The JCT 05 Standard Building Contract

Law and Administration

The JCT 05 Standard Building Contract

Law and Administration

Second edition

Issaka Ndekugri
Michael Rycroft

AMSTERDAM • BOSTON • HEIDELBERG • LONDON • NEW YORK • OXFORD
PARIS • SAN DIEGO • SAN FRANCISCO • SINGAPORE • SYDNEY • TOKYO
Butterworth-Heinemann is an imprint of Elsevier

Butterworth-Heinemann is an imprint of Elsevier
Linacre House, Jordan Hill, Oxford OX2 8DP, UK
30 Corporate Drive, Suite 400, Burlington, MA 01803, USA

First published by Arnold as The JCT 98 Building Contract: Law and Administration, 2000
Reprinted 2006
Second edition, 2009

British Library Cataloguing in Publication Data
A catalogue record for this book is available from the British Library

Library of Congress Cataloguing in Publication Data
A catalogue record for this book is available from the Library of Congress

ISBN: 978-1-8561-7629-3

For information on all Butterworth-Heinemann publications
visit our web site at www.elsevierdirect.com

Typeset by Macmillan Publishing Solutions
(www.macmillansolutions.com)

Printed and bound in United Kingdom

09 10 11 12 13 10 9 8 7 6 5 4 3 2 1

Contents

Preface to Second Edition

In the nine years since the publication of the first edition of this book, much has changed for those contracting with the Joint Contracts Tribunal's forms of contract. Developments in both statutory and common law and innovation in procurement methods have had some effect; but the greatest impact has been from the re-drafting by the JCT in 2005 of their entire suite of contracts, giving participants in the building industry a new set of forms of contract for a wider range of procurement options. Nevertheless, the Standard Building Contract, the subject of this book, remains the JCT's flagship form and, despite the development of procurement systems such as partnering, and framework arrangements, it still appears to be the model of administrative and risk allocation norms in not only the building sector, but also the wider construction industry.

This book is published as a second edition to our earlier book, *The JCT 98 Building Contract: law and administration*, simply because the 2005 version of the standard contract, despite the change of title, is in essence a re-presentation of it's predecessor JCT 98. Whilst the 2005 edition is a dramatic change in form, the allocation of risk has remained largely the same. Indeed, the JCT have gone to considerable lengths to assist those familiar with the 1998 contract to adapt to the new contract, particularly with the provision of tables showing how clauses in the new contract map onto equivalent clauses in JCT 98. We have, therefore, tried to keep chapters and section numbers of our book the same as those in the first edition wherever possible. There are some obvious changes, such as the introduction of chapters on third party rights and CDM obligations, but they conveniently fill the gaps left by the exclusion of the provisions on Nominated Sub-contractors and Suppliers.

We have concentrated on the With Quantities version of the JCT Standard Building Contract, as that form contains the widest variety of features. We considered, at length, the short title for use throughout the book. Strictly, JCT 05 describes the whole suite of JCT contracts, and the particular contract form we deal with is designated 'SBC/Q' by the JCT. However, 'SBC/Q' is not a term used much in the industry, where simply 'JCT 05' is often used to describe the Standard Building Contract, which distinguishes it from 'JCT 05 Design and Build', or 'JCT 05 Major Projects' etc. We therefore hope and pray that our readers will treat our use of 'JCT 05' as the short title with indulgence.

In the preface to the first edition, we noted that disputes appeared to result largely from lack of understanding of the standard terms then current, or confusion arising from poorly prepared contract documents. Our observations suggest that, in this respect, little has changed, and that considerable risk remains for any of the participants in the contracting process to become embroiled in dispute. Accordingly, our book is still aimed at all construction participants in developer, management, design, or construction organizations involved in the planning, execution or administration of building contracts and, also, at undergraduate and postgraduate students studying towards careers in the industry. Feedback from the first edition suggests that even experienced construction lawyers are likely to find it very useful.

We are aware that readers will use this book in different ways; some will approach it chapter by chapter, but many will dip into a particular topic from the index to research the problem in hand. For that reason, there is a certain amount of duplication of the general coverage of some topics. For example, the principles underlying liquidated damages can be found in the chapter dealing with time and liquidated damages, and also in the chapter dealing with preparation of the contract documents. Whilst we cover the effect of the provisions in the JCT contract in considerable detail, our aim is also to put the contract into context by explaining the statutory and common law background to each topic. This inevitably results in some overlap, and we have tried to cross-reference to other chapters where wider reading is necessary.

The last nine years have seen a dramatic increase in access to case law through the Internet. In particular, the publication on-line of most House of Lords, Court of Appeal and High Court judgments, together with Commonwealth Courts' and other common law decisions has produced a plethora of examples to demonstrate particular points. However, readers should bear in mind that only a few cases produce new principles; many of the cases cited simply demonstrate established principles in action, applied to the particular facts, and often influenced by unique agreements between the parties.

At the publication of the first edition in 2000, the 1996 Construction Act was still in its infancy. Our chapters on payment and dispute resolution in particular were, therefore, written with the benefit of only a few cases. Whilst we now have the judicial help of several hundred cases, a few may become redundant as, at the time of going to print, a long-awaited amendment to the 1996 Act under the title 'Local Democracy, Economic Development and Construction Bill, Part 8 Construction Contracts' was working its way through Parliament. When it becomes law the JCT will be obliged to amend their contract to implement the changes. Although we have dealt with payment, dispute resolution and parties' attempts to amend the standard contract by reference to the 1996 Act, we have also outlined the proposed statutory amendments to give an indication of the changes that may come about in 2010.

On the matter of style, for the reasons given in the Preface to the first edition, we have again followed the JCT in their use of 'he' for reference to contractual dutyholders. We have also retained the use of capital letters when referring to specific JCT terminology (e.g. the Architect, the Employer, Variation, Parties), and lower case where the reference is general (architects, employers, variations, parties). In many places we have referred to judges, some of whom later went on to higher positions. Where we cite a judge's rank (e.g. HHJ, LJ, etc) it is the rank at the time of the relevant judgment but we have tried to avoid the repetitive use of 'as he/she then was'.

We owe thanks to the University of Wolverhampton for the use of its extensive facilities, to Hill International Midlands office for the kind use of research materials, and to Mr John Wood, a director of Quantex Consulting who again gave time freely to bench test the user-friendliness of our style. Thanks also go to the production team for their contribution. Finally, but by no means least, we are indebted to our wives Mariama and Judith, without whose patience and support this book would never have been finished.

We have tried to state the law at January 2009; any errors or omissions are, of course, ours.

Issaka Ndekugri
Mike Rycroft
February 2009

Preface to First Edition

Periodic surveys of standard building contract forms suggest that the Joint Contracts Tribunal stable of forms is in widespread use. In particular its flagship, the Standard Form of Building Contract, in its various versions is popular where works are of more than minimal complexity; and to the extend that the industry can be said to adopt a norm, the administrative principles of the JCT Standard Form appear to be the nucleus of that norm. However, use of the Standard Form has been dogged by a high incidence of disputes. It is an open secret that many of these disputes have been attributed either to lack of understanding of the terms, or to confusion arising from cavalier preparation and alteration of documents. Our personal observations indicate that the problem affects all participants in the contracting process, including Employers' professional teams and Contractors' site or office management. In recent years there has been a huge increase in the law and procedures content of academic built environment courses; this is matched by a similar increase in continuing professional development courses dealing with contractual issues. Our book results from our considerable involvement in both types of course, together with the realization that difficulties faced by practising construction professionals and individual students are commonplace.

This work is aimed at all construction professionals, whether they be in design or construction organizations, who are involved in the planning, execution or administration of building contracts; it is also aimed at undergraduate and post-graduate students studying towards careers in the industry. Of course we must not forget the lawyers; we hope the form and content of this book will assist in particular, those encountering their first construction cases, who may be unfamiliar with the jargon or traditions of the construction world.

We have not dealt with all versions of JCT Standard Form, but have concentrated on the Private Edition with Quantities; comparison of the major differences from other versions are noted in the text. To comply with tradition we have used JCT 98 as the short title. The book was commenced when the 1980 Edition with Amendments 1 to 17 was current, and was necessitated by a dearth at that time of up-to-date texts. The publication of Amendment 18 in April 1998, followed, in November 1998 by the 1998 Edition, seriously delayed drafting; no event could have highlighted more dramatically for us the ongoing problem faced by students and practitioners alike in trying to keep up with change. The inevitable Amendment 1 was issued by the JCT in June 1999. The amendment deals with tax matters which we had previously decided would be outside the scope of the book, and consequently it receives only passing mention in the commentary on preparing documents, and payment obligations. Just before publication, JCT 98 became available on CD with the facility for completing the form including incorporating the standard amendments. This must ease the onerous task described in Chapter 1.

One change in the construction industry which is not reflected in the book, but which deserves special mention, is the welcome increase in the numbers of female practitioners throughout all disciplines. We gave considerable thought to getting around the inadequacy of the English language, which thwarts any attempt to neutralise contractual duty holders. Rather than reduce everyone to 'it', or resort to the clumsy 's/he' or 'he/she', we reluctantly decided on the use of 'he' throughout; this was principally because recurring reference needs to be made to the contract terms, and 'he' is the term used exclusively by the JCT.

Change is not confined to the construction industry, and in April 1999 new Court Civil Procedure Rules came into force as a result of Lord Woolf's *Access to Justice* Report. The main effect of the Rules on this book is on the terminology. Where reference is made in a general sense to the rights or obligations of a party bringing a case to the Court, we have used the term 'claimant' in accordance with the Rules. Where we have used the term 'plaintiff' it refers to the plaintiff in a case preceding the new Rules.

On the matter of style, at the risk of confusing the reader we have adopted the use of capital letters when referring to specific JCT terminology (e.g. the Architect, the Employer, Variation), and lower case where the reference is general (architects, employers, variations).

We owe much to Mr John Wood, a director of James R. Knowles, who gave his time freely to act as a sounding board for our ideas and to bench test the user-friendliness of our style. We are also indebted to the editorial and production team who waited so patiently while we struggled to catch up with the Joint Contracts Tribunal.

Finally, we have tried to state the law at September 1999; any errors or omissions are ours.

Issaka Ndekugri
Michael Rycroft

January 2000

Glossary of Latin Terms

ad hoc	for this special purpose
caveat	warning; in relation to a document used to describe a conditional term
caveat emptor	let the buyer beware
consensus ad idem	all in agreement as to the same thing
contra proferentem	the doctrine that the least favourable construction of words will be adopted against the person putting them forward
de facto	actual, as opposed to legally recognized
de jure	by right
ex gratia	as a favour
force majeure	acts of God or man-made events beyond people's control
functus officio	a person's powers have ended
inter alia	among other things
in terrorem	to frighten
nemo dat quod non habet	you cannot give away what you do not own
obiter	by the way
obiter dicta	something said by the way; observations made by a judge in passing, but not binding as precedent
pari passu	with equal pace; together
pro forma	(in legal and business sense) a standard format
pro rata	in proportion
quantum meruit	as much as he has earned; (used in legal sense as a synonym for *quantum meruit* and *quantum valebat*)
quantum valebat	as much as they were worth
ultra vires	beyond someone's powers

Abbreviations

Law Reports

ABC LR	Arbitration, Building & Construction Law Reports
AC	Appeal Cases, from 1891
ALJR	Australian Law Journal Reports
ALL ER	All England Law Reports
App Cas	Appeal Cases, up to 1890
BCL	*Building and Construction Law*
Beav	Beavan's Reports
BLISS	*Building Law Information Subscriber Service*
BLR	Building Law Reports
Burr	Burrow, 1756–1772
CA	Court of Appeal
CB (NS)	Common Bench, New Series
CILL	*Construction Industry Law Letter*
CL	*Current Law*
CLC	Commercial Law Cases, from 1994
CLD	*Construction Law Digest*
ConLR	Construction Law Reports
Const LJ	*Construction Law Journal*
CSIH	Court of Session, Inner House
CSOH	Court of Session, Outer House
DLR	Dominion Law Reports
EG	*Estates Gazette*
EGLR	Estates Gazette Law Reports
ER	English Reports
EWCA	England & Wales Court of Appeal
EWHC	England & Wales High Court
Ex	Exchequer Reports

F & F	Foster & Finlayson, 1858–1867
FSR	Fleet Street Reports
GWD	Green's Weekly Digest
H & C	Hurlstone & Coltman, 1862–1866
HL	House of Lords
HL Cas	House of Lords Cases, 1847–1866
ICR	*International Construction Law Review*
JP	*Justice of the Peace and Local Government Review*
KB	King's Bench
LGLR	Local Government Law Reports
LGR	Local Government Reports, 1992–1995
LJ Ch	Law Journal Reports – Chancery, 1831–1849
Lloyd's Rep	Lloyd's Reports, from 1951
LR CP	Law Reports – Common Pleas Cases, 1865–1875
LRQB	Law Reports Queen's Bench
LT	Law Times Reports, 1859–1947
M & W	Meeson & Welsby, 1836–1847
NE 2d (Ind App)	North Eastern Reporter 2d series, Indiana Appeals
NSWLR	New South Wales Law Reports
NSWSC	New South Wales Supreme Court
NTSC	Northern Territory Supreme Court
NZLR	New Zealand Law Reports
ORB	Official Referee's Business
PNLR	Professional Negligence & Liability Reports
QB	Queen's Bench
Salk	Salkeld, 1689–1712
SARL	South African Law Reports
SCLR	*Scottish Construction Law Review*
ScotCS	Scottish Court of Session
SJ	*Solicitors' Journal*
SLT	Scots Law Times
Stark	Starkie, 1814–1823
TCC	Technology and Construction Court
TCLR	Technology and Construction Law Reports
TLR	Times Law Reports
UKHL	United Kingdom House of Lords
WASC	Western Australia Supreme Court
WL	*Westlaw*
WLR	Weekly Law Reports

General

AA 1996	Arbitration Act 1996
ACOP	Approved Code of Practice (Health & Safety Commission)
AI	Architect's Instruction

CDM	Construction (Design and Management) Regulations 1994–2007
CDP	Contractor's Designed Portion
CIArb	Chartered Institute of Arbitrators
CIMAR	Construction Industry Model Arbitration Rules
CIS	Construction Industry Scheme
CLCA	Civil Liability (Contribution) Act 1978
COW	Clerk of works
CPM	Critical path method
CPR	Civil Procedure Rules
CRTPA	Contract (Rights of Third Parties) Act 1999
FIDIC	Fédération Internationale des Ingénieurs Conseils (Association of National Associations of Consulting Engineers)
HASAWA	Health and Safety at Work Act 1974
HHJ	His/Her Honour Judge
HMSO	Her Majesty's Stationery Office
HSE	Health and Safety Executive
ICE	Institution of Civil Engineers
IRS	Information Release Schedule
JCT	Joint Contracts Tribunal
LC	Lord Chancellor
LCCP	Law Commission Consultation Paper
LDs	Liquidated damages
LJ	Lord Justice
LOSC	Labour only sub-contractor
LPA 25	Law of Property Act 1925
LPCDIA	Late Payment of Commercial Debts (Interest) Act 1998
MR	Master of the Rolls
NEDO	National Economic Development Office
NJCBI	National Joint Council for the Building Industry
P&T	Purchasers and Tenants
PIC	Person-in-Charge
RIBA	Royal Institute of British Architects
RICS	Royal Institution of Chartered Surveyors
ROT	Retention of Title
SCL	Society of Construction Law
SMM7	Standard Method of Measurement 7th edition
UCTA	Unfair Contract Terms Act 1977
VAT	Value Added Tax

Publications

Hudson's	*Hudson's Building and Engineering Contracts*, 11th edn I N Duncan Wallace QC (Sweet & Maxwell, 2003)
Keating	*Keating on Construction Contracts*, 8th edn, The Hon Sir Vivian Ramsey; Stephen Furst QC (Sweet & Maxwell 2008)

Contracts

DOM/2	*Standard Form for Domestic Sub-Contract* (The Construction Confederation, 1998)
GC/Works/1	*General Conditions of Government Contract for Building and Civil Engineering Works* (HMSO, 1973)
ICE 7th	Institution of Civil Engineers, The Association of Consulting Engineers, The Civil Engineering Contractors Association, *ICE Conditions of Contract 7th edition* (Thomas Telford Services Ltd, 1999)
SFA	*RIBA Standard Form of Agreement for the Appointment of an Architect*

JCT Contracts, guides and warranties

IFC 98	*Intermediate Form of Building Contract for works of simple content, 1998 Edition* (RIBA Publications)
JCT 05	(For the purposes of this book) *Standard Building Contract (2005 edition); see also SBC*
JCT 39	*RIBA Standard Form of Building Contract (1939 Edition) 1957 Revision*
JCT 63	*Standard Form of Building Contract 1963 Edition*
JCT 80	*Standard Form of Building Contract 1980 Edition*
JCT 98	*Standard Form of Building Contract 1998 Edition*
CWa/F	*Contractor Collateral Warranty for a Funder*
CWa/P&T	*Contractor Collateral Warranty for a Purchaser or Tenant*
SCWa/E	*Sub-Contractor Collateral Warranty for Employer*
SCWa/F	*Sub-Contractor Collateral Warranty for a Funder*
SCWa/P&T	*Sub-Contractor Collateral Warranty for a Purchaser or Tenant*
DB 2005	Design and Build Contract (2005 Edition)
DB/G	Design and Build Contract Guide
MP 2005	*Major Project Construction Contract (2005 Edition)*
SBC	*Standard Building Contract (2005 Edition)*
SBC/AQ	*Standard Building Contract With Approximate Quantities (2005 Edition)*
SBC/G	*Standard Building Contract Guide*
SBC/Q	*Standard Building Contract With Quantities (2005 Edition)*
SBC/XQ	*Standard Building Contract Without Quantities (2005 Edition)*
SBCSub/C	*Standard Building Sub-Contract Conditions (2005 Edition)*

1

Style of JCT 05 Standard Building Contract, contract execution and related problems

1.1 Standard form contracts generally

Any transaction between two business concerns will inevitably place a demand on the fixed overhead resources of the parties. Whether the contract is a simple exchange of letters or a lengthy document signed by each party, there is some administrative input; the extent of the input is determined largely by the effort required to negotiate terms and to commit them to writing. The nature of some companies' work may be such that transactions are numerous; other companies' transactions may be few but complicated. In either case it is an advantage to any organization to develop personal terms of trading. A contract which sets out such terms is widely known as a standard form contract. The advantages to the parties issuing such terms are many (e.g. familiarity from use, understanding, saving on drafting time, saving on negotiating time, or saving on senior staff time when transactions can be concluded by clerical staff). They are all good reasons for the issuing parties to enter into their own standard form contract. This type of standard form is normally issued unilaterally.

However, for the party intending to use the standard form of another (unilateral standard form) there are clearly some disadvantages. Savings on time, both in drafting and negotiating, will certainly remain an advantage, and are perhaps the principal reasons for the use of standard forms; but clearly there is a danger that the terms will also contain onerous provisions for some users. In any event, the main benefit of standard forms is still lost to one party since, unless the company enters into a particular contract form regularly, it will not develop familiarity or understanding. As a result, unilateral standard forms are often looked upon with suspicion by those with previous bad experiences. The way out of such situations for some is to insist on their own standard terms. There are then two sets of unilateral terms and the scene is set for either a negotiation or a battle[1] with the party in the stronger position dictating the basis of the contract.

There is another option; that is to reap the benefits of standard form contracts whilst avoiding the disadvantages. Many industries have developed a different type of standard form contract, a type which is more worthy of the general title 'standard', and which could be called 'industry standard forms'. The terms are settled through negotiation by representative bodies of commercial interests in a particular industry, and are based on mutual experience of the practices of the industry. Such forms, as a class, have been described by the courts as being widely adopted because experience has shown that they facilitate the conduct of trade, and that there is a strong presumption that their terms are fair and reasonable because they are used widely by parties whose bargaining power is evenly matched.[2]

The main virtues of an 'industry standard form contract' can be summarized as:

- being a device for allocating contingent risk whilst saving time and assisting bargaining at arm's length;
- being a device for avoiding writing terms for each transaction;
- having the benefit of providing understanding by familiarity and experience in practice;
- being less likely to protect the interests of only one party, having been negotiated by independent bodies representing all interests in an industry;
- producing savings in transaction costs, avoiding the need to negotiate each contract;
- removal of unwanted discretion from individuals, enabling a structured approach to negotiations;
- enabling allocation of risk to be anticipated and provided for in calculations;
- providing a familiar structure for payment, varying work, and dispute resolution;
- enabling necessary quotations from others such as sub-contractors and suppliers to be obtained with greater accuracy.

Disadvantages of industry standard form contracts as a general class appear to be few, since they go some way towards removing the principal objection to unilateral forms (that is a stronger party imposing its will on a weaker party). However, within an industry the individual industry standard forms in common use may draw some criticism, and the building industry is no exception.[3]

1.2 The Joint Contracts Tribunal Limited

One of the most prolific producers of contract forms for the building industry is the Joint Contracts Tribunal Limited (formerly the Joint Contracts Tribunal for the Standard Form of Building Contract), commonly called the JCT.

The constituent members are representative bodies of various commercial interests in the building industry, who settle the terms of the JCT stable of contract forms; at the time of writing they are:

- British Property Federation Limited;
- Construction Confederation;

---

1 See discussion on 'the battle of forms', Section 1.7 in this chapter.
2 *Per* Lord Diplock, *Schroeder (A) Music Publishing Company Ltd* v. *Macauley* (1974), 2 All ER 616.
3 See Section 1.3.3 in this chapter.

- Local Government Association;
- National Specialist Contractors Council Limited;
- Royal Institute of British Architects;
- The Royal Institution of Chartered Surveyors;
- The Scottish Building Contract Committee Limited.

In 1931 the JCT was formed comprising the Royal Institute of British Architects and the National Federation of Building Trades Employers (now the Construction Confederation). In 1939 the JCT published a major standard form of contract revising and replacing an earlier form[4] which had been known for many years as 'the RIBA Form'. Although the JCT was self-appointed, its forms were in common usage and in 1963 a further new edition was published which was again revised in 1977. By this time the constituent bodies had grown in number to include the Royal Institution of Chartered Surveyors, Association of County Councils, Association of Metropolitan Authorities, Association of District Councils, the Greater London Council, Committee of Associations of Specialist Engineering Contractors, Federation of Specialists and Sub-Contractors, the Association of Consulting Engineers, British Property Federation, and Scottish Building Contracts Committee.

In 1980 a substantial new edition, called the *Standard Form of Building Contract 1980 Edition* (JCT 80), was published in several Private and Local Authority versions – With Quantities (Q), Without Quantities (WQ), and With Approximate Quantities (AQ).

JCT 80 incorporated changes from the earlier *1963 Edition*, *1977 Revision*, notably a decimal numbering system, reference to all paragraphs and sub-paragraphs as clauses, a definitions clause, and separately published fluctuations clauses. Between 1983 and April 1998, JCT 80 was changed by a further approximately 200 items in 18 published amendments and 4 reprints. The most significant of these changes is Amendment 18 issued in April 1998. Amendment 18 was prompted by the need for construction contracts in writing to comply with the Housing Grants Construction and Regeneration Act 1996, Part II.[5] However, the JCT went a step further. The 1996 Act followed some, but not all, of the recommendations contained in a report by Sir Michael Latham called *Constructing the Team*.[6] The JCT took the opportunity to incorporate the spirit of some of the other recommendations, and to introduce provisions in their contracts that reflected what was then seen as modern practice; Amendment 18 was used as the vehicle. Unfortunately Amendment 18 together with its guidance note totalled 60 pages containing 22 items, 2 bond agreements, and 2 misleadingly titled adjudication agreements (perhaps 'adjudicator agreements' would be more accurate). Amendment 18 created a daunting task for those wishing to use JCT 80 as a working document; earlier criticism of JCT 80[7] was reinforced. The JCT seemingly realized the size of the problem, for in December 1998 they published a new edition of JCT 80 incorporating all amendments, under the title *Standard Form of Building Contract 1998 Edition*. Following precedent the 1998 edition became known as 'JCT 98'. JCT 98 was published as both Private and Local Authorities versions, 'With Quantities', 'With Approximate Quantities', and 'Without Quantities', and is the subject of the first edition of this book.[8]

---

4 An early twentieth-century version of a form used in the late nineteenth century; a copy of the latter is included in *Hudson on Building Contracts*, 3rd edn (Sweet & Maxwell, 1906) vol. 2.

5 The 1998 Act (widely known as the Construction Act) came into force on 1 May 1998.

6 Published in 1994.

7 See criticisms at Section 1.3.3 in this chapter.

8 Issaka Ndekugri and Michael Rycroft, *The JCT98 Building Contract: Law and Administration*, Butterworth-Heinemann, 2000.

In July 2005, the JCT amended and republished the Standard Form again, this time under the title *Standard Building Contract* (*SBC 2005*), as part of a large suite of standard forms. The new 2005 edition, the subject of this book, is published in With Quantities, With Approximate Quantities, and Without Quantities versions. Each version is suitable for either Private or Local Authorities use.

Again following precedent, the 2005 edition rapidly became known colloquially as 'JCT 05', although this can cause some confusion as the whole suite of contracts is also referred to as JCT 05.

In April 2007, in Amendment 1, the JCT published the first amendments to their new forms to reflect changes to CDM Regulations, and to introduce third party rights or warranties for the Employer. The JCT then published JCT 05 Revision 1 in June 2007, incorporating Amendment 1, together with a few minor corrections. A further amendment was issued in February 2008 to allow execution as deed by a single Director and witness, following changes in legislation.

In this book, reference to 'JCT 05' means the Standard Building Contract, SBC2005, Revision 1, June 2007, amended February 2008.

1.3 JCT 05: Use, style and criticism

1.3.1 Use of JCT 05

The JCT suite of contract forms contains a wide variety of options, and it can be a daunting task for even the most experienced exponents of the industry to select the most appropriate form for the circumstances.

A note is printed inside the front cover of each of the JCT forms explaining where a particular contract form would be appropriate, and circumstances where it can be used. The same note appears against the title of each form in the JCT Catalogue *JCT 05 Contracts and Documentation*. A superficial comparison of the forms can be achieved by scanning all the various forms in the catalogue, but the JCT helpfully have produced Practice Note:[9] *Deciding on the appropriate JCT contract*.

The Practice Note classifies projects by design and management duty (e.g. design by owner's consultants or by Contractor), and by means of evaluation (e.g. Lump Sum, Measurement, and Cost Reimbursable). Reference is made to 'larger Works', 'major Works', 'large-scale projects', 'Works simple in character', and 'Works of a simple content'. Each of these terms is relative, depending on experience, and no indication is given as to the meaning of large-scale[10] or simple. The Practice Note does, however, emphasize wisely that the guidance given is not a substitute for professional advice.

The Standard Building Contract is published in three forms (i.e. with quantities, without quantities, and with approximate quantities) which are dealt with separately in the Practice Note. Understandably the criteria for use of the three forms contain many similarities.

---

9 First published August 2006, Revised February 2007 and February 2008. This note replaces Practice Note 5 Series 2, *Deciding on the appropriate JCT form of Main Contract*, published by JCT for use with the 1998 editions.

10 Compare with JCT Practice Note 5, Series 2 which proposed a value range.

All versions: The JCT suggest that they can all be used where the Contractor, under a predominantly building contract, is nevertheless required to design discrete parts of the Works,[11] and where the Work is divided into Sections;[12] all are suitable for use by both private and local authority employers. Another similarity is that they are all appropriate where the Conditions will be administered by an Architect/Contract Administrator and Quantity Surveyor (in practice the Standard Contract cannot be operated without them).

All three are stated to be appropriate for 'larger Works designed and/or detailed by or on behalf of the Employer, where detailed contract provisions are necessary …' Design duty apart, comparison of this with the user profile for the Major Project Construction Contract[13] seems to suggest that some parties to construction contracts need to be given a structured procedure to follow. In theory, JCT 05 can be used for projects of any size and complexity, where the work is designed predominantly by the owner. However, in practice, the criteria of size, value and complexity are not the only factors influencing use of the form. They are all criteria related to the project; but contracts are more to do with relationships than with projects. It has to be emphasized that some construction industry standard forms, including JCT 05, are highly procedural.[14] The one factor that is most likely to affect the successful use of the form on any project, of whatever value or complexity, is the parties' ability to administer the procedural requirements. If the procedures are not understood, or if insufficient administrative resources are allocated, then the parties' relationship is vulnerable to deterioration through misunderstanding. The important criterion then, when deciding suitability, should be the administrative competence of both the contract administrator (whether he is an architect or from another discipline) and the likely contractor. If the nature and complexity of the contract form is such as to impinge on the success of what could otherwise be a project designed and constructed economically in a time to the satisfaction of all parties, then that contract form is not the correct form for that particular relationship. The parties should then consider using one of the more simple JCT forms.[15]

The differences between the three versions of the Standard Building Contract (With Quantities, Without Quantities, and With Approximate Quantities) lie in the Contractor's risk in what is included in the scope of work. The risk is identified by describing items of work with their respective quantities.[16]

Without Quantities: In essence, the 'Without Quantities' form is a pure lump sum contract in which the Contractor provides a single price for building what is shown on drawings and described in specifications prepared by the building owner or his consultants. The tenderer is required to bear the cost of his own errors in coming to his price, including underestimation of the amount of work involved; conversely over-estimation of the work will be to his benefit. The level of risk is summed up by the JCT in their observation that Contractors may be reluctant to tender 'where the project exceeds a certain size or

---

11 See Chapter 4.
12 See Section 1.5.2 in this chapter referring to the Sixth Recital, and introduction to Chapter 3 for effect of dividing work into Sections.
13 Suitable for regular users, where the Contractor is experienced, and where procedures are already in place.
14 The Practice Note February 2008 does not emphasize complexity, but contains a useful comparison of main contract provisions in the frequently used JCT forms.
15 For example, the JCT Intermediate Building Contract, or JCT Minor Works Building Contract.
16 See also Section 1.3.4 in this chapter.

complexity'.[17] Certainly, size and complexity are relevant factors, but so too is the completeness of the information provided by the employer. The larger the gaps in design or knowledge that the Contractor has to fill, the greater the risk he takes, and the greater must be the skill of his estimator.

With Quantities: The 'With Quantities' version is also a lump sum contract, but in order to provide a basis for comparing tenders, the building owner commissions a Quantity Surveyor to prepare a bill of quantities describing the work required, which the tenderers then price.[18] Where bills are used, the tenderers may rely on the contents at face value in the knowledge that the cost arising out of any errors should be corrected at the building owner's expense.

With Approximate Quantities: The 'With Approximate Quantities' form is a 'remeasurement' contract. The Bills are used to provide a market level of pricing and basis for tender comparison; the final price is determined by measuring the work done and applying the Bill rates. The provisions are very similar to the 'With Quantities' version, except instead of adjusting the original price up or down to reflect changed work, the quantities in the Bills are totally remeasured. This version is most suitable where drawings are prepared but have not been sufficiently detailed to prepare accurate quantities. However, in practice, owners in such a position will be tempted to pass the risk of completing the design to the Contractor, rather than take the risk themselves by firming up the quantities.

1.3.2 Style of JCT 05

JCT 05 is published in three forms: Standard Building Contract With Quantities (SBC/Q), Standard Building Contract Without Quantities (SBC/XQ), and Standard Building Contract With Approximate Quantities (SBC/AQ). Each version is suitable for use by private companies or individuals, or by Local Authorities.

The need for some building owners to take over parts of their new facility in sections is accommodated within the contract by choice of options in the Contract Particulars. The Contract Particulars section, that is the set of variables containing the data and risk choices.relevant to the particular contract (and formerly called the Appendix in JCT contracts) forms part of the Agreement at the front of the printed form.

Design by the Contractor is also incorporated by use of an optional Contractor's Designed Portion, again triggered by an entry in the Contract Particulars.

Annexures containing optional procedures, standard collateral documents, and lengthy options, are included as 'Schedules' operated by clauses in the Conditions. The Schedules include a 'Contractor's Design Submission Procedure', a 'Schedule 2 Quotation' procedure, three insurance options, a 'Code of Practice' for use when opening up work, provision for 'Third Party Rights' when the parties intend that identified third parties should be able to enforce benefits under the Contract, standard 'Forms of Bonds' for use where payment is made in advance, for off-site materials, or without retentions, and three 'Fluctuations Options'.

---

17 See Practice Note – Deciding on the Appropriate JCT Contract, para 26.

18 See Sections 1.4.3 and 1.5.4 in this chapter for general comment on bills of quantities.

The rationale of JCT 05 is well described by the JCT themselves in their Guide SBC/G.[19]

> The JCT's intention is that the content of SBC 2005 should, as far as possible, be familiar to users of JCT 98. Certain points have been clarified but there has been no material alteration in risk allocation and many of JCT 98 procedures have been retained.

Provisions are grouped in sections related to topics. For instance, payment issues such as certification, evaluation, loss and expense and fluctuations, all appear in Section 4 – Payment; similarly issues related to carrying out the Works, such as possession, completion, supply of information, design procedures, all appear in Section 2 – Carrying out the Works. The sections are listed in Section 1.4.1.3 in this chapter.

A brief introduction to the contract is provided by the JCT in their publication *Standard Building Contract Guide* (SBC/G). The Guide is a mixture of explanation of the provisions, and a comparison with JCT 98 which this contract replaces. There is a useful schedule for those moving from JCT 98 to JCT 05 at the back of the Guide, listing new clause numbers against the corresponding clause number in JCT 98.

1.3.3 Criticism of JCT 05

JCT 05 is a derivative of JCT 98 and the JCT and RIBA forms which preceded it. JCT contracts have attracted much general criticism in the past,[20] aimed more at form than at risk allocation. It is on the form that draftsmen seem to have concentrated in the production of JCT 05.

Form is important; it helps to realize one of the principal purposes of any standard form contract – the way in which it is used. The Standard Building Contract, like any other written contract, is a combination of several types of statement. Statements as to obligations, entitlements and liabilities concerning the goods and services to be supplied and the price to be paid, are intertwined with statements of procedure (i.e. how the job will be administered). This last set of statements (procedures) is the code by which those drafting a form direct the various parties how to go about their daily business. The intention may be admirable; unfortunately, often the execution is not. Historically, the JCT standard terms have been convoluted and prolix; they have been totally unsuitable for the day-to-day requirements of many building sites. The difficulty is that contract forms are required to act, not only as a legal record of the binding agreement, but also as a working tool. This requires clauses to be simple statements wherever possible, with minimal subdivision (but see JCT 98, with its bewildering clause number 13.4.1.2.A7.1.1). Unfortunately many forms, including those in the JCT lineage, are written with a mind to dissection in the event of dispute; they are not written in language or in a style capable of being understood, with any justifiable confidence, by those using them.[21] Even the courts have had difficulty. In 1967 Lord Justice Sachs said:[22] 'The difficulties arise solely because of the

19 SBC/G para 3.
20 See Ndekugri & Rycroft, *The JCT98 Building Contract: Law and Administration*, para 1.3.3.
21 But see provisions of JCT Major Project Construction Contract 2005 (Reference *MP2005*) drafted using plain English and avoiding excessive cross-referencing.
22 *Bickerton* v. *NW Metropolitan Hospital Board* [1969] 1 All ER 977 at 979.

unnecessary amorphous and tortuous provisions of the RIBA contract: those difficulties have for a number of years been known to exist.'[23]

In the 2005 edition, the JCT has gone some way towards meeting its critics. It has clearly tried to move away from the presentational rut of earlier revisions, to create a more user-friendly document, albeit maintaining the principles, procedures and risk allocation familiar to users.[24] The grouping of clauses relating to like topics has obviated the need for much of the confusing cross-referencing which made JCT 98 so difficult to follow. Similarly, the use of defined terms has cut down unnecessary repetition. The compression of annexes to conditions, and annexes to appendices, fluctuations clauses, design and quotation procedures, into a single set of schedules helps to simplify navigation around the contract. In particular, the introduction of Contractor's design, and work divided into sections, incorporated by triggers in the Recitals, replacing the messy amendments required under JCT 98, must bring a sigh of relief to practitioners preparing contract documents. All this goes towards making a more user-friendly document than previous editions of the standard contract.

Nevertheless, JCT 05 is not without its faults. There are some missed chances. For example, disputes still abound over the meaning of practical completion and its effect. The JCT Major Projects form published in 2003 contained a helpful definition of practical completion; it is surprising that a similar definition does not appear in JCT 05. The absence of clear time limits on the Contractor's obligations creates an uncertainty regarding the effect of practical completion, which the courts have addressed with conflicting views.[25]

Whilst considerable effort has been made to simplify the clause structure, newcomers to the contract may find the lack of a subject index irritating.[26] There are several provisions where users are required to provide descriptions (e.g. descriptions of Sections where the Works are divided, descriptions of acceptable electronic communication, lists of warranty providers). In an attempt to provide maximum flexibility, users are given the option of providing information in separate documents.[27] Unfortunately, that has the effect of increasing the likelihood at operational level, of documents being split or even ignored. A major introduction is the positive use, albeit limited, of third party rights legislation by entries in Part 2 of the Contract Particulars. In the Guide, the JCT describe the provisions in Part 2 as 'inevitably complex' as a result of the number of variables. The number of variables does indeed make completion of the form complicated, but Part 2 is not simply complex; it is confusing. The layout and numbering system of JCT 05 generally is clear and uniform, wherever it is opened, except in Part 2 of the Contract Particulars. Part 2 is unstructured, and not consistent with the rest of the form. It gives the appearance of having been drafted in a vacuum, without any regard for the need one day to cross-refer to individual entries in it.

Turning briefly to the sub-contracting arrangements in JCT 05, a frequent source of misunderstanding (and dispute) has been removed by excluding the nominating procedures,[28] whereby the Architect could impose sub-contractors onto the Contractor, who had only limited opportunity to complain.

---

23 See also the judgment of Salmon LJ in *Peak Construction Ltd* v. *McKinney Foundations Ltd* (1970) 1 BLR 111 at 114.

24 See intention stated by the JCT in Standard Building Contract Guide (SBC/G), para 3.

25 See Chapter 6, Section 6.7.2.

26 Compare with engineering forms of contract such as ICE 7th Edition.

27 See Section 1.5.4 in this chapter for list of potential annexed documents.

28 Described in *Scobie & McIntosh Ltd* v. *Clayton Bowmore Ltd* (1990) 49 BLR 119 at 128 as being 'of inordinate and needless length'.

The simplification of terms in JCT 05 is unlikely to affect the successful administration of projects where historically success has been marred by the extent of procedures. Contractors or professionals who fail to provide sufficient administrative backup will still find the Standard Building Contract very procedural. Indeed with the introduction of the Contractor's design submission procedures to counter the loss of nomination, there is probably as much paperwork as with JCT 98.

It was observed in the JCT 98 edition[29] that JCT forms invariably work in practice, due largely to the operators' wish to make them work. This is achieved often by ignoring the strict terms, cutting procedural corners, and by applying some pragmatism and common sense. The JCT seemed to recognize this particular issue in JCT 98 by ratifying common practices such as Contractors' applications for payment, Contractors' evaluation of variations, priced schedules to assist in interim valuations, and by introducing a fixed starting point for the interim payment timetable. Evaluation of variations by the Contractor (the Alternative A procedure in JCT 98) has fallen by the wayside, on grounds of under-use. Nevertheless, the likelihood is that Contractors still value variations and submit them, but not using the somewhat convoluted Alternative A procedures. The other features are carried forward into JCT 05. However, whilst pragmatism and common sense may lead to a high success rate for the Standard Building Contract, when it fails the ensuing dispute can be very messy. The procedures may be complex, but generally they work if followed. Once either party starts to cut procedural corners, ignore notices, time-scales, requirement for communication or confirmation in writing under the pressure of maintaining the programme, he risks recrimination and a breakdown in relationships if things go wrong.

The Standard Building Contract is still complex, albeit its form and simplified clauses make it more likely than previous editions to be used as a tool. Whether simplification will assist in the retrospective legal analysis of a dispute, only time will tell, but simplification does not always lead to clarity.

For too many of the building industry's practitioners, JCT 05 – like JCT 98 – is probably still a document to be signed, put in a drawer and forgotten until things go wrong. If that is so, JCT 05 will fail as a set of working rules, not least because the parties will examine the document only to remind themselves of their rights; their duties and obligations will be forgotten until it is too late.

1.3.4 Scope of this book – JCT 05 Standard Building Contract, With Quantities

The commentary in this book is based on 'JCT 05 Standard Building Contract With Quantities (SBC/Q), Revision 1, 2007' as amended by 'Attestation Update' in February 2008. The 'With Quantities' version is chosen solely on the grounds of its comprehensive nature. The main differences between this form and the 'Without Quantities' and 'With Approximate Quantities' versions relate to evaluation of the Works and calculation of interim payments.

The 'Without Quantities' version contains no reference to Bills. The Contractor is required to price a schedule of Work or specification (provided by the Employer), or to give a price backed by an analysis. The Contractor bears the risk for the quantities. The principles

29 Ndekugri & Rycroft, *The JCT98 Building Contract: Law and Administration*, Butterworth-Heinemann, Section 1.3.3.

of valuing variations are similar to those applying to the 'With Quantities' form, but using rates from a priced schedule. As there are no Bills, there is no reference to the Standard Method of Measurement. A significant effect of this is to avoid an implied warranty by the Employer that information provided to the Contractor is accurate and sufficient.[30]

The 'With Approximate Quantities' version is a remeasurement contract. The 'Tender Price' provides an initial indication of the contract value. The final price, called the 'Ascertained Final Sum', is calculated by applying evaluation principles (similar to the 'With Quantities' version) to measurement of the whole Works.

1.4 Documents forming the Contract

1.4.1 JCT 05: the printed form

1.4.1.1 Contents

JCT 05 contains a contents section at the front of the document listing all the clause headings, but unlike some of its engineering industry counterparts,[31] there is no subject index.

1.4.1.2 Articles of Agreement – Recitals, Articles, Contract Particulars, Attestation

The Articles of Agreement section is the most important part of the contract. It is the core statement of what the parties have agreed; without it, and in the absence of some other contractual arrangement, there is no contract. The Articles are also probably the least read section of the contract documents, except by legal advisers and those preparing documents for execution. This is both strange and disappointing. The Articles not only set out the obligations of the parties, but can also be looked upon as a mission statement.

The section begins by recording the names and addresses of the parties, who are introduced immediately as the Employer and the Contractor. The parties' names are not used again except to identify signatories. The parties may be companies or individuals.

Recitals: The recitals put the agreement into context, explaining the facts on which the contract is based. There are seven recitals describing what is required and what events have taken place:

- *The First Recital* describes the work required by the Employer, identifies the site and indicates that a design and bills of quantities have been prepared on his behalf.
- *The Second Recital* confirms that the Contractor has supplied the Employer with a priced copy of the said bills of quantities, and also where applicable with a priced Activity Schedule (see Chapter 15, Section 15.4.2). The Bills are identified by signature as the Contract Bills.
- *The Third Recital* identifies the Contract Drawings by number and states that the Drawings have been signed.

---

30 See Section 1.4.5 in this chapter dealing with the *Bryant* case.
31 For example, ICE, *Conditions of Contract 7th Edition* (Thomas Telford Ltd, 1999).

- *The Fourth Recital* identifies, by reference to the Contract Particulars, whether the Employer is considered to be a Contractor for the purposes of tax deduction under the Construction Industry Scheme.
- *The Fifth Recital* is optional and refers to an Information Release Schedule given by the Employer to the Contractor (see also Chapter 11, Section 11.4.6).
- *The Sixth Recital* identifies whether the Works are to be carried out in Sections.

The Seventh to Tenth Recitals apply only where the Contractor is to carry out some design under the provisions for a Contractor's Designed Portion.

- *The Seventh Recital* identifies the work to which the Contractor's Designed Portion (CDP) applies.[32]
- *The Eighth Recital* introduces the Employer's Requirements being documents supplied by the Employer showing his requirements relating to work, where the Contractor is to complete the design.
- *The Ninth Recital* states that the Contractor has responded to the Employer's Requirements, and has provided proposals, together with an analysis of the price related to the Contractor's Designed Portion.
- *The Tenth Recital* confirms that the Employer has inspected the Contractor's Proposals, and is satisfied that they appear to meet the Employer's Requirements. (The relevance of this provision is dealt with in Chapter 6, Section 6.12.6.5).

Articles: The Recitals are followed by the Articles, setting out what is agreed:

- *Article 1* states the overriding obligation of the Contractor; subject to the Contract Documents, he will carry out and complete the Works. It is this obligation which prevents the Contractor walking off site unless the terms of the Contract provide.[33]
- *Article 2* states the principal obligation of the Employer – to pay the Contract Sum or such other sum as becomes payable under the Contract. JCT 05 is a lump sum contract, subject only to adjustment as provided in the terms. The main adjustment is likely to be in Variations and correction of errors (see Chapter 6), and in loss and expense (see Chapter 13). This still seems to come as a surprise to those professionals who rely on copious hidden approximate quantities in the Contract Bills and to Contractors who see wholesale remeasurement of the work as an easy option.
- *Article 3* identifies the person to carry out the duties of the Architect (or the Contract Administrator)[34] under the Contract, and gives the Employer the power and the obligation to maintain someone in the role of the so named duty-holder. The Employer is obliged to replace the person named as Architect as soon as reasonably practicable and no later than 21 days of his ceasing to be the Architect for whatever reason (Clause 3.5 describes the procedure). Except where the Employer is a Local Authority and the replacement duty-holder is an official of it, the Contractor can object within 7 days. The Contractor's objection does not have to be reasonable; the test is that the Contractor's reasons are 'considered to be sufficient' by the Employer or a person appointed under

---

32 See Chapter 4.
33 Subject also to statutory rights to suspend performance under the Housing Grants Construction and Regeneration Act 1996, Part II.
34 See also Section 1.5.2.3 in this chapter, Article 3.

the disputes resolution procedures in the Contract. It seems Adjudication, Arbitration or Litigation is a pre-requisite to the Contractor's right to influence the replacement name. It is sometimes tempting for an Employer (particularly lay clients who are not familiar with the building industry) towards the end of a project to discharge the Architect to save on fees, and with that in mind strike out the express obligation to re-appoint. Such temptations should be resisted since it has been held that failure to re-appoint, even in the absence of an express obligation, is still a breach of contract.[35] Likewise it is tempting to discharge the Architect for making decisions with which the Employer does not agree, and either to carry out the function himself, or to re-appoint with a replacement who concurs with the Employer.[36] Again the temptation should be resisted. In the case of *Scheldebouw BV* v. *St James Homes (Grosvenor Dock) Ltd*,[37] it was emphasized that an Employer (because of self-interest) could not appoint himself into a role required by the contract to be that of an independent certifier, unless it was agreed by the Contractor, or the Contractor had tendered on that understanding. Even the appointment of a sympathetic replacement will not assist the Employer with past decisions since Clause 3.5.2 expressly provides that the replacement Architect cannot disregard or overrule any certificate, opinion, decision, approval or instruction of the replaced Architect.

- *Article 4* identifies the person to carry out the role of Quantity Surveyor under the Contract. The obligation to replace, and the right of objection, is the same as for replacement of the Architect under Article 3 and Clause 3.5. However, there is no corresponding express obligation to maintain the replaced Quantity Surveyor's decisions, except in relation to measurement and valuation. This is not surprising since the Quantity Surveyor has no power under the Contract to make decisions except in relation to measurement and valuation. What little authority he had in JCT 98 (under the Variations procedures Alternative A) has been removed with the deletion from JCT 05 of the Contractor's Price Statement system.
- *Article 5* identifies the name of the CDM Co-ordinator under the CDM Regulations, if it is not the Architect,[38] and provides for the Employer to appoint a replacement if necessary.
- *Article 6* identifies the name of the Principal Contractor[39] under the CDM Regulations, if it is not the Contractor, and provides for the Employer to appoint a replacement if necessary. In the absence of any other name, the duty-holder role of 'Principal Contractor' will default to the Contractor, leaving the Employer and the professional team (whose job it is to check) in a vulnerable position if the Contractor does not have the competence or sufficient resources required under the CDM Regulations. In practice it is likely that the Contractor will usually be suitable, but there are situations where the Contractor is chosen because of his trade specialism, rather than competence as a CDM duty-holder. A typical situation would be where a trade Contractor such as

---

35 *Croudace Ltd* v. *London Borough of Lambeth* (1986) 33 BLR 20 CA.

36 The Employer will be in breach of the Contract if interfering with the Architect in the exercise of duties that require the Architect to form an opinion, or to act fairly; see Chapter 2 on the duties and obligations of the Employer and the Architect.

37 [2006] BLR 113.

38 See comment on appointment of CDM Co-ordinator in Section 1.5.2.3 in this chapter, Article 5; and Chapter 10, Section 10.4.

39 See comment on appointment of Principal Contractor in Section 1.5.2.3 in this chapter, Article 6; and Chapter 10, Section 10.6.

a shopfitting firm, or a mechanical and electrical services Contractor, with little or no experience as Principal Contractor, is the main Contractor because of the high specialist content of the overall Works.

- *Article 7* reminds the parties that they have the right to refer any dispute or difference arising under the Contract to adjudication under Clause 9.2. This complies with the Housing Grants, Construction and Regeneration Act 1996, Part II, s.108, and identifies the rules to be applied to such adjudication as being the statutory default rules in the Scheme for Construction Contracts (England and Wales) Regulations 1998.[40]
- *Article 8* applies only if the Contract Particulars entry against Article 8 has been deleted to make clear that Article 8 applies. It constitutes an Arbitration Agreement within the provisions of the Arbitration Act 1996. With the exception of matters referred to in the Article, the parties must refer any dispute or difference arising under or in connection with the Contract to arbitration. Whilst the parties are given a choice of forum for dispute resolution in addition to adjudication, the default position is the Arbitration Agreement. If Article 8 is not expressly stated to apply in the Contract Particulars, the forum for dispute resolution defaults to legal proceedings under Article 9.
- *Article 9* applies as a default if the Contract Particulars entry against Article 8 has not been deleted clearly to activate Article 8.

Contract Particulars: JCT 05 is a standard form which is required to suit different situations, requirements and circumstances. The Contract Particulars section is a schedule for the variables[41] in the Agreement and Conditions and contains two main types of entry. Some of the information in the Contract Particulars is data (e.g. the length of Rectification Period and Date of Possession), whereas some entries trigger which of alternative clauses in the Articles of Agreement or the Conditions are to apply (e.g. whether the parties have chosen the Court (by default), or arbitration (Article 8) as the proper tribunal to hear disputes).

The concept of the Contract Particulars entries being no more than a set of variables is important, and it is essential that the variables are compatible with the relevant operating mechanism in the Agreement, Conditions and Schedules. In practice disputes may arise when either party or the professional team applies Contract Particulars information without referring properly to the operating clause. The case of *Bramall & Ogden Ltd* v. *Sheffield City Council*[42] provides a salutary example. The contract, like JCT 05, contained a partial possession clause[43] under which the Contractor's total liability for liquidated damages based on the sum in the appendix of that contract would be reduced *pro rata* the value of work taken over by the Employer in advance of the remainder. The appendix expressed liquidated damages at a rate of £20 per week for each uncompleted dwelling. Whilst it was possible to make a calculation of damages from the appendix data, it was not possible to insert that data into the operating clause, since there was no sum to be reduced *pro rata*. The court decided the provisions should be construed strictly, and the parties should not strain to make the calculation work. Since the appendix entry was not consistent with the operating clause, the clause was held void for uncertainty and the Employer lost his right to damages.

---

40 Statutory Instruments, 1998 No. 649.
41 See Section 1.5.2.4 in this chapter for comment on each entry.
42 (1983) 29 BLR 73.
43 JCT 98 Clause 18 is of the same effect as JCT 05 Clause 2.33.

The Contract Particulars are split into two parts. Part 1 contains general information and choices referred to throughout the Conditions. Part 2 is reserved for information about the rights of named third parties to enforce benefits granted to them under the Contract.

Detailed commentary is given on the entries required against each clause (see Sections 1.5.2.4 and 1.5.2.5 in this chapter below).

Attestation: The Attestation concludes the Articles of Agreement, confirming that the contract has been concluded in the presence of witnesses. Space is left for the parties either to sign the Contract under hand as a 'simple' contract, or to sign as a deed (also known as a 'specialty' contract or a 'contract under seal')[44] (see also Section 1.5.2.6 in this chapter, Attestation, and Section 1.5.5.6 dealing with amending the limitation period).

Main Differences between 'Simple'[45] and 'Specialty' contracts (deeds): There are three main differences between simple contracts and deeds. First, a gratuitous promise is binding under a deed, whereas under a simple contract a promise must be supported by consideration. Thus in a simple contract a promise to do work is usually supported by a promise to pay money. In JCT 05 the promise can be seen in Article 1: 'The Contractor shall carry out and complete the Works ...', and in Article 2: 'The Employer shall pay ...'. Whilst consideration is not strictly necessary in a deed, a nominal consideration is often included (e.g. 'In consideration of the payment of one pound (£1) ...').[46] This is sometimes incorporated to ensure that a simple contract will exist in any event if the document is flawed as a deed by, say, failure to complete the signing formalities properly.[47]

The second difference is that the facts stated in a deed contract cannot be denied. Thus, if JCT 05 is signed as a deed, statements such as those in the recitals (e.g. Fifth Recital: 'the Employer has provided the Contractor with a schedule' cannot be denied later by the Contractor). Clearly it is essential that the parties check factual statements, particularly if the contract is to be executed as a deed.

The third difference is the most significant and affects the parties' exposure to legal action. Under the Limitation Act 1980 s. 5 an action for breach of contract under a simple contract cannot be commenced later than 6 years after the breach occurring; in the case of a deed, under s. 8 the period is 12 years. These periods start when 'the cause of action accrues' (i.e. the date of the breach of contract). In the case of defective work under a building contract where the obligation is to complete construction work, the point at which the Contractor's breach occurs is thought, for most breaches, to be at practical completion (i.e. the date of the last opportunity to correct the defect before handing it over as being complete).[48] It follows that an action in respect of defective work will normally have to be commenced within 6 years or 12 years of practical completion.[49]

---

44 The term 'under seal' is still used in some quarters to describe a deed, although since 1989 it is no longer necessary to apply a company seal. See also Section 1.5.2.6 in this chapter, Attestation.

45 Sometimes referred to as contracts 'under hand'.

46 For example, see JCT Collateral Warranty CWa/F, p.4.

47 See Section 1.5.2.6 in this chapter, Attestation; and Section 1.6.4, Failure to complete formalities.

48 See *William Tomkinson & Sons Ltd* v. *Parochial Church Council of St Michael-in-the-Hamlet* (1990) 6 Const LJ 319.

49 For the position where the contract is amended to limit or extend the statutory limitation period, see Section 1.5.5.6 in this chapter.

The limitation period for legal action should not be confused with the Rectification Period. The 6-year and 12-year periods do not increase the parties' liability; they merely set the time during which the parties are exposed to action for a breach. Thus in the absence of any breach the parties would be free from contractual liability indefinitely. However, the longer the limitation period, the more time is allowed for breaches committed during the course of the contract to emerge. As a result the prudent Contractor when signing a deed will ensure that major sub-contractors are also engaged under deeds. In this way he may avoid being in a position where he is pursued after 6 years for a sub-contractor's defect, but is unable to pursue the sub-contractor through being 'statute barred'.

1.4.1.3 Conditions

The Conditions set out the terms, or qualifications, attached to the Agreement. Related topics are grouped into Sections 1 to 9:

Section 1: Definitions and Interpretation
Section 2: Carrying out the Works
Section 3: Control of the Works
Section 4: Payment
Section 5: Variations
Section 6: Injury, Damage and Insurance
Section 7: Assignment, Third Party Rights and Collateral Warranties
Section 8: Termination
Section 9: Settlement of Disputes.

Commentary on the principal provisions of the Conditions is dealt with by topic elsewhere in this book.

1.4.1.4 Schedules

Earlier editions of JCT contracts have been criticized for their complexity, partly caused by mixing optional provisions and detailed procedural rules with the general clauses. This was complicated further by a plethora of annexures to the appendix, annexures to the Conditions, and supplemental provisions. JCT 05 is less cluttered. Detailed procedures and options appear in a single set of schedules, and are activated by entries in the Contract Particulars, or by reference in the Conditions. There are seven schedules.

Schedule 1: Contractor's Design Submission Procedure: The Contractor's Design Submission Procedure is a set of rules, applying only to Contractor's Designed Portion Works, giving the Architect opportunity to comment on the Contractor's design. It is the default procedure where the parties have not agreed some other means for the Architect to comment. The submission procedure is dealt with in detail in Chapter 4, Section 4.2.6.

Schedule 2: Schedule 2 Quotation: Clause 5.3 of the Conditions entitles the Architect to instruct the Contractor to give a quotation for a Variation, using the rules set out in Schedule 2. The rules proscribe the timescales for quoting and responding, together with a description of the content required and the effect following acceptance or rejection. The quotation procedure is dealt with in detail in Chapter 6, Section 6.13.

Schedule 3: Insurance Options: Schedule 3 sets out three optional insurance provisions, Option A – New Buildings – All Risks Insurance by the Contractor, Option B – New Buildings – All Risks Insurance by the Employer, and Option C – Existing Structures – Insurance by the Employer. Clause 6.7 identifies which option applies by reference to the Contract Particulars, which in turn register the parties' choice by use of 'strike-out' options. The three options are dealt with in detail in Chapter 7 (see also Section 1.5.2.4 in this chapter, regarding entries in the Contract Particulars).

Schedule 4: Code of Practice (for identification of defects): Clause 3.18.4 deals with the opening up and testing of work that the Architect considers may not comply with the Contract. The difficulty for the Architect once he has identified a defective piece of work is to discover whether the defect is present elsewhere without demanding wholesale demolition of good work. The Code of Practice sets out the criteria to be considered by the Architect, the aim being to help in the fair and reasonable operation of Clause 3.18.4.

Schedule 5: Third Party Rights: JCT 05 is one of the few standard form contracts to make positive use of the Contracts (Rights of Third Parties) Act 1999,[50] by granting enforceable rights to purchasers, tenants and funders as an option under Clauses 7A and 7B of the Conditions. Application is triggered by relevant entries in Part 2 of the Contract Particulars. Schedule 5, Part 1, sets out rights, with limitations and conditions, promised by the Contractor to purchasers and tenants, and in Part 2, the rights, with limitations and conditions, promised by the Contractor to funders. Sub-contractors providing similar rights through collateral warranties are also identified here (see Section 1.5.2.5 in this chapter for commentary on the Contract Particulars entries, and Chapter 9 for commentary on the rights set out in Schedule 5).

Schedule 6: Forms of Bonds: Clauses 4.8, 4.17 and 4.19 of the Conditions provide for optional early payment for the Works, or in respect of materials not yet delivered to the site, and for interim payments in full, without reduction for retention monies. Schedule 6 contains model associated agreements. They are the terms of three bonds. All three forms of bond are agreed between the JCT and the British Bankers Association, and all are intended to be executed as a deed[51] between the Surety and the Employer; the Contractor is not a party to the bond.

The Contract does not provide a choice of bond other than the model.[52] The individual Bonds are dealt with in Chapter 15, Section 15.10.

Schedule 7: Fluctuations options: Fluctuations in price due to the effect of inflation are introduced in Clause 4.21 of the Conditions. Three options are given: A – fluctuations in Employer's contribution, levies and taxes calculated by reference to the Contractor's records and actual cost incurred, B – fluctuations in the cost of labour and materials and

---

50 See Chapter 9 for commentary on effect of the 1999 Act.

51 See comment on Attestation in Section 1.4.1 in this chapter, Simple and specialty contracts, and Section 1.5.2.6.

52 Model bonds were provided in JCT 98 but the Seventh Recital made clear that other forms could be used provided copies were given to the Contractor before entering into contract.

taxes calculated by reference to the Contractor's records and actual cost incurred, and C – a price adjustment formula calculation using monthly published indices. The clauses governing calculation under the three options are set out in Schedule 7. The choice of option is identified in the Contract Particulars by use of a 'strike-out' entry.

For commentary on the application of the fluctuations options see Chapter 14.

1.4.2 Contract Drawings

In Clause 1.1, the definition of 'Contract Documents' includes the 'Contract Drawings'. In turn the 'Contract Drawings' are defined as those referred to in the Third Recital, where they are listed.[53] The Contract Drawings, prepared on behalf of the Employer, show the extent of Architect-designed work to be carried out by the Contractor under the Contract, but subject to further description of work in the Contract Bills[54] and also subject to the issue of further details.[55]

The latitude allowed by the Contract for the Architect to issue further information as the work progresses has historically been interpreted by some Architects and other designers to mean the design need be in no great detail at the contract signing stage. Often the result is *ad hoc* design as site work proceeds, leading to disruption of the Contractor's progress, and dispute.

A feature of JCT 05 is the Information Release Schedule to inform the Contractor when information will be issued by the Architect. The Fifth Recital stating the Employer has provided such a schedule is optional, and the schedule is not required to be annexed to the Contract.

The term 'Contract Drawings' misleadingly gives the impression that the drawings listed as such represent the total of drawings in the Contract describing the entire work. However, there are other drawings in the Contract Documents; they are hidden away in other groups of documents. When part of the Works are required to be designed by the Contractor, the Employer's Requirements (prepared by the Employer or his Architect) may contain further drawings, as too may the Contractor's Proposals (prepared by, or on behalf of, the Contractor).

1.4.3 Bills of Quantities

In Clause 1.1, the definition of 'Contract Documents' includes the 'Contract Bills'. In turn the 'Contract Bills' are defined as those referred to in the Second Recital. Bills of quantities are quantitative lists of descriptive items which, together with CDP Documents where applicable, make up the entire work to be done by the Contractor. The bills are usually prepared by a Quantity Surveyor from drawings and specifications and are measured and described in accordance with a recognized code or 'method of measurement'.

---

53 Not all contracts contain drawings as contract documents in their own right. For example a contract under the JCT Design and Build Contract may incorporate the Employer's drawings but they must be contained with other information in a Contract Document called 'Employer's Requirements'. This principle is replicated in SBC2005 in respect of CDP Works.

54 See Clause 2.1; for commentary on discrepancies in or divergences between documents see also Chapter 6, Section 6.14.

55 See Clause 2.12.

JCT 05 Clause 2.13.1 identifies the method used as the 'Standard Method of Measurement', which is defined in Clause 1.1 as the current edition (unless agreed otherwise) of the Standard Method of Measurement of Building Works, published by the Royal Institution of Chartered Surveyors and the Construction Confederation. Each tenderer is presented with a copy of the bills, and is required to insert unit rates or lump sums. These are then extended by multiplying such rates by the relevant quantities in the bills. The aggregate of all the extensions becomes the amount of the tender and, in the case of the successful Contractor, ultimately is transcribed to Article 2 to become the Contract Sum; the bills become the Contract Bills. Strangely, the important connection between the Contract Bills and the Contract Sum is not to be found in the Articles, the Recitals, or even the Definitions. Instead it is implied, tucked away in Clause 4.1: 'The quality and quantity of the work included in the Contract Sum shall be deemed to be that which is set out in the Contract Bills and … the CDP Documents'. Strictly, if the total amount in the Contract Bills (and the CDP Analysis) and the amount inserted as the Contract Sum in Article 2 were not the same, there is no machinery in the contract for making adjustment; under the Contract rules the Contract Sum could still be adjusted for Variations and other matters without any more complication than exists if the two amounts are the same. However, if there are any overt pricing errors in the tender bills, it is recommended practice[56] for the Employer's team to notify the Contractor before accepting the tender to give him the chance to correct or stand by the error.

Bills of quantities are a common feature throughout all contractual levels of the United Kingdom's building industry. Such is the place of bills of quantities in the education of building industry professionals that even under principal contracts not containing Contract Bills, the Contractor will invariably produce his own bills of quantities, either for his own cost control purposes, or for use in sub-contracts. In the early 1960s the National Federation of Building Trades Employers agreed with its members that they would not tender for work over £8000 in value unless the contract incorporated bills.[57] However, the wide range of present-day standard form contracts suggests that the use of bills as the basis of the contract is on the wane. This is probably due partly to changing techniques in project procurement led by the development of design build contracts,[58] partly to time constraints preventing Architects from preparing full drawings sets, partly by recognition of practical shortcomings of bills,[59] and partly because Employers are not prepared to pay for Quantity Surveyors' full services. The two main advantages of using bills of quantities are (1) a means of comparing tenders like for like, and (2) a pricing structure for valuing variations and calculating the amounts for interim payments, and where applicable, inflation costs by the formula method.[60] In practice, there is a problem in the use of bills for calculating the value of variations, particularly when the bills have been produced before the design is complete.

---

56 In *McMaster University* v. *Wilchar Construction* (1971) 22 DLR (3d) 9 Can, The Court of Ontario held that the employer was prevented from maintaining a contract based on its purported acceptance of a tender knowing that the tender contained an error.

57 In re *Birmingham Association of Building Trade Employers' Agreement* [1963] 1 WLR 484, this agreement was held to be against the public interest under the Restrictive Trade Practices Act 1956.

58 Following suggestions in the 'Banwell Report' that design and construction processes should be less segregated: *The Placing and Management of Contracts for Building and Civil Engineering Work* (HMSO, 1964).

59 For detailed criticism of bills of quantities see Duncan Wallace, I. N., *Construction Contracts: Principles and Policies in Tort and Contract* (Sweet & Maxwell, 1986), Chap 26.

60 See commentary on Fluctuations – Option C in Chapter 14.

As long ago as 1944, the 'Simon Report'[61] described the purpose of bills as 'to put into words every obligation or service which will be required in carrying out the building project'. There is no need, therefore, for a separate 'specification' document under JCT 05 'With Quantities'; indeed there is no place for it, since the specification ought to be contained in the Contract Bills. Except for adding and omitting quantities for variations the bills are supposed to describe the work to be done, but often, due to time constraints and incomplete design the bills are little more than an informed guess as to the content of the work. Since the bills are prepared by the Employer's team the risk in their accuracy is expressly placed on the Employer; thus errors are required to be corrected as though they were a variation.[62] This provides a buffer which enables the tender enquiry documents to be prepared more quickly to meet demanding time targets. Unfortunately when such tender bills become the Contract Bills the strict and limited rules for their adjustment can become unmanageable, and particularly so when there are numerous variations to be adjusted in addition to the task of correcting the original quantities to match the developing design.

Other benefits of bills are not so affected by their content. Even an inaccurate bill will at least give a uniform basis for tendering, enabling the Quantity Surveyor to make adjustments in his cost advice to his client; and provided the descriptions contained in the bills are representative of the type of work required, the unit rates are likely to be suitable for valuing variations. Of greater effect is the practice adopted by some Contractors of inserting unit rates either to suit anticipated cashflow (i.e. 'front loading' to increase the value of interim payments for early trades), or to 'gamble' on the items likely to be increased or omitted,[63] in the hope of increasing the quantity of high profit items.

One feature of bills which has attracted debate is the tendency of Quantity Surveyors to incorporate 'special conditions', sometimes conflicting with the JCT Conditions. A typical example is the incorporation of design duties for specified work other than in accordance with Contractor's design provisions in the Conditions. The effectiveness of such provisions is considered later, in Section 1.5.5 in this chapter, when dealing with amendments and potential conflict between the Bills and Clause 1.3.

1.4.4 Documents related to Contractor's Designed Portion (where applicable)

1.4.4.1 Employer's Requirements

In Clause 1.1, the definition of 'Contract Documents' includes the 'Employer's Requirements'. In turn the 'Employer's Requirements' are defined as those referred to in the Eighth Recital and the Contract Particulars. The Employer's Requirements are accurately named, being a statement of what the Employer wants the Contractor to design for him, as part of the overall Works. The Employer's Requirements may include specifications and drawings in addition to those in the Contract Bills and Contract Drawings (see Chapter 4, Section 4.2.2 for commentary on the Employer's Requirements, their content and form).

---

61 *The Placing and Management of Building Contracts*, HMSO.

62 Clauses 2.14.1 and 2.14.3.

63 See Chapter 6, Section 6.12.2.3 for further comment, and *dicta* of Lord Justice Pearce in *The Mayor Aldermen and Burgesses of the Borough of Dudley* v. *Parsons and Morrin Ltd.*

1.4.4.2 Contractor's Proposals

In Clause 1.1, the definition of 'Contract Documents' includes the 'Contractor's Proposals'. In turn the 'Contractor's Proposals' are defined as those referred to in the Ninth Recital and the Contract Particulars. The Contractor's Proposals are the Contractor's response to the Employer's Requirements, and may include specifications and drawings in addition to those in the Contract Bills, the Contract Drawings, and the Employer's Requirements (see Chapter 4, Section 4.2.3 for commentary on the Contractor's Proposals, their content and form).

1.4.4.3 CDP Analysis

In Clause 1.1, the definition of 'Contract Documents' includes the 'CDP Analysis'. In turn the 'CDP Analysis' is defined as that referred to in the Ninth Recital and the Contract Particulars. The CDP Analysis is a breakdown of the portion of the Contract Sum relating to the work to be designed by the Contractor under the Contractor Designed Portion provisions (see Chapter 4, Section 4.2.4 for commentary on the CDP Analysis, its form and function).

1.4.5 Other documents: Standard Method of Measurement

Clause 2.13.1 states 'the Contract Bills are to have been prepared in accordance with the Standard Method of Measurement', which is defined in Clause 1.1 as 'the *Standard Method of Measurement of Building Works*, 7th Edition, published by the Royal Institution of Chartered Surveyors and the Construction Confederation'. This document is known throughout the industry as 'SMM7'.

One purpose of SMM7 is to standardize the way in which work is measured and described, so that Contractors and Employers' professional teams understand what is included, and what is not. In this way, in theory, disputes over issues such as whether a unit rate in the Contract Bills includes all associated items of labour can be avoided.

Another purpose of SMM7 is to ensure that the tenderer whose task it is to price the bills knows the extent of the work for which he is tendering. This squares with the description of bills of quantities in the Simon Report:[64]

> to put into words every obligation or service which will be required in carrying out the building project.

As a result, probably the most important section of SMM7 is Rule 1.1: 'Bills of Quantities shall fully describe and accurately represent the quantity and quality of the works to be carried out. More detailed information than is required by these rules shall be given where necessary in order to define the precise nature and extent of the required work'.[65]

SMM7 also influences the Contractor's entitlement to extension of time and to recovery of loss and expense. The rules are mainly concerned with accurate measurement. However, they also provide for the inclusion of approximate quantities,[66] and provisional

---

64 See commentary on bills of quantities in Section 1.4.3 in this chapter.

65 See *C. Bryant & Son Ltd* v. *Birmingham Hospital Saturday Fund* [1938] 1 All ER 503. In that case the equivalent clause using similar words in the Standard Method of Measurement of the time, when read with the Contract which stated the Standard Method had been used, was construed by the Court to be a warranty by the Employer that information provided to tenderers was both accurate and sufficient for their needs. In other words such information could be relied on at face value.

66 See Clauses 2.29.4 and 4.24.4 (extension of time and loss and expense in respect of inaccurate forecast of approximate quantities).

sums,[67] in circumstances where work is not designed, or cannot be properly measured for any reason. General Rules 10.1 to 10.6 of SMM7 provide that:

- where work is capable of description but quantities cannot be calculated accurately, the quantities are to be estimated and described as approximate;
- where work cannot be described and given in items it must be given as a provisional sum and identified as being for defined or undefined work;
- where a provisional sum is identified as being for defined work the Contractor will be deemed to have made allowance for the work in his programme and in pricing Preliminaries.

Clauses 2.29 and 4.24 state that the adjustment of approximate quantities and provisional sums for undefined work are grounds for entitlement to extension of time, and to loss and expense if suffered.

1.4.6 Other documents: Adjudication Agreements

The JCT have prepared 'Adjudication Agreements' for use when appointing an adjudicator. The use of the JCT forms is not mandatory, and is referred to in JCT 05 only by use of a *Footnote [23]*. There are two forms of agreement; one for use where the Adjudicator is named in the contract and one for use either where a name is agreed by the parties or where a name is nominated by the nominating body in the Contract (see Chapter 17 for further comment).

1.4.7 Other documents: Model Arbitration Rules

Where arbitration is chosen in the Contract as the forum for dispute resolution, Article 8 refers to the JCT 2005 edition of the Construction Industry Model Arbitration Rules (CIMAR) (see Chapter 17 for comment).

1.4.8 Other documents: Sundry descriptive documents

Several clauses in the Recitals and Contract Particulars require a detailed entry or a reference to separate identified documents annexed to the Contract. A brief description of such optional descriptive documents is provided in Section 1.5.4 in this chapter.

1.5 Preparing the Contract Documents

1.5.1 Generally

JCT 05 is drafted for execution as an express agreement using the printed form. There is no standard alternative such as that found in some forms[68] where provision is sometimes made for a letter of acceptance together with the Tender to create the contract. Nevertheless, and unfortunately, there is a common trait, particularly amongst industrial and occasional

67 See JCT 05 Clauses 2.29.2.1 and 4.24.2.1 (extension of time and loss and expense not allowed in respect of provisional sum for defined work).
68 For example, ICE *Conditions*, 7th Edition which incorporates the Tender as a contract document.

Employers, to forego using the printed form. Instead, an owner's own order form is issued, making reference to the standard JCT form. The result is often confusion when an optional clause has not been chosen. Some of the problems resulting from such misuse of the standard form are dealt with later in this chapter.

The principal documents forming the contract are described earlier in this chapter, and it is essential that they are all constantly kept in mind as a group when preparing the JCT form for execution. Indeed, it is well worth having the various documents available as a complete set so that they can be checked against one another to reduce risk of discrepancy and misunderstanding.

The sections of the form requiring completion are the Articles of Agreement and the Contract Particulars, but there may also be loose printed amendments[69] to be inserted.

As an alternative to using the printed paper form, an electronic version of JCT 05 is available from the JCT,[70] with on-line update support. Entries are made electronically by working through the document, and responding to questions, thus reducing the risk of missing an entry. On completion of entries the contract can be printed for use in the normal manner.

Whether the document prepared for execution is in printed paper form, or an electronic form printed after completion, the entries described in the following sections in this chapter are the same.

Note: reference to '*Footnote [ ]*' in the following Sections 1.5.2 to 1.5.4, is reference to the relevant footnote printed in JCT 05 printed form.

1.5.2 Agreement, Contract Particulars, Schedules

1.5.2.1 Articles of Agreement

The Parties: The names and addresses of the parties are entered in the appropriate spaces. The date is normally inserted after the second party has completed the attestation at the end of the Articles of Agreement.

Footnote [1] reminds the parties to delete the references to 'Company No' and 'registered office' if they are not a company registered under the Companies Acts or registered in another country. Companies incorporated outside England and Wales should insert particulars of the place of incorporation.

There may be difficulty when the Employer is an organization such as a club or religious group, where there may be no corporate identity. It will be necessary to identify who will enter into the contract, and take legal responsibility. The Contractor may, in the interests of prudence, feel it wise to require the contract to be signed personally by a number of the organization's representatives to spread the risk in the event of legal action.

1.5.2.2 Recitals

First Recital: A brief description of the intended works and address of the site is entered.

---

69 JCT 05 Revision 1 2007 incorporates Amendment 1, April 2007. If using a previous version, Amendment 1 is a loose sheet to be inserted. In February 2008, 'Attestation Update' was published, which requires deletion of the section in the Contract '*Execution as a Deed*' and insertion of new Notes and Attestation pages.

70 JCT Contracts Digital Service.

Second Recital: Ensure a fully priced copy of the Contract Bills has been supplied to the Employer and copies are available. The copy for the Contract Documents should be signed or initialled by or on behalf of each party.

If a priced Activity Schedule has been provided, a copy should be annexed to the Contract, after checking that the items priced total the Contract Sum excluding Provisional Sums and the value of Approximate Quantities. Whilst there is no express requirement in the recital to sign or initial the schedule, it helps if the schedule later becomes detached from the Contract Documents.

If a priced Activity Schedule has not been provided by the Contractor, reference must be struck out, as *Footnote [3]*.

Third Recital: The numbers of Contract Drawings are to be inserted. It is important to ensure the Drawings numbers and revisions referred to are the same as those referred to in the Contract Bills, and that the copies for incorporation in the bundle of Contract Documents are copies of the specified Drawings. It is not unknown for the latest edition of a Drawing to be appended in error to the contract in place of the earlier revision specified.

The Contract Drawings are required to be signed or initialled by or on behalf of each party for identification at the same time as they sign the Attestation.

Fourth Recital: Reminds the parties that an entry is required in the Contract Particulars as to the status of the Employer as a 'Contractor' for the purposes of the Construction Industry Scheme.

Fifth Recital: If an Information Release Schedule has not been provided by the Employer, reference must be struck out, as *Footnote [5]*. If a schedule has been provided, there is no express requirement to annex it to the Contract.

Sixth Recital: If the Works are divided into Sections, possibly with different start and finish dates, they will be shown in the Contract Bills, the Contract Drawings or in other documents identified in the Contract Particulars. The care needed in completing this entry cannot be overstressed. The term 'Section' here does not refer to division of the site for site-mapping purposes. If Sections are shown on Contract Drawings or in the Contract Bills it is not sufficient simply to mark different areas without notation, hoping that they will be impliedly imbued with special status. Dividing the Works into Sections for the purpose of the Conditions has far-reaching effect on administration of the contract, and the rights and obligations of the parties, including extension of time and liquidated damages. It should be made clear that the reference is to Sections for application of the 'Sections' provisions in the Contract. The scope of work within each Section needs to be described sufficiently for the Parties (and the Architect) to know, in due course, whether a Section is complete, or whether there is more work to do. The effect of poor definition can be seen from the case of *Taylor Woodrow Holdings Ltd & Another* v. *Barnes & Elliott Ltd*.[71] In that case, sections were described, but none of the sections specifically incorporated the roads, garages and other common services. It was held that it was not possible to determine whether any particular section was complete, and it followed that the liquidated damages provisions in respect of those sections were inoperable; there was no effective trigger for the liquidated damages clause to operate.

71 [2004] EWHC 3319 (TCC).

Seventh Recital: The nature of work, not only to be constructed but also to be designed, by the Contractor under the Contractor's Designed Portion provisions (if applicable) is described here. This recital should be deleted if the Contractor's Designed Portion does not apply.

Eighth Recital: If the Contractor's Designed Portion applies, the Employer must have supplied his requirements (Employer's Requirements) to the Contractor (see Tenth Recital). This recital should be deleted if the Contractor's Designed Portion does not apply.

Ninth Recital: If the Contractor's Designed Portion applies, the Contractor must have supplied to the Employer his proposals (Contractor's Proposals), and an analysis of the relevant part of the Contract Sum (CDP Analysis) (see Tenth Recital). This recital should be deleted if the Contractor's Designed Portion does not apply.

Tenth Recital: If the Contractor's Designed Portion applies, the Employer declares in this recital that he has examined the Contractor's Proposals, and is satisfied that they appear to meet his requirements (see commentary in Chapter 6, Section 6.12.6.5 dealing with conflict between documents).

The Employer's Requirements, the Contractor's Proposals and the CDP Analysis must be signed or initialled by or on behalf of each party, and each document is described in the Contract Particulars. This recital should be deleted if the Contractor's Designed Portion does not apply.

1.5.2.3 Articles

For general commentary on the Articles see Section 1.4.1 in this chapter above. Only Articles that require action other than an entry in the Contract Particulars are dealt with here. Redundant parts of Articles containing options triggered by an entry in the Contract Particulars do not need to be deleted.

Article 2: The Contract Sum must be inserted using figures from the Contract Bills and the CDP Analysis. Since it is common for last minute amendments to be made to the Bills and to Contractor designed work, it is wise to transcribe the figure from the copy of the Contract Bills and CDP Analysis which have been prepared for signing. If the figure does not match the total anticipated after negotiations, amendment may be required to the Bills and Analysis.

Article 3: The name and address of the Architect or Contract Administrator should be inserted, but care should be exercised to avoid striking out the irrelevant part of the term 'Architect/Contract Administrator'. Striking out to leave 'Architect' would require the named person to be entitled to be called an Architect in accordance with the Architects Registration Acts 1931 to 1969 amended by the Housing Grants, Construction and Regeneration Act 1996, Part III. As printed, the term applies adequately to either an Architect or a Contract Administrator, and is used throughout the Contract.

Article 4: The name and address of the Quantity Surveyor should be inserted. There are no restrictions on the use of the title for the purposes of the Contract.

Article 5: The name and address of the CDM Co-ordinator should be inserted here only if the Architect is not going to fill the role. Care should be taken before the Architect is appointed automatically, to ensure that the Client has satisfied himself that the Architect is not only competent but has sufficient resources to be a duty-holder under the CDM Regulations. This article may be deleted only when the Works are not notifiable under the CDM Regulations 2007.

Article 6: The name and address of the Principal Contractor should be inserted here only if the Contractor is not going to fill the role. However, it is essential that the Contractor is assessed by the Employer or his professional team (an obligation under the CDM Regulations), for competence and availability of resources to fulfil the role of Principal Contractor.[72]

[***Further articles inserted by the Parties***: *the standard form does not provide expressly for the insertion of more articles. However, there is no bar to such insertions, and although amendment should be approached with great caution, there will be times when circumstances demand that amendments be made to the standard form. Examples include the introduction of special conditions, or the need to comply with the appointment of duty holders under new legislation. To avoid misunderstanding and dispute over priority, we suggest the proper place for recognition of such amendment is by introduction of further articles here – see Section 1.5.5 dealing with amendments.*]

1.5.2.4 Contract Particulars – Part 1: General

The Contract Particulars entries are variables, inserted to enable the relevant operating clause in the Articles of Agreement, the Conditions, and the Schedules to have effect. The Contract Particulars entries are not substitutes for the relevant operating clause; nor do they operate in isolation as terms of the Contract. It is essential therefore, that each entry is compatible with its operating clause.[73]

Entry *– Fourth Recital and Clause 4.7*: The alternative standard entries should be deleted as appropriate to identify whether the Employer is or is not a 'Contractor' for the purposes of the tax regulations.

Entry *– Sixth Recital*: If the Works are to be treated as a number of Sections, and such Sections are not shown in the Contract Drawings or Contract Bills, the description of the Sections is entered here. If the description appears in a separate document annexed to the Contract, a reference to the annex is sufficient.

However, in a note to the entry, there is an invitation to insert the identifier of a document containing the description, which from *Footnote [11]* impliedly need not be annexed to the Contract. It is not stated whether this option is an intention of the JCT, but considering the commercial effect of dividing the Works into Sections, it is important that there is no lack of clarity as to the extent of those Sections. In the interests of clarity and dispute avoidance, it is suggested that a separate document should always be signed and annexed to the Contract.

---

72 See comment on Article 6 in Section 1.4.1 in this chapter.

73 See example in *Bramall & Ogden* v. *Sheffield City Council* given in Section 1.4.1 in this chapter, dealing with Contract Particulars.

***Entry** – Eighth Recital*: The reference or other identifiers of the Employer's Requirements (where applicable) are entered here (see also commentary on 'Entry – Sixth Recital' above).

***Entry** – Ninth Recital*: The reference or other identifiers of the Contractor's Proposals and the CDP Analysis (where applicable) are entered here (see also commentary on 'Entry – Sixth Recital' above).

***Entry** – Article 8*: The relevant standard entry should be deleted if disputes are to be decided by Arbitration. If the standard entry is not deleted, dispute will be determined by legal proceedings. Note, the parties rights to refer a dispute to adjudication are not affected.

***Entry** – Clause 1.1 Base Date*: The Base Date is to be inserted here. The date is usually, and logically, either the date, or close to the date, of the Contractor's tender. Alternatively, if the Employer wishes, he may insert a date in his tender inquiry documents to provide a date upon which the tender will be based, albeit the tender may be submitted before or after that date. In the event of protracted negotiations the Parties may agree a later date to fairly reflect the pricing level, or relevant conditions. The Parties are reminded by the JCT *Footnote [34]* that the Base Date is relevant to changes in Statutory Requirements and to Fluctuations Options for use in calculating reimbursement in respect of inflation. It is also useful for determining which version of a document applies when it is incorporated by reference, and is not annexed or specifically described. Examples given by the JCT are the Standard Method of Measurement and definitions of prime cost of dayworks, but would also include standard specifications and codes of practice, if dated editions are not identified.

***Entry** – Clause 1.1 CDM Planning Period*: The period for the Contractor and others to carry out CDM planning is inserted here. The style of the optional entries allows for start and finish dates, or for a period with a start or a finish date.

***Entry** – Clause 1.1 Date for Completion of the Works, and of Sections*: The date related to the Works, to be inserted here is the date fixed by the parties at the time the contract is executed. It is not a fixed date in the sense that it cannot be adjusted. Time is not 'of the essence'[74] since express provision is made for adjustment in Clauses 2.26 to 2.29.

Relating to Sections, the Section description and the relevant Date for Completion of each Section is entered.

***Entry** – Clause 1.7*: The Parties insert their respective addresses to which notices must be sent. The issue of a notice under JCT 05 is usually associated with a strict timetable for the recipient. It is in both parties' interest therefore to insert an address that will enable prompt action. The need for a 'quick response' address is emphasized by the inclusion of spaces for fax numbers.

***Entry** – Clause 1.8*: The alternative standard entry should be deleted as appropriate to identify either details of electronic communications that satisfy requirements in the

---

74 'An essential condition or stipulation in a contract without which the contract would not have been entered into': *Osborn's Concise Law Dictionary*, 10th edn (Sweet & Maxwell, 2005).

Contract for communication in writing, or the description of a separate document containing that information. Any continuation of the entry onto further sheets should be signed or initialled by or on behalf of each Party and annexed to the Contract. A note reminds the Parties that if no description is entered here, all communications 'are to be in writing …'. To that extent the use of electronic communication and its status seems clear. Unfortunately, the JCT cloud the issue by adding '… unless agreed otherwise', which suggests that an agreement to accept something other than in writing need not be set out in the Contract Particulars. This inevitably introduces an element of uncertainty as regards priority when a notice or other communication is challenged.

Whilst communications in writing may be agreed to be satisfied by electronic means generally, the Parties need to keep in mind that Clause 1.8 is expressly subject to specific requirements described elsewhere in the contract. Third party notices (schedule 5) and notice of termination (Clause 8.2.3) require actual service or by post, so in order to avoid unnecessary possible conflict when relationships may be fragile, any agreement here should exclude these notices, and any others that the Parties feel are so serious as to demand special attention.

A common habit throughout the industry in sending emails, often with attachments, is to address communications direct to individuals, rather than to a company central address. Too often, the effect is to personalize the message, and deprive the company of the knowledge that the message has arrived. This can create expensive problems if the intended recipient is on leave, unless the company has a system in place for someone else to open, action, or forward all emails. It is therefore essential, particularly regarding communications that could delay the Works if not acted upon, for agreements to contain conditional terms reliant on direction to a central address, albeit copied to an individual if necessary.

Entry – *Clause 2.4*: *Date of Possession of the Works*, *and of Sections*: The date related to the Works, to be inserted here is the date fixed by the parties at the time the contract is executed.

Relating to Sections, the Section description, and the relevant Date of Possession of each Section is entered. It should be noted that dates relating to Sections are treated individually, and are not connected to possession or completion dates of other Sections, unless a connection is clearly stated. For example, if the entries read 'Possession of *Section 1: 1 August, and Section 2*: *1 September*,' a delay to the start of Section 1 will not change the start date for Section 2. Similarly, if the date of completion of Section 1 is the day before the date for possession of Section 2, an excusable delay on Section 1 will not affect the Contractor's entitlement to take possession of Section 2 on time. The two sets of dates are not linked. The Contract and the Contract Particulars refer to dates, not periods, so it is not sufficient simply to enter periods in entries against Clause 1.1 or 2.4. If the Parties wish the timing of one Section to follow on from completion of another, it is necessary to create a link by describing the connection clearly.[75]

Entry – *Clause 2.5*: The alternative standard entry should be deleted as appropriate to identify whether Clause 2.5 applies (entitlement of Employer to defer possession of the site).

---

75 In *Liberty Mercian Ltd* v. *Dyball Construction Ltd* [2008] EWHC 263 (TCC), the date of possession for sequential Sections were entered as 'upon completion of (the previous section)' so that a culpable delay in Section 1 ran through later sections, attracting liquidated damages to each.

If Clause 2.5 does apply, a maximum period other than 6 weeks may be entered provided it is less than 6 weeks.

Alternative provision is made for the same information relating to Sections where applicable, with the facility to have different periods of deferment for different Sections.

Entry – *Clause 2.19.3*: Where the Contractor's Designed Portion applies, the Parties may agree a limit to the Contractor's liability for loss of Employer's use, loss of profit and other consequential losses in the event of defective design (other than the cost of correcting the defective design and damages for delay). The agreed limit (if any) is entered as a sum in respect of the Employer's loss over all Sections. There is no facility in the printed form for a limit per Section. Employers should be wary of attempting to specify separate limits, unless they are confident that they can particularize their losses sufficiently to set losses on a particular Section against the relevant limit.

Entry – *Clause 2.32.2*: The entry for liquidated damages must be a rate (£) per unit of time to be compatible with Clause 2.32.2, and with Clause 2.33 providing for partial possession by the Employer (see Section 1.4.1.2 in this chapter, Contract particulars: regarding confusion between the concepts of partial possession and sectional completion in so far as they affect the Contract Particulars entry for liquidated damages).[76]

The Employer (or his team) must avoid the temptation to treat the liquidated damages entry as a means of encouraging the Contractor to meet deadlines. An amount appearing to be a threat may be construed as a penalty,[77] and face challenge when attempts are made to enforce it. The general principles applying to penalties and liquidated damages are discussed in Chapter 11, Section 11.1. There is a danger when calculating what amount of liquidated damages to put in a contract, to decide upon a figure that is unconscionably high compared with the greatest loss that may be incurred. This does not mean that any figure higher than the actual anticipated is unenforceable, but it may be considered unreasonable and unenforceable if the discrepancy is substantial[78] between the damages stipulated and the damages likely to be suffered. Also, there is a tendency amongst those drafting liquidated damages entries in JCT contracts, to refer to a sum 'per week or part thereof', presumably to ensure that damages apply from the beginning of delay, and do not apply only to complete weeks. It may be that the anticipated loss for a day is the same as for a week, but there is a risk that it is not, and could be construed as unconscionably high. Some contracts[79] avoid the risk by specifying an amount per day.

It is not unknown for a Contractor to accept a high level of liquidated damages on the presumption that he can, at a later date, challenge it, but it is a high risk strategy. Whilst there are circumstances in which liquidated damages can be challenged at the time of their attempted recovery, often on grounds of establishing them as a penalty, success is by no means certain. The approach of the courts was summed up neatly in *Alfred McAlpine Capital Projects Ltd* v. *Tilebox Ltd*, when it was said:[80]

> Because the rule about penalties is an anomaly within the law of contract, the courts are predisposed, where possible, to uphold contractual terms which fix the level of damages

---

76 Referring to *Bramall & Ogden* v. *Sheffield City Council* (1983).
77 *Dunlop Pneumatic Tyre Company* v. *New Garage and Motor Company Ltd* (1915) AC 79.
78 See *Alfred McAlpine Capital Projects Ltd* v. *Tilebox Ltd* [2005] BLR 271, at 279, para 48.
79 See ICE 7th edition.
80 [2005] BLR 271, at 280, para 48.3.

> for breach. This predisposition is even stronger in the case of commercial contracts freely entered into between parties of comparable bargaining power.

It was then emphasized that the parties' pre-contract approach was relevant:[81]

> A figure of (£…) was considered not only by the parties, but also … by their legal advisers. The fact that (the provisions) survived such scrutiny is further evidence that … the liquidated damages provision was reasonable.

There will be occasions when a tenderer cannot afford the risk of high liquidated damages compared with the contract value. In such cases he may need to discuss a reasonable level with the Employer, or load his tender to cover his risk. Where the Employer rejects an approach from the tenderer and is not prepared to discuss at all the level of damages, he may at a later date face a challenge from the Contractor, for which the courts have greater sympathy. To avoid the risk of legal proceedings, it seems clear that the most sensible approach is to sort out the amount of damages acceptable from the outset; the only cost to the parties is the loss of somewhat uncertain rights to challenge the amount.

Alternatively, the tenderer may find it more palatable if the Employer considers relying on actual damages to be proved, instead of liquidated damages. This situation can occur where the actual damages would fluctuate wildly depending on circumstances at the time. If the Employer does not wish to apply liquidated damages at all, but instead wishes to rely on his common law entitlement to recover unliquidated damages (i.e. such damages as he can prove) it is not sufficient for the Employer to simply insert 'Nil' or 'n/a' in the Contract Particulars. In *Temloc* v. *Errill Properties*[82] the Court of Appeal decided the liquidated damages provision was exhaustive of the employer's rights to damages and was capable of being operated at the rate of '£ nil' per week, but in any event, a '£ nil' entry indicated to the Contractor at the time of entering the contract that the employer would suffer no damage. The Court held that the employer lost his right to any damages, liquidated or unliquidated. Similarly in *Chattan Developments Ltd* v. *Reigill Civil Engineering Contractors Ltd*[83] it was held that 'n/a' referred to in a letter confirming an oral contract on JCT terms deprived the Employer of both liquidated and unliquidated damages. However it was observed that a written contract with the whole liquidated damages clause struck out would probably have enabled a claim for unliquidated damages. Despite the *Chattan* case, the effect of an 'n/a' entry is probably still uncertain, depending upon the facts and the parties' construed intentions. In the Australian case of *Silent Vector Pty Ltd* v. *Squarcini*,[84] it was held that 'n/a', in this instance, did not deprive the Employer of the right to claim unliquidated damages. The particular use of 'nil' and 'n/a' in several places in the contract suggested that the parties did not intend to throw away their normal rights, but that the relevant clause should not apply. Nevertheless, the uncertainty in this area was also emphasized, and the cause summarized as being a failure by parties to delete, amend, or add clauses in a clear and consistent manner. If the Employer wishes to create certainty in order to rely on actual damages rather than liquidated damages, he should amend the Contract by deleting the operating clause 2.32.

---

81 *Ibid* at 286, para 95.5.
82 (1987) 39 BLR 30.
83 [2007] EWHC 305 (TCC).
84 *Silent Vector Pty Ltd t/a Sizer Builders* v. *Squarcini* [2008] WASC 246 (30 October 2008).

If the Works are divided into Sections, a separate sum must be inserted for each Section adjacent to the Section description. The principles regarding a single sum apply also to each sum related to Sections. Care needs to taken in deciding upon each amount, for there is a danger that, if damages were claimed for all the Sections, the aggregate could include duplication, and produce an extravagant amount compared with the greatest loss that could have been anticipated.[85] It may be that the greatest anticipated loss in respect of the whole project will be incurred by delay to any one of the individual Sections, in which case it may be justifiable to insert that amount against each; but there is potential, where Sections overlap, for the maximum loss to be recovered against each of the Sections in respect of the same period of concurrent delay,[86] giving rise to a foreseeable duplication.

In the interests of certainty, each Section should be treated as though it were a separate project, and the likely loss for each Section estimated separately. In *Liberty Mercian Ltd* v. *Dean & Dyball Construction Ltd*,[87] it was observed that the entry of different sums against each Section to indicate the different losses that would flow, was strong support for the proposition that they were a genuine pre-estimate, rather than a penalty.

Entry – *Clause 2.37*: If the Works are divided into Sections, the value of each Section is entered here adjacent to the Section description. The purpose of the value is to enable calculation of proportionate reduction of liquidated damages in the event of part of a Section being taken over by the Employer before the whole of the Section is complete.

Entry – *Clause 2.38 – Rectification Period for the Works, and for Sections*: The length of the Rectification Period must be entered here, either for the whole of the Works, or if Sections apply, for each Section. This is the period during which any defects appearing must be notified to the Contractor, to enable him to carry out rectification work. That is both his obligation and his right. The end of the period does not signify an end to the Contractor's liability; it is simply an end to the Employer's obligation to allow the Contractor to put right his own work (see Chapter 3, Section 3.5 regarding liability for defects and defects appearing after the Rectification Period). A period of 6, 9 or 12 months is common. Many building projects have a high mechanical and electrical services or process plant content, in which defects do not become apparent until many months after Practical Completion; it is common, therefore, for the services elements of the Works to have a Rectification Period of at least 12 months. If no period is inserted the Contract Particulars state that the period is 6 months.

Relating to Sections, the Rectification Period may differ from Section to Section, and the periods are entered here adjacent to the Section description. The period starts from the date of practical completion of the relevant Section.

Entry – *Clause 4.8 – Advance Payment*: The alternative standard entry should be deleted as appropriate to identify whether the Employer wishes to make advance payment. If he does then he must enter details of the amount (£) or, if applicable, the percentage of the

---

85 In *Jeancharm Ltd* v. *Barnett FC* (CA) [2003] EWCA Civ 58, a rate of 260% as interest was held to be a penalty and unenforceable.

86 In *Liberty Mercian Ltd* v. *Dean & Dyball Construction Ltd* [2008] EWHC 263 (TCC), it was held that a delay running through several Sections would attract all the different rates of liquidated damages related to them, but in that case the Sections were not concurrent, one Section starting on completion of another.

87 [2008] EWHC 263 (TCC), at para. 25.

Contract Sum to be paid. He must also enter the date or the time when payment will be made and the time(s) at which the advance payment will be reimbursed by the Contractor.

A note reminds the Parties that in the case of Local Authority Employers, Advance Payment will not be applicable.

Entry – *Clause 4.8 – Advance Payment Bond*: If an advance payment bond is not required the appropriate alternative entry must be deleted. Failure to complete this entry will default to a bond being required.

Entry – *Clause 4.9.2*: The parties have three options for identifying the payment certification timetable: (1) Specified start date followed by the same date each month or the nearest business day, or (2) Periods of no longer than one month after the Date for Possession, or (3) The last day of each month, following the last day of a defined month. Option (3) requires amendment to the standard entry (see JCT *Footnote [17]* for suggested words). If no entry is made, the default position is the second of the above options.

Entry – *Clause 4.17.4*: If an off-site materials payment bond is not required for 'uniquely identified' listed items, the standard entry should be deleted. If a bond is required the amount of the bond should be entered.

Entry – *Clause 4.17.5*: If provision for payment for 'not uniquely identified' listed items off site does not apply, the standard entry should be deleted. If provision for payment does apply, a bond is required and the amount of the bond should be entered.

Entry – *Clause 4.19 – Contractor's Retention Bond*: If a bond is to be provided by the Contractor in return for interim payment in full (i.e. 100% of the value of the Works without abatement for retention) the relevant alternative entry should be deleted. A note reminds the Parties that this does not apply to Local Authority Employers.

Clause 4.19.2 of the Conditions requires the use of the form of bond published at Part 3 of Schedule 6, and Clause 2 of that bond requires a maximum aggregate sum to be stated. The sum to be stated in the bond is entered here. Since the bond is intended to replace the retained fund which would have been available for the Employer to use if the Contractor failed to perform, the amount of the bond that the Employer may call upon must be a sum similar to the amount of retention that would have been held. As the retention would have been an accumulating amount based on the value of work executed including Variations, the amount of the bond is an accumulating sum with a maximum equal to the entry in the Contract particulars. The amount entered therefore should take into account the amount of Variations that the Employer may order, to ensure sufficient protection.

Clause 6.3 of the bond requires statement of the date on which the surety is released from liability if a Certificate of Making Good has not been issued, or if the maximum aggregate has not been reached. The relevant date is entered here.

Entry – *Clause 4.20.1*: An entry is only required if the Retention Percentage is not 3%. The Conditions do not prevent any rate higher or lower than 3%.[88]

88 The previous edition of this contract specified 5% or less.

Entry – *Clause 4.21 and Schedule 7*: The alternative standard entries should be deleted as appropriate to identify which of the three optional fluctuations clauses applies. See comment under Section 1.4.1 in this chapter dealing with 'Schedule 7: Fluctuations options'. See also Chapter 14 for commentary on the three options.

A percentage addition to amounts payable under Fluctuations Options 'A' and 'B' should be entered.

The Base Month for the purposes of Rule 3 of the Formula Rules, where Fluctuations Option 'C' applies, should be entered (see commentary on Option 'C' in Chapter 14).

The alternative standard entry should be deleted as appropriate to identify whether formula adjustment is to be made under the Work Category Method (i.e. Part I) or the Work Group Method (Part II) of Section 2 of the Formula Rules. Note the method used should be the method stated in the tender documents.

Entry – *Clause 6.4.1.2*: The amount inserted is the minimum cover required in respect of the Contractor's liability for injury or death to persons and injury or damage to property.

Entry – *Clause 6.5.1*: The alternative standard entry should be deleted as appropriate to identify whether or not a Joint Names Policy insurance *may* be required. The amount inserted is the amount of insurance cover to be provided if the Employer requires the Contractor to insure in joint names against injury or damage to property which is the liability of the Employer. *Footnote [20]* reminds the Parties to amend the printed entry if the amount stated is to be an aggregate, and not for any one occurrence or series of occurrences.

Entry – *Clause 6.7 and Schedule 3 – Insurance of the Works*: The alternative standard entries should be deleted as appropriate to identify which of Insurance Options 'A', 'B' or 'C' applies. See also comment under Section 1.4.1 in this chapter dealing with 'Schedule 3: insurance options' (see Chapter 7 for commentary on use of alternative clauses).

Strictly, there is no need to delete the redundant alternatives in Schedule 3, although in practice it is often done in an attempt to avoid confusion. Unfortunately it sometimes has the reverse effect. It is not unknown for the wrong alternative to be deleted leaving a discrepancy in the Conditions.

Footnote [21] notes that Terrorism Cover will require an additional premium, or may be difficult to obtain. The note advises that where such difficulty arises the Parties should involve their insurance advisers. The advice in *Footnote [21]* applies equally to any matter of insurance where the Contractor's normal insurance policies, which may have to cover the requirements of a diverse range of standard form contracts, do not exactly match the requirements of the JCT Contract.

Entry – *Clause 6.7 and Schedule 3 – Percentage to cover professional fees*: The percentage entry is the percentage to be added to insurance cover to allow for professional fees required to reconstruct the work. If no percentage is entered the rate defaults to 15%.

Entry – *Clause 6.7 and Schedule 3 – Annual renewal date*: The annual date supplied by the Contractor for renewal of insurance.

Entry – *Clause 6.11 – CDP Professional Indemnity Insurance*: The minimum amount of indemnity required by the Employer is entered here. The amount may relate to individual claims or a series of claims, or to an aggregate during a yearly period, the choice of which is selected by deleting the standard entry as appropriate. If the Parties choose the period option, and require a period other than one year, the length of the period must be stated, albeit there is no specific space provided. If no amount of indemnity is stated, a note informs the Parties that Clause 6.11 insurance is not required.

Space is provided to insert the separate level of cover required for pollution/contamination claims. If no figure is entered, the level is the full amount of indemnity stated.

The period prior to expiry of PI Insurance is selected by deleting the standard entry as appropriate, or by inserting a period if other than 6 or 12 years. A note guides the Parties to choose a period not exceeding 12 years. If no period is selected, the period defaults to 6 years.

***Entry** – Clause 6.13*: The alternative standard entry should be deleted as appropriate to identify whether or not the Joint Fire Code applies.

If the Joint Fire Code applies, the alternative standard 'yes/no' entry should be deleted as appropriate to identify whether or not the insurer under Schedule 3, Insurance Option A, B or C has specified that the Works are a 'Large Project'.

Footnote [22] reminds the parties that where Option A applies, these entries are made on information supplied by the Contractor. The printed footnote applies to both entries (perhaps in error), but the choice of whether the fire code will apply or not should be a matter for the Employer to decide at the tender enquiry stage; it is not a fact for the Contractor to supply.

***Entry** – Clause 6.16*: The alternative standard entry should be deleted as appropriate to identify who bears the cost of any amendments made to the Fire Code after the Base Date.

***Entry** – Clause 7.2*: The alternative standard entry should be deleted as appropriate to identify whether or not an assignee to whom the Employer has assigned his benefits under the Contract after Practical Completion may commence proceedings in the Employer's name to enforce such benefits. If neither entry is deleted, Clause 7.2 will apply.

If Clause 7.2 applies, and the Works are divided into Sections, the parties may choose whether assignees rights apply to each Section (by leaving the standard entry), or not (by deleting the standard entry). It may be that only parts of the project are intended to be assigned by the Employer, in which case Clause 7.2 may apply to specific Sections only. If that is the case the relevant Sections have to be identified.

***Entry** – Clause 8.9.2*: The period for which the Employer is entitled to suspend the Works as a result of his own culpable events without giving cause for the Contractor to give notice of termination should be entered here. If no entry is made, a standard period of 2 months applies.

***Entry** – Clauses 8.11.1.1 to 8.11.1.5*: The period for which the Employer is entitled to suspend the Works as a result of *force majeure*, default by statutory undertakers, Specified Perils, civil commotion, threat of terrorism or terrorism activity, or the UK Government exercising statutory power without giving cause for the Contractor to give notice of termination should be entered here. If no entry is made, a standard period of 2 months applies.

***Entry** – Clause 9.2.1*: the Parties may agree upon and name an Adjudicator in the space provided (see commentary in Chapter 17, Section 17.4.2 regarding advantages and disadvantages of naming an Adjudicator). If no name is entered, the alternative standard entries should be deleted as appropriate to identify which Adjudicator nominating body is selected. If no selection is made here by the parties, the referring party to an adjudication may choose the nominator, but it must be one of the standard entries. If all the standard entries are deleted, the referring party will be entitled to ask any nominator[89] to select an Adjudicator.

If the parties wish to agree on another nominating body they are free to do so. Whilst *Footnote [23]* draws attention to standard Adjudicator agreements prepared by the JCT, their use is not mandatory.

Entry – *Clause 9.4.1*: The alternative standard entries should be deleted as appropriate to identify which Arbitrator nominating body is selected. If no selection is made here by the parties the standard selection is stated to be the RIBA.

1.5.2.5 Contract Particulars – Part 2: Third Party Rights and Collateral Warranties

Part 2 is divided into five sections (A to E), in three topic groups:

- Purchasers and Tenants Identities, and rights from the Contractor (A, B);
- Funder's identity, and rights from the Contractor (C, D);
- Collateral Warranties from Sub-Contractors to Purchasers, Tenants, Funder, or Employer (E).

Purchasers', Tenants' and Funders' Rights

Sections A to D relating to Purchasers, Tenants and Funders and their rights need to be completed in full only if rights or warranties to others are required from the Contractor, and they are not set out in a separate document.

If rights or warranties are indeed set out separately, the relevant document must be identified in the space provided at the head of Part 2. The document should be signed or initialled by or on behalf of each party, and annexed to the Contract (*Footnote [27]*). If required rights or warranties are not set out separately, Part 2 needs to be completed.

Part 2 identifies benefits and beneficiaries other than the parties to the Contract. Benefits may be provided either through collateral warranties or by statement in the Contract enforced under the Contracts (Rights of Third Parties) Act 1999. The use of a collateral warranty is triggered by reference to Clause 7C of the Conditions, and reliance on the 1999 Act is triggered by reference to Clause 7A of the Conditions.

Section (A) – Identity of Purchasers and Tenants, and whether rights are conferred as third party rights, or by collateral warranty: If the parties wish to confer rights on purchasers or tenants, the recipients must be identifiable, although they need not even exist when the Contract is entered into. If rights are intended to be enforceable under the 1999 Act, this section must be completed to overcome the limitation provisions in Clause 1.6. A name, a description, or a class is sufficient, so long as a beneficiary wishing to enforce his rights can be identified as an intended beneficiary. The Guide gives as an example, 'all first purchasers' and 'all original/first lessees'. A list of intended beneficiaries must be entered in the first column, and a brief description of the Works relevant to each intended

89 See Scheme for Construction Contracts (England and Wales) Regulations 1998, s. 2(1)(c).

beneficiary must be inserted in the second column. Failure to complete these columns nullifies both Clauses 7A and 7C. The adjacent entries in the third column state against each beneficiary whether Clause 7A or Clause 7C applies (i.e. whether the rights will be conferred by third party rights, or by collateral warranty). If no choice is inserted, the default method is by Clause 7A, third party rights.

There is room for confusion in an anomaly between the side-heading which refers to Clauses 7A, 7C and 7E, and the third column referring only to a choice between Clauses 7A or 7C. Clause 7E refers to Collateral Warranties provided by sub-contractors, and it is likely that an identified Purchaser or Tenant will be intended to receive benefits from a sub-contractor. The third column of this section is stated in the headnote to refer only to rights provided by the Contractor. Rights from sub-contractors are dealt with in section (E). However, if the Employer requires third party rights to be provided by a sub-contractor he will need first to enter that intention in this section, but he will also need to ensure that the relevant sub-contract between sub-contractor and the Contractor contains appropriate provision. This is fraught with difficulty, as it requires the Employer's team to 'police' and control the Contractor's supply chain; a clear requirement to provide rights in a sub-contract would be of little comfort to a beneficiary such as a Tenant, if the sub-contract in fact did not in turn contain those rights. In the Guide, the JCT wisely suggest 'it would be unrealistic and overly prescriptive to require the Contractor in every case to use a JCT sub-contract', and that 'it appears … impractical at that level to provide for third party rights …'

Section (B) – Rights from the Contractor to Purchasers and Tenants: Entries in this section relate to paragraphs in Schedule 5 or the corresponding clause in the relevant JCT Collateral Warranty.[90]

Entry – *Paragraph/Clause 1.1.2*: The alternative standard entry should be deleted as appropriate to identify whether the Contractor accepts liability, in principle up to a stated maximum, for losses incurred by a Purchaser or Tenant (other than the cost of actual remedial work) in the event of remedial work being necessary. This typically covers losses such as loss of income or profit, but subject to proof of actual loss.[91] The maximum must be inserted as a sum in the space provided. Paragraph 1.1.2 will not apply unless both the entry in respect of 1.1.2 and a maximum is inserted. The type of maximum liability (i.e. per breach or an aggregate limit) is identified by deleting the irrelevant standard option.

Entry – *Paragraph/Clause 1.3.1*: Space is provided to insert the names of 'Consultants' who agree to give undertakings such as those given by the Contractor reducing the Contractor's liability. A note states that if no names are specified, they shall be the Architect and Quantity Surveyor.

Entry – *Paragraph/Clause 1.3.2*: Space is provided to insert the names of sub-contractors who have agreed to give undertakings in relation to design of Sub-Contract Works for which the Contractor is not liable under the Contract. A note states that if no names are

90 Rights described and paragraph numbers in Schedule 5 are the same as those provided in the corresponding clauses of the JCT Collateral Warranty.

91 See Chapter 12 for general principles applying to damages.

specified, they shall be the sub-contractors that have agreed to give third party rights or collateral warranties to any Purchasers or Tenants.

Section (C) – Identity of Funder: If the parties wish to confer rights on the Funder, he must be identifiable, although he need not even exist when the Contract is entered into. The principles of identification are similar to those described for section (A) above. A note reminds the Parties that if no-one is identified, the Contractor will not be required to provide rights, either by third party rights or by collateral warranty.

Section (D) – Rights from the Contractor to Funder: Entries in this section relate to paragraphs in Schedule 5 (Part 2), or the corresponding clause in the JCT Collateral Warranty.[92]

Entry – *Nature of Funder Rights*: The alternative standard entry should be deleted as appropriate to identify whether rights from the Contractor are to be provided as third party rights, or by collateral warranty.

Entry – *Paragraph/Clause 1.1*: Space is provided to insert the names of Consultants and sub-contractors who have agreed to give undertakings in respect of consultants services or sub-contractor's design. A note states that the names shall be the same as those specified under Section (B), unless otherwise stated.

Entry – *Paragraph/Clause 6.3*: An entry is necessary only if the period required for the Contractor to give warning to the Funder of his intention to terminate the Contract is other than 7 days. The Parties need to consider carefully whether to increase or decrease the standard 7 days warning. The advance warning to the Funder applies to any termination notice intended by the Contractor to the Employer, so if a notice of intention to terminate is followed by a notice actually terminating employment under the Contract, there is a delay in each case before the Contractor can act. The effect, using the default period of 7 days, is to add 14 days (or other period depending on this entry) to the termination process. A longer warning period gives the Funder more time to resolve the problem if he can, but of course it also delays the Contractor from enforcing his rights under the standard terms of the Contract.

Collateral Warranties from Sub-Contractors

Section (E) – Collateral Warranties from Sub-Contractors: If collateral warranties are required from sub-contractors, particulars including names, beneficiaries, rights and terms should be set out in a separate document. If the terms of a warranty are required other than one of the standard JCT Collateral Warranties,[93] it is essential that details are specified in the separate document. A description of the document for identification is entered here. The document should be signed or initialled by or on behalf of each party, and annexed to the Contract (*Footnote [27]*). If sub-contractors' names or types of warranty are not entered here, the second part of Section (E) needs to be completed.

Condition (i) to Section (E) identifies recipients entitled to warranties from sub-contractors; unless otherwise stated, they are any Purchaser, Tenant or Funder identified in Section (A) or (C), and the Employer.

---

92 Rights described and paragraph numbers in Schedule 5 are the same as those provided in the corresponding clauses of the JCT Collateral Warranty.

93 SCWa/P&T (for a Purchaser or Tenant), SCWa/F (for a Funder), SCWa/E (for an Employer).

Condition (v) specifies the warning periods in respect of termination notice by reference to Section (C) entry for 6.3.

Entry – *Clauses 3.7 and 3.9 of the Conditions (first column)*: *Identity of Sub-Contractors from whom Collateral Warranties are required*: Names of sub-contractors, or category, should be listed here only if not entered in the separate document. Section (E), *Footnote [31]*, advises Employers to be selective in listing sub-contractors. It is unusual for sub-contractors having no design input (e.g. an earthmover) to be required to give a warranty, but it is common, and advisable, for a sub-contractor whose work includes a specialist installation designed by the sub-contractor (e.g. curtain walling, building management system).

Entry – *Clauses 3.7 and 3.9 of the Conditions (second column) – Type of warranty required*: Entries here, against each adjacent name, should identify only standard JCT Warranties.[94] If non-standard terms of warranty are required they should be specified in the separate document.

Whatever the form of warranty, liability is limited if a maximum liability is specified in respect of the Contractor in Section (B).[95] The specified maximum will apply also to a sub-contractor's warranties, unless a lower amount is specified. Similarly, if liability is limited in respect of contribution by Consultants, the Consultants are those listed in Section (B).[96]

Entry – *Clauses 3.7 and 3.9 of the Conditions (third column) – Level of Professional Indemnity insurance*: The level of Professional Indemnity insurance (PI), if required, must[97] be entered here as a sum. Such insurance will not be necessary if the sub-contractor has no design input. Condition (ii) of Section (E) provides that the level and basis of insurance should match the entry for Clause 6.11 in the Contract Particulars. However, many sub-contractors who manufacture equipment as part of a specialist installation have Product Liability insurance, and may be reluctant to take out individual PI policies for individual projects. The interface between Professional Indemnity and Product Liability insurances is outside the scope of this book, but in any event, Employers and their professional teams should involve their insurance brokers or representatives in deciding who is best placed to bear the risks, and who should pay the premiums.

1.5.2.6 Attestation

The Attestation contains two clearly headed alternative sections, together with useful user notes, to enable the parties to execute the Agreement either 'under hand', or 'as a Deed' (see commentary on Attestation in Section 1.4.1.2 in this chapter).

Under hand: If the Contract is to be signed as a simple contract under hand, each party, or his authorized representative, and witness are required to sign where indicated.

---

94 See Chapter 9, Section 9.4.
95 Part 2 Section (E) Condition (iii).
96 Part 2 Section (E) Condition (iv).
97 Part 2 Section (E) Condition (ii).

As a Deed: If the Contract is to be signed as a deed the parties are required to enter their full names (i.e. company or individual etc) and affix their common seal[98] and sign, or simply sign, in the respective spaces. There are four options:

- where a party is a company not using its common seal, and the document is signed by a Director and the Company Secretary, or two Directors – option (A);
- where a party is a company or organization using its common seal, affixed in the presence of, and signed by, a Director and the Company Secretary, or two Directors – option (B);
- where a party is a company not using its common seal, and the document is signed by a single Director in the presence of a witness, who attests the signing with his own signature and his name and address – option (C);[99]
- where a party is an individual, and the document is signed by that individual in the presence of a witness, who attests the signature with his own signature and his name and address – option (D).

The Notes [1] to [6] printed in the Attestation provide detailed guidance on application of the alternatives.

The reason for the alternative spaces for companies, is that under the Companies Act 2006 a company need not use a seal.[100] A document without a seal but stated to be intended as a deed, and signed in accordance with option (A) or (C), will have effect as a deed on delivery. The document is presumed to be delivered when it is executed, provided there is no evidence of intention that delivery should take place later or upon some condition being met.

Under the Law of property (Miscellaneous Provisions) Act 1989, if a party is an individual[101] it is not necessary for him to affix a seal. Provided the document makes clear that it is intended as a deed[102] and it is signed by the individual in the presence of, and signed by, a witness, the document will have effect as a deed.

Local Authorities must use their seal, so only option (B) is available to them.

Care needs to be taken to ensure both parties complete all the relevant formalities. This includes signing any Contract Documents which require signing (e.g. the Contract Drawings, Contract Bills), and initialling all amendments. If the relevant formalities are not completed the contract may not have effect as a deed. However, that is not to say that no contract exists; it is likely that a simple contract will still be concluded.[103]

One situation to be avoided is where one party has signed the Contract, and the other party has either sealed the Contract or signed as a deed. The Contract is still binding, but the party who has signed will be bound under a simple contract, whereas the party signing under seal or as a deed will be bound under a deed (specialty). The result is that the party

98 Under the Companies Act 2006, a company need not have a common seal at all – s.45(1); but if a company does have a seal the name of the company must be clearly engraved on it – s.45(2). Failure to comply with s.45(2) is an offence – s.45(3).
99 Companies Act 2006, s.44(2)(b).
100 Companies Act 2006, s.44(1)(b).
101 Law of Property (Miscellaneous Provisions) Act 1989, s.1(3).
102 Law of Property (Miscellaneous Provisions) Act 1989, s.1(2).
103 See commentary in Section 1.4.1 in this chapter, Attestation, regarding consideration in simple contracts and specialties.

under the simple contract will be vulnerable to pursuit for a period of 6 years, and the party bound under the deed will be exposed for 12 years.[104] For this reason, a prudent practical measure when documents are being prepared for execution is to delete the alternative not being used before a signature is entered in error.

1.5.2.7 Conditions

Only clauses in the Conditions that require action other than an entry in the Contract Particulars are dealt with here. Redundant parts of clauses containing options triggered by an entry in the Contract Particulars do not need to be deleted.

Clause 1.1 Definitions – Public Holiday: The meaning of Public Holiday should be amended if appropriate. See *Footnote [35]*.

Clause 1.12: This clause should be amended if the law applicable to the Contract is not the law of England. See *Footnote [36]*.

1.5.3 Drawings and Bills of Quantities

1.5.3.1 Drawings

The Contract Drawings must be collected together and checked to ensure the drawing numbers correspond exactly with the numbers in the Third Recital. If, for example, drawing No. 04B is listed, then drawing No. 04B should be annexed, not No. 04A or 04C. This point is so obvious that it should not need to be made, but the error is common, often as a result of late changes. Another necessary check, which is a result of growing flexibility of computer-aided design and electronic communication, is to ensure that amendments to drawings are registered and 'frozen' on a drawing revision. The danger is that 'advance copy' drawings may be communicated freely between design disciplines, or designer and Quantity Surveyor, without control. It is not uncommon to find two or more copies of a drawing in circulation, bearing the same reference, but containing different information; which version is the basis of the contract can easily be forgotten.[105]

The correct drawings must all be copied and signed by the parties to comply with the statement in the Third Recital.

1.5.3.2 Bills of Quantities

The Contract Bills should be checked to ensure that:

- the Bills are arithmetically correct;
- the total price, when added to the price in the Contractor's Proposals (if any) is the price transcribed into Article 2 (the Contract Sum);
- any drawing numbers listed as being the drawings on which the Bills have been prepared are the same numbers as those listed in the Third Recital;

---

104 See commentary in Section 1.4.1 in this chapter, Attestation, regarding statutory limitation periods.
105 See Section 1.6 in this chapter, 'When things go wrong'.

- data given in the Bills relating to the Contract particulars does not conflict with the entries in the Contract (whilst the Contract Particulars will override the Bills, it should be remembered that in day-to-day use, members of the project team are more likely to refer to the Bills than the Contract; it is wise to avoid unnecessary potential dispute by removing the likely causes);
- if the Bills provide that certain work must be carried out by persons listed in the Bills, such a list must contain not less than three names to comply with Clause 3.8 of the Conditions;
- if the Employer wishes to pay for materials off site, that the list supplied to the Contractor showing uniquely identified and not uniquely identified materials in pursuance of Clause 4.17 of the Conditions is annexed to the Contract Bills.

The Contract Bills must be signed by the parties to comply with the Second Recital.

1.5.4 Optional documents

There are a number of documents referred to in the Contract which are optional, or result from optional provisions, and which require to be annexed to the Contract:

Activity Schedule: If an Activity Schedule has been provided by the Contractor (see Second Recital) it should be checked to ensure that each activity listed is priced, and that the sum of those prices equals the Contract Sum excluding Provisional Sums and the value against Approximate Quantities. The schedule should be signed or initialled by or on behalf of each party to comply with the Third Recital.

Information Release Schedule: If an Information Release Schedule has been provided by the Employer (see Fifth Recital) there is no express requirement in the Contract for it to be annexed. However, in the interests of avoiding misunderstanding and dispute, the schedule should be signed or initialled by or on behalf of each party and annexed to the Contract.

Description of Sections: If the Works are to be treated as divided into Sections, and if the description of those Sections is not included in the Contract Drawings or Contract Bills, a separate document will be needed to identify the Sections to comply with the Sixth Recital. The document should be signed or initialled by or on behalf of each party, and annexed to the Contract.

List of electronic communications: If the Parties have agreed that certain communications will be accepted as being 'in writing', the details must be written in a separate document or listed in the Contract Particulars (see Entry – Clause 1.8). The separate document, or if applicable, a continuation onto further sheets of the list written out in the Contract Particulars, must be signed or initialled by or on behalf of the Parties and annexed to the Contract.

Dates of Possession of Sections (Entry – Clause 2.4); Maximum period of deferment of Sections (Entry – Clause 2.5); Section Sums (Entry – Clause 2.37); Rectification Periods of Sections (Entry – Clause 2.39): The printed form provides space for entries relating to only three Sections. If there are to be more, the Parties will need to continue these entries in the Contract Particulars onto further sheets. The further sheets must be signed or initialled by or on behalf of each Party and annexed to the Contract.

Contractor's Designed Portion Documents: If a Contractor's Designed Portion applies (see Seventh to Tenth Recitals) the Employer's Requirements, the Contractor's Proposals, and the CDP Analysis must all be signed or initialled by or on behalf of each party, and annexed to the Contract. Checks should be carried out similar to those listed for Bills of Quantities (see Section 1.5.3 in this chapter above). In addition it is important for both parties, in the interests of avoiding later dispute, to ensure that:

- the Contractor's Proposals and the Employer's Requirements and the CDP Analysis are compatible (see commentary in Chapter 4, Sections 4.2.1 to 4.2.4, and Chapter 6, Section 6.12.6.5 dealing with problems in conflict between documents);
- any Provisional Sums in the Contractor's Proposals are transferred to the Employer's Requirements (see commentary in Chapter 4, Section 4.2.3);
- unnecessary information provided with the Contractor's Proposals not intended to make the Contractor's Proposals condition, is removed;
- any drawings or other non-public documents to which the Employer's Requirements or the Contractor's Proposals refer, are annexed to the relevant document;
- if any assumptions[106] are expressed in the Contractor's Proposals, the intended effect is clear in the event that such assumptions do not materialize. In the interests of certainty, if assumptions are intended to make the Contractor's Proposals conditional, they should be transferred into the Employer's Requirements (see Chapter 6, Section 6.12.6.5 dealing with conflict between documents);
- where optional proposals have been submitted by the Contractor for the Employer's consideration, the rejected proposal is deleted. Where an accepted option differs from the Employer's Requirements, the Employer's Requirements should be modified.

Third Party Rights/Collateral Warranty Documents: The Contract Particulars, Part 2, if applicable, refer to two optional documents setting out:

- rights or warranties required from the Contractor (Part 2 heading paragraph);
- rights or warranties required from sub-Contractors (Part 2, section (E)).

Each document must be signed or initialled by or on behalf of each party and annexed to the Contract to comply with the recommendation in Part 2, *Footnote [27]*.

1.5.5 Amending the Standard Contract Form

1.5.5.1 Generally

The standard form is sometimes amended by the parties, either through choice, or because it is necessary.

106 In *Ove Arup & Ptnrs International Ltd & Anr* v. *Mirant Asia-Pacific Construction (Hong Kong) Ltd & Anr* [2005] EWCA Civ 1585, paras 91-94, the Court of Appeal held that a foundations designer who had made assumptions, had a duty to ensure that further information was acquired to verify those assumptions, and to notify his client of that need.

Whilst JCT 05 is published by an organization which includes representatives of various construction industry client bodies, individual building owners do not necessarily see the entire standard form as their own. Many will introduce amendments or additions to the form to bring it into line with their own commercial practices. The normal received wisdom is that the parties should avoid tinkering with a standard form. This is partly for fear of destroying the delicate balance of risk allocation; it is also to avoid legal uncertainty created by changing individual clauses which form part of a complex web of interacting cross-referred terms. Amendment by choice has attracted judicial criticism: [107]

> A standard form is supposed to be just that. It loses its value if those using it or, at tender stage those intending to use it, have to look outside it for deviations from the standard.

However, reallocation of risk is a matter for the parties if that is what they are prepared to agree upon, and one can only advise caution on the manner of its reallocation.

There are occasions, albeit rare, when amendment becomes necessary, usually as a result of new legislation. The JCT are quick to bring all of their standard forms into line with new legislation, but there is a risk of a standard form becoming non-compliant for a short period, while the JCT catch up. However, the risk is more acute when the parties, for whatever reason, use an out of date edition of the printed contract.

The following sections, 1.5.5.2 to 1.5.5.6, deal with enforceability of non-standard amendments to JCT 05, introduced by the parties.

1.5.5.2 Amendment by introduction in the Bills and other documents

Employers (or their professional teams) often attempt to introduce changes to obligations through description in the Contract Bills or other descriptive schedules. A typical example is the tendency sometimes to incorporate design duty and liability in the absence of express provisions in the Conditions. It is argued by many Contractors that the purported imposition of design duty and liability in the bills is an attempt to modify the Agreement, and that Clause 1.3 prevents it.[108] Clause 1.3 states 'Nothing contained in the Contract Bills or the CDP Documents shall override or modify the Agreement or these Conditions'. However, in *Moody* v. *Ellis*[109] the Court of appeal held that wording to a similar effect in the 1963 edition of the JCT Standard Form did not prevent provisions in the Bills from being incorporated. An example can be seen in the case of *Haulfryn Estate Co. Ltd* v. *Leonard J. Multon & Ptnrs and Frontwide Ltd*,[110] where it was held that an item in the Bills[111] placing design obligations on the Contractor, did not override or modify the interpretation of the Conditions which contained no provision for design by the Contractor; rather they 'added to but were consistent with the obligations imposed by the Conditions'. In other words, design and build is an extension of build only, not a contradiction; the concepts are not mutually exclusive.

---

107 See *Royal Brompton NHS Trust* v. *Hammond & Others (No. 9) [2002]* EWHC 2037, at para. 60.

108 The authority used for this view is invariably *John Mowlem & Co. Ltd* v. *British Insulated Callenders Pension Trust Ltd* (1977) 3 ConLR 63.

109 [1983] 26 BLR 45.

110 ORB; 4 April 1990, Case No. 87-H-2794.

111 The contract was JCT Minor Works form which contained words very similar to clause 1.3 of JCT 05.

Similarly, in *Royal Brompton NHS Trust* v. *Hammond & Others (No. 9)*,[112] it was held that the imposition of programming obligations in the Bills did not offend against the Conditions in JCT 80 (Clause 2.2.1 of JCT 80 is a forerunner of JCT 05, Clause 1.3).

These decisions will please those who view the Bills as a specific description of the work required on a particular project, and that should in any event have priority over the standard terms which are not project specific.[113]

However, such amendments are not always successful. Another example of an attempt to increase the Contractor's obligation can be found in *M. J. Gleeson (Contractors) Ltd* v. *London Borough of Hillingdon.*[114] In this case the Bills, by a hand-written amendment, contained obligations to complete various parts of the Works by specified dates; the Conditions (JCT 63) contained only a single Date for Completion, together with a clause containing similar words to those in JCT 05, Clause 1.3. It was held that the higher obligation as to time in the Bills did not override that in the Conditions. Here the concept in the Conditions of one completion date did contradict the concept of several dates in the Bills, so the equivalent of Clause 1.3 could be operated, and the Bills were over-ridden.

A cavalier approach to inserting special terms in the Bills should be avoided if success is to be certain; it is better to amend the Conditions, if amendments are necessary at all.

1.5.5.3 Amendments to the printed standard form

It is beyond the scope of this book to consider all the possible circumstances that can give rise to amendment to the standard form. However, there are some matters that deserve mention.

Amendment to the Agreement or Conditions required to comply with new legislation: The JCT are usually fairly prompt in bringing their standard forms in line with legislation.[115] Nevertheless, there may be some delay, and the Parties then need to deal with the issues themselves.

A typical example is the coming into force of the Site Management Plans Regulations 2008, which require the Client on certain projects to appoint a principal Contractor to perform specified administrative duties. It may be that the Contractor is already appointed as Principal Contractor under the CDM Regulations, but that is irrelevant, and the two functions and titles must not be confused. An additional article in the Articles of Agreement is the logical place to deal with an appointment, if the Contractor is to take on the role.

A more far reaching example is the amendments to JCT 05 that would become necessary if Parliament passed amendments to the Construction Act (see Section 1.5.5.4 below). In December 2008, a long awaited 'Construction Contracts Bill' was published,[116] which if enacted would require changes to the provisions in JCT 05 dealing with payment and adjudication (see Chapters 15 and 17 for comment on the issues dealt with in the Bill).

---

112 [2002] EWHC 2037.

113 In the Scottish case of *Barry D. Trentham* v. *McNeil* (1995) GWD 26-1366, it was held that written amendments prevailed over Clause 2.2.1 of JCT 80.

114 23 April 1970; (1970) 215 EG 165.

115 For example, Amendment to Attestation provisions following changes in the Companies Act.

116 Local Democracy, Economic Development and Construction Bill [HL], Part 8 Construction Contracts.

Amendment to the Agreement to recognize other amendments: Amending the standard form should always be approached with caution and reluctance. If amendments to terms are introduced, in the interests of clarity and readability, it is far better to amend the printed form than to simply refer to a separate set of changes. However, if a set of separate special conditions are necessary, they need to be placed in the contractual hierarchy. An additional article in the Articles of Agreement is a logical place for the amendments, and their priority, to be identified.

The risks in amending the standard form: It is also beyond the scope of this book to consider all the possible problems which may arise when the standard form is amended by the parties, save to suggest that if amendments are necessary at all, they should always be done with professional guidance. There are three principal areas of risk. First is the danger of affecting multiple clauses where they interact through cross-references; only careful diagnosis and tracking will avoid ambiguity or confusion.[117] Next is interface with the common law. In *Peak Construction* v. *McKinney Foundations*[118] it was held that the Contractor was entitled to payment for inflation up to practical completion, even though he may not be entitled to extension of time, where the printed text of the extensions clause had been amended.[119] It was also held that if the Employer obstructs the Contractor by his act or omission, and there is no corresponding ground in the contract to grant extension of time, the Contractor's obligation is to do no more than complete in a reasonable time in all the circumstances; the Employer then loses his right to recover liquidated damages. Clearly, deleting standard clauses, particularly those relating to extension of time, can be costly for the Employer.

The third area of risk is conflict, or apparent conflict, with statute, particularly (1) the Housing Grants Construction and Regeneration Act 1996, Part II, (2) the Unfair Contract Terms Act 1977, and (3) The Limitation Act 1980, which are dealt with in turn below.

1.5.5.4 Amendments to the Conditions in breach of the 'Construction Act' 1996[120]

Generally: It comes as a great surprise to many building owners new to the construction industry that the freedom to contract in construction agreements is limited by statute. Amendment of standard form contracts may fall foul of the Housing Grants Construction and Regeneration Act 1996, Part II (commonly known in the construction industry, and throughout this book as 'The Construction Act') if provisions protected by the Act are removed. It comes as an equal surprise to prospective Employers from other industries that the restrictive parts of the Act are not restricted to housing or to grants, but encompass most types of construction including heavy engineering. The main purpose of the 'Construction Act' is to influence the culture of the construction industry, to ensure contracts incorporate at least minimum provisions with regard to payment and dispute resolution. It was enforcement

117 For example of difficulty see Section 1.4.1 in this chapter dealing with the Contract Particulars and referring to *Bramall & Ogden* v. *Sheffield City Council*; the partial possession clause in the Conditions conflicted with the appendix entry for liquidated damages, rendering the clause void and unenforceable, and deprived the Employer of his rights to damages.

118 (1970) 1 BLR 111.

119 JCT 05, Schedule 7, Clauses A9.2, B10.1, C6.1, incorporate this principle (see Chapter 14, Section 14.5).

120 Housing Grants Construction and Regeneration Act 1996, Part II.

of the Act from 1 May 1998 which prompted the JCT to make radical amendments to the Standard Building Contract through Amendment 18 of JCT 80, so that the following JCT 98, and subsequently JCT 05, are fully compliant. Further amendment, albeit agreed between the parties, may contravene the requirements of the Act, triggering the application of statutory 'default provisions' contained in a statutory instrument generally known as the 'Scheme for Construction Contracts'.[121] The Scheme contains a Schedule in two parts. Part I provides rules for the selection of an adjudicator with powers to make enforceable decisions; Part I will apply where the contract contains no adjudication provisions, or where the provisions do not comply with the Act. Part II provides payment rules which will apply either where the contract fails to make provision at all, or where the parties have failed to agree on details such as timing of payments.

The Act requires that, unless one of the parties is a residential occupier, a contract in writing[122] for a 'construction operation' must contain rules sufficient to enable the party receiving payment to know in advance when payments will become due, how each payment is calculated, and the latest date by which payment must be received. In addition, the remedy of suspending performance is provided for late or under payment, and the contract must expressly give either party the right to refer disputes to an adjudicator.[123] Further, a clause making payment conditional on payment from a third party[124] is unenforceable unless the third party is insolvent. These limitations on the parties' agreements are dealt with in more detail in the following paragraphs.

Contracts in which the Parties' amendments may be effective: Amendments to JCT 05 may be effective where the Act does not apply. Section 105(2) of the Act lists exclusions from the definition of 'construction operations' (see below). Contracts for supply and delivery only are not affected automatically by the Act, and nor are contracts for:

- drilling for, or extraction of, oil or natural gas;
- extraction of minerals;
- assembly, installation or demolition of plant or machinery, and its support or access steelwork, where the main activity of the site is nuclear processing, power generation, water/effluent treatment, manufacture or storage (other than warehousing) of chemicals, pharmaceuticals, oil, gas, steel, food or drink;
- wholly artistic works.

If a JCT 05 contract for one of the excluded operations is amended or added to, even though it is done in a way that appears to contravene the substantive provisions of the Construction Act (i.e. s.108 to s.116), the amendment will be not be prohibited or replaced by the Act. This is because the contract would not be construed as a 'construction contract', for the purposes of the Act, and the Act will not apply.

121 Scheme for Construction Contracts (England and Wales) Regulations 1998; Statutory Instrument 1998 No. 649.

122 The requirement for a contract to be 'in writing' has caused considerable difficulty in the courts, and at the time of writing is being reviewed. Amendment removing the requirement for a contract to be in writing is proposed in The Construction Contracts Bill.

123 See Chapter 17.

124 Commonly known as 'pay when paid'.

Contracts in which the parties' amendments may not be effective: Amendments to JCT 05 may not be effective where the Act applies. The scope of the Act is wide, applying to contracts for 'construction operations' which are defined in s.105. A lengthy list of operations is set out in s.105(1), including amongst other things, construction, maintenance and demolition of permanent and temporary buildings and structures, works ranging from power lines to railways, from heating installations to communications systems, from site clearance to dismantling scaffolding. In short, the Act applies to contracts for most of the activities for which JCT 05 would be used (except exclusions listed in s. 105(2) – see above). If the contract is one to which the Act applies, any amendments which have the effect of avoiding compliance with the Act will cause the relevant default provisions of the Scheme for Construction Contracts to bind the Parties. This can, in some cases, result in 'cherry picking' individual details from the Scheme (e.g. the length of the payment period where the contract does not provide one), or it may mean the complete replacement of a non-compliant adjudication clause with the adjudication provisions in the Scheme.

Mandatory contractual provisions where the Act applies – Payment: Where the Act applies, unless the duration of the work is less than 45 days, a Contractor is entitled to interim stage or periodic payments, together with notification of amounts due under a compliant notification procedure and timetable (ss.109 to 111) (see Chapter 15 for commentary on payment procedure and timetable).

To the extent that amendment of clauses in Section 4 of JCT 05 (dealing with payment) removes any of the payment provisions covered by the Act, the amendment will not be effective, and will be replaced by the relevant corresponding provision in the Scheme for Construction Contracts, Schedule Part II.

Mandatory contractual provisions where the Act applies – Suspension: The Act does not expressly require a construction contract to provide for suspension in the event of late payment, although JCT 05 contains such a provision at Clause 4.14. However, s.112 of the Act does give statutory rights to suspend performance of obligations where a payment due is not made in full by the final date for payment.

An alteration to Clause 4.14 of JCT 05 (dealing with suspension), that removes any of these express provisions from the Contract, will still leave the Contractor with a statutory right.

Under the Act s.112(4), a party who suspends is also entitled to an extension of time for the period of the actual suspension, so deletion of Clause 2.29.5 (the relevant event of suspension) will not deprive the Contractor totally of his right to an extension. However, the statutory right to extension under the Act is not so wide as Clause 2.28 of the Contract; the Act provides only for the net period of the suspension, whereas the right in Clause 2.28 relates to delay caused by a suspension, which may incorporate time lost bringing back resources. The Construction Contracts Bill, if passed, would change this position by amending the Act to entitle the Contractor to suspend only part of the Works, and to an extension of time in consequence of his exercising his rights, thus allowing time for re-mobilizing.

Under the Act there is no statutory right to recover loss and expense. It follows that if Clause 4.24.3 (i.e. entitlement to loss and expense as a result of suspension) were to be deleted, a Contractor delayed by his own rightful statutory suspension would need to demonstrate a breach of contract by the Employer, and claim damages at common law.

Again the position would change if the Construction Contracts Bill were to become law, by amending the Act to entitle a Contractor, who correctly exercises his rights to suspend, to recover a reasonable amount in respect of his costs and expenses reasonably incurred.

Mandatory contractual provisions where the Act applies – Adjudication: Adjudication is discussed in detail in Chapter 17. The Act s.108(1) gives express right to the parties to refer any dispute arising under the Contract for adjudication, and sets out, in ss.108(2) to (4), the matters that must be covered in an adjudication clause in the contract. These matters are dealt with in Chapter 17. However, the essential point here is that both s.108 (5) of the Act and para. 2 of the Scheme state that adjudication provisions in Part I of the Schedule to the Scheme will apply (i.e. by default) where the requirements of s.108(1) to (4) are not met. JCT 05 provides for adjudication in Article 7 and Clause 9.2. Under Clause 9.2 the rules governing an adjudication are stated to be the provisions of the statutory Scheme for Construction Contracts (England and Wales) Regulations 1998, with minor amendments in Clause 9.2.2.

If Clause 9.2.2 were to be deleted by the Parties, the Scheme would apply but without amendment. If the contract is construed as a 'construction contract' under the Act, and if the whole of Clause 9.2 were to be deleted by the Parties, the Scheme without amendment would still apply by operation of the Act.

Ineffective contractual provisions where the Act applies – Conditional Payment: 'Pay when paid' clauses are often associated with sub-contracts, but are not unknown in main contracts. Where an Employer relies on a third party for funding the project, he may wish to insert a clause into JCT 05 making payment to the Contractor conditional on his own receipt of the necessary funds. When the Act applies to the contract, such a clause would be unenforceable under the Act, unless the third party on whom the Employer relies becomes insolvent.

The position would be different, and the inserted clause would be enforceable if, instead of relying on payment from a third party, a conditional payment clause relied on certification under a different contract. This is common in sub-contracts where entitlement to payment by the Contractor may be related to certification under the head contract that the work done is satisfactory. Criticism from the construction industry has led to inclusion of a provision in the Construction Contracts Bill, with the aim of making such so-called 'pay when certified' clauses unenforceable.

Ineffective contractual provisions where the Act applies – Adjudication Costs: One of the benefits of adjudication, as a dispute resolution forum, is that it is intended to be accessible to both parties. Any provision in the contract that requires the costs of a dispute to be borne by the losing party, or indeed by one specified party whether he wins or loses, may deter a small, or an impecunious party from referring a dispute to adjudication. This will apply particularly where it is likely that the responding party will engage expensive legal and technical support. The Act does not expressly bar such clauses if inserted in the contract.

In the interests of promoting equal access to justice, the Construction Contracts Bill, if passed, would make ineffective any clause inserted in the contract regarding the parties' costs relating to an adjudication, and would set limitations on post contract agreements.

1.5.5.5 Amendments and additions in breach of the Unfair Contract Terms Act 1977

The imposition of liability on the other party may sometimes seem unfair, but the general position is that if two parties wish to make a bad bargain, the courts will not seek to make it good. The Unfair Contract Terms Act 1977 (UCTA) does not always help; the title is misleading. UCTA does not govern general fairness in contracts between two businesses, but it does control some exemption and limitation clauses which attempt to reduce liability for negligence and breach of contract. The Act applies to contracts where one party is a consumer, and to some contracts where both parties are businesses.

A clause attempting to exclude or limit liability for death or personal injury caused by negligence is unenforceable under s. 2(1). Section 2(2) makes any clause limiting any loss or damage caused by negligence unenforceable, except where such a clause can pass a test of reasonableness.[125] Under s. 3, where a party is either a consumer or deals on the other's standard terms, the other cannot rely on a term which attempts to exclude or limit liability for his breach, unless the term 'satisfies the requirements of reasonableness'. The guidelines in UCTA for application of the 'reasonableness test' include strength of bargaining position,[126] knowledge of the relevant term, and the practicalities of complying with any conditions on which liability would rely. Under UCTA s.11(5) it is for the party alleging a term is reasonable to demonstrate that it passes the test.[127]

Putting UCTA into context, amendment of JCT 05 to exclude or limit liability for death or personal injury will always be unenforceable. Other amendments may be subject to the reasonableness test, if they are considered to be one party's own terms. JCT 05 is generally considered to be a consensus contract, since it is published by an organization made up from bodies representing both parties' interests. However, when amendments are introduced, they may have the effect of converting the contract into the unilateral terms of one of the parties. Sometimes amendment will be by genuine agreement after negotiation, but it is not unusual for Employers unilaterally to impose new or changed terms. These may appear in a tender enquiry in which the Employer also states that qualification or amendment by the tenderer will disqualify the entire tender. The terms may then be the Employer's own, and their effectiveness will depend, amongst other things, on whether the parties have equal bargaining power. However, for the purposes of UCTA, the size of a company is not relevant to equality of bargaining power, since legal advice is available to both. Indeed it can be very difficult to establish inequality, since there is no obligation to enter into the contract at all,[128] unless there is already a legal commitment known to the other party and on which the other party then relies to impose its terms.[129]

---

125 In *Regus (UK) Ltd* v. *Epcot Solutions Ltd.* [2007] EWHC 938 (Comm), a clause was held unenforceable because it left no remedy for a breach. The Court of Appeal reversed the decision [2008] EWCA Civ 361, but on the grounds that the clause in question did not deny all remedies. The principle held.

126 In *Chester Grosvenor Hotel Co. Ltd* v. *Alfred McAlpine Management Ltd* (1992) 56 BLR 115, McAlpine's own form limited liability, but it was held the parties had equal bargaining power so the term was not unreasonable.

127 In *AEG (UK) Ltd* v. *Logic Resource Ltd*, CA 20 October 1995, terms were held unreasonable since the plaintiff failed to plead its case in such a way as to adduce evidence showing that its terms were reasonable.

128 In *Denholm Fishselling Ltd* v. *Anderson*, 2 November 1990, 1991 SLT 24, it was held that the buying power of the parties was equal even though the buyer could not purchase on any other terms locally, since he was not obliged to buy from any individual seller.

129 For example, in *Northern Construction Ltd* v. *Gloge Heating and Plumbing* (1984) 6 DLR (4th) 450, a sub-contract tenderer was held to be in breach of a collateral contract when it revised its tender, knowing the Contractor was committed under a main contract won on the basis of the sub-contractor's original figures.

UCTA provides some protection to the parties, but in practice the protection is limited to a very narrow band. In the context of JCT 05, the tendering procedure under which the contract is normally brought about means that the relevance of UCTA is more likely to be protection for the Contractor than for the Employer. However, a Contractor who knowingly enters into a contract containing terms which he considers to be onerous does so at his peril.

1.5.5.6 Amendments to alter periods under the Limitation Act 1980

The Limitation Act 1980 provides a defence against legal proceedings in respect of a breach of contract starting after a specified period. The period is 6 years in the case of a simple contract (s. 5), and 12 years in the case of a deed (s. 8) (see also Sections 1.4.1.2 and 1.5.2.6 in this chapter above). If proceedings are started after that time, the defence, which must be raised by the defendant in order to have effect, is that the claim is made too late.

However, there are occasions when the Parties may wish to amend the statutory period, either to lengthen the period of exposure to legal attack, or to shorten it. For example, such amendment may be necessary when a contract is to be backed by a collateral warranty, and the overall limitation period would result in exposure to legal action for a period longer than that created by the contract. Similarly, a Contractor who is required to enter into a deed may seek to reduce the period of his exposure, and consequently, his Professional Indemnity insurance premiums, when the contract includes some design.

The limitation periods set out in the statute are not statutory entitlements; they are long-stops, after which an action cannot normally be brought. However, the Parties may, by agreement, change those periods; they may extend them, shorten them, or do away with them altogether. Such an agreement will be binding.[130] Express agreement to amend the period may be included in the contract, or in a collateral warranty, or in a separate contract. The position is summed up succinctly by the Law Commission in Consultation Paper 151, when stating the present law:[131]

> 3 Contracting out of, or waiving, the statutory limitation period
> 9.7 The defendant may agree by contract not to plead the statutory limitation period (or that the limitation period should be extended, postponed or suspended) (Lade v. Trill).[132] Such a contract will be valid provided that it is supported by consideration (or effected as a deed).

and further, in Law Commission Report No. 270:[133]

> (3) Contracting out of, or waiving, the statutory limitation period
> 2.96 The limitation period may be excluded by agreement, express or implied. Similarly a party may be estopped by his own conduct from asserting a limitation defence.

In the interests of certainty, the most convenient method of adapting the Contract in order to specify a limitation period, different from that provided in the Limitation Act, is to add a further Article to that effect in the Articles of Agreement.

---

130 See Chitty on Contracts, General Principles (27th edn 1994), paras 28-078 to 28-085.
131 Law Commission Consultation Paper (LCCP 151), 1998, Section A: The Current Law, para. 9.7.
132 *Lade* v. *Trill* (1842) 11 LJ Ch 102 .
133 Law Commission Report (Law Com No. 270) Limitation of Actions, Item 2, 2001, Part II: An Outline of the Present Law, para. 2.96.

1.5.6 Custody of Contract Documents

Clause 2.8.1 requires the Contract Documents to remain in the custody of the Employer, but to be available for inspection by the Contractor. A copy of the full set must be provided to the Contractor, at no charge, together with two further copies of the Contract Drawings and unpriced bills of quantities. Strangely, if the Contractor's Designed Portion applies, there is no obligation to provide the Contractor with further copies of the Employer's Requirements which are likely to contain further drawings, albeit they are not part of the Contract Drawings.

1.6 When things go wrong

1.6.1 Two versions of a drawing revision, the wrong version of which is signed

If this occurs the Contractor is both obliged and entitled to carry out the work on the signed, or annexed, drawing. The general position is that there is an error of intended scope of work which both parties have accepted in signing the Contract, and which neither is entitled to have corrected unless the other is in agreement. The position might be different if one of the parties were aware of the error and took advantage of what he saw as a bargain. In that instance the remedy of rectification may be available.[134]

1.6.2 Contract Drawings listed do not correspond with Drawings signed

An example would be the listing of Drawing No. 123–27B, and the incorporation in the signed drawings of Drawing No. 123–27C (or even No. 123–27A). The difficulty is in identifying which drawing revision number represents the contractual obligation. Reference to the Third Recital seems to indicate the Contract Drawings are the drawings signed by the parties: 'the drawings are numbered … annexed to this Contract ('the Contract Drawings') … have … been signed … by'. Thus the drawing to be followed is the signed drawing rather than the number listed, which may be a drawing in existence and in the possession of the Contractor but which may have been discarded temporarily. This situation can occur easily when last minute changes are made, particularly when they have been made for budgetary purposes and are later reintroduced.

The difficulty in practice is that the Contract Documents are often separated,[135] although the Contract at Clause 2.8.1 requires them to be retained by the Employer and available for inspection.[136] Contractors often seem to forget their right, or are reluctant, to insist on viewing the original, particularly when they have already been provided with a copy of the

---

134 For full discussion on equitable remedy of Rectification see *Keating on Building Contracts*, 8th edn (Sweet & Maxwell, 2006), paras 11-011 to 11-017 (hereafter *Keating*).

135 See Section 1.5.6 in this chapter, Custody of Contract Documents.

136 This is an improvement introduced by the JCT from JCT 98, which required the documents to be split between Employer and Architect.

drawings listed in the Contract Bills or the Recitals. Under Clause 2.8.2 the Contractor is provided with a certified copy of the Contract Documents immediately after execution of the Contract, but there is often delay in the copying process; if the Contractor's copy is prepared and handed over at the time of signing, transcription errors are likely to be discovered at that point. The prudent Contractor should have checked the Contract Documents thoroughly before signing in any event; but unfortunately the euphoric atmosphere on signing, together with the strain on resources in getting the job under way at the start of a new project, can easily push final checks to the back of the mind.[137]

1.6.3 Failure to complete the Contract Documents

The documents for a contract under JCT 05 require considerable effort to bring to a state ready for signing, leading to documents not being completed in time or errors in, and dispute over, the parity of the documents. This is particularly so when there are many documents to be annexed to the Contract (see Sections 1.5.3 and 1.5.4 in this chapter above). Many building owners seek to buy time, perhaps while finalizing a design or completing contract details, or even while concluding financing arrangements; their method is by use of a device called a 'letter of intent'. The popularity and notoriety of letters of intent justify more than passing comment, so a section devoted to the topic is included in Section 1.8 in this chapter.

Sometimes the parties do not sign the contract at all, but a contract may still exist. JCT 05 contains a form of agreement and, unlike some standard forms,[138] does not refer to a letter of acceptance; but nevertheless an unequivocal acceptance by the building owner of a Contractor's valid tender will create a binding agreement (i.e. a contract). The terms will be the terms of the tender which in most cases are likely to be the terms of the Employer's enquiry, but if the Contractor has qualified his tender in any way, even if the bills, as they often do, state the tender must be unqualified, then the qualification will be binding.

An example can be seen in the engineering case of *Yorkshire Water Authority* v. *Sir Alfred McAlpine & Son (Northern) Ltd.*[139] The tender was based on the ICE *Conditions 5th edn* 1973, reprinted 1986, which provides for a programme to be submitted by the Contractor after entering into contract. A programme and method statement were attached to the tender. The tender was accepted by the Employer by letter, but no formal contract was signed. A dispute arose over the method of carrying out the works when site circumstances prevented the Contractor from following the method statement. The court held the method statement was incorporated in the contract and was the specified method of construction; the Contractor was entitled to have the change of working method treated as a variation.

There are occasions when work commences while the final details of the contract are still being negotiated, but agreement is reached at a later date. In those circumstances it has been held the agreement acts retrospectively and the terms bind both parties in their performance of the contract prior to their agreement. In *Trollope and Colls Ltd* v. *Atomic Power Construction Ltd.*[140] work started in June 1959, but agreement was not reached until April 1960. It was suggested in the judgment in that case that the parties would have

137 See Section 1.8.1 in this chapter, referring to cavalier approach to letters of intent.
138 For example, Mechanical and Electrical Engineering Form MF/1, *ICE Conditions*, 7th Edition.
139 (1985) 32 BLR 114.
140 (1962) All ER 1035.

said as a matter of course, if they had been asked: 'This contract is to be treated as applying, not only to our future relations, but also to what has been done by us in the past since the date of the tender in the anticipation of the making of this contract'.

1.6.4 Failure to complete the formalities

There are three common situations in which one or both of the parties regularly fail to complete the formalities of signing a contract:

Orders or agreements 'subject to contract': The words 'subject to contract' are usually used in relation to the sale of land, but occasionally the term is used in construction agreements to signify that a formal contract is to follow. It is a term normally used to give comfort, without committing to being contractually bound. However, there are a few instances where all the terms are agreed and the parties commence performance. In those circumstances there can be a binding contract based on the agreement as it stands, irrespective of whether the formal document has been completed. In *Stent Foundations* v. *Carillion Construction*[141] the Court of Appeal held that an agreement stated to be 'subject to contract' could nevertheless be binding, depending on the conduct of the parties; in this case it was said that everyone behaved as though the works contract was in place.

Failure to sign the proposed contract agreement after starting work: When work starts before the Contract is signed, the incentive to sign may wane, which occurs regularly in association with letters of intent.[142] This is particularly so when problems in performing proposed obligations provide one of the parties with hindsight as to the risks under the Contract. However, failure to sign does not necessarily mean that no contract exists. If the essential matters are agreed, or if negotiations cease without clear dissent on some point, then a contract may be concluded if work is started or continued. An example can be seen in the approach taken by the judge in *Birse Construction Ltd* v. *St David Ltd.*[143] Whilst the decision in this case was not supported later by the Court of Appeal[144] following a jurisdiction challenge, the reasoning was not criticized. The Contractor received documents referring to the standard form JCT 80 after work had started. Negotiation on final matters including programme came to an end at a meeting a week later, but the documents were never signed and returned. It was held that a contract had been concluded both by offer and acceptance, and by conduct:

> The offer was accepted by the plaintiff's confirmation that the dates were feasible ... at the meeting ... Even if there had been no such agreement ... then alternatively I conclude that there was acceptance of (various matters) by conduct by one party or the other in either permitting or continuing the execution of work.

It is often a mixture of fact and law that determines whether the parties have reached agreement, and the judge in the *Birse* case saw the legal analysis as artificial; he recognized that there was no distinct offer and acceptance, but that the parties were building their contract

---

141 (2000) 78 ConLR 188; see also *Bryen & Langley Ltd* v. *Martin Boston* [2005] EWCA Civ 973.
142 See Section 1.8.1 in this chapter.
143 [1999] BLR 194.
144 [2000] BLR 57 [CA].

like the pieces of a jigsaw, the last piece of which had been put in place. The continuance of work without dissent was also crucial: 'given that work was continuing on a basis apparently acceptable to both parties, it would require clear evidence to displace the inference that it was not being carried out pursuant to some contract'.

In *Harvey Shopfitters Ltd* v. *ADI Ltd*[145] a formal contract was being prepared. Work started when the Architect wrote on behalf of his client, authorizing work, and stating that his client intended to enter into contract based on the tender, which was in turn based on IFC84. The letter continued by promising that a *quantum meruit* would be paid if the contract should fail to proceed and be formalized. The formal IFC84 contract never materialized. The Court of Appeal looked at the manner in which the parties had conducted themselves throughout the project. The parties had acted as though they were working in compliance with IFC84. It was held that the intention to formalize a contract did not prevent a contract existing, and that the parties in this case had agreed all the important matters. The letter was simply a part of the overall developing relationship. Conversely, in *Haden Young Ltd* v. *Laing O'Rourke Ltd*,[146] work at a football stadium proceeded as though a contract existed, but it was held that the parties had failed to agree on essential terms, or a means by which agreement could be reached, so no contract existed.

If work is being done (or allowed to be done) while negotiations continue, an express statement that it is in the absence of a contract seems necessary if either party wishes to be certain of avoiding agreement.[147]

Failure to complete the formalities of a deed: Sometimes contracts signed as a deed will be completed by the parties together, but it is not unusual for one party to sign, then return the document to the other who does not get around to signing. In those circumstances a confusing situation can develop. The party signing as a deed will be bound to his promises under the deed, whereas the other party who did not sign will only be bound if a simple contract has been created in place of the deed. The main effect of a deed is to bar actions after 12 years, whereas the limitation period on legal actions on simple contracts is 6 years (see Section 1.4.1.2 in this chapter, Attestation, on simple contracts and 'specialties'). Where only one party has completed a deed, only that party will be exposed to action for its breaches for the longer 12-year period. The other party will be exposed under its simple contract for 6 years.

1.6.5 Discrepancies

A fruitful area for misunderstanding and dispute is in ambiguity or conflicting requirements built into and between the various parts of the contract documentation. Simple examples would be, say, a specification requiring carpets throughout whilst a drawing showed floor tiles in toilet areas; or drawings indicating use of engineering bricks where the bills describe facings. The issue is seeded during the preparation of the contract documents, and it normally emerges in the context of Variations, triggered by the Contractor

145 [2003] EWCA Civ 1757, CA; 13 November, 2003.

146 [2008] EWHC 1016 (TCC); see para 73 for a list of factors to consider, citing Keating (8th Edition).

147 See also Section 1.7 in this chapter, with reference to *Con Kallergis Pty* v. *Calshonie Pty Ltd* for another example on this point.

asking for extra payment or resisting a credit Variation. Treatment of discrepancies and inconsistency is dealt with in detail in Chapter 6, Sections 6.12.6.5 and 6.14.

1.7 Forming a contract and the 'Battle of Forms'

In English law a contract is simply a legally binding agreement.[148] An agreement is made (a) when a statement of agreement is signed, as in the case of signing JCT 05 Articles of Agreement, or (b) when one party makes an unambiguous offer capable of being accepted, and the other accepts it unequivocally, or (c) when negotiations cease without obvious remaining disagreement.[149] Whichever route is used, it will be based on 'offer and acceptance' in some form, following an enquiry. In the context of JCT 05, the contractor's tender is usually an offer even if it does not comply with the enquiry, just as long as it fulfils the legal requirements of a valid offer. A contract would be formed if the building owner were to say to the tendering contractor 'I accept your tender'. However, if the owner were to respond 'I accept on the basis that the completion date in the enquiry document applies – not as stated in your tender', that would not be an unequivocal acceptance and no contract would be formed; it may simply signal the common desire of many in business to contract on their own terms.

A qualified acceptance of an offer in law is not acceptance at all, and no contract is concluded; but provided the response fulfils the criteria for an offer, it may be construed as a counter-offer which would in turn kill the previous offer and be open for acceptance by the other party. The procedure has become known as the 'battle of forms'. Whether or not a contract is formed will depend on the facts and the courts may apply the rules of contract construction in order to decide whether one party has accepted the other's terms by words or conduct. The difficulty which arises is that the counter-offer may itself be countered by a further counter-offer reverting to the terms of the original offer or introducing new terms. Such a situation was examined by the Court of Appeal in the case of *Butler Machine Tool Co Ltd* v. *Ex-Cell-O Corporation (England) Ltd.*[150] The dispute involved the sale of machine tools quoted on the seller's terms, which included a price variation clause. The buyer purported to accept the quotation by a standard purchase order incorporating its own standard conditions which stated that the price was on a fixed basis. The order contained a tear-off acknowledgement slip accepting the terms which the seller duly signed and returned with a covering letter. The seller's covering letter stated that the order was accepted in accordance with the seller's quotation. The machine was then delivered and a price increase was claimed in accordance with the variation clause. A dispute arose over whose terms formed the basis of the contract. In the Court of Appeal it was held that the seller's covering letter, although referring to the initial quotation, did so merely to identify the subject matter. Lord Denning said:

> In some cases the battle (of forms) is won by the man who fires the last shot. He is the man who puts forward the latest terms and conditions: and, if they are not objected to by the other party, he may be taken to have agreed to them.

148 See *Courtney & Fairburn Ltd* v. *Tolaini Bros (Hotels) Ltd:* CA; (1975) 2 BLR 97.

149 See Lord Hoffman in *Investors Co-operative Scheme Ltd* v. West *Bromwich Building Society* [1998] 1 WLR 893, at 912 for a modern approach to interpretation of agreements.

150 (1979) 1 All ER 965.

Clearly, 'shots' can only include those responses capable of being construed as offers; in this case the seller's covering letter was decided by the court to be something short of a counter-offer, thus leaving the acknowledgement of the buyer's counter-offer as acceptance of the buyer's terms.

The *Butler Machine Tool* case involved relatively simple facts compared with situations arising on building projects. The exchange of communication was short and the obligation was fulfilled by one delivery. In building contract negotiations the communication can extend to many letters and meetings over a period of many months leading to numerous changes to documents which become annexed to the proposed contract conditions.

The battle of forms is used by many to get a foot in the door, backed by a general philosophy that the exchange of standard terms or tender qualifications prevents acceptance of the other party's terms, but without real expectation that either's terms will be agreed. The Contractor is then relying on receiving some sort of fair payment,[151] in the belief that it will be better than signing a contract with all its obligations. In short, the 'last shot' doctrine becomes a powerful tactic in the risky game of contract avoidance. Unfortunately for those employing the tactic, there is a rule of contract construction which may prove fatal if ignored; that is that a party can impliedly accept the other's terms by its conduct. Application of the rule is not without its difficulties.

If, having responded to an offer by issuing a counter-offer, a Contractor immediately commences work, such conduct cannot be construed as acceptance since a Contractor cannot accept his own offer. He starts work at risk. Nevertheless, the conduct of the person receiving the offer, in allowing the work to continue without dissent, may be construed as acceptance, but in the short intervening period more correspondence may have passed changing the terms yet again. In the building industry such situations occur frequently, exacerbated by the desire to commence preliminary work to meet building or fiscal deadlines, and cushioned by the use of letters of intent (some of which may be construed as further counter-offers).

The risks of acting hastily are clear; so too are the risks of delaying. Adoption of the 'last shot' doctrine requires fine timing. In the event of a dispute the relationships in contract will depend entirely on the sequence of 'shots' and how it relates to the point at which one of the parties conducts himself in a manner signifying his acceptance of the other's terms. One example can be seen in the case of *Chichester Joinery Ltd* v. *John Mowlem and Co plc.*[152] Chichester's quotation for joinery was submitted under cover of a letter which referred to standard conditions printed on the reverse. Some months later Mowlem sent out to Chichester and others an invitation to tender referring to printed conditions. Following a telephone conversation during which further terms were agreed, Chichester confirmed they would do the work in accordance with their original quotation and the oral agreement. Mowlem then sent an order referring to its own terms (different from those in their own invitation). It was held that Mowlem's order and Chichester's subsequent acknowledgement containing their earlier terms were each counter-offers killing the other side's previous offers. The judge went on to reject the suggestion that Chichester accepted Mowlem's terms by their conduct when they started preparing for manufacture, deciding the evidence fell far short of that necessary to establish acceptance by conduct. On considering the later conduct of Mowlem in accepting delivery of the joinery on site,

---

151 See Section 1.8.2 in this chapter dealing with *quantum meruit* in the absence of a contract.
152 (1987) 23 ConLR 30.

he found it constituted acceptance of the terms in Chichester's counter-offer (the acknowledgement of order). The alternative would have been that no contract existed at all.

The significance of this case is cited purely as an example. It is not new law, but the application of established principles in the analysis of situations, with the possibility of unexpected results. The risk for the parties was recognized in the judgment:

> This case highlights the risk facing parties who seek to impose their own respective conditions rather than some well established form of contract commonly used. ... Unless such a form of contract is used, the courts may be faced (as here) with what is essentially an artificial state of affairs.

The artificial state of affairs referred to is the urging by legal advisers of their clients to respond to each document from the other party with an endorsement or annexation of their own terms. In other words, the judge was openly criticizing the promotion of the 'battle of forms'.

For those engaged in the battle, however, it is not so simple. The insertion of such matters as completion dates, which a tenderer knows during negotiations cannot be met, can be resisted only with counter-offers or express rejection. The former is seen by many as an opportunity to secure work, albeit without agreeing terms, but in the hope that the other side will eventually agree either expressly or by conduct. Express rejection carries with it the risk of the other party taking its custom elsewhere, so rejection is frequently withheld until sufficient work is done to make it too late to involve another Contractor without the owner incurring significant extra cost. An example can be found in *Kitsons Insulation Contractors Ltd* v. *Balfour Beatty Buildings Ltd.*[153] In that case the Contractor sent a standard form sub-contract to the sub-contractor, inviting him to sign the document and return it. Certain amendments which the sub-contractor then saw for the first time were not acceptable to him. Work had already started but acceptance of the document would have retrospective effect. The sub-contractor wrote to the Contractor refusing to sign the contract since it did not contain a number of terms from their offer concerning payment, and further contained a scope of work different from that in the offer. The main issue in the case concerned an early letter of intent and whether or not it gave rise to a contract. That letter was potentially one of the 'shots' in a series of offers and counter-offers which culminated in the issue of formal documents. It was held there was no contract; the parties had never reached agreement. Where there is no contract as a result of a battle of forms ending in stalemate, the Contractor's entitlement to payment is governed by the same principles as those applying where letters of intent are used.[154]

A similar result can be seen in *Midland Veneers Ltd* v. *Unilock HCP Ltd*[155] in which negotiations, a letter of intent, an order and an acknowledgement of order all came to nought. Negotiations between the two companies for the supply of veneered heater fittings resulted in a letter of intent. It was a pure letter of intent stating no more than an intention and confirming costs incurred would be met if no order was placed. Some goods were then supplied. A month after the letter Unilock placed an order referring to the terms of the main contract, Unilock's own terms on the back, and other matters including liquidated damages and a guarantee. Midland Veneers responded immediately stating the order had

---

153 ORB 1461; 17 January 1991; (1991) 8-CLD-05-04.
154 See Section 1.8.2 in this chapter dealing with *quantum meruit.*
155 CA; 12 March 1998, QBENF 96/1652 CMSI; [1998] ABC LR 03/12.

been passed to the production unit for processing and they would send an order acceptance shortly. Further letters were exchanged ending with a letter from Unilock confirming an agreement on the damages and guarantee, and also that the rest of their order was acceptable to Midland Veneers. On receipt of this letter Midland Veneers made the remaining deliveries which were accepted. Disputes developed over delivery dates and the court was asked to rule on the basis of the contract, if any. It was held no contract was concluded, and Unilock appealed. The Court of Appeal could find no evidence of agreement; it held that the acknowledgement letter from Midland Veneers was no more than that, and it was clear that an acceptance was intended to follow from another department in the future. Thus the letter itself could not be an acceptance of the order. It was also held that the conduct of the parties in delivering and accepting the goods did not signify agreement; indeed the only thing agreed was a price. In short, there was no contract. The amount to be paid was not at issue in this case because Unilock had always maintained there was a contract based on an agreed price. The dispute was over the liquidated damages, which in the absence of a contract could not be deducted. The court was only asked to decide whether or not a contract existed.

In the Unilock case neither express nor implied agreement could be found, but nor was there any express rejection of the other's terms. The possible effect of the absence of express rejection can be seen in the Australian case of *Con Kallergis Pty Ltd* v. *Calshonie Pty Ltd.*[156] A supplier quoted to an electrical Contractor for luminaires, referring to the Contractor's drawing. The Contractor sent an order attaching a different drawing and requesting samples. The supplier responded with a document describing itself as 'cost variation subsequent to our original offer' and sent samples. The Contractor then issued a Variation, stating it was to be valued by negotiation based on information to be provided by the supplier. The supplier claimed its 'cost variation' was a counter-offer which had been accepted when the Contractor accepted the samples. The court looked at the conduct of the parties, and held it showed they had reached a concluded agreement in the terms of the Contractor's order. A letter from the supplier signifying that the Contractor's order was not accepted would probably have protected the supplier against a contract being formed on the other's terms.

Again, the issue of an order was the subject of dispute in the American case of *ICC Protective Coatings* v. *A.E. Staley Manufacturing Co.*[157] In response to an invitation from Staley to quote for protective coatings to plant, ICC submitted a proposal. The proposal contained a price and terms dealing with correction of defects. Staley issued a purchase order referring to their own terms on the back, and stating that any additional or different terms proposed by ICC were rejected unless expressly accepted in writing. A dispute arose when Staley did not give ICC the opportunity to put right their own work. The court held that ICC had accepted Staley's terms without question, and they were the basis of the contract, to the exclusion of ICC's terms.

Timing is critical in the battle of forms. For example, in the *Kitsons* case the letter rejecting the sub-contract documents is central to the conclusion that there was no contract. If the timing of commencing substantive work had been different, the result of that case could have been reversed. If the sub-contractor had not started work so early, but instead had waited until after receiving the sub-contract document, and then had started

156 25 March 1997; (1998) 14 BCL 201–214; Building Case Law Digest 1998, Pub. LLP 1999, p. 67.
157 695 N.E.2d 1030 (Ind. App. 1998); *Construction Claims Monthly* 20 (9) September 1998, 2.

before commenting on the amendments, the court may have construed a contract, as it did in the *Con Kallergis* case. Whether or not that is the desired result will depend on the reason for engaging in the battle; for both Kitsons and Midland Veneers it could have been fatal, since one of their main objectives in bringing the action was to avoid contractual obligations as to time.

The battle of forms is an academic means of identifying the basis of a relationship, and it sometimes conflicts with reality. The analysis can only conclude that either one party's terms apply, or that there is no contract at all. A case in the Court of Appeal indicating a move towards recognition of the commercial cut and thrust of contract negotiation is *Hertford Foods Ltd and Another* v. *Lidl UK GmbH*,[158] in which Lord Justice Chadwick said:

> 24. The judge found as a fact that (the parties) intended to and did reach contractual agreement by the end of their telephone conversation. On that basis, knowing that they had not, and … probably could not, reach agreement as to the applicability of either set of terms, the only inference that can be drawn is that neither set of standard terms would be applicable. That conclusion seems to me at least as likely to accord with reality as a conclusion that they reached no binding agreement at all, or that either agreed to contract on the standard terms of the other.

This conclusion was followed in *Leicester Circuits Ltd* v. *Coates Bros plc*,[159] in which it was held that neither party had expressly agreed to the other's terms, nor indeed had either given relevant thought or consideration to them, so neither party's terms applied. In those circumstances, a contract was concluded, but the terms were to be implied.

1.8 Letters of intent and the right to payment

1.8.1 Letters of intent: meaning and effect

1.8.1.1 Generally

Strictly, letters of intent are not a feature required in the formation of contracts using the JCT forms, but their use prior to concluding a JCT contract is so common that they deserve a section to themselves.

Parties who are negotiating or are about to negotiate a contract will often start work towards fulfilling the obligations of that contract before all the terms of contract are resolved or a contract is signed. They do so usually in the knowledge and with the intention that the contract, when it is signed, will act retrospectively. In other words all their actions when performing within the scope of the envisaged contract will become actions carried out under and subject to the terms of the actual contract.[160] In these circumstances one of the parties will commit resources and expenditure before the other; since it is unlikely that payment will be made before any reciprocal performance, the first to commit himself will be usually the supplier. As there is no contract the purchaser would have a distinct advantage – receipt of work that he wants to be done, with no clear express contractual obligations as to either

158 [2001] 27 BLISS 11.
159 [2002] EWHC 812 (QB).
160 *Trollope & Colls Ltd* v. *Atomic Power Constructions Ltd* (1963) 1 WLR 33.

the amount or timing of payment for it, whilst at the same time being able to avoid the pressure of negotiating unwanted terms in a tight timescale. In short, the purchaser gains time. Such time enables him to prepare lengthy documentation without delaying commencement of the work and to hold out in negotiations for his most favourable terms. The supplier, too, is given time to negotiate, but has the disadvantage of having committed himself to cost; for him the attraction of the situation is the start of work against an anticipated order which he hopes will become more certain with the passage of time.

A device often used to create such situations is the 'letter of intent'. A principal may send a letter stating that he intends to enter into a contract with the recipient. In its purest form that is all a letter of intent comprises. When the recipient knows that he is one of a number of tenderers in competition, the letter is taken often as a sign that the uncertainty is over, the contract is secured and that it is simply a question of completing formalities; the choice of Contractor has been made and the work can get under way. If a contract is subsequently concluded, then the common intention has been fulfilled and clearly any dispute which arises will be a dispute over rights and obligations under the Contract. The difficulty arises where the formalities are not concluded, but the parties have proceeded as though they were, and a dispute then arises. Often the dispute will concern the rights or obligations of one of the parties under the terms of the contract that was envisaged. Whether or not that envisaged contract ever came into existence becomes a critical issue and represents a significant cause for concern when letters of intent form the basis of a relationship.

The traditional approach of the English courts has been to treat letters of intent as a means of postponing legal liability. It is a fundamental requirement in English law that for a contract to be concluded there must be *consensus ad idem*;[161] the parties must have reached agreement and negotiations on all but minor terms must be finished.[162] A statement from one party to another that there is an intention to enter into a contract is a statement that the contract is not yet concluded; it is a thing for the future.[163]

Even more speculative is the letter which states 'it is our intention to place an order'. Since an order is often no more than an offer[164] open for acceptance by the recipient, there is no relationship whatsoever created by a mere statement of intention to make an offer. Accordingly there is no legally binding obligation on either party in so far as such obligations are seen as fulfilling terms of the envisaged contract.

Whilst this is the position when a letter of intent is in its simple and pure form, complications arise in those instances where the letter contains more information or instructions. The issuing party will often refer to what the terms will be with the possible effect of turning what purports to be a letter of intent into an offer. Whether this effect is deliberate or not is immaterial. If the letter contains the constituent elements of an offer capable of unequivocal acceptance by the recipient, then it may be construed as an offer.[165] In *Ove Arup & Ptnrs International Ltd & Anr* v. *Mirant Asia-Pacific Construction (Hong Kong) Ltd & Anr*,[166]

---

161 (Lat. agreement as to the same thing).

162 *Per* Lord Denning MR, *Courtney & Fairburn Ltd* v. *Tolaini Bros (Hotels) Ltd* (1975), 2 BLR 97 at 101–102.

163 *Peter Lind* v. *Mersey Docks and Harbour Brd* (1972) 2 Lloyd's Rep. 234.

164 *Butler Machine Tool Co* v. *Ex-Cell-O Corporation (England) Ltd* (1979) 1 All ER 965.

165 *Per* Bigham J, *Crowshaw* v. *Pritchard and Renwick* (1899), 16 TLR 45; cited by Powell-Smith, V. and Furmston, M.P., *A Building Contract Casebook*, 2nd edn (BSP Professional Books, Blackwell Scientific Publications Ltd, 1990) p. 17.

166 [2003] EWCA Civ 1729, at para. 9.

a letter of intent contained sufficient provisions for its acceptance to be construed as a contract for design. The statement of intention does not cut across this concept. Indeed, a statement of intention to enter into a contract contained in an offer, merely serves to reinforce the presumed intention to create legal relations.[167]

The significance of the interpretation of letters of intent can be seen by comparison of two cases. In *Kitsons Insulation Contractors Ltd* v. *Balfour Beatty Buildings Ltd*[168] the court was required to decide whether a letter of intent formed the basis of a contract. The letter sent by Balfour Beatty to Kitsons stated an intention to enter into a contract using a standard form sub-contract called DOM/2, with amendments to be forwarded in due course; the letter also requested Kitsons to accept the letter as authority to commence the sub-contract works. The formal contract was drawn up some months later, but it was never signed and certain terms were rejected by Kitsons. Meanwhile work continued and Balfour Beatty made payment consistent with the payment rules of the standard DOM/2 terms. Notwithstanding the actions of the parties, when a dispute arose the court had to decide a preliminary point as to whether the letter of intent created a contract. Regarding intention, it was said:[169] 'I am satisfied that Balfour Beatty were not prepared to enter into a concluded sub-contract with Kitsons on 23 March 1988, although they were anxious to get the work started'.

On the question of a contract being concluded later the evidence showed that the documentation differed from that referred to in the letter and was rejected by Kitsons. The judge consequently found that the rejection precluded any possibility of a contract: 'it remained one of the matters upon which the parties never agreed'. In this case, the letter of intent gave rise to no contractual relationship whatsoever, and moreover had by its terms fixed Kitsons' expectations, which resulted in their subsequent rejection of the formal documents.

In *Harris Calnan Construction Co. Ltd* v. *Ridgewood (Kensington) Ltd*,[170] an accepted letter of intent was held to contain sufficient for a contract to have been concluded:[171]

> In the present case, the letter of intent made plain that there was complete agreement as to the parties to the contract; as to the contract workscope (because it was contained in what was described as 'Tender Documents dated 2nd November, 2005'); as to an agreed lump sum of £200,787.75; as to an agreed set of contract terms (namely the JCT 2005 Standard Form, Private with Quantities), with 5 percent retention and £5,000 per week liquidated damages; and as to a contract period of sixteen working weeks.

The content of most letters of intent lies somewhere between that in the Kitsons and Harris Calnan cases, and there has been a move by the courts to find some contractual relationship, albeit of limited extent. In the case of *C. J. Sims Ltd* v. *Shaftesbury plc*,[172] a dispute arose over the amount of remuneration. Sims had made an offer based, as requested, on standard JCT 80 terms. In response Sims were sent a letter of intent and were asked to start work immediately. The letter provided for payment of reasonable costs

---

167 In the case of a business there is a presumption that there is an intention to create a legal relationship when it makes an offer; when a private individual makes an offer there is a presumption that there is no such intention. In either case the presumption is reversed by an express intention.

168 1989, ORB: 1461; (1991) 8-CLD-05-04; Judgment 17 January 1991.

169 *Per* his Honour Judge Thayne Forbes, Judgment 17 January 1991, at 9, para. 3.

170 [2007] EWHC 2738 (TCC).

171 At para 11.

172 (1991) 25 ConLR 72.

in the event that the contract did not proceed, but also that such costs were to be 'substantiated in full to the reasonable satisfaction of [the defendant's] Quantity Surveyor'. Work proceeded while precise terms were negotiated. No agreement was reached and a dispute arose over entitlement to payment. It was agreed by the parties that a contract existed although they also agreed that the terms were to be found strictly in the letter of intent. The court found that the letter created a contract and that the requirement for costs to be substantiated to the satisfaction of the Quantity Surveyor was a condition precedent to the entitlement to any payment. In contrast with the Kitsons case, the court in the Sims case found a relationship in contract based on the wording of the letter of intent together with the subsequent actions of the parties in reliance on that wording.

A developing willingness of the courts to find a contract of some sort, albeit not for the whole project, can also be seen in other cases. In *Durabella Ltd* v. *J. Jarvis & Sons Ltd*[173] it was held:

> It is now well established that where a letter of intent authorises work ... pending the conclusion of some further agreement it will, if accepted, constitute a contract ... for what it requires.

In *A.C. Controls Ltd* v. *British Broadcasting Corporation*[174] a letter of intent was signed by both parties. The letter authorized certain advance survey work to enable the BBC to decide on the final specification before entering into a formal contract for the whole project. It was said by the Court: '... the letter is, in effect, a mini contract with a defined scope of work which ACC is to implement in full'. Again, in *Tesco Stores Ltd* v. *Costain Construction Ltd, Peter Hing and Jones (a firm), Vale (UK) Ltd, and Whitelight Industries Ltd*,[175] a letter of intent was signed and returned by the Contractor. It was held that a contract was formed and that the only document incorporated into the contract was that letter; the only express terms were that the Contractor would start work before execution of a formal contract, and that payment in the event that the formal contract was not executed would be in accordance with the terms of the letter.

Whilst it seems from these cases that the courts may find a contract in a letter of intent, it is not a foregone conclusion. The judgment in *Twintec Ltd* v. *GSE Building and Civil Engineering Ltd*[176] is a reminder: '... there is no settled law on the meaning and effect of letters of intent. The court must decide each case on its facts. ... I am reminded of the basic legal requirements of offer, acceptance, intention to create legal relations, and consideration ...'

Clearly, the words of a letter of intent are critical if the nature of a relationship is to be certain before commitments are made, but where a letter of intent containing more than a simple statement of intention is accepted, either expressly or by conduct, and where the basic legal requirements are fulfilled, there is a likelihood that at least some binding obligation is created.

1.8.1.2 Letters of intent and standard form contracts

In both the *Kitsons* and the *Sims* cases reference had been made to standard form contracts. In both cases the result turned on the content of the letter of intent. It must

173 (2001) 83 ConLR 145.
174 (2002) 89 ConLR 52.
175 [2003] EWHC 1487 (TCC).
176 [2002] EWHC 605 (TCC).

therefore be questioned what would have been the result of either case if the terms contained in the letter had been precise, clear and complete. The mere fact of the letter being a pre-contractual document does not of itself prevent a contract existing. In *Charles Church Developments Ltd* v. *Pacific Western Oil Corporation*,[177] Lord Templeman stated: 'A preliminary arrangement which contemplates a formal contract may itself constitute a binding agreement'.

Likewise if a written communication, albeit headed as a letter of intent or containing words to that effect, contains all the necessary elements required to constitute a valid offer, then it may be construed as an offer, irrespective of its title. The consequent risk for issuer and recipient alike is that a letter of intent is seldom clear, but is nevertheless capable in some instances of being an offer, acceptance of which, including implied acceptance by conduct,[178] will be binding.

This point raises yet another aspect of risk. Where standard forms are contemplated, an Architect, engineer, or project manager with apparent authority will be involved in the pre-contractual stages. It is possible therefore that a letter of intent, if sent to a Contractor, will be issued by the Architect or other professional; for example, in the Birse case referred to above, a letter authorizing work was sent by the owner's Architect. However, the person sending the letter may be acting without any authority. In *GPN (In receivership)* v. *O2 (UK) Ltd*,[179] it was held that a Quantity Surveyor who had authority to negotiate contract terms on behalf of the owner had no ostensible authority to bind his client in a contract with the Contractor. Another salutary example can be seen in the case of *A. Monk Building and Civil Engineering Ltd* v. *Norwich Union Life Insurance Society*.[180] The judge was required to consider various preliminary issues including the status of a letter of intent and the authority of the issuer to make a binding contract. He decided on the facts that although the project managers had been given express authority to send the letter of intent, notwithstanding its content they had no authority to commit their client, Norwich Union, to a binding agreement. So even if the letter gives every appearance of being a valid offer capable of acceptance, the question of authority to bind a building owner may reverse an otherwise clear situation long after the parties have committed themselves to commercial courses of action.

The letter of intent is a useful instrument to get work under way while documents are prepared; this is one of the classic purposes of such letters, albeit there are risks attached. Reciprocally, letters of intent can be made clearer and their issue may be facilitated by referring to what the terms will be. This creates greater confidence in what might otherwise be no more than a bare promise to negotiate. The publicized nature of standard forms such as JCT 05 assists incorporation into any contract by mere reference to them; indeed in *Killby & Gayford Ltd* v. *Selincourt Ltd* [181] the Court of Appeal held that the words in a letter 'subject to the normal standard form of RIBA contract' were sufficient to incorporate the then current JCT terms into the contract. Consequently, a recipient of a letter of intent referring to industry standard terms is given illusory comfort, first by the stated intention, and second by the passing reference to a standard form contract; this may be

177 (1980) Unreported, Case No. 914 (Transcript: Association), 8 July 1980.
178 See Section 1.7 in this chapter, Forming a contract and the 'Battle of Forms'.
179 [2004] EWHC 2494 (TCC).
180 62 BLR 107.
181 (1973), 3 BLR 104.

particularly so when such reference is made by a respected professional, albeit he may have no authority. Not wishing to appear mistrustful of a potential customer, the recipient is consequently induced into commencing preparatory work, design or even installation.

From the principal's point of view, some delays in getting a complex project under way are almost unavoidable, and a letter of intent is a useful tool; but a letter of intent is also a facility which can by itself promote delay. It eases the pressure, and there seems little doubt that once a letter of intent (particularly one referring to standard terms) has been acted upon, both the incentive and enthusiasm to pour resources into preparing and concluding a formal contract must surely wane. The result is a period of uncertainty; but often the effects of the uncertainty are appreciated only after it is too late to remedy the situation, and a dispute has arisen.

Much depends on the circumstances of any ensuing dispute and the entitlement to be paid. An example can be seen in the High Court decision in *British Steel Corporation* v. *Cleveland Bridge and Engineering Co Ltd.*[182] The defendant Cleveland Bridge sent British Steel a letter stating an intention to enter into a sub-contract for the supply and delivery of castings. The price was stated and authority was given to commence work pending preparation of sub-contract documents. Before the completion of documents a schedule of delivery dates was given to British Steel and all but one of the deliveries were made. Disputes arose over quality and delivery dates, Cleveland claiming that deliveries had been made late compared with the schedule. The court decided that no contract existed but that the content of the letter gave rise to an entitlement to be paid a *quantum meruit* (see below). The decision was significant for British Steel, for had it been decided that a contract was concluded, then the terms may well have included the delivery schedule, and a counterclaim from the defendant for faulty performance might have succeeded.

Similarly in the earlier unreported case of *Hammond and Champness Ltd* v. *Hawkins Construction (Southern) Ltd*,[183] it was argued by the defendant Contractor that if the court found there was no contract and therefore no completion date, then at least the sub-Contractor's work should be carried out in a reasonable time; there was, they said, an implied obligation to that effect. The court disagreed:

> If a party does work not pursuant to an agreement but at the request of another party he is free to stop work … when he wishes and that is a consideration which would militate against there being any obligation to do a specified body of work or to do it within a reasonable time, or any other time.

It is clearly an attractive proposition for any Contractor to be in a position where the difficult requirements of a contract can be avoided, but payment can still be claimed. The attraction is greater still if the requirement which can be avoided is related to progress and completion or to quality.[184] After all, those are the areas where Contractors are most at risk of having contractual counterclaims or damages levied against them, and where they have the greatest difficulty in complying with strict procedural requirements of standard form contracts. It follows that if the potential confusion and extreme administrative requirements of forms like JCT 05 can be avoided and payment can be received for

182 (1984) 1 All ER 504; (1981) 24 BLR 94.

183 Unreported, Judgment 24 April 1972; *Construction Law Yearbook* (Sweet & Maxwell, 1995), pp. 181–88.

184 In *Tesco Stores Ltd* v. *Costain Construction & Others* [2003] EWHC 1487, terms to use reasonable care and skill were implied into a limited contract.

work done simply by working in reliance on a letter of intent, it may be wondered why a Contractor should ever wish to enter into a contract, standard or otherwise. One reason is the ability to enforce rights through adjudication; another is the question of calculating remuneration, and its payment.

It can be seen from the *British Steel* and *Hammond and Champness* cases that the recipient of a letter of intent may be entitled to payment in the absence of a contract. However, even though there may be valid authority to carry out work, there is some uncertainty as to the amount of payment that may be due. It is the absence of an agreed price, and the lack of any rules to determine one, that many recipients of letters of intent do not recognize or simply choose to disregard. Yet many Contractors accept letters of intent almost without question. Perhaps the rationale behind such an apparently cavalier approach is summed up in the suggestion:[185] 'Businessmen compare a letter of intent with the alternative, which is to provide no guarantee at all and therefore see it as better than nothing and significant in forming a limited relationship'.

1.8.2 Letters of intent: right to payment

1.8.2.1 Right to payment: introduction

Whenever a Contractor, in order to keep to a tight programme, starts work before a contract is concluded, he does so in anticipation of being paid eventually under that contract. A problem arises when the contract has not come into existence but the Contractor having carried out work wishes to realize his expectations; often he will claim the price he would have recovered under the expected contract. Likewise the building owner will often act as though the contract existed and make payments in accordance with the proposed terms. If a contract for a limited scope of work has been formed, as occurred in the *A.C. Controls* case and the *Tesco Stores* case referred to above, payment will depend on the terms of the letter. Tender rates, reasonable costs, costs plus reasonable overheads and profit, are all typical provisions in letters of intent. If the letter is silent, then a reasonable value should be paid.

The difficulties in the use of letters of intent usually start when one of the parties considers the other is not fulfilling all its obligations. A typical situation would be where the programme of work is not being maintained by a Contractor and as a consequence the building owner suffers losses which he then attempts to set off against payments being made. Since the set-off is in respect of what would be an alleged breach of contract, it is a convenient defence for the Contractor to revert to the true position and argue that since there is no contract there can be no breach; consequently there can be no entitlement to set-off damages. The question which then arises, if there is no contract, is whether there is an entitlement to be paid anything for the work done, and if so, how much.

The position in these circumstances was considered in *William Lacey (Hounslow)* v. *Davis*:[186]

> In neither case was the work to be done gratuitously, and in both cases the party from whom payment was sought requested the work and obtained the benefit of it. In neither case did the parties actually intend to pay for the work otherwise than under the supposed

185 Ball S.N., Work Carried Out in Pursuance of Letters of Intent – Contract or Restitution? (1983) 99 LQR, 572.
186 (1957) 1 WLR 932; *per* Barry J.

> contract .... When the beliefs of the parties were falsified, the law implied an obligation – to pay a reasonable price for the services which had been obtained.

Similarly in *British Steel* v. *Cleveland Bridge*[187] it was said:

> Both parties confidently expected a formal contract to eventuate ... one then requested the other to expedite the contract work, and the other complied with the request ... if, contrary to their expectation, no contract was entered into, then the performance of the work is not referable to any contract ... and the law simply imposes an obligation on the party who made the request ... such an obligation sounding in quasi-contract or, as we now say, in restitution.

It should be noted at this point that whilst most Contractors probably think subjectively in terms of an 'entitlement to receive' in the absence of a contract, the courts clearly place emphasis on an 'obligation to give' by the recipient of the service. The two concepts do not necessarily produce the same results, but it seems something becomes due and the rationale is to be found in what is variously referred to as *quasi-contract* or now more commonly, 'restitution'.

1.8.2.2 Quasi-contract and restitution

Historically the basis of liability and entitlement in arrangements of a contractual nature when no actual contract exists has been the subject of two main theories, i.e. that there is some sort of implied contract; alternatively that a person should not be unjustly enriched at the expense of another. The former theory, endorsed by the idea that a contract was necessary to enable a claim to be classified,[188] gives rise to what is known as a 'quasi-contractual' claim and covers various circumstances such as work done under a void or illegal contract, or when the contract had been terminated for breach, but excludes work done voluntarily. The particular area relevant to work on a letter of intent, however, may be seen as an exception to the general rule that work voluntarily done does not confer a benefit for which payment must be made. The exception occurs where there is an express or implied request for services to be rendered. This will include the situation where the performance of services is requested in anticipation of a contract which does not materialize, such as a request or authority in a letter of intent. The second theory falls into the area known as 'restitution'. Restitutionary remedies may be applied throughout all areas of law and the nature of restitution has been defined as[189] 'the law relating to all claims, quasi-contractual or otherwise, which are founded on the principle of unjust enrichment'. The idea that a person ought not to gain at another's expense is broadly in accord with the general views of many in the building industry who find themselves in the position of working without a contract. It is seen by many as being 'fair'. This is not so far from the rationale behind the concept. In *Moses* v. *Macferlane*,[190] Lord Mansfield said:

> The gist of this kind of action is that the defendant under the circumstances of the case, is obliged by the ties of natural justice and equity to refund the money.

---

187 (1981) 25 BLR 94.

188 Classification in civil actions in tort or in contract; see *Halsbury's Laws of England*, 4th edn (Butterworth), vol. 9, para. 636.

189 Goff, R., and Jones, G., *The Law of Restitution*, 3rd edn, p. 3.

190 (1760) 2 Burr, 1005 at 1012; cited by Owen-Conway, S., *Restitution and Quantum Meruit (1985–86)* (Sweet & Maxwell, 1896); 16 UWA Law Rev., 155.

Unfortunately for many Contractors, they are prone to apply their concept of natural justice and equity to their own actions of providing a service and a consequent obligation to pay, omitting to take into account the circumstances of the situation. The omission is crucial. Restitution is about refunding a gain (or in many circumstances paying for it) but is an exercise for the recipient of the service and taken from his standpoint. The point is emphasized succinctly by Lord Hope of Craighead in *Sempra Metals Ltd* v. *Inland Revenue*:[191]

> ... the remedy of restitution differs from that of damages. It is the gain that needs to be measured. The gain needs to be reversed if the claimant is to make good his remedy.

What then is a fair value of a Contractor's work? There are a number of possible answers. The choice is between (1) the actual cost to the Contractor, (2) the cost, plus a reasonable profit for the Contractor, (3) the usual contract price between the parties, (4) a typical price in the local market place, (5) the benefit or gain by the owner, or (6) a mixture of all these choices. Each may be a *quantum meruit* in different circumstances, but which type applies in any particular case could depend on the reason for claiming a *quantum meruit*.

1.8.2.3 What is quantum meruit?

The term *quantum meruit* is a synonym meaning 'as much as he has earned' and 'as much as they were worth';[192] but in the context of restitution the latter should provide the more accurate guide. Unfortunately the courts appear reluctant to assist, for rarely do they provide any rules or general guidance as to the calculation of a *quantum meruit* in the absence of contract. The furthest they appear prepared to go is to identify the various parts of a service provided and to attach obligation in principle by looking at the nature of the enrichment. An example can be seen in *Marston Construction Co. Ltd* v. *Kigass Ltd*,[193] from which several principles emerge. The court considered the provision of various estimates and drawings by a Contractor. The work was done in expectation of being awarded a design and build contract for a factory. The judge found that all the work was carried out in reasonable contemplation of payment (to be made through the future contract)[194] and much of the work was requested either expressly or impliedly by the owner. He determined that the owner had to pay for the drawings prepared for planning approval which had realizable value in that they were capable of being used, even though in fact the project was unlikely ever to proceed. On the other hand, the owner was not obliged to pay for the work carried out in the preparation of revised tenders which was of value only to the Contractor for the purpose of gaining further work.

The *Marston* judgment suggests that in order to succeed in a *quantum meruit* claim in the absence of any contract, there must be an express or an implied instruction to do the work, together with actual expectation of payment in circumstances where it is reasonable to expect payment; and the principal must be left with something of value to him that is

---

191 *Sempra Metals Ltd. (formerly Metallgesellschaft Ltd.)* v. *Her Majesty's Commissioners of Inland Revenue & Anr* [2007] UKHL 34, at para 28.

192 *Osborn's Concise Law Dictionary, 10th edn*: *Quantum meruit is a synonym for quantum meruit et quantum valebant.*

193 (1989) 46 BLR 109.

194 This aspect of the decision is open to the criticism that it was not reasonable for the Contractor to expect payment, since in this case the owner had said the development relied on a successful insurance claim.

realizable. The criteria in the *Marston* case have since been refined, although the practical effect for those seeking a restitutionary payment probably remains much the same. In *R (On the application of Charles Rowe)* v. *Vale of White Horse District Council*[195] the Court identified four essential ingredients:

> i) a benefit must have been gained by the defendant; (ii) the benefit must have been obtained at the claimant's expense; (iii) it must be legally unjust, that is to say there must exist a factor (referred to as an unjust factor) rendering it unjust, for the defendant to retain the benefit; (iv) there must be no defence available to extinguish or reduce the defendant's liability to make restitution.

Unfortunately in neither of the above cases was there any need to deal with value, but the approach in determining liability is nevertheless indicative of value-based rather than cost-based calculations (i.e. returning the benefit). This idea must sound warning bells to those who rely with blind faith on recovering a *quantum meruit* when all else fails (e.g. all those who work in reliance on letters of intent).

Most Contractors assume that quantification of their work on a *quantum meruit* should be done on a time and materials basis (cost plus), in order to compensate them for the resources provided. That approach can be seen in the judgment in the Australian case of *ABB Engineering and Construction Pty Ltd* v. *Abigroup Contractors Pty Ltd*:[196]

> whatever approach might have been made at an evidentiary level to support a finding of reasonable remuneration, the appropriate approach was to look at the costs that ABB had actually incurred in doing that work and then apply an appropriate margin for overheads and profit.

However, there are three main situations giving rise to *quantum meruit*, and the elements making up its value may vary. First, there is a situation where a contract has been broken and remuneration is being sought for work done. Second, there is the position where there is a contract, but there are no precise provisions as to price or payment. The third situation is where work has been done on request, in the absence of a contract, but in the knowledge that the work is not being done gratuitously.

***Quantum meruit* – breach of contract**: The breach of contract requires an assessment of damages. 'Time and materials', or cost recovery claims, are essentially a form of damages recoverable by the injured party from the party in breach of a contract. Thus time and materials claims could logically be associated with a 'quasi-contractual' relationship (but not a claim based on restitution).

***Quantum meruit* – contract exists, but no precise provisions**: This situation also gives rise to a type of contractual claim, but based on an implied promise to pay 'a reasonable sum'. It has been said that a reasonable sum in these circumstances is 'the value between a willing seller and a willing buyer'[197] where the court is required to take up a bargaining stance on behalf of both parties and to mentally enact the negotiation that never took place. Thus in *British Bank for Foreign Trade Ltd* v. *Novinex Ltd*[198] it was considered that

---

195 [2003] EWHC 388 (Admin) (07 March 2003), at para 11.
196 [2003] NSWSC 665.
197 *Per* Lord Denning MR in *Seager* v. *Copydex (No 2)* (1969) 1 WLR 809 at 813.
198 (1949) 1 All ER 155.

a commission basis was a proper mode of reasonable remuneration in line with the normal practice of the particular trade in question. Clearly, in the evaluation of *quantum meruit* in this second analysis the calculation must take into account the circumstances not only of both parties, but also external influences. It is this type of *quantum meruit* that could apply if acceptance of a letter of intent created a contract, without identifying the price.

Quantum meruit – requested works known not to be done gratuitously: *Quantum meruit* in these circumstances is relevant to letters of intent and to the battle of forms if no contract is formed, and may give rise to a restitutionary claim. The aim is 'to strip the defendant of a benefit which he has unjustly gained at the plaintiff's expense'.[199] Here the value of the *quantum meruit* should be determined by viewing the position from the 'outcome' of the arrangement (i.e. the value of the benefit conferred on the recipient). The method of calculation, however, is a problem, since there is little unequivocal guidance from the courts.

The problem was addressed by the Court of Appeal in *Crown House Engineering Ltd* v. *Amec Projects Ltd.*[200] The issue concerned circumstances which are common in the building industry. Crown House had carried out work on instruction in the absence of a contract. Amec had made payments on certificates as though a contract existed, but on the last certificate had set off sums in respect of counterclaims which included a charge for delays, allegedly resulting from Crown's tardy performance. Amec in defending, argued that if there was no contract and a *quantum meruit* claim could be made against them, then at least regard should be had to any necessary cost incurred by them as a result of receiving the service later than they had required. This, they said, had the effect of reducing the value of the benefit. At trial the judge said the value should be the 'objective value of the job that has been done'; he looked for genuine reduction which could be made in the objective valuation of the job, but found none. He considered tardiness could only be related to a contractual obligation; consequently, in the absence of a contract, Amec's counterclaim had no basis. An appeal from Amec was rejected, but in his speech Lord Justice Bingham said:

> Crown has argued … that these matters [tardiness] are wholly irrelevant to the assessment of what reasonable remuneration Crown should recover. It may well be that they are right … But the answer does not seem to me to be obvious … The doctrine of unjust enrichment … does no doubt require [that] … a customer should not take the benefit of a Contractor's services rendered at his request without making fair recompense … but it does not so obviously require that assessment should be made without regard to the acts or omissions of the Contractor when rendering those services which have served to depreciate or even eliminate their value to the customer.

The message appears to be clear; fair value should take into account the worth of the service provided to the recipient. This seems particularly relevant when the service has little or no value, albeit that the Contractor may have incurred substantial costs.

In *J.S. Bloor Ltd* v. *Pavillion Developments Ltd*,[201] a road, installed without prior instruction, was held to represent no overall benefit, and thus no value. Whilst the road was

199 See Goff & Jones, *Law of Restitution*, 3rd edn, pp. 19, 148.
200 (1989) 48 BLR 32.
201 [2008] EWHC 724 (TCC).

retained by the Employer, he argued successfully that, once built, he had no alternative but to accept it, but that the disadvantages outweighed any benefits. Benefits and disadvantages weighed in the balance included the use of the road for a while, against deprivation of the opportunity to have the road designed to the employer's own specification, and his inability to obtain collateral warranties.

The principle may be applied both ways. A service which is inexpensive to provide may be of great value. An example of this can be seen in the next case, where the challenge of defining a *quantum meruit* is met head on.

In *Costain Civil Engineering Ltd and Tarmac Construction Ltd* v. *Zanen Dredging and Contracting Co. Ltd*[202] there was a sub-contract for a tunnel under a river which required a casting basin as temporary works. It was decided to convert the casting basin into a marina. It was held that the work to the marina was not part of the original contract and it was not a variation. No contract existed, but the sub-Contractor was entitled to a *quantum meruit.* In deciding what a *quantum meruit* included the court had to consider two unusual circumstances. Since the sub-contractor was on site it did not have to incur substantial setting-up costs, and since the conversion to a marina was a beneficial afterthought by the client, the Joint Venture Contractor received additional profit. If the value was to be based on costs, the sub-contractor would not be out of pocket, but it would be a bargain for the Contractor since the costs were much lower than any other sub-contractor would have incurred; if the value was to be based on benefit, the inclusion of the profit gained by the Joint Venture could put a value on the sub-contractor's work higher than if done under a sub-contract. The court held that a *quantum meruit* in the absence of a contract should be enrichment-based and therefore valued as a benefit. The difficulty always lies in valuing, and it is convenient to relate to costs, even when the calculation is not theoretically cost-based. Here Zanen was awarded a 'top up', to make the costs comparable with what another sub-contractor would have had to charge as setting-up costs. Zanen was also awarded a split of the Joint Venture's profit. It is important to remember that what is being paid here is the value from the recipient's point of view, not the value to the provider.

In the Zanen case the *quantum meruit* was probably higher than a corresponding contractual price would have been; but it could in other circumstances be lower, and the relevant circumstances are often outside the control or even the knowledge of the provider. A difficulty for the Contractor is in establishing the value, but as in the Zanen case a logical starting point is the actual cost. Even here there can be difficulty as the Contractor discovered in *Bentley Construction Ltd* v. *Somerfield Property Company Ltd and Somerfield Stores Ltd*[203] in which the judge observed:

> The unhappy position in which I am left concerning … claims in respect of preparatory work … is that while I am satisfied on the evidence that some work was done … I am unable to say what … would be appropriate compensation for such work as was done. For that reason the claims … fail.

However, a more robust approach can be seen in the House of Lords' decision in the *Sempra* case referred to above, in which interest on the repayment of tax was claimed in restitution against the Inland Revenue. The relationship between the Inland Revenue and

---

202 (1996) 85 BLR 77.
203 [2002] 82 ConLR 163.

the Government made it difficult to calculate the actual benefit of the use of the claimant's money. It was held that it would not be unjust to award compound interest, which would reflect the commercial realities.

One significant area of risk for the Contractor lies in the limits that are sometimes inserted in letters of intent. Typical wording may be along the lines: '... in the event of the project not proceeding you will be paid your reasonable costs incurred up to a maximum of £50,000 ....' The introduction of a maximum is likely to be binding, whether it is considered as a limit on the Contractor's reasonable expectation of payment, or as a payment term in a limited contract. In the case of *Mowlem plc* v. *Stena Line Ports Ltd*,[204] a series of letters from Stena, gradually increasing the maximum, ceased at £10 million. Mowlem carried on working after reaching the maximum, and spent more than £10 million. The Court held that it would go against commercial sense if Mowlem retained the right to payment for the excess when Stena had stipulated how much they were prepared to pay.

In contrast, in *ERDC Group Ltd* v. *Brunel University*[205] it was agreed that a limited contract had been created by the letter of intent, but that work carried out after the date when authority in the letter expired was done in the absence of a contract. The Court held that the later work should be paid for at reasonable rates. Reasonable rates in this instance were said to be the rates being paid under the earlier contract, on the grounds that rates that were reasonable did not suddenly become unreasonable.

Similarly, a limit on the value of authorized work also featured in *Diamond Build Ltd v. Clapham Park Homes Ltd.*,[206] in which it was held that a monetary limit in the letter of intent capped the value of a limited contract created in the letter, but the cap did not cover additional work, not included in the tender; such extra work was to be paid for on a *quantum meruit* basis.

The only certainty about working in expectation of a *quantum meruit* in the absence of a contract is the uncertainty. Contractors who deliberately work without a contract are always at risk. In many of the cases cited in this section it was clear that the court looked upon the failure to achieve agreement as an unfortunate and unexpected event. Generally, the parties had gone about their business in anticipation of a contract being concluded, albeit that stage had not yet been reached. However, if a Contractor, on receipt of a letter of intent or an unacceptable counter-offer, deliberately decides to 'hook' the building owner with no intention of ever concluding a contract,[207] it must be questioned whether restitution as a just (i.e. equitable) entitlement would be available to him at all; the Contractor may be entitled to nothing. An example of this principle can be seen in the case of *R* v. *Vale of White Horse District Council*,[208] in which a restitutionary claim for payment in respect of council services was partly disallowed. There was no contract between the council and householders, but the council provided sewerage services for 6 years without charging separately, because they were uncertain as to their rights. The council eventually

---

204 [2004] EWHC 2206 (TCC).

205 [2006] EWHC 687 (TCC).

206 [2008] EWHC 1439 (TCC); [2008] CILL 2601.

207 In *Barclays Bank plc and Another* v. *Hammersmith and Fulham LBC*, [1991] 12 CL 64, CA Neill LJ suggested one of the factors to be taken into account might be the state of knowledge of the parties. In *Sabemo* v. *North Sydney Municipal Council* [1997] 2 NSWLR 880, judgement was given for the plaintiff contractor where the failure to conclude a contract was caused by the defendant's action carried out in bad faith.

208 *R* (*On the application of Charles Rowe*) v. *Vale of White Horse District Council* [2003] EWHC 388 (Admin).

became aware of their entitlement to charge after receiving legal advice, but for a further six months they did nothing about it. Letters were then sent to all householders informing them that they would be charged in future, and they would also be charged the arrears for the service they had received over the previous 6 years. It was held that the claimant householder had a general obligation to pay for the services provided, but that it would not be unjust for him to retain the past benefit (i.e. it was not unfair for him to avoid paying the arrears). The Court observed that: '... it was folly and short-sighted [on the part of the council] to place and deliberately leave the residents ... in the dark'. The council's approach lost them their entitlement.

In summary, it appears that there is no clear formula on which to calculate a reasonable sum to be paid for work carried out in the absence of a contract, and despite the ascendancy of restitutionary principles over 'damages' principles, neither is likely to oust the other completely. The resultant calculation emerges as an amalgam. The 'time and material' (cost plus) basis is a convenient starting point, as in the *Zanen* case, but costs should be modified by any diminution or enhancement of value to the recipient of the service. Costs should then be moderated further, if applicable, by external factors such as the general market rate or the type of costs which would be included in a general market price. One thing is clear, however; that is the amount to be received cannot be certain, and the principles behind the calculation provide an extremely unstable bedrock from which to build the foundations of any relationship.

2

Participants in the project and their roles under the Contract

Execution of a project let on JCT 05 requires the performance of distinct roles by a variety of participants. In most cases, their roles and duties are determined by a contract with the Employer. The contractual and administrative relationships among them are shown in Fig. 2.1. In parallel with their contractual duties, and depending on their contribution to the project, some of the participants may also assume duties of a statutory nature – e.g. duties as a 'client', 'designer', 'contractor', 'CDM Co-ordinator', or 'principal contractor' under the Construction (Design and Management) Regulations 2007 (CDM 2007). It is important to note that these labels for the participants refer to holders of certain duties rather than types of professional or business organizations and that it is possible for one organization or professional to discharge the role of more than one participant. For example, it is possible, although not advisable, for the Architect also to be the Employer's Representative, a designer and the CDM Co-ordinator.

The aim in this chapter is to examine their roles under the Contract that are not expressly covered by CDM 2007. Their duties as CDM dutyholders are explained in Chapter 10.

2.1 The Employer

The Employer plays easily the most dominant role in the whole project because it is he who initiates it and secures the necessary funding. This section considers only his role and duties during the construction phase. Readers are directed to textbooks on project management for an examination of his wider role in the total procurement process.

2.1.1 Express contractual duties

At the post-contract stage, the Employer generally stays in the background and allows the Architect, specialist design consultants, the Quantity Surveyor and the Clerk of Works,

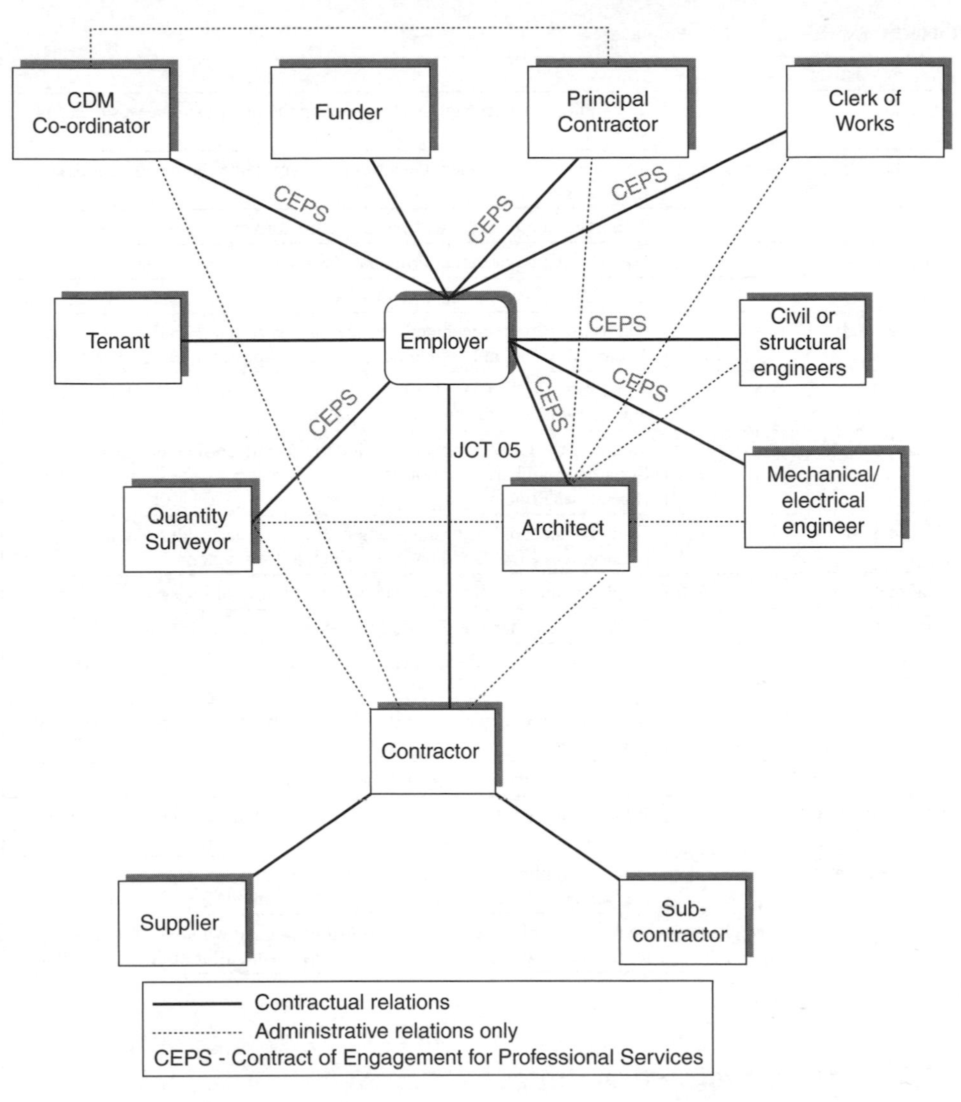

Fig. 2.1 Project participants

all of whom he employs, to act on his behalf. However, the Contract expressly requires the Employer to perform certain duties as summarized in Table 2.1. The term 'duty', as used in this context, denotes a task that the Employer must perform in any circumstances and is generally prefaced by 'the Employer shall…'. The table does not therefore include steps in procedures that the Employer may only have to take depending upon whether the relevant procedure is invoked (e.g. notice to refer to adjudication or arbitration and notice to withhold payment). The most important of the Employer's duties is his obligation, stated in Article 2, to pay the Contractor the Contract Sum or other sum payable under the Contract at times and in a manner explained in detail in Chapter 15.

Table 2.1 Express duties of the Employer

Article/Clause:	Duty
Article 3	To appoint a replacement Architect when the incumbent ceases to act as such
Article 4	To appoint a replacement Quantity Surveyor when the incumbent ceases to act as such
Article 5	To appoint a replacement CDM Co-ordinator whenever necessary
Article 6	To appoint a replacement Principal Contractor whenever necessary
Clause 2.4	To give possession of the Site to the Contractor on the Date of Possession
Clause 2.8.4	Not to use certain documents for any purpose other than the carrying out of the Works and not to divulge the rates and prices in the Contract Bills
Clause 3.25.1	To ensure that the CDM Co-ordinator and the Principal Contractor carry out their duties under the CDM Regulations
Clause 3.25.2	To notify the CDM Co-ordinator and the Architect of changes to the Construction Phase Plan notified to the Employer by the Contractor in his capacity as Principal Contractor
Clause 3.26	To notify the Contractor in writing of the name and address of a replacement CDM Coodinator or Principal Contractor
Clause 4.13.1	To pay on any Interim Certificate within 14 days of its issue
Clause 4.13.3/4.15.3	To serve a Payment Notice after the issue of every payment certificate
Clause 4.15.4	To pay on the Final Certificate within 28 days of its issue if the amount stated on it is payable by the Employer
Clause 7.1	Not to assign any right under the Contract without the written consent of the Contractor
Paragraph 2 of Schedule 2	To consider Schedule 2 Quotations from the Contractor
Insurance Option B	Where it applies, to take out and maintain insurance against damage to the Works
Insurance Option C	Where it applies, to take out and maintain insurance against damage to the Works and existing structures and their contents
Paragraph 23(2) of the Scheme for Construction Contracts	To comply with the decision of an Adjudicator to whom a dispute is referred until it is resolved finally by agreement, arbitration or litigation

2.1.2 Implied contractual duties

In addition to the express duties, duties normally implied into construction contracts may apply unless, in respect of the particular duty, the Contract expressly provides to the contrary. Examination of most of the express duties will show that they are no more than the expression of some of the principles flowing from an employer's duty, long established by common law, to cooperate with the contractor in the carrying out of the works. Flowing from this general duty to cooperate is an implied term that neither party will do anything to prevent the other from performing his side of the contract.[1] The positive form of this term is an implied obligation to do whatever is necessary for the other to perform his obligations. However, it would appear that the positive duty only applies to contracts that specify an outcome that cannot be achieved unless both parties cooperate to bring it

1 *McKay* v. *Dick* (1881) 6 App Cas 251; *Barque Quilpé Ltd* v. *Brown* [1904] 2 KB 264.

about. In *London Borough of Merton* v. *Leach*[2] Vinelott J. held that all these terms were implied into the JCT 63 Conditions. As JCT 05 does not depart from JCT 63 in respect of general philosophy and approach, it is submitted that the implied duty to cooperate in general and its derivative duties also apply to it.

2.1.3 Duty not to interfere with the discretion of the Architect or Quantity Surveyor

As explained in Section 2.4 in this chapter, where the Architect or Quantity Surveyor is not acting as an agent of the Employer but as an independent professional holding the balance fairly between the Employer and the Contractor (e.g. carrying out of valuation of variations or certification), the Employer does not warrant that those duties will be performed with reasonable skill and care. For example, the Employer is not liable at common law for the financial losses of the Contractor flowing from any under-certification by the Architect.[3] The remedy is for the Contractor to invoke the appropriate dispute resolution mechanism to have the certificate revised.

However, the Employer is under an implied warranty that the Architect and the Quantity Surveyor will act fairly independently and that he will not interfere with their free exercise of their professional judgement. *Hickman & Co.* v. *Roberts and Others*[4] illustrates unlawful interference with the independence of an architect under a building contract. In response to the architect's letter to the employer that the contractor was entitled to a certificate for interim payment and that he intended to issue it, the employer replied that he was not to do so until the contractor had supplied an account of all extras. The architect wrote advising the contractor that he had been instructed to withhold interim payment certificates and that the contractor was to resolve the problem with the employer. There were also letters from the employer instructing the architect to cut down the contractor's claims as much as possible as the employer had financial difficulties. The House of Lords decided that the employer's conduct amounted to wrongful interference.

Case law suggests that the Employer may even be under a duty to stop any unfairness to the Contractor of which the Employer becomes aware.[5]

2.1.4 Rights of third parties

The nature of third party rights introduced into JCT 05 is described in Chapter 9. Depending on the way the Contract Particulars are completed, the Contractor may have

---

2 *London Borough of Merton* v. *Stanley Hugh Leach* (1985) 32 BLR 51 (hereafter *Merton v Leach*); see also *Holland Hannen & Cubitts* v. *Welsh Health Technical Services Organisation* (1981) 18 BLR 80 in which the implication of these terms was approved by Judge Newey QC.

3 See Sections 2.4.6 and 2.4.8 in this chapter and Chapter 15, Section 15.9, for more detailed discussion on this issue.

4 [1913] AC 229; on this implied duty see also *Sutcliffe* v. *Thackrah* [1974] AC 727; *Merton* v. *Leach*; *John Mowlem & Co. Ltd* v. *Eagle Star Insurance Co. Ltd and Others* (1992) 62 BLR 126. Under Clause 8.9.1.2 interference with the issue of certificates under the Contract is a specified default by the Employer (i.e. the Contractor is entitled to terminate his own employment under the Contract) (see Chapter 16, Section 16.9.2).

5 *Panamena Europea* v. *Leyland* [1943] 76 Lloyd's Rep. 114; *Perini Corporation* v. *Commonwealth of Australia* (1969) 12 BLR 82 (Supreme Court of New South Wales). See Chapter 15, Section 15.3, for discussion of negligent certification of which the Employer is aware.

obligations to grant rights in relation to the Contract to third parties. These rights are triggered by the Employer's receipt of a notice from the Contractor, who has no obligation to notify the beneficiaries of the rights. Unfortunately, third party beneficiaries will be unable to enforce their entitlements against the Contractor unless they become aware of their existence. It is for the Employer to notify them, whether or not he has express obligations under separate contracts with the third parties.

2.1.5 The Employer as client under the Site Waste Management Plans Regulations

The Employer may have statutory duties in respect of a client under the Site Waste Management Plans Regulations 2008,[6] which came into force on 6 April 2008. They apply to construction and demolition projects with an estimated cost greater than £300,000 (excluding VAT) planned for after 6 April 2008. They also apply to projects planned for earlier but which started after 1 July 2008. Regulation 5 requires that any client intending to carry out a qualifying project should prepare a site waste management plan (SWMP) for it before construction begins. The client must also appoint a principal contractor where there is the intention to use contractors to carry out the project. The principal contractor here may be different from the dutyholder of the same name under the CDM Regulations although, in the interest of effective communication, it would be advisable for one person to perform both roles.

For a project estimated to cost less than £500,000, the prescribed content of the SWMP must include:

- the identity of the client;
- the identity of the principal contractor;
- the identity of the person who drafted it;
- the estimated costs of the project;
- decisions taken before the drafting of the plan as to how quantities of waste from the project would be minimized;
- description of the type of waste expected to be produced from the site;
- estimates of the amount of each type of waste expected to be produced;[7]
- the waste management plan for each of type of waste, including reuse, recycling and lawful disposal;
- declarations by the client and the principal contractor that they will take reasonable steps to ensure compliance with the requirements of the Regulations.

The SWMP is expected to be a living document. For projects of value of £500,000 or less, the principal contractor, whenever waste is removed from the site, must record on the plan the:

- identity of the carrier;
- types of waste removed;
- site to which the waste is removed.

---

6 They were imposed by Statutory Instrument SI 2008 No. 314.

7 It is expected that, as a minimum, the classification system should recognize three types of wastes: inert, non-hazardous, or hazardous.

There are more onerous obligations where the value of the project is greater than £500,000. For example, the information on the plan must include the waste carrier's registration number and copies of relevant licenses, permits and waste transfer notes. The principal contractor must keep the plan under periodic review to ensure that it accurately reflects progress on the project. In any event, the frequency of review must not be less than every 6 months. The review must maintain a record of types and quantities of waste produced and their ultimate outcomes. Within three months after completion of the project, the principal contractor must close out the plan by recording on it:

- confirmation that the plan was properly monitored and updated;
- comparison of the estimated quantities of each type of waste against actual quantities;
- explanation of any differences between the estimates and the actual waste;
- an estimate of cost savings flowing from compliance with the Regulations.

The Employer under JCT 05 would be a client for the purposes of these Regulations and must therefore ensure compliance. JCT 05 form needs to be amended to address these additional duties (this had not been done as at the time of writing). Until this is done, the Employer should seek legal advice on appropriate amendments to the Contract and their other contractual arrangements.

The enforcement agencies are local authorities and the Environment Agency. Failure to comply with the SWMP is a criminal offence punishable by a fine not exceeding £50,000. Parties who may be liable include individuals and companies acting as clients for the purposes of the Regulations. Where a company or other corporate entity commits an offence under the Regulations with the consent or connivance of a director, manager, secretary or other like officer of the company, that person would also be liable for the breach. A person in breach of the Regulations may be given the opportunity to discharge liability for the breach by payment of a fixed penalty of £300.

2.2 The Employer's Representative

It is common practice for the Employer's role to be delegated to an in-house project manager or an external project management firm. JCT 05 recognizes this practice by providing in Clause 3.3 that the Employer is entitled to do this by written notice identifying the name of the individual to act for him and the extent of the delegation. The delegation takes effect from the date of the notification. It is important to note that the Employer's Representative must be a named individual and not that of the firm he may be working for. A footnote advises against the Architect or the Quantity Surveyor being named as the Employer's Representative on grounds of possible confusion arising from the duality in their roles that would result.

The duties to be performed by the Employer's Representative are a matter for his contract with the client, the Employer. Case law suggests that, in the absence of terms to the contrary, the Employer's Representative may be considered a watchdog on behalf of the Employer, with responsibilities for reporting any failings of the Architect, the Quantity Surveyor or other consultants to the Employer. In *Chesham Properties Ltd* v. *Bucknall Austin Project Management Services Ltd*,[8] his Honour Judge Hicks, QC, deciding a

8 (1996) 82 BLR 92.

preliminary issue, held, among other things, that a project manager on a contract in the terms of JCT 80 owed a duty in contract and in tort to advise and/or inform the Employer of actual or potential deficiencies in the performance by the Architect, the Quantity Surveyor and other consultants of their contractual duties to the Employer. It is submitted that this also applies to the Employer's Representative under JCT 05.

2.3 The Contractor

Article 1 states the Contractor's primary obligation as being to carry out and complete the Works in accordance with the Contract Documents. This obligation is repeated in Clauses 2.1–2.3 but with more detail on the quality of materials and standards of workmanship to be achieved and any Contractor's Designed Portion (CDP).[9] The Conditions contain express and specific duties within this general obligation for purposes of certainty, monitoring and remedial action as appropriate.

2.3.1 Materials and workmanship

The Contractor's responsibility to the Employer for workmanship and materials at common law is next explained for the purpose of putting in context the provisions of JCT 05 on the subject. The Supply of Goods and Services Act 1982 has put the common law principles discussed in statutory form.

2.3.1.1 The common law position

The common understanding of 'workmanship' in the construction industry concerns the skill and care exercised by a contractor in the physical execution of work. However, to the extent that choice of materials is left to the contractor, this may also mean design (i.e. suitability of the materials for the purpose for which they have been used). In the absence of terms to the contrary, or in directing the contractor as to the detailed manner in which the work is to be done, the law has, for a long time, implied a term that the work will be done in a proper and workman-like manner (i.e. with the skill and care of an ordinary competent contractor).[10]

From general principles of the law of contract, a contractor is under an obligation to ensure compliance with the specification where materials are specified in the contract. If the specification states a brand name or a particular supplier of the material, the contractor would still be under a warranty that such materials are of good quality when used. This obligation is absolute (i.e. it is no defence to the liability for a defect that it was not discoverable even with the most careful examination). A warranty of fitness of purpose may also apply if the circumstances indicate that there was reliance on the contractor's skills regarding suitability of the materials (e.g. leaving the choice of the type of material

9 The Contractor's Designed Portion is covered in detail in Chapter 4.
10 *Duncan* v. *Blundell* (1820) 3 Stark. 6; *Pearce* v. *Tucker* (1862) 3 F & F 136; *Test Valley Borough Council* v. *Greater London Council* (1979) 13 BLR 63.

to the contractor). The leading authority on quality and fitness for purpose of materials in construction contracts is *Young & Marten Ltd* v. *McManus Childs Ltd*:[11]

> MC, developers and the main contractor[12] under a building contract, sub-let the roofing to YM. The sub-contract called for the use of a type of tile called 'Somerset 13', which was available from only one manufacturer. Soon after construction, some of the tiles began to disintegrate. This problem was traced to a batch of the tiles that contained defects which could not have been discovered even with reasonable inspection. MC sought to recover the costs of re-roofing from the sub-contractor. YM argued that, since they had not been relied upon regarding the type of tile and they could not reasonably have discovered the defects, they were not liable. The House of Lords decided that, unless the circumstances of a particular case are such as to exclude it, there will be implied into a contract for work and materials a term that the materials will be of good quality and a further term that the materials will be reasonably fit for the purpose for which they were used. It was held that, in this particular case, although there was no obligation for fitness for purpose, there was still a duty to ensure that the materials were of good quality because, as YM had a choice whether or not to accept the batch of defective tiles, there was reliance upon them to inspect them before acceptance. YM were therefore liable.

However, where the circumstances of the particular contract indicate that there is no reliance on the skill and care of the contractor on the issue of quality (e.g. the terms of the contract compel the contractor to accept materials from a particular supplier), the contractor will not be liable for defects in them.

> *Gloucestershire C.C.* v. *Richardson*:[13] the respondents, main contractors on a building contract let on JCT 39 (1957 revision), were instructed by architects to accept and use concrete columns from nominated suppliers. After the columns had been incorporated into the works, defects were discovered. A PC sum in the Bill of Quantities merely required the contractor to erect columns to be supplied by nominated suppliers. All the circumstances of the contract indicated that the contractor had no choice but to accept the supply even though the supplier's terms negotiated by the architect contained exclusion clauses. The House of Lords held that, in those circumstances, the contractor was not subject to a warranty that the columns would be of good quality.

As their Lordships emphasized, generally the principles discussed above apply to the work and materials of sub-contractors and suppliers even if the employer nominated them.[14] Two reasons are normally given for this. First, main contracts often state expressly that the contractor is to be responsible for the work and materials of all sub-contractors and suppliers exactly as if they are his own. Second, there is a need to maintain a chain of liability from the employer down to the manufacturer. Without such a chain of liability, a sub-contractor or supplier can, at best, recover only nominal damages from the next party down the chain (wholesaler or manufacturer). Generally, the law follows the view that society is not well served by allowing those causing loss or damage to escape liability whilst those who suffer

---

11 (1969) 9 BLR 77 (hereafter *Young & Marten*).

12 The main contractors were Richard Saunders Ltd but the pleadings showed that they were acting as agents for the developers.

13 [1968] 1 AC 480; 2 All ER 1181; this decision was made by the same Law Lords and given the same day as *Young & Marten*.

14 *Rumbelows Ltd* v. *A.M.K.* (1980) 19 BLR 25.

the loss are denied any remedy. Wherever possible, the courts therefore interpret contracts in such way that the chain of liability is maintained from the ultimate consumer right down to the manufacturer. In *Young & Marten*, Lord Reid explained the need, wherever possible, to construe contracts so as to maintain the chain of liability in these words:

> There are, in my view, good reasons for implying such a warranty if it is not excluded by the terms of the contract. If the contractor's employer suffers loss by reason of the emergence of the latent defect, he will generally have no redress if he cannot recover damages from the contractor. But, if he can recover damages, the contractor will generally not have to bear the loss; he will have bought the defective material from a seller who will be liable under s. 14(2) of the Sale of Goods Act, 1893, because the material was not of merchantable quality. And, if that seller had in turn bought from someone else, there will again be liability, so that there will be a chain of liability from the employer who suffers the damage back to the author of the defect.

2.3.1.2 The contractual position

On the quality of materials and workmanship, the Contract states:

> 2.1 The Contractor shall carry out and complete the Works in a proper and workmanlike manner and in compliance with the Contract Documents, the Construction Phase Plan and the Statutory Requirements and shall give all notices required by the Statutory Requirements.
>
> 2.3.1 All materials and goods for the Works, excluding any CDP Works, shall, so far as procurable, be of the kinds and standards described in the Contract. Materials and goods for any CDP Works shall, so far as procurable, be of the kinds and standards described in the Employer's Requirements or, if not there specifically described, as described in the Contractor's Proposals or documents referred to in Clause 2.9.2. The Contractor shall not substitute any materials or goods so described without the written consent of the Architect, which shall not be unreasonably delayed or withheld but shall not relieve the Contractor of his other obligations.
>
> 2.3.2 Workmanship for the Works, excluding any CDP Works, shall be of the standards described in the Contract Bills. Workmanship for any CDP Works shall be of the standards described in the Employer's Requirements or, if not there specifically described, as described in the Contractor's Proposals.

The requirement to work to the Contract Documents implies that, in respect of materials and workmanship specified in detail in the Contract Bills, the Employer's Requirements or the Contractor's Proposals, it is no defence to liability for non-complying work or materials that the Architect or the Clerk of Works (COW) approved them or even included them in payment certificates. The same point is expressly stated in Clause 3.6. These two parties are on the site solely for the benefit of the Employer. In the performance of their duties of inspection, they have no responsibility to the Contractor to discover his mistakes.[15] They do owe that responsibility to the Employer and if they fail to discover defects that they should, with reasonable skill and care, have discovered, they will be liable to him for the resulting loss.

---

15 See Section 2.4.8 in this chapter for a detailed discussion of the Architect's liability to the Contractor.

From the general principles already explained, the Contractor is not under a warranty of fitness for purpose in respect of materials specified in the Contract Bills. Some specifications contain lists of different types of materials for the same purpose, with the choice left to the Contractor as to which type to use. Such specifications raise the issue whether the Contractor is liable if his choice is not fit for purpose whilst the others on the list are. This issue came before the Court of Appeal in *Rotherham Metropolitan Borough Council* v. *Frank Haslam Milan & Co. Ltd and M.J. Gleeson (Northern) Ltd*[16] which arose from a contract let on JCT 63. Clause 6(1) of that form of contract stated that 'All materials, goods and workmanship shall so far as procurable be of the respective kinds and standards described in the Contract Bills'. Hardcore to be used as fill to the underside of the ground floor slab of a building was specified in the Contract Bills as:

> Granular hardcore shall be well graded or uncrushed gravel, stone, rock fill, crushed concrete or slag or natural sand or a combination of any of these. It shall not contain organic materials susceptible to spontaneous combustion, materials in a frozen condition, clays or more than 0.25 of sulphate ions as determined by BS 1377.

The Contractor used steel slag that expanded after completion, causing heaving of the ground floor with resulting cracking in reinforced concrete floors. It was argued for the Employer that as the choice of type of hardcore was left to the Contractor, he warranted the fitness of his choice for purpose as hardcore. This was unanimously rejected in the Court of Appeal.

The Contractor's obligation to use materials of the kind and standard specified in the Contract Bills, the Employer's Requirements or the Contractor's Proposals is qualified by 'so far as procurable': Clause 2.3.1. It is to be noted that the Contract does not specify the geographic limits of procurability. This omission could give rise to arguments whether materials available only abroad are procurable. If after execution of the Contract, any item of material ceases to be procurable, the Architect may issue a Variation altering the kind or standard of the item. The Architect may also consent to the use of substitutes. The Contractor's obligation on standards of workmanship is not qualified by 'so far as procurable'. This means that it is no defence to the obligation that the human skills or equipment to achieve that standard of workmanship are not procurable.

Where the quality of materials and standards of workmanship are left to the opinion of the Architect, they must be to his reasonable satisfaction: Clause 2.3.3. This is an objective standard. The Architect has no power to demand quality or workmanship of the highest standards without a Variation. A decision of the Court of Appeal suggests that the Architect may be entitled to take into account the competitiveness of the Contractor's prices for the relevant materials or work in deciding whether they are to his reasonable satisfaction. In *Cotton* v. *Wallis*[17] the specification in the contract required all materials and workmanship to be the best of their kind and to the full satisfaction of the Architect. The contract also provided that 'the contractor shall carry out and complete the works in accordance with this contract in every respect with the direction and to the reasonable satisfaction of the architect'. It was held by a majority that it was reasonable for the Architect to accept work and materials not of the best quality because the contract price was very low.[18]

---

16 (1996) 78 BLR 1.

17 [1955] 1 WLR 1168.

18 See also Chapter 4, Section 4.2.2, dealing with the use of superlatives to describe quality in the Employer's Requirements.

In respect of quality standards that must meet the Architect's reasonable satisfaction and with which he is dissatisfied, Clause 3.20 requires the Architect to express any dissatisfaction within a reasonable time from execution of the unsatisfactory work. The Contract does not state the effect of the Architect's failure to do this. A possible construction is that the Architect cannot thereafter reject the work no matter how bad it is and that he must therefore include it in Interim Certificates. However, the Architect may advise the Employer to challenge the quality of the work before expiry of 28 days after the issue of the Final Certificate. After the 28 days, as explained under Chapter 3, Section 3.9.4, the Contractor is no longer answerable for work expressed as subject to the reasonable satisfaction of the Architect. An alternative, and preferred, construction is that the Architect can still reject the work but the delay in expressing his dissatisfaction is a breach of Contract by the Employer for which the Contractor would be entitled to damages.

To the extent that quality of materials and goods and standards of workmanship are neither described in the Contract nor left to the reasonable satisfaction of the Architect as already explained, Clause 2.3.3 requires such quality and standards to be appropriate to the Works or the CDP Works as the case may be. This is probably a different standard from the 'good and workmanlike manner' standard normally implied at common law. For example, the standard of finishes appropriate to farm buildings intended for use by animals would not be appropriate for a five-star hotel. As the Architect must decide whether such work has been properly executed before including it in Interim Certificates under Clauses 4.10 and 4.16, it is his decision whether the workmanship used by the Contractor is appropriate for the Works. The practical difference between this benchmark and the reasonable satisfaction of the Architect is that the Final Certificate is not conclusive evidence that the quality of materials and goods and standard of workmanship are appropriate.

Under Clause 2.3.4, in respect of materials and goods specified in the Contract by reference to the Architect's satisfaction with their quality, the Contractor must, upon request, provide the Architect with evidence that any such materials and goods used were to his satisfaction. There is no specified time limit within which the Architect may make such a request. A prudent Contractor would therefore obtain the Architect's written statement of his satisfaction depending on the importance of the relevant items. Also, lack of response to the Contractor's request to the Architect to provide reasons for expressed dissatisfaction as required by Clause 3.20 could constitute such evidence. There is a corresponding obligation to provide evidence that materials and goods not specified in detail anywhere are appropriate to the Works. Some of the relevant evidence may be vouchers but it is to be noted that the Contract no longer limits the evidence that may be requested to vouchers.

Liability in respect of quality of materials and standard of workmanship specified in detail in the Contract lasts for either 6 or 12 years, depending upon whether the Contract is executed as a simple contract or a deed (limitation of actions is discussed in detail in Chapters 1.4.1.2 and 1.5.5.6). However, where quality of materials and standards of workmanship are left to the opinion of the Architect, the Employer cannot challenge any decision of the Architect that the materials or workmanship are to his reasonable satisfaction unless proceedings are commenced within 28 days after the issue of the Final Certificate.[19]

19 See Chapter 3, Sections 3.9.3 and 3.9.4 for detailed discussion of this issue.

2.3.1.3 *Substitution of materials required under the Contract*

Clause 2.1 prohibits the Contractor from substituting any materials or goods required by the Contract without the written consent of the Architect, which is not to be unreasonably delayed or withheld. It is debatable whether straight substitution (i.e. there is no alteration in the kind and standard of the item)[20] is within the Clause 5.1 definition of a Variation. However, it is submitted that, as a Variation is defined by inclusive phraseology, straight substitution is likely to be within the definition. The Architect, in deciding whether or not to give his consent to a requested substitution, must therefore bear in mind the possibility of the Contractor later claiming that the consent amounts to a Variation. The Architect can close off this possibility by stating in the clearest possible terms that the consent is subject to the condition that it does not take effect as a Variation or in any other way change the Contractor's other obligations under the Contract.[21]

A common practice in drawing up specifications for materials and goods is to add to the description of relevant items in the Contract Bills the phrase 'or equivalent approved by the Architect'. The construction of a similar phrase in *Leedsford Ltd* v. *Bradford Corporation*[22] suggests that such a strategy gives the Architect absolute discretion whether or not to accept materials of similar specification offered by the Contractor. This construction is unlikely to apply to JCT 05 as the Architect is required expressly under Clause 2.3.1 not to delay or withhold unreasonably his consent to substitute materials and goods.

2.3.2 The Contractor's duty to give notices

The Contractor is required to give a variety of notices to the Architect and/or the Employer. They are summarized in Table 2.2. Where, in respect of a particular requirement for a notice or a document, the Contract prescribes the manner of giving or serving it, the Contractor must comply with the prescription.[23] Clause 1.7 deals with situations where the Contract does not state the manner of notice. The Contractor must give the notice or serve the document by any effective means to the address stated in the Contract Particulars or any agreed address. The notice or document is deemed effectively given if addressed and given by actual delivery or pre-paid postage to the applicable address. If no address is agreed, the notice or document is deemed given to or served on a registered company if delivered or properly posted to its registered or principal office. In the case of a non-registered body, there is a need to deliver or post it properly to the last known principal business address.

The notice requirements serve a number of purposes, including to:

1. enable the Architect to deal with the matter notified or to take timely and appropriate remedial action to minimize its negative effects;
2. alert the Architect to monitor the situation and to collect contemporaneous information with a view to avoiding future disagreements concerning what actually happened;

---

20 An example is where the Contract Bills specify a brick from a particular manufacturer and there is the same kind and standard of brick obtained from another.

21 See Chapter 6, Section 6.6 for detailed discussion on sanctioned Variations under Clause 3.14.4.

22 (1956) 24 BLR 45.

23 For example, Clauses 8.2 prescribes the manner of serving notices of default and termination.

Table 2.2 Notices to be served by the Contractor

Clause	Matter to be Notified
Clause 2.1	Notice required by any Statutory Requirement (the appropriate body)
Clause 2.6	Proposed use or occupation of any part of the site or the Works by the Employer and additional premium payable for such use or occupation
Clause 2.12.3	Time for provision of information by the Architect of which he appears to be unaware
Clause 2.15	Errors, discrepancies and divergences within or between documents listed under that clause
Clause 2.17	Divergence between the Statutory Requirements and a Clause 2.15 document or a Variation instruction
Clause 2.18	Emergency work to comply with the Statutory Requirements
Clause 2.27.3	Material circumstances of actual or likely delay to the progress of the Works
Clause 2.27.3	Material changes to estimates of particulars of delay and additional information reasonably necessary to keep the Architect updated
Clause 3.25.2	Amendment to the Construction Phase Plan (this applies only if the Contractor is also the Principal Contractor)
Clause 5.3.1	The Contractor's disagreement with an AI stating that Schedule 2 Quotation procedure applies to a proposed Variation
Clause 8.5.2	Occurrence of certain insolvency events in relation to the Contractor
Clause 8.9.1	Specified default by the Employer (optional)
Clause 8.9.3	Termination of the Contractor's employment for a specified default or suspension event
Clause 8.10.1	Termination by the Contractor of his own employment for insolvency of the Employer
Clause 8.11	Termination by the contractor of his own employment for serious suspension by events listed under that clause
Clause 3.22.4	Fossils, antiquities and other objects of interest or value found on the site
Paragraph A.4.1, B.3.1 & C.4.1 of Schedule 3	Loss or damage to any part of work executed or Site Materials
Option A and B of Schedule 7	Occurence of events for which the Contract Sum is to be adjusted for fluctuations

3. allow the Architect properly to keep the Employer informed about the project;
4. allow the Employer to make necessary arrangements to deal with the notified matter (e.g. making arrangements for additional funds);
5. assist the Quantity Surveyor in valuing the Works.

Failure to serve any notice in accordance with the stipulations of the Contract is technically a breach of contract for which the Employer may claim damages or even terminate the Contract at common law for a fundamental breach. Also, the Contractor may not be able to enforce his contractual rights in relation in the matters he failed to notify. For example, under Clause 4.23, it is a precondition for ascertainment of loss and/or expense under the Contract that the Contractor has made a written application that regular progress is being, or will be, disturbed. In any event, to the extent that the failure to notify prevented operation of the appropriate contractual machinery, he cannot enforce such operation because it is a long standing common law principle that 'no person can take advantage of the non-fulfilment of a condition the performance of which has been hindered by himself'.[24]

24 *Roberts* v. *Bury Commissioners* (1870) LR 5 CP 310.

2.3.3 The Contractor's design responsibility

This issue begs the question what is 'design'? Although it is not possible to produce a universally acceptable definition, there is no doubt that the term is wide enough to include decision-making during the drawing up of specifications for the Works, determination of dimensions, determination of reinforcement needs and choice of working methods.[25] Generally, in a 'construct' contract (i.e. one in which a full design has been produced on behalf of an employer), with the contractor only implementing what has been designed, the contractor bears no design responsibility towards the employer. This proposition is based on a presumed common intention that the contractor is not to be responsible for a matter on which the employer never placed reliance upon his skills.

> *Lynch* v. *Thorne*:[26] a builder put up a house in accordance with plans and specifications annexed to the contract. The specifications required use of nine-inch bricks in an external wall without any rendering. The litigation concerned the builder's liability for water penetration through the wall. The Court of Appeal held that, since the builder had complied exactly with the specifications and drawings, he was not liable.

The wording of Article 1 and Clauses 2.1 and 2.2 of JCT 05 suggests that, except where a CDP is included in the Works, it is a construct contract as the Contractor is simply to carry out the works in accordance with the Contract Documents. *John Mowlem & Co. Ltd* v. *BICC Pensions Trust Ltd & S. Jampel & Partners*[27] was decided on the unchallenged assumption that a Contractor under JCT 63, which is virtually indistinguishable from JCT 05 without a requirement for any CDP on this issue, had no responsibility for design.

A section of the Contract Bills contained a performance specification which stated:

> retaining walls forming external walls to buildings and basement slabs are to be constructed so that they are impervious to water and damp penetration, and the contractor is responsible for maintaining these in this condition'. Clause 12(1) of JCT 63, which is similar to Clause 1.3 of JCT 05, provided that nothing in the Contract Bills was to modify or override the printed conditions. The action concerned cracking and water penetration that occurred. It was held that the performance specification imposed a design responsibility on the contractor and that it was therefore ineffective against Clause 12(1).

Whilst the whole tenor of JCT 05 without a CDP is that the Contractor carries no design responsibility, some of the responsibilities of the Contractor require the exercise of a certain amount of design skills, e.g. workmanship and selection of materials to the extent that these matters have not been covered in the specification compliance with Building Regulations and other statutory obligations, and reporting of errors, departures, divergences and discrepancies.[28]

---

25 See *Hudson's Building and Engineering Contracts*, ed. Wallace, I.N.D. 11th edition (Sweet & Maxwell, 1995), paras 4.064 and 4.065 (hereafter *Hudson's*).

26 [1956] 1 WLR 303.

27 (1977) 3 ConLR 63. For a contrasting view of the effect of Clause 1.3 see Chapter 1, Section 1.5.5.

28 But see also Chapter 4, Sections 4.3 and 4.4.4 regarding work specified by performance in the Contract Bills.

2.3.4 The Contractor's CDP obligations

Clause 2.2 provides that if the Works includes a Contractor's Design Portion the Contractor must: (i) complete the design of the CDP; (ii) comply with the directions of the Architect for the integration of the Contractor's design of the CDP with the design of the rest of the Works; (iii) comply with Regulations 11, 12, 18 of CDM 2007. The CDP provisions, in effect, embed a design and build contract into what would otherwise be a construct only contract. However, whilst a design build contractor is generally under an implied duty to achieve fitness for purpose, Clause 2.19.1 purports to limit the contractor's duty in relation to the design component of the CDP to one of skill and care of a reasonably competent architect. These matters are covered in Chapter 4, Section 4.4.

2.3.5 The Contractor's duty to warn[29]

Case law suggests that there are circumstances in which a contractor may be held liable under a construct contract for design defects on the basis of breach of an implied duty to warn the owner, or his representatives, of design defects. In the Canadian case of *Brunswick Construction* v. *Nowlan*,[30] the contractor was to build a house in accordance with drawings provided by the owner but without any supervision from the architect who had carried out the design and produced the drawings. The Supreme Court of Canada decided that, in those circumstances, the owner must have relied upon the contractor regarding obvious defects in the design and that a duty to warn the owner of such defects was implied. In *Equitable Debenture Assets Corp. Ltd* v. *William Moss*[31] Judge Newey QC decided that a contractor owed a duty to warn of design defects of which the contractor was actually aware. In *Victoria University of Manchester* v. *Hugh Wilson*[32] the same Judge held that it is an implied term of a construct contract that the contractor should warn the employer of defects in the design that he believed to exist. He explained that 'belief' in this context required more than mere doubts as to the correctness of the design but less than actual knowledge of errors. These cases arose from contracts let on versions of JCT 63 that did not contain express terms requiring the Contractor to warn the Architect of defects that he actually discovered. However, in *University of Glasgow* v. *William Whitfield and John Laing*,[33] which arose from a subsequent version of JCT 63 that contained express obligations[34] to warn of discovered defects, Judge Bowsher said *obiter* (Lat. by the way) that there was no implied duty to warn of design defects. He distinguished the earlier cases on the grounds that whilst those cases involved reliance by owners upon contractors regarding design defects, there was no evidence of such reliance in the case before him.

A contractor has a duty under the Supply of Goods and Services Act 1982 to exercise reasonable skill and care in the performance of the construction contract. It is

---

29 See Nicholls, H., 'Contractors Duty to Warn' (1989) 5 Const LJ 175; Wilson, S., and Rutherford, L., 'Design Defects in Building Contracts' (1994) 10 Const LJ 90.

30 (1974) 49 DLR 93.

31 (1984) 1 Const LJ 131.

32 (1984) 1 Const LJ 162.

33 (1988) 42 BLR 27; this decision was preferred by Judge Esyr Lewis, QC in *Chancellor, Masters and Scholars of the University of Oxford (trading as Oxford University Press)* v. *John Stedman Design Group and Others* (1991) 7 Const LJ 102.

34 The express provisions worked against implications of additional obligations in the sense that the parties are deemed to have defined completely the Contractor's duty to warn.

arguable that the limitation of a contractor's duty to warn of only defects actually discovered may not be effective where a defect is so glaring that any contractor complying with this statutory duty ought reasonably to have discovered it. This argument is more applicable to undiscovered defects that present serious health and safety risks than aesthetic defects or those that only affect the pocket of the employer. Judge Newey adopted such an approach when the issue came before him again in *Edward Lindenberg* v. *Joe Canning and Others.*[35] In that case the defendant was engaged by the claimant developer to carry out preliminary work in the basement of a block of flats that the claimant wished to convert into flats. Drawings produced by the claimant's surveyor showed 9-inch internal walls, including the chimney breast wall, as non-load-bearing. The defendant demolished the walls without props or other suitable precautions, resulting in deflection of the floor of the flat above. The defendant contended that, in assuming that protective measures were unnecessary, he had simply followed the drawings provided by the claimant and that, therefore, he was not liable for the damage. Judge Newey QC held that the defendant should have had grave doubts about the correctness of the information on the drawing, not only from the fact that so obviously an important structural member as the chimney breast wall was indicated as non-load-bearing, but also from the fact that 9-inch walls should be required for a non-load-bearing function. Following his previous decisions, he held that the defendants were in breach of an implied term of the agreement that the defendant should exercise the skill and care expected of an ordinary competent builder.

The existence of a contractor's duty to warn of defects eventually got to the Court of Appeal in *Plant Construction plc* v. *Clive Adams Associates and JMH Construction Services Ltd.*[36] Plant Construction was engaged by the Ford Motor Company as the main contractor to design and construct two pits for engine mount rigs in Ford's research and development centre. Plant sub-contracted the substructure works to JMH. Plant also engaged Clive Adams as consulting structural engineers to design and monitor the substructure works. The substructure works involved removal of part of the base to an existing steel column supporting part of the roof to the centre. Ford's engineer instructed that, pending provision of underpinning of the base, the roof trusses were to be supported by Acrow props at specified positions. JMH were contractually obliged to comply with the engineer's instructions. The site agent of JMH and the consulting engineer thought the Acrow props were inadequate and discussed their concerns with a representative of Plant.

The roof supported by the props collapsed when there was heavy rainfall, resulting in Plant being liable to Ford for over £1 million. Plant sought to recover this amount plus its other losses consequent upon the collapse from JMH and the consulting engineer. The claim against the consulting engineer was settled and the action continued against only JMH. It was contended on behalf of Plant that JMH owed it a duty to exercise care and skill to warn Plant of the inadequacy of the propping. The Court of Appeal decided unanimously that a contractor was under an implied obligation to exercise the skill and care of an ordinarily competent contractor and that such an implied obligation carried with it a duty to warn of danger perceived by the contractor. The Court further decided that, considering the seriousness of the danger, JMH should have protested against complying with the instruction in more vigorous terms. The factors considered by the Court to be crucial were that the works were obviously dangerous and that JMH knew them to be so.

---

35 (1992) 62 BLR 147.
36 [2000] BLR 137.

Lord Justice May, who delivered the judgment of the court, expressly reserved the court's opinion on the question of liability where: (a) the contractor did not know but arguably ought to have known of the danger posed by the design; and (b) the contractor knew or ought to have known of a design flaw that was not dangerous. The case law supportive of liability is therefore of only persuasive authority on these issues.

In *Aurum Investments Ltd* v. *Avonforce Ltd (in Liquidation) and Others*[37] the defendant Contractor entered into a Contract with the claimant to design and construct a basement and garage extension to a building. The second defendant was a specialist underpinning sub-contractor engaged by the Contractor to underpin the flank wall of the adjoining property. The design of the underpinning works was provided on behalf of the Contractor by structural engineers and entailed the construction of a mass concrete wall beneath the footing of the flank wall. On completion of the underpinning works and necessary backfilling by the sub-contractor, the Contractor started excavation for the basement in the area near the flank wall without lateral support to the mass concrete underpinning wall. This resulted in lateral movement of the wall and collapse of the excavation. Aurum brought a claim against the Contractor who brought in the specialist for an indemnity or a contribution. The Contractor's case against the specialist was that it had failed to discharge its duty to warn the Contractor of the need to provide lateral support during the excavation for the basement. Dyson J. held that there had been no such duty on the specialist sub-contractor. He distinguished that case from *Plant* v. *Adams* on two factors: (a) the danger was not in the work being done by the party alleged to be in breach of duty but the work of others done after that party's work; (b) the underpinning sub-contractor was not aware of any danger as it did not know of the method to be used for the subsequent excavation.

One of the difficulties associated with this issue, is that contractors may be placed in an invidious position. A contractor who, because of his expertise, is actually aware of a danger may warn. If the architect responds that he should 'get on with it', the question then arises whether he should refuse on the grounds of danger, thereby risking termination of his employment. It is suggested that he would have great difficulty in defending prosecution or even civil action following a design failure causing injury by claiming that he did the work, knowing it to be dangerous, because he had been told to do it.[38]

2.3.6 Duty to comply with Architect's Instructions (AIs)

Clause 3.10 of the Conditions requires the Contractor to comply with all instructions of the Architect issued under the Contract. Some of the most problematical issues in contract administration arise from this duty. For that reason, AIs are treated as a separate topic in Chapters 5 and 6.

2.3.7 Duties relating to the Contractor's master programme

A properly prepared and updated programme is a very powerful tool for managing a project, particularly where a method statement is provided with and cross-referenced to it. Such a programme not only alerts the project participants to the timetable for performance of their various obligations but also allows analysis and prediction of the impact of any

37 (2001) 17 Const LJ 145.
38 This point is also dealt with in Chap 5.3.1.

event likely to cause delay for appropriate control action. A master programme of some sort is normally submitted with the contractor's tender. Clause 2.9.1.2 provides that, unless he has already done so, the Contractor must provide the Architect with two copies of his master programme as soon as possible after execution of the Contract. The Contractor is to amend the programme to reflect any extension of time granted or any Pre-agreed Adjustment[39] of the Completion Date and supply the Architect with two copies of the revised programme within 14 days after the granting of extension of time or the agreement.

JCT 98 did not impose any obligation on the Contractor to produce or update a master programme. The imposition of these obligations by JCT 05 is therefore to be welcomed. However, the provisions on the Contractor's programme still suffer from a number of shortcomings:

1. There is no indication of the nature of the programme required. It could be a simple list of activities, a bar chart, a linked bar chart or a Critical Path Method (CPM) network diagram. Technically, the Contractor would be complying with Clause 2.9.1.1 if he supplies any form of programme;
2. The programme does not have to be to the satisfaction of the Architect. This means that the Architect has no effective way of getting the Contractor to amend an unrealistic programme;
3. The Contractor is not required to amend his master programme for reasons other than extension of time or a Pre-agreed Adjustment (e.g. for delay due to his own inefficiency); neither is the Contractor required to give copies of the revised programme to the Architect if he produces one for such other reasons;
4. The Contract provides no sanction against failure of the Contractor to comply with his programming obligations.

These shortcomings are often cured by having appropriate additional provisions in the Contract Bills. Such amendment of the Contract is unlikely to be invalidated by Clause 1.3. In *Royal Brompton Hospital National Trust* v. *Hammond (No. 9)*,[40] HHJ LLoyd QC stated that the equivalent clause in JCT 80 did not have that effect, as such additional provisions in the Contract Bills, rather than contradicting them, only supplemented lax programming provisions in the Conditions. Many contractors choose to follow the practice intended by these amendments even where they are not imposed contractually because they consider it good practice.

2.3.7.1 The Society of Construction Law's Protocol

In response to a high incidence of disputes concerning delays, extension of time and liquidated damages, a committee of the Society of Construction Law (SCL), developed and published the Society of Construction Law Delay and Disruption Protocol,[41] which is intended for use as guidelines in dealing with these issues on construction projects. Great benefit by way of understanding of the relevant issues and avoidance strategies can be derived from reading it although some aspects are not directly applicable to JCT 05. For

39 'Pre-agreed Adjustment' is defined under Clause 2.26.2 as 'the fixing of a revised Completion Date for the Works or a Section in respect of a Variation or other work referred to in Clause 5.2.1 by the confirmed Acceptance of a Schedule 2 Quotation'.

40 [2002] EWHC 2037 (TCC).

41 Society of Construction Law, *The Society of Construction Law Delay and Disruption Protocol*, October 2002.

the principles advocated in the Protocol to apply to a project on which the adopted form of contract is JCT 05, they must be incorporated into the Contract.

It is recommended in the SCL Protocol that, in all but the simplest projects, the master programme should be prepared as a CPM network using commercially available project planning software.[42] Information to be provided on it includes:

- all relevant activities relating to design, manufacturing, and procurement;
- on-site construction activities;
- the critical path;
- resource requirements;
- all major items of information the contractor requires from the contract administrator.

The contractor should be required to update the programme monthly, or even more frequently where the project is very complex. Electronic copies of the programme, the method statement, their updates and reports explaining the modifications made in the updates are to be downloaded and archived for future reference.

According to the Protocol, it is good practice for the contract administrator to be under a contractual duty to review and approve the contractor's programme, method statements and their updates. Although the Institution of Civil Engineers' Conditions of Contract, 7th edition, requires that the Contractor's programme be accepted by the Engineer before it is implemented, there are many in the construction industry who are of the view that how a contractor intends to carry out the works is a matter entirely for the Contractor and that it would be inappropriate for the contract administrator to have to police this obligation, thereby accepting some responsibility for the Contractor's working methods and arrangements. The fact that the Architect under JCT 05 has no vetting powers over the Contractor's programme does not mean that he cannot comment on the programme if he has any concerns.

Other recommendations in the Protocol considered in this work concern the keeping of site records,[43] extension of time procedures[44] and analysis of concurrent delays.[45]

2.3.7.2 The contractual status of the Contractor's programmes

The contractual status of a contractor's programme has been considered in the case law. In *Glenlion Construction Ltd* v. *The Guinness Trust*[46] the Contract Bills in a JCT 63 contract required the Contractor to provide a programme. The programme provided contained a completion date earlier than the Completion Date in the contract documents. It was held by Judge Fox-Andrews that: (a) although the Contractor was entitled to earlier completion, it was not a contractual obligation; (b) there was no obligation on the Employer or the Architect to do all the things necessary to enable completion according to the Contractor's programme. They were obliged to do only the things necessary to enable completion by the Completion Date in the contract documents.

Clause 2.9.1 provides that descriptive schedules, the master programme and its updates and similar documents are not to be considered as imposing any contractual obligations

---

42 Most of the software packages available allow barcharts to be produced from the network.
43 See Chapter 11, Section 11.15.2.
44 See Chapter 11, Section 11.15.3.
45 See Chapter 11, Section 11.12.
46 (1987) 39 BLR 89; 11 ConLR 126 (hereafter *Glenlion* v. *Guinness*).

beyond those imposed by the Contract Documents. Apart from restating the common law position highlighted in *Glenlion* v. *Guinness*, this provision also avoids the problem highlighted by the litigation in *Yorkshire Water Authority* v. *Sir Alfred McAlpine & Son (Northern) Ltd.*[47] In that case, which arose from a contract in the terms of the ICE *Conditions of Contract (5th Edn)*, the contract signed incorporated the contractor's method statement. Skinner J held that the contractor was entitled to a variation order (with the normal consequences on entitlement to recovery of extra costs and extension of time for delays) to effect a change in the construction methods necessitated by impossibility of the method statement. This outcome was in stark contrast to the customary practice in the construction industry whereby the costs and schedule implications of a contractor's chosen methods of construction are matters for which the contractor is entirely responsible. Clause 2.12.2 further provides that the Architect is only obliged to supply the Contractor with drawings and other information not listed in the Information Release Schedule in sufficient time to allow completion by the Completion Date (i.e. there is no obligation on the Architect to go out of his way to supply the information at times necessary for the Contractor to finish earlier than the Completion Date under the Contract).

However, a decision in *London Borough of Merton* v. *Stanley Hugh Leach Ltd*[48] suggests that, in certain circumstances, a contractor's programme could constitute compliance by the contractor with an obligation to supply specific information or serve a particular type of notice. This case also arose from a JCT 63 contract, which provided as a Relevant Event:

> the contractor not having received in due time necessary instructions … drawings, details or levels from the architect for which he specifically applied in writing provided that such application was made on a date which having regard to the Completion Date was neither unreasonably distant from or unreasonably close to the date on which it was necessary for him to receive the same.

On a preliminary issue, Vinelott J held that a programme annotated with dates by which the architect was to supply instructions, drawings, or other information could constitute the specific application contemplated in that Relevant Event provided the two requirements of being neither unreasonably distant nor unreasonably close were met.

There is now no requirement for the Contractor to have applied for the information. As explained under Section 2.4.1 in this chapter, the onus is rather on the Architect to determine when it is necessary for the Contractor to receive the information. The Contractor's role here is only to advise the Architect of his need for information where he is aware, *and* has reasonable grounds to believe, that the Architect is unaware of them. It is to be noted not only that the test of when the Contractor should advise the Architect is subjective but also that the advice is not required to be in writing. By analogy with the *Merton* v. *Leach* decision, it is submitted that the Contractor will have complied with his obligation to advise the Architect in this respect if he submits up-to-date programmes with the appropriate level of detail. The Contractor does not even have to do this because of the subjective nature of the test of when he should advise the Architect. This balance of obligation is likely to be considered unfair by many non-contractors in the construction industry. All this suggests a policy on the part of JCT to discourage use of the form without a comprehensive Information Release Schedule.

---

47 (1985) 32 BLR 114.
48 (1985) 32 BLR 51 (hereafter *Merton* v. *Leach*).

2.3.8 Statutory Requirements

It would be implied into a building contract that the contractor must comply with Building Regulations and the like. This obligation is within the express terms of Clause 2.1, which requires the contractor to comply with, and give any notices required by any of the Statutory Requirements. The Statutory Requirements are defined under Clause 1.1 to include the following:

- any Act of Parliament;
- any instrument, rule or order made under any Act of Parliament;
- any regulation or bye-law of any local authority with jurisdiction over, or in connection with the Works;
- any regulation or bye-law of any statutory undertaker with jurisdiction over, or in connection with the Works.

Failure to comply with such regulations constitutes an actionable breach of contract. Furthermore, such failure may constitute some form of illegality that could affect the Contractor's right to payment because, as a general principle from common law, a contractor cannot recover payment for an illegal contract even if he did not know of the illegality.[49] This principle applies to a contract which, by their very nature, contravenes statutory provisions (such as the Building Regulations and Town and Country Planning legislation).[50] In such a situation, not only is the Contractor not entitled to be paid, but sums already paid by the Employer could also be refundable.[51]

A distinction has to be made between inherent illegality (e.g. where there is no planning permission for the development) and illegality resulting from the way the Contractor performed his contractual obligations. Depending upon the circumstances, the Contractor may be entitled to payment even with the latter type of illegality. Factors to be taken into account include whether the illegality can be cured by remedial work and whether or not the enforcement authority is prepared to waive the contravention.

Townsend (Builders) Ltd v. *Cinema News & Property Management Ltd*:[52] the location of water closets in a building being constructed by the contractor contravened building byelaws. The contract was in the terms of the then current predecessor of JCT 05. The Contractor became aware of the contravention only when the construction had reached an advanced stage. On the assurance of the architect that he would resolve the problem with the local authority, the contractor completed the work without a variation to deal with the contravention. Although the local authority condemned the work, they allowed it to remain subject to conditions. The Court of Appeal held that, since the local authority had allowed the work to remain, the contract was not illegal and that the contractor was therefore entitled to payment. It was also held that failure of the contractor to comply with a term similar to JCT 05 Clause 2.1 was a breach for which the employer was entitled to damages, effectively the cost of remedial work necessary to eliminate the contravention.

Clause 2.21 provides that the Contractor shall pay and indemnify the Employer against liability for fees and charges in connection with the Statutory Requirements that are

---

49 Re *Mahmoud and Ispahani* (1921) 2 KB 716; *Bostel Bros Ltd* v. *Hurlock* (1949) 1 KB 74.

50 *Steven* v. *Gourley* (1859) 7 CB (NS) 99; *Townsend (Builders) Ltd* v. *Cinema News & Property Management Ltd* (1959) 20 BLR 118.

51 *Smith & Son Ltd* v. *Walker* [1952] 2 QB 319.

52 (1959) 20 BLR 118.

legally payable pursuant to the relevant statutory provision. Where there is a Provisional Sum in the Contract Bills for such fees and charges, they are to be dealt with in accordance with Clauses 4.3.2.1 and 4.3.3.3. In any other case, the amount of such fees and charges is to be added to the Contract Sum unless: (i) they are priced in the Contract Bills, or (ii) they arise solely in respect of a CDP.

2.3.8.1 Divergence in contractual requirements[53]

The Conditions expressly provide for the situation where the Contractor or the Architect discovers a divergence between the Statutory Requirements and any of the documents listed under Clause 2.15.[54] By Clause 2.17.1, the Contractor or the Architect who discovers any such divergences must immediately give written notice of it to the other. As explained in Section 2.3.5 in this chapter, it is also implied into the Contract that, in the carrying out of the Works, the Contractor will exercise the skill and care of a reasonably competent contractor. It would therefore be a breach of Contract if the Contractor fails to serve notice of contraventions of the Statutory Regulations that a reasonably competent contractor ought to have discovered.

If the divergence is between the Statutory Requirements and any CDP Document, the Contractor must also advise the Architect of how the Contractor proposes to amend the relevant document to remove it. Within 7 days of receiving the Contractor's notice of a divergence or the Architect's independent discovery of the divergence, the Architect must issue appropriate instructions. He has a longer period of 14 days to deal with a Contractor's notice of proposed amendments to a CDP Document. In so far as any instruction given under Clause 2.17.2 in respect of a divergence not involving a CDP Document requires a Variation, it is to be considered a Variation under Clause 5.1, with all the implications for entitlements to extensions of time and recovery of direct loss and/or expense. For a detailed commentary on divergences affecting any CDP Document, see Chapter 6, Section 6.14.

The Contractor's duty to comply with the Statutory Requirements and to give notice of any divergence between them and any other requirement of the Contract is subject to Clause 2.17.3 which states:

> Provided that the Contractor is not in breach of Clause 2.17.1, the Contractor shall not be liable to the Employer under this Contract if the Works (other than the CDP Works) do not comply with the Statutory Requirements to the extent the non-compliance results from the Contractor having carried out the work in accordance with the documents referred to in Clauses 2.15.1 to 2.15.4 or with any instruction requiring a Variation (other than a Variation in respect of the Contractor's Designed Portion).

The intention behind this clause appears to be that, as long as the Contractor gives the notices on discovered divergences, he is not to be liable to the Employer for non-compliance with the Statutory Requirements to the extent that the non-compliance is due to the carrying out of the Works in accordance with the other requirements of the Contract. Whether the courts will give effect to this intention is debatable. Clause 2.17.3 is therefore an exclusion clause that purports to limit the contractual liability that would otherwise apply to the

53 See also Chapter 6, Section 6.14.5.1.

54 The Contract Drawings, Contract Bills, Architect's instructions other than any requiring a Variation, Drawings and other documents issued by the Architect to the Contractor under Clauses 2.9 to 2.12 and CDP Documents.

Contractor in relation to breach of the Statutory Requirements which is the direct consequence of carrying out the Works in accordance with the design provided by the Employer. The question is whether it is effective against legislation on exclusion clauses in contracts. Section 7 of the Unfair Contract Terms Act 1977 requires reasonableness if the clause is to be effective against an Employer who, for the purposes of the Contract, deals as a consumer. It is submitted that, subject to serving notices of contraventions of the Statutory Requirements actually discovered or which a reasonably competent contractor ought to discover, the clause meets the requirements of reasonableness.

There is the related question whether the Contractor is liable to the Employer in tort in circumstances where he is exonerated by Clause 2.17.3 from liability under the Contract. There is now little doubt that the effect of the 'contract structure' defence against imposition of a duty of care is that the contractor would not be liable in the tort of negligence. Whether there is liability for breach of statutory duty is a question of whether the relevant statute provides for civil liability. Under s. 38 of the Building Act 1984, breach of the Building Regulations gives rise to civil liability but that section is not yet in force. Where the Contract requires 'work for or in connection with the provision of a dwelling' the contractor owes a duty to the employer under s. 1(1) of the Defective Premises Act 1972. His obligation is 'to see that the work that he carries on is done in a workmanlike manner, with proper materials … and so as to be fit for the purpose required'. However, s. 1(2) exonerates from liability 'a person who takes on any such work for another on terms that he is to do it in accordance with instructions given by or behalf of that other … to the extent to which he does it properly in accordance with those instructions'. This section is clearly Clause 2.17.3 in relation to dwellings.

2.3.8.2 Unauthorized variations to comply with Statutory Requirements[55]

Under normal circumstances, the Contractor must not vary the Works without a Variation order from the Architect. However, sometimes it happens that the contractor has to carry out Variations to comply with the Statutory Requirements in circumstances where it is not practicable to wait for Variation instructions. For example, the Health and Safety Executive (HSE) may make orders requiring immediate remedying of a particular hazard. The Conditions recognize the possibility of such situations arising by allowing the Contractor to supply any materials and do any work necessary without waiting for instructions from the Architect. However, the Contractor must inform the Architect of such emergency and the steps taken. Where the emergency work is made necessary by a divergence between the Statutory Requirements and Contract, it should be treated as if the Architect had given an instruction pursuant to a notice of the divergence from the Contractor.

2.3.9 Duty to indemnify the Employer

The Contractor undertakes to indemnify the Employer against any expense, liability, loss, claim or proceedings in connection with any of the following matters:

- statutory fees/charges (Clause 2.21);
- the Contractor's infringement of patent rights in the carrying out of the Works described in the Contract Bills (Clause 2.22);

---

55 See also Chapter 6, Section 6.14.5.

- personal injury or death (Clause 6.1);[56]
- damage to property (Clause 6.2).[57]

As a cause of action under an indemnity accrues only when the Employer's loss from the relevant matter is established,[58] the contractor may incur liability under the indemnities many years after expiry of the normal limitation period applicable to the Contract. To illustrate how onerous the indemnities are, consider a third party action in tort against the Employer for latent damage to neighbouring property caused by subsidence for which the Contractor is responsible under Clause 6.2. Under the Latent Damage Act 1986, such an action may be brought against the employer 15 years after the date on which the damage occurred. The limitation period for the indemnity, which would be 12 years if the Contract was executed as a deed, may therefore begin to run from the fifteenth year after the damage, giving rise to potential liability for up to 27 years after commission of the relevant wrong.

2.3.10 Insurance obligations

The Contractor is required to take out and maintain insurance policies against certain risks. Details on the risks and the insurance requirements are provided in Chapter 7.

2.3.11 Duties with respect to assignment and sub-contracting

These matters are covered in Chapters 8 and 9.

2.3.12 Obligations in respect of other contractors of the Employer

It is not uncommon for the Contract to be for only a part of a larger project. Clause 2.7 is designed to facilitate other work to be carried out by the Employer or his other contractors in parallel with the carrying out of the Works. It provides that where such other work is identified in the Contract Bills with sufficient information to enable the Contractor to carry out and complete the Works, the Contractor must allow such other work to be carried out. Where the other work is not so identified, consent of the Contractor must first be obtained but such consent is not to be unreasonably delayed or withheld if requested.

Delay caused by such other work would amount to 'impediment, prevention or default' by one of the Employer's Persons, a Relevant Event under Clause 2.29.6. It is therefore doubtful whether the risk of such delay in itself is sufficient to entitle the Contractor to withhold his consent. In contrast, withholding consent for health and safety reasons would be considered reasonable.

---

56 See Chapter 7, Section 7.2.2 for a detailed commentary on this indemnity.
57 See Chapter 7, Section 7.2.3 for a detailed commentary on this indemnity.
58 *Post Office* v. *Norwich Union Fire Insurance Ltd* [1967] 1 QB 363; *County & District Properties* v. *C. Jenner & Son Ltd* [1967] 2 Lloyd's Rep. 728; *R. & H. Green Silley Weir Ltd* v. *British Railway Board* (1980) 17 BLR 94.

2.3.13 The principal contractor under the Site Waste Management Plan Regulations

It is explained in Section 2.1.5 in this chapter that the Employer may be under a duty as client under the Site Waste Management Plans Regulations 2008 to appoint a principal contractor for the purposes of those Regulations. The Contractor would have this additional responsibility where he is appointed to that role. Where the Contractor has not been given the opportunity to price for the role before execution of the Contract, such additional responsibility may be treated as a Variation. Alternatively, it is arguable that, to be able to impose the duties of a principal contractor under these Regulations on the contractor, the Contract itself has to be varied.[59]

2.4 The Architect

Traditionally the Architect has been the leader of the design team. His relationship with the employer is, in the main, governed by the contract between them. In most cases, the terms of this contract are those of the RIBA's Standard Form of Agreement for the Appointment of an Architect (SFA).[60] However, it is not uncommon for the architect's appointment to be informal. Where this is the case, a contract in those terms may be implied by a previous course of dealing between the particular parties. In the absence of such history of dealings, the courts may not imply an intention to contract on those terms.[61]

The SFA covers the full diet of services offered by architects, ranging from feasibility studies, through design and supervision on site to commissioning. However, the Architect is referred to in the Conditions mainly in his capacity as the administrator of the Contract. His role as designer is not dealt with expressly. Similarly, the role of specialist design consultants is not explicitly referred to in the Conditions although they play a major role during construction. From the Contractor's standpoint, the generic role of designer is treated as part of the Architect's. Whilst this arrangement has the advantage of administrative convenience, it is artificial in some respects because decisions often have to be made in the same proceedings as to whether a mishap or other undesirable event is the responsibility of the Architect, the Contractor or a specialist design consultant. For this reason, the role of the Architect wearing only his administrator's hat is covered in this section. The generic role of designer, which covers both any specialist design consultant and an architect who also contributed to the design of the Works, is dealt with separately in Section 2.6 in this chapter. This means that where the Architect also provided or provides some design input, the reader needs to read both sections to develop a full picture of his role.

JCT 05 expressly provides that the Architect must perform certain duties. They are summarized in Table 2.3. Such provisions, in themselves, do not impose on the Architect obligations to perform them because the Architect is not a party to the Contract. However, he would be under such obligations through express or implied terms in his contract of

---

59 See Chapter 6, Section 6.2, for comments on variations 'to' or 'under' the Contract.

60 Royal Institute of British Architects, *A Guide to the RIBA Forms of Appointment 1999*, RIBA Publications Ltd, 1999.

61 *Sidney Kaye, Eric Firmin & Partners* v. *Bronesky* (1973) 4 BLR 1.

Table 2.3 Duties of the Architect

Clause	Duty
1.9	To issue any certificate due under the Contract to the Employer but with a duplicate copy to the Contractor
2.10	To determine levels and provide to the Contractor dimensioned drawings and such other information to enable the Contractor set out the Works
2.11	To ensure that 2 copies of any item of information referred to in the Information Release Schedule are supplied to the Contractor at the time stated in the Schedule
2.12.1	To provide the Contractor with 2 copies of any instruction, further drawings or detail necessary to explain or clarify the Contract Documents
2.14.1	To correct any errors in the Contract Bills (although the clause does not state expressly that this is the duty of the Architect, it is so inferred from the deeming of the correction as a Variation under clause 2.14.3)
2.15	To issue instructions to resolve discrepancies within, or divergences between, contractual documents
2.17.2	To issue instructions to resolve any divergence between the Statutory Requirements and any other contractual document within 7 days of the Contractor's notice of it or its independent discovery by the Architect; to issue necessary instruction in relation to any notified amendment to any CDP Document within 14 days
2.28.1	To grant extension of time in accordance with the clause
2.28.5	To review extension of time within 12 weeks after practical completion
2.30	To issue the Practical Completion Certificate/Sectional Completion Certificate when the conditions for its issue have been satisfied
2.33	To issue a written statement to the Contractor identifying part(s) of the Works taken into possession by the Employer before issue of the Practical Completion Certificate
2.35	To issue a Certificate of Making Good in respect of any part taken over when the Contractor has made good all defects in the part for which the Contractor is responsible
2.38	To issue a schedule of defects not later than 14 days after expiration of the Rectification Period if defects, shrinkages or other faults appear within that Period
2.39	To issue a Certificate of Making Good when the Contractor has made good all defects in the Schedule of Defects
3.12.1	To issue all instructions in writing
3.13	To specify the clause authorizing any instruction that is challenged by the Contractor in writing
3.16	To issue instructions to the Contractor in regard to the expenditure of Provisional Sums in the Contract Bills or in the Employer's Requirements
3.23	To issue instructions in regard to what is to be done concerning fossils, antiquities and other objects or interest or value found on the site
3.24	To ascertain or to instruct the Quantity Surveyor to ascertain loss and/or expense for disturbance to progress of the Works by discovery of fossils and the like
4.4	Take any addition to, or deduction from, the Contract Sum into account in the computation of the next Interim Certificate
4.9	To issue Interim Certificates in accordance with Clause 4.9
4.15	To issue the Final Certificate
4.23	To ascertain or instruct the Quantity Surveyor to ascertain loss and/or expense for which the Contractor has applied

engagement to that effect.[62] In the performance of his duties, the architect wears two hats but not both at the same time. With respect to some of the duties, he acts as an agent of the employer. Any default by the architect in the performance of such duties may therefore be treated by the contractor as a default on the part of the employer (for example, failure to provide necessary information or to certify interim payment). With other duties, the architect is an independent professional and the employer is not in breach of contract for defective performance of duties such as assessment of claims and quantification of the amount payable in certificates. This duality was described by Lord Reid in *Sutcliffe* v. *Thackrah*[63] in these terms:

> It has often been said, I think rightly, that the architect has two different types of function to perform. In many matters he is bound to act on his client's instructions, whether he agrees with them or not; but in many other matters requiring professional skills he must form and act on his own opinion. Many matters may arise in the course of the execution of a building contract where a decision has to be made which will affect the amount of money which the contractor gets… The building owner and the contractor make their contract on the understanding that in all such matters the architect will act in a fair and unbiased manner, and it must therefore be implicit in the owner's contract with the architect that he shall not only exercise due care and skill but also reach such decisions fairly, holding the balance between his client and the contractor.

In *Costain Ltd and Others* v. *Bechtel Ltd and Others*[64] Jackson J. (as he then was) referred to these statements as having been accepted for the last 30 years by the construction industry and the legal profession in the UK as correctly stating the duties of architects, engineers and other certifiers under conventional forms of construction contract.

The Employer's responsibility to the Contractor for a decision of the Architect depends on the role in which it was taken. In *Merton* v. *Leach*[65] Vinelott J. explained the difference in relation to the Architect's decisions under JCT 63 in these terms:

> It is to my mind clear that, under the standard conditions, the Architect acts as a servant or agent of the building owner when supplying the Contractor with the necessary drawings, instructions, levels and the like and in supervising the progress of the work and in ensuring that it is properly carried out. … To the extent that the Architect performs these duties the building owner contracts with the Contractor that the Architect will perform them with reasonable diligence and with reasonable skill and care. The contract also confers on the Architect discretionary powers which he must exercise with due regard to the interests of the Contractor and the building owner. The building owner does not undertake that the Architect will exercise his discretionary powers reasonably; he undertakes that, although the Architect may be engaged or employed by him, he will leave him free to exercise his discretions fairly and without improper interference by him.

---

62 In *Townsend* v. *Stone Toms & Partners* (1984) 27 BLR 26 the Court of Appeal decided, *inter alia*, that it was an implied term of an architect's contract of engagement in relation to contract administration that he would administer the contract in accordance with its terms.

63 [1974] AC 727.

64 [2005] EWHC 1018 (TCC).

65 *London Borough of Merton* v. *Stanley Hugh Leach* (1985) 32 BLR 51, at pp. 136–37.

2.4.1 Duty to provide drawings, instructions and other information

One of the most common types of claims made by contractors against employers is often based on an allegation by the contractor that he suffered delays or incurred additional cost on account of delay by the contract administrator in supplying necessary drawings, instructions or other information. The main provisions on the Architect's duty to provide drawings and other information are in Clauses of 2.11 and 2.12. The former concerns information listed in the Information Release Schedule (IRS)[66] where the Contract incorporates such a document whilst the latter deals with the supply of drawings and other details not in the Schedule. It is stated in Clause 2.11 that, to the extent that the default of the contractor or of a person for whom the Contractor is responsible prevented release of information in accordance with the IRS, the Employer is not responsible for the consequences of the Architect's failure. It is not clear why there is no equivalent provision in relation to failure to comply with Clause 2.12.

2.4.1.1 Items in the Information Release Schedule

Clause 2.11 requires the Architect to ensure that two copies of any information referred to in the IRS are released at the specified time. The Contractor and the Employer may agree to vary the times in the IRS. Such agreement is not to be delayed or withheld unreasonably. If the Architect cannot comply with any particular deadline for release of any relevant item of information he must explain the situation to the Employer and get him to negotiate a variation of the release date with the Contractor. Without the authority of the Employer the Architect cannot agree directly such a variation with the Contractor. Whilst the Employer and the Contractor may revise the IRS without involving the Architect, such action would work against the Employer's interests in the event that the Architect is unable to release any information earlier than required by the original IRS.

2.4.1.2 Information not listed in the Information Release Schedule

Clause 2.12 places obligations on the Architect to supply the Contractor with two types of information. First, the Architect must provide free of charge to the Contractor two copies of 'any further drawings as are *reasonably necessary* to explain and amplify the Contract Drawings' where such drawings or details are not included in the IRS. The 'reasonably necessary' qualification is highly significant. The Contractor, by assuming responsibility to carry out and complete the Works in accordance with the Contract Documents, warrants that he possesses the skills necessary to interpret the Contract Drawings and Contract Bills and translate them into working or installation drawings that describe exactly what needs to be done in operational terms. Thus, as remarked by HHJ LLoyd QC in *Royal Brompton Hospital National Health Trust* v. *Hammond (No. 9)*,[67] the Contractor is entitled to receive only further drawings and instructions without which a reasonably competent contractor would not be able to carry out and complete the Works. There is no

66 See Chapter 1, Section 1.5.2.2 for commentary on the fifth Recital, which concerns this Schedule.
67 [2002] EWHC 2037 (TCC).

right to be spoon-fed with information that the Contractor should be capable of producing for himself.[68]

The second type of information consists of 'such instructions…as are necessary to enable the Contractor to carry out and complete the Works in accordance with this Contract'. The obligation to issue instructions as to the expenditure of Provisional Sums is referred to expressly. The qualification to the Contractor's entitlement to be issued with an instruction is that it is necessary rather than 'reasonably necessary'. This distinction may be explained on the basis that, whilst the need for further drawings depends on the quality of the design and the Contractor's in-house ability to prepare them, the instructions contemplated are those that any Contractor would need in any event.

The further drawings and instructions are to be provided 'at the time it is reasonably necessary for the Contractor to receive them' (Clause 2.12.2). The Contractor's intentions are a relevant factor in determining this baseline. Such intentions may have been documented in the contractor's master programme, minutes of pre-contract meetings, minutes of site meetings or written requests for the specific information. Regarding the relevance of the master programme, it is to be borne in mind that sometimes a drawing issued by the Architect cannot be put to immediate use on an operational activity because the Contractor has to prepare working or installation drawings from it first. In such a case, or where long delivery materials are required, there would be a lag between when the contractor actually needs the drawing and the start date of the relevant activity. It is therefore arguable that the Architect should always take into account such lead times where the Contractor's needs are expressed by reference to a programme. If in doubt, he would be acting prudently to seek confirmation from the Contractor.

Many in the construction industry would contend that, where the Contractor made a specific request for a drawing or other information by a certain date, the Architect should be entitled to assume that the Contractor had taken into account his internal requirements for producing working drawings. However, the Court of Appeal in *Royal Brompton Hospital Health Service Trust* v. *Hammond (No. 5)*[69] stated that the contractor's own view of when the information was necessary was not decisive. The quality of drawings actually provided would be a relevant factor as the contractor may have expected them to be of such high quality that they could be put to immediate use with minimal prior requirements for the production of working drawings. Should a requested drawing turn out to be of poor quality the time when the contractor should have received it is to be determined on an objective basis taking into account what needed to be done before it could be put to use.

Actual progress on site is a further factor to consider. If the Contractor was so behind with the Works that he would not have been in a position to use the information even if it had been provided as initially expected, the time when it was necessary for him to receive the information would be postponed accordingly. This proposition follows from the decision in *Merton* v. *Leach* that a contractor's master programme annotated with dates for submission of information would not be an effective notice of when the information is required if performance has not been in accordance with the programme. This proposition

68 In practice, the dividing line between details that the Contractor can work out for himself and outstanding design information within the Architect's remit is often blurred, resulting in contractors erring on the side of safety by waiting for the Architect.

69 [2001] BLR 297, CA; [2001] EWCA Civ 550 (hereafter *Royal Brompton No. 5*).

also received the support of the Court of Appeal in *Royal Brompton No. 5*. At para. 79 Aldous LJ stated:

> I believe that [*counsel for the defendants*] is correct in his submission that Clause 5.4 [*materially the same as JCT05 Clauses 2.11 and 2.12*] is a term which must be construed as imposing an obligation to provide the drawings when necessary in the sense that the drawings must be provided when actually necessary as opposed to when they were perceived to be necessary. No doubt in most cases the perceived need of a contractor will coincide with actual need, but this may not be such a case.

Clause 2.12.2 addresses expressly a common source of contention which occurs where the Contractor's actual progress is such that the works are likely to be completed earlier than the Completion Date. In such a situation when it was reasonably necessary to provide the drawing or instruction is to be judged by reference to the contractual Completion Date and not the earlier date towards which the Contractor was working. This focus on the Completion Date is in line with the position already reached at common law. In *Glenlion Construction Ltd* v. *The Guinness Trust*,[70] which arose from a contract in the terms of JCT 63, it was decided that the Architect did not have to go out of his way to supply the Contractor with information at times that would allow the Contractor to complete earlier than contractually necessary.

If the Contractor is likely to be delayed by lack of information not included in the IRS, his role in all this is to advise the Architect of the need for the information but only if he is aware that the Architect is ignorant of this need. He does not even have to do so in writing. The onus is therefore on the Architect to find out when the Contractor needs what information. This requires close examination of the Contractor's programme and actual progress. It may be prudent practice to impose on the Contractor, through appropriate provisions in the Contract Bills, a duty to provide a detailed schedule of information requirements for the whole contract and to table at every site meeting an updated schedule of information requirements for the next period. In particular, the Contractor should be required to highlight any information needed long before the scheduled commencement of the relevant activity.

2.4.1.3 Information from other designers

The source of some of the drawings and other information required by the Contractor is often a specialist design consultant engaged directly by the Employer (e.g. building services and structural engineers). Such specialist designers are also under a duty to exercise skill and care in providing some of the information needed by the Architect to supply to the Contractor in accordance with Clauses 2.11 and 2.12. Correspondingly, the consultants need to be advised on the timetable for providing the information to the Architect. On large projects the flow of information between designers and the Contractor calls for very careful coordination. Under normal circumstances, the responsibility of such coordination is the architect's. However, where the Employer has also engaged a project manager, the coordination role can be blurred unless the allocation of this responsibility is clarified and communicated to all concerned.

---

70 [1987] 30 BLR 89.

2.4.1.4 Claims for delayed drawings and other Information

JCT 98 made specific provision for claiming extension of time[71] and recovery of loss and/or expense[72] on account of delay by the Architect in supplying necessary drawings and other information. Such express provisions have not been carried forward into JCT 05. However, the Contractor may still successfully claim for the effect of such occurrences. Delayed drawings or information may amount to 'impediment, prevention or default' by the Architect, the Relevant Event in Clause 2.29.6, and the Relevant Matter under Clause 4.24.5.

2.4.2 Quality control duties

This section is to be read in conjunction with Section 2.3.1 in this chapter, which discusses the Contractor's obligations in respect of quality of materials and standards of workmanship.

Quality control is easily the most important part of the Architect's supervisory role, the aim of which is to detect any non-conforming work or materials and to have the non-conformance remedied by using powers given him under the Contract. The Architect may do any of the following in relation to his quality control function during the carrying out of the works:[73]

- request evidence of the quality of certain materials and goods (Clause 2.3.4);
- carry out inspection and testing of materials or excuted work (Clauses 3.17 and 3.18.4);
- instruct removal of non-conforming work (Clause 3.18.1);
- accept non-conforming work after due consultation with the Contractor and obtaining the Employer's agreement (Clause 3.18.2);
- issue appropriate Variation and other instructions after consultation with the Contractor (Clauses 3.18.3 and 3.19);
- omit non-conforming work when preparing Interim Certificates (Clauses 4.10 and 4.16);
- exclude from the site any person employed on it (Clause 3.21);
- withholding the issue of the Practical Completion Certificate.

In support of these powers, the Architect is entitled under Clause 3.1 to be allowed access at all reasonable times to the Works and workshops and other locations where work is being done for the Contract. The Contractor is to ensure that the Architect has similar rights of access to the work and workshops of sub-contractors. The Contractor is to include dovetailing terms in sub-contracts.

2.4.2.1 Evidence of Architect's satisfaction and appropriate materials and goods

Clause 8.2.1 of JCT 98 authorized the Architect to request vouchers covering any items of materials and goods for the Works. There is no similar specific provision on vouchers in JCT 05. Instead, under Clause 2.3.4, the Architect may request evidence that the quality of materials, goods and workmanship are as required by the Contract. As vouchers often contain information on quality (e.g. specifications, prices and names of suppliers), the Architect may still request them. However, it is submitted that the Architect is not to insist

---

71 See Clauses 25.4.6.1 and 25.4.6.2.

72 See Clause 26.2.1.1 and 26.2.1.2.

73 For discussion of the Architect's powers in relation to defects after the issue of the Practical Completion Certificate see Chaper 3, Sections 3.5 and 3.6.

on being supplied with vouchers, if the Contractor can establish the appropriate quality standard by other types of evidence.

2.4.2.2 Inspection and testing

The Architect may issue instructions requiring the Contractor to carry out any test on executed work or materials for the Works either himself or by third parties (Clause 3.17). Where the work has already been covered up, he may instruct that it be opened up for inspection and such testing. Where the inspection and testing show the relevant work or materials to be in accordance with the Contract, the cost of compliance with the instruction is to be added to the Contract Sum unless there is provision for the opening up, inspection, or testing of work or materials in the Contract Bills.[74] The exception is probably designed to avoid double payment for the same work (pursuant to the provisions in the Contract Bills and under Clause 3.17).

Clause 3.18.4 is intended to deal with situations where the Architect, as a result of discovery of actual non-compliance with the Contract, has reasonable grounds to suspect other instances of further or similar non-compliance. He may issue further instructions for the opening up or testing of other work as reasonably necessary to deal with those concerns. For example, consider where, on excavating one of a row of column footings of identical design, it is discovered that its dimensions are wrong. The Architect may insist on excavating the other footings to check that the same or similar mistakes have not been made. Reasonableness of the Architect's actions pursuant to Clause 3.18.4 is governed by a Code of Practice annexed to the Conditions as Schedule 4.

It is stated in Clause 3.18.4 that the Contractor is not entitled to any additional payment for compliance with an architect's instruction issued pursuant to that clause, provided that the instruction was reasonable in the circumstances. There is no entitlement even if the inspection or test shows that the work and materials are in accordance with the Contract. As this clause applies only where, as a consequence of prior discovery of non-compliance, the Architect needs to be satisfied that there is no further non-compliance of a similar nature, it is only fair that the Employer does not pay for a situation brought about by the Contractor's failure to comply with the Contract. The Relevant Event under Clause 2.29.2 is 'instruction of the Architect for the opening up for inspection or testing any work, materials or goods under Clause 3.17 or 3.18.4 (including making good) unless the inspection or test shows that the work, materials or goods are not in accordance with this Contract'. There is therefore entitlement to extension of time if the Contractor is delayed by instructions for further opening up and testing necessitated by non-compliant work if the work, materials or goods are found to be in accordance with the Contract. Some Employers would question the justification for giving the Contractor more time for dealing with a consequence of his own breach of contract.

2.4.2.3 Removal of non-conforming work and materials

If the Architect expresses his opinion that work done ought to be taken down or that specific materials should not be used, the Contractor will normally comply. However, there is always the danger that once the Architect's back is turned, an unscrupulous contractor may do whatever he likes, particularly where the relevant work is to be covered up soon afterwards. Clause 3.18.1 is designed to avoid this risk by allowing the Architect to require

74 Such provision will normally be in the form of a Provisional Sum.

the Contractor to remove the non-conforming work or materials from the site altogether. For the obligation to remove work or materials to arise the Architect must expressly and unequivocally require the Contractor to do so. Merely condemning the work or materials as non-conforming and requiring the Contractor to ensure compliance with the Contract would not amount to instructing their removal.[75] Here again, the Architect may advise the Employer to consider exercising his right under Clause 3.11 to have the work or materials removed at the Contractor's expense if the contractor fails to comply with the architect's instruction. As an ultimate sanction, the Employer may terminate the contractor's employment for refusal to comply with such instruction, which constitutes the specified default under Clause 8.4.1.3 where the Works are materially affected by such refusal.

2.4.2.4 Acceptance of non-conforming work

The Architect may, after consultation with the Contractor and the agreement of the Employer, allow non-conforming work or materials to be left and to make an appropriate deduction from the Contract Sum to reflect the non-conformance (Clause 3.18.2). The Architect is required to confirm such acceptance of non-conforming work to the Contractor in writing. The Clause states expressly that such allowance does not amount to a Variation under the Contract. Such express provision avoids uncertainty of the kind that arose in *Simplex* v. *The Borough of St Pancras*.[76] In that case a contract administrator accepted an alternative proposed by the contractor to deal with defects. The contractor's contention that the alternative amounted to a variation, with the usual entitlement to extension of time and recovery of loss and/or expense, was accepted by the court.

There is no mechanism for determining the amount of the 'appropriate deduction'. There is even no requirement for the Quantity Surveyor to be involved except to adjust the Contract Sum by the amount of the appropriate deduction although, in practice, such a role goes unchallenged by the Employer and the Contractor. The Guidance Notes to Amendment 5 to JCT 80, which introduced the equivalent of Clause 3.18.2, took a different line on the role of the Quantity Surveyor. They stated that the determination of the amount was a matter for him. However, there are at least two arguments against this construction. First, a duty to include an amount in the final account does not necessarily mean a duty to determine its quantum. In many similar situations, the Quantity Surveyor is expressly required to determine the quantum (e.g. valuation of Variations under Clause 5.2.1), even though Clause 4.3 also requires him to take account of the amount in the final accounts. Second, and more importantly, the whole arrangement is subject to the agreement of the Employer. He is entitled to withhold agreement unless he considers the amount involved to be the 'appropriate deduction'. Generally, with matters for the Quantity Surveyor's determination, the Employer cannot interfere in this way.

It was cautioned in the Guidance Notes that 'The Tribunal considers that in no circumstances should the deduction take the form of a "penalty" on the Contractor nor should it equate or relate to the cost of making the non-conforming work conform to the Contract which the Contractor has saved because the non-conforming work is being allowed to remain'. These arguments are relevant only where the Employer agrees to the Architect allowing the non-conforming work to remain. If the Employer refuses, the Contractor

---

75 *Holland Hannen & Cubitts (Northern) Ltd* v. *Welsh Hospital and Technical Services Organisation* (1981) 18 BLR 80.

76 (1958) 14 BLR 80.

would have to make good the non-conformance, particularly as the Architect has several ways of ensuring this. For example, the Architect may instruct removal of the non-conforming work and advise the Employer to exercise his right under Clause 3.11 to have it done if the Contractor fails to comply. It would therefore appear that the Employer is in the driving seat. However, he must act reasonably because, to the extent that the cost of demolition and reinstatement or other remedial measures is disappropriate to the benefit to be obtained, it may not be recoverable.[77]

In any case it has to be noted that, according to the Court of Appeal's decision in *TFW Printers Ltd* v. *Interserve Project Services Ltd*,[78] guidance notes published by contract drafting bodies may not be admissible in court as aides to the construction of the contracts they are published to accompany. At best they only describe recommended good practice.

2.4.2.5 Variations and other instructions

Under Clause 3.18.3, the Architect may issue any Variation instruction made reasonably necessary by an AI to remove non-conforming work and materials or acceptance of non-conforming work. Before the Architect can do this, he must consult the Contractor. Under Clause 3.19, the Architect may, after similar consultation, issue an instruction to deal with the Contractor's 'failure to comply with Clause 2.1 in regard to the carrying out of work in a proper and workmanlike manner and/or in accordance with the Construction Phase Plan'. This is irrespective of whether he has instructed removal of the work or allowed it to remain.

An instruction under Clause 3.19 may, but does not have to, be a Variation. If the Contractor fails to comply with the instruction, the Employer may exercise his right under Clause 3.11 to give effect to the instruction at the Contractor's expense. To the extent that an instruction under Clause 3.18.3 or 3.19 is made necessary by the non-conformance, it is not a ground for extension of time or addition to the Contract Sum. It is only fair that the Contractor should not gain from his own breach of contract.

Whether a contract administrator has the power to issue an instruction requiring the contractor to make good defects during the carrying out of the works is often contested by contractors. A common argument by contractors is that they have up to the date of practical completion to make good any defects in the works[79] and that an order to do so earlier amounts to an instruction to carry out the work in a specific order, a variation under most construction contracts.[80] JCT 05 is much clearer on this point than JCT 98. Where the defect must be made good without delay in the interest of health and safety or other strong reason, the instruction would be reasonably necessary. The Architect may therefore issue such an instruction under Clause 3.19 and the Contractor would not to be entitled to extension of time or adjustment of the Contract Sum on that account.

It is implicit from the last sentence of Clause 3.19 that the Architect may issue instructions for immediate making good of defects even where immediate action is not reasonably necessary as a consequence of the breach of contract. The Contractor would be entitled to more time or money if immediate compliance causes delay or additional costs.

---

77 *Forsyth* v *Ruxley Electronics and Construction* (1995) 73 BLR 1; *McGlinn* v. *Waltham Contractors Ltd and Others* [2007] EWHC (TCC) 149. See also Section 2.6 in this chapter.

78 [2006] EWCA Civ 875; [2006] 2 CLC 106; [2006] BLR 299; 109 ConLR 1; (2006) 22 Const LJ 481.

79 This argument often relies on the 'temporary disconformity' concept articulated by Lord Diplock in *Kaye Ltd* v. *Hosier & Dickinson Ltd* [1972] 1 WLR 146.

80 For example see Clause 5.1.2.4 of JCT 05.

Table 2.4 Instructions under Clauses 3.17, 3.18 and 3.19

Instruction	Outcome	Costs of compliance	Extension of time	Loss and/or expense
Under Clause 3.17 only	Materials/work are found to be in accordance with the Contract	To be added to the Contract Sum unless provided for in the Contract Bills	Relevant Event under Clause 2.29.2.2	Relevant Matter under Clause 4.24.2.2
	Materials/work are found not to be in accordance with the Contract	Not to be added to Contract Sum – *See discussion*	No entitlement	No entitlement
Under Clause 3.17 but pursuant to Clause 3.18.4	Materials/work are found to be in accordance with the Contract	Not to be added to the Contract Sum if instruction was reasonably necessary	Relevant Event under Clause 2.29.2.2	No entitlement
	Materials/work are found not to be in accordance with the Contract	Not to be added to the Contract Sum if instruction was reasonably necessary	No entitlement	No entitlement
Clause 3.18.1		Contractor's cost	No entitlement	No entitlement
Under Clause 3.19	Compliance with the instruction	Contractor's cost	No entitlement	No entitlement

The difference in consequence is probably the justification for the express direction to the Architect to consult the Contractor before issuing instructions under Clause 3.19. As the differentiating factor is the necessity of the instruction, the consultation should consider this question together with the Contractor's entitlement to time and money.

It is important to bear in mind that there are five different types of instruction involved here: those under Clauses 3.17, 3.18.1, 3.18.3, 3.18.4, and 3.19. Each type may entail: (i) additional costs incurred directly in carrying out the tasks ordered; (ii) loss and/or expense incurred as a result of material disturbance to the progress of the Works; (iii) delays to completion of the Works. The Contractor's entitlements in respect of these consequences are summarized in Table 2.4. In view of the differences in the Contractor's entitlements, the Architect must always state the empowering clause in the relevant instruction itself. Any disputes regarding which of the clauses is applicable to any given instruction may be resolved in accordance with the provisions on the settlement of disputes.[81]

2.4.2.6 Defective work and Interim Certificates

Under Clause 4.16.1.1 the Architect is to include in Interim Certificates only work properly executed. This qualification represents probably the most potent weapon by which the Architect can enforce compliance with the required quality standards. It is to be noted that the Architect has no discretion to include non-conforming work in Interim Certificates. In *Townsend and Another* v. *Stone Toms & Partners*[82] the Court of Appeal held that it was a breach of his contract of engagement for an architect to include defective work of which he was aware in interim certificates under the then current edition of the JCT's prime cost

81 See Chapter 17.
82 (1984) 27 BLR 26.

contract. It was stated that the fact that there was enough retention to cover the cost of making them good, if the contractor failed to do so, was not a proper ground upon which the architect could ignore the terms of the contract to pay for only work properly executed.

2.4.2.7 Exclusion of persons

The Architect may issue instructions excluding from the site any person employed on it: Clause 3.21. This power is probably designed to avoid recurrent problems with consistently sub-standard work by particular individuals or firms, but is not to be exercised 'unreasonably or vexatiously'.[83]

2.4.2.8 Withholding the Practical Completion Certificate

Under Clause 2.30 it is a condition precedent for the issue of the Practical Completion Certificate that the Architect reaches the opinion that practical completion of the Works has been achieved. He may therefore withhold the issue of the Certificate until defects, other than those of a trifling nature, are made good. Furthermore, he may issue a Non-Completion Certificate, one of the triggering events for the Employer's entitlement to liquidated damages, if the defects remain up to the Completion Date. Considering the advantages of the Practical Completion Certificate, and disadvantages of the Non-Completion Certificate to the Contractor, the powers over their issue are probably the most powerful weapons in the Architect's arsenal for ensuring that the Works are handed over without serious defects.

2.4.2.9 Construction Skills Certification Scheme

This is a registration scheme administered by the Construction Industry Training Board aimed at promoting development and verification of skills and competence in a relevant trade and awareness of health and safety issues. To acquire registration, an operative must possess membership of an equivalent skills registration scheme or relevant formal qualifications and/or successfully complete recognized training and assessment. A registered operative is given a CSCS card which lasts for three or five years. Many construction organizations will not employ on their construction sites any operative without a CSCS card.

Clause 2.3.5 requires the Contractor to take all reasonable steps to encourage the Contractor's Persons to hold CSCS Cards or equivalent. The provision lacks serious contractual bite.[84] Auditable practice of monitoring possession of the cards linked to skills-based recruitment, engagement of sub-contractors and training towards registration are likely to be enough.

2.4.3 The extent of the Architect's duty to supervise/inspect

Performance of the duties already outlined requires some amount of the Architect's presence on the site. The question is what the requisite amount should be. It is clear from the case law[85] that exercise of the appropriate standard of skill and care does not require the

---

83 See Chapter 16, Section 16.19 for explanation of this phrase.

84 This provision was introduced through Amendment 5 (July 2003) to JCT 98 with the stated aim of encouraging contractors to use personnel for building works who are registered CSCS cardholders or equivalent.

85 *East Ham Corporation* v. *Bernard Sunley & Sons Ltd* (1966) AC 406 (hereafter *East Ham* v. *Sunley*); *Corfield* v. *Grant and Others* (1992) 59 BLR 102.

Architect to be permanently on the site. In *Corfield* v. *Grant and Others*[86] Judge Bowsher had this to say on the subject:

> What is adequate by way of supervision and other work is not in the end to be tested by the number of hours worked on site or elsewhere, but by asking whether it was enough. At some stages of some jobs exclusive attention may be required to do the job in question … at other stages of the same jobs, or during most of the duration of other jobs, it will be quite sufficient to give attention to the job only from time to time. The proof of the pudding was in the eating. Was the attention given enough for this particular job?

It would appear that the experience and competence of the particular contractor is also a factor to be taken into account. In the *East Ham* v. *Sunley* case, Lord Upjohn said that where an architect knows the contractors sufficiently well and can rely on them to do a good job, it would be proper to relax the supervision of detail in favour of other matters. It was also suggested in *Sutcliffe* v. *Chippendale and Edmondson*[87] that the degree of supervision should be higher where the experience or competence of the contractor is questionable. However, where a job requires more inspection by an architect than is possible because of the architect's other responsibilities, the architect may still be liable.[88]

From the above discussion, the mere fact that there are undiscovered defects does not necessarily mean that the Architect was negligent in his supervision of the Works. Indeed, in *Gray and Others* v. *Bennett & Sons and Others*,[89] Sir William Stabb QC, sitting on Official Referee's business, decided, on a preliminary issue, that an architect was not liable for defects that the contractor's workmen had deliberately concealed. Similarly, in *Department of National Heritage* v. *Steensen Varming Mulcahy and Others*[90] Judge Bowsher rejected an employer's argument that defective work was itself evidence of inadequate supervision by consulting engineers who carried supervisory responsibility similar to that of the Architect under JCT 05. In *Consarch Design Ltd* v. *Hutch Investments*[91] the same judge stated that an architect acting as contract administrator does not guarantee that his inspections will reveal or prevent all defective work. In that case he rejected a claim against the architect for failing to notice that a limestone floor had been laid badly. However, in *McGlinn* v. *Waltham Contractors Ltd and Others*[92] HHJ Coulson QC stated that the mere fact that defective work was carried out and covered up in between the inspector's visits may not be an effective defence where the circumstances were such that the inspector should reasonably have contemplated this problem and made precautionary arrangements.

To establish that defects are attributable to negligent inspection, it has to be shown that the frequency and duration of inspections actually carried out were less than the amount necessary, taking into account the nature of the works and the perceived competence and trustworthiness of the contractor. In the *McGlinn* v. *Waltham Contractors* case, the architect from whom the employer was claiming damages for negligent inspection had carried out inspections on only days of site meetings. The Judge stated that carrying out inspection piggyback with site meetings alone would not ordinarily be sufficient. It did

86 See Note 85.
87 (1971) 18 BLR 149. This case went to appeal under the name of *Sutcliffe* v. *Thackrah* [1974] AC 727 but not on this issue.
88 *Corfield* v. *Grant and Others* (1992) 59 BLR 102.
89 (1987) 43 BLR 63.
90 60 ConLR 33; (1998) CILL 1422.
91 [2002] PNLR 31.
92 [2007] EWHC 149 (TCC).

not help the defence of the architect in that case that no record of the findings of his inspections had been made.

The SFA deals with the frequency of the architect's site visits expressly. Clause 3.1.1 of that form requires him to make such visits as he reasonably expected to be necessary when he was appointed. This test of sufficiency of the visits is therefore subjective, a contrast with the objective standard under the common law. The architect is under a duty to confirm this expectation to his client in writing. It is submitted that a duty to do so within a reasonable time of his engagement would be implied. Reasonableness should reflect the state of the client's brief and the complexity of the project. However, there is no need to delay the confirmation until he is certain of the frequency required because Clause 3.1.2 anticipates changes in his expectation by providing that, if he does revise them, he must so inform the employer in writing. It is therefore clear that constant presence of the architect on site is not the norm. Clause 3.3 of the SFA reflects this by requiring the architect to advise the employer in writing of any need for staff (e.g. resident architects, representatives of designers and clerks of works) to assist him in the supervision and administration of the contract. With the exception of the provision in Clause 3.4 for a Clerk of Works, JCT 05 does not provide for the roles of such additional staff because, as stated in the SFA, all site staff so appointed are to be under the direction and control of the Architect.

In recognition of the limitation to the amount of site presence required of the architect, there is a view that he is more properly referred to as an 'inspector' rather than a supervisor. Indeed, the current edition of the SFA refers to inspections rather than supervision, the latter requiring much closer oversight of the contractor's work.

2.4.4 Certification duties

The Architect is responsible for issuing various certificates or certifying specified matters. They are to be issued to the Employer but with duplicate copies immediately sent to the contractor (Clause 1.9). The certificates or matters are:[93]

- the Practical Completion Certificate (Clause 2.30);
- the Certificate of Making Good (Clause 2.39);
- where the Employer took possession of a part of the Works before the issue of the Practical Completion Certificate, a certificate stating that defects, shrinkages or other defaults in that part have been made good (Clause 2.35);
- Non-completion Certificate (Clause 2.31);
- Interim Certificates (Clause 4.9);
- Final Certificate (Clause 4.15.1).

A certificate is not issued until it is sent to the Employer and the Contractor. Under Clause 1.7.1 a document is 'deemed to be duly given or served if addressed and given by actual delivery or sent by pre-paid post to the Party to be served at the address stated in the Contract Particulars …' unless in relation to the particular document the contract provides otherwise. Thus, the date of signing and dating a certificate is not necessarily the date of its issue.[94]

93 Some of these certificates may be issued for Sections where sectional completion applies.

94 See *Penwith Dictrict Council* v. *VP Developments Ltd* [1999] EWHC 231 (TCC) which concerned this issue in relation to JCT80.

It is to be noted that the certificates fall into two categories. First, there are Interim Certificates, which the architect must issue on specific dates.[95] Failure of the architect to issue this type of certificate by the specific date would therefore be a breach of the Contract by the Employer. Generally, such certificates involve statements of quantum and the employer is not vicariously liable for any negligence of the architect in arriving at the quantum. The appropriate remedy is for the contractor to challenge the contents of the certificate by invoking the applicable dispute resolution procedure. The second category consists of decisions of the Architect arrived at through exercise of his professional skills that certain events have occurred (e.g. practical completion of the works or completion of making good defects). The employer is not liable to the contractor for any delay in issuing or failure to issue this type of certificate. Here too, the appropriate remedy is through invocation of the relevant dispute resolution procedure to determine whether or not the certificate should be issued. Any negligence by the Architect in the performance of duties to certify is a breach of his contract of engagement for which the Employer would be entitled to damages.[96]

2.4.5 Duty to report on failings of other members of the design team

The Architect is often under an express duty under his contract of engagement to coordinate the work of all members of the design team not only during design but also during the construction phase. Such a term may even be implied as a custom of the construction process. A first instance decision suggests that the Architect may owe the Employer a duty in contract and in tort to advise him of actual or potential deficiencies in the performance by the Quantity Surveyor and other consultants of their contractual duties to the Employer.[97] However, it was also held in that case that there is no duty to warn in respect of the Architect's own deficient performance or that of the Employer's project manager. It may even be argued that, depending on the type of error and all the surrounding circumstances on site, the Architect may even be under a duty to correct design errors by other designers. An example is where the Architect should reasonably be expected to possess relevant design expertise and it is clearly in the interests of the Employer that a design change is implemented immediately without inviting or waiting for a response from the original designer. However, the Architect would be prudent to follow the formal process of informing the Employer and getting his express approval to the change.

2.4.6 Standard of skill and care expected of the Architect

At common law, the standard of skill and care to be exercised by the Architect in the performance of his duties is that required of every professional person (i.e. the reasonable skill and care an ordinary competent architect would exercise in the circumstances of the particular job).[98] Clause 1.2.1 of SFA provides that, in the provision of the services he was engaged

---

95 See Chapter 15, Section 15.2.1 for how these dates are fixed.

96 See Section 2.4.7 of this chapter and Chapter 15, Section 15.9.

97 *Chesham Properties Ltd* v. *Bucknall Austin Project Management Services Ltd* (1996) 82 BLR 92.

98 *Bolam* v. *Friern Hospital Management Committee* [1957] 1 WLR 582; 2 All ER 118; the *Bolam* standard was applied to the supervisory duties of architects by the Court of Appeal in *West Faulkner Associates* v. *London Borough of Newham* (1994) 71 BLR 1.

to deliver, 'the architect shall exercise reasonable skill and care in conformity with the normal standards of the architect's profession'. Where there is no such express term in the particular contract of engagement, it will be implied unless there is express provision to the contrary. Even where there is exclusion of the duty of skill and care, it will almost certainly be struck down by the Unfair Contract Terms Act 1977. Case law[99] suggests that he must possess a good general understanding of the law in relation to building contracts. In particular, he would be expected to have good understanding of the terms of this standard form and relevant authorities. Although he is not expected to possess the expert knowledge of a lawyer, he must be able to recognize when to advise his client to seek appropriate legal advice.

2.4.7 Liability of the Architect to the Employer

Where the Architect fails to discharge his role as architect with the required standard of skill and care, he would be liable to the Employer in contract and tort. For example, the House of Lords held in *Sutcliffe* v. *Thackrah*[100] that an architect was liable in contract for negligent certification. In *West Faulkner Associates* v. *London Borough of Newham*,[101] which arose from a JCT 80 contract, the Employer wished to terminate the Contractor's employment on the grounds of alleged failure to proceed regularly and diligently with the carrying out of the Works, as they were entitled to do, but could not do this because the Architect refused to issue the Contractor with a notice of the default, a precondition of the Employer's right to terminate. The Court of Appeal found that, on the facts, the Contractor had committed the default and that the Architect should have issued the notice. The Architect was therefore held liable for the Employer's loss caused by not terminating when he wished to do so.

Unless his contract of engagement provides to the contrary, the Architect would be under a duty to advise the employer on the general operation of the Contract and on the Employer's responsibilities within it. Where the Architect does not possess the required special knowledge, he should advise the Employer to consult a lawyer or other expert. In *Pozzolanic Lytag Ltd* v. *Brian Hobson Associates*[102] it was decided on a preliminary issue that a project manager appointed to administer a contract let on JCT 80 With Contractor's Designed Portion Supplement was under a duty to check that the scope of the Contractor's insurance arrangements met the requirements of the Contract. To the project manager's contention that his role in that respect was simply to collect evidence on the insurance arrangements and to pass it to the Employer for consideration, Dyson J said:

> I cannot agree with the opinion of [the project manager]. If a project manager does not have the expertise to advise his client as to the adequacy of the insurance arrangements proposed by the contractor, he has a choice. He may obtain expert advice from an insurance broker or lawyer. Questions may arise as to who has to pay for this. Alternatively, he may inform the client that expert advice is required, and seek to persuade the client to obtain it. What he

99 See *Townsend (Builders) Ltd* v. *Cinema News & Property Management Ltd* (1959) 20 BLR 118; *B.L. Holdings* v. *Robert J. Wood & Partners* (1979) 12 BLR 1; *West Faulkner Associates* v. *London Borough of Newham* (1994) 71 BLR 1; *Pozzolanic Lytag Ltd* v. *Bryan Hobson Associates* [1999] BLR 267; 1 TCLR 233; 63 ConLR 81;15 Const LJ 135; *Royal Brompton Hospital NHS Trust* v. *Hammond* (No. 9) [2002] EWHC 2037 (TCC).

100 [1974] AC 727.

101 (1994) 71 BLR 1; (1995) 11 Const LJ 157.

102 For a similar finding see *William Tomkinson* v. *The Parochial Church Council of St Michael* (1990) 6 Const LJ 319.

> cannot do is simply act as a 'postbox' and send the evidence of the proposed arrangements to the client without comment.

It is submitted that these comments apply to the Architect under JCT 05 except where the circumstances show that the Employer never relied on the Architect in relation to these matters.

2.4.8 Liability of the Architect to the Contractor and other third parties

As there is no contract between the Architect and any of these parties, any liability would have to be found in tort. The general principles governing this type of liability are outside the scope of this book. However, three points have to be made. First, there is no legal responsibility on the Architect to protect the Contractor's interests by intervening in the manner in which the Contractor is carrying out the Works even if the Architect notices that the Contractor is working inefficiently or even dangerously. In *Oldschool* v. *Gleeson*[103] in which a contractor contended that supervising engineers were liable to them for the collapse of a wall, Sir William Stabb, QC, sitting on Official Referee's Business,[104] stated:

> The duty of care of an architect or of a consulting engineer in no way extends into the area of how the work is carried out. Not only has he no duty to instruct the builder how to do the work or what safety precautions to take but he has no right to do so, nor is he under any duty to the builder to detect faults during the progress of the work. The architect, in that respect, may be in breach of his duty to his client, the building owner, but this does not excuse the builder for faulty work.

Second, as explained in some detail in Chapter 15, Section 15.9, it is doubtful whether the Architect will incur liability to the Contractor for his loss arising from under-certification of payment. Third, there is little doubt that the Architect may be found liable in tort for personal injury and damage to property. For example, in *Clay* v. *Crump & Sons Ltd*[105] an architect was found jointly liable with a contractor for injury suffered by contractor's workers as a result of the collapse of a wall that the architect had instructed the contractor to leave standing on the site. The architect was found negligent because had he inspected the wall, which he failed to do, he would have found that the wall was too unstable to be left standing. There may also be liability for pure economic loss under the *Hedley Byrne* principle[106] where the Architect goes out of his way to advise the Contractor and the requirements for that type of liability are met.[107]

---

103 (1974) 4 BLR 103; see similar comments in *Clayton* v. *Woodman & Son (Builders)* [1962] 2 All ER 33; [1962] 1 WLR 585.

104 Now renamed the 'Technology and Construction Court'.

105 [1963] 3 WLR 866.

106 The basic principle here is that where a person gives advice knowing that it will be relied upon and the person advised does rely on it and suffers loss as a consequence, the adviser may be held liable for the loss if his advice was given negligently. The liability could still arise even where the advice was given free of charge.

107 For an example that pre-dates the development of *Hedley Byrne* liability, see *Townsend (Builders) Ltd* v. *Cinema News & Property Management Ltd* (1959) 20 BLR 118 in which the architect was held liable to the contractor for costs of remedial work to comply with bye-laws on account of advice to the contractor during construction that he would resolve the contravention with the local authority and that the contractor should therefore ignore it.

2.5 The Quantity Surveyor

The Contract states the duties of the Quantity Surveyor as to:

1. value Variations in accordance with Clauses 5.6–5.10 or, in the case of a Variation relating to CDP, Clause 5.8;
2. ascertain loss and/or expense that has been, or is being, incurred by the Contractor on account of a Relevant Matter if instructed by the Architect to do so (Clauses 4.23 and 3.24);
3. carry out valuations for interim certificates (where Fluctuation Option C applies the Quantity Surveyor must always do so but where it does not apply he must do so only if instructed by the Architect) (Clause 4.11);
4. prepare the final accounts for the Contract (Clause 4.5).

In the performance of his duties under the Contract, the standard of skill and care to be exercised by the Quantity Surveyor is the normal *Bolam*[108] standard applicable to all professional people. A pre-*Bolam* case suggests that occasional arithmetical or clerical errors may be consistent with exercise of the proper standard of skill and care. In *London School Board* v. *Northcroft*[109] it was held that a quantity surveyor was not liable for two clerical errors made by his assistant in the computation of the contract price for completed buildings. However, it is doubted in *Hudson's Building and Engineering Contracts* whether such leniency would apply today.[110] The Quantity Surveyor is required by s. 13 of the Supply of Goods and Services Act 1982 to exercise reasonable skill in the delivery of his professional services. It is therefore arguable that, although he has no duties with respect of quality matters, he owes the Employer a duty to draw to the attention of the Architect obvious defects that he notices in the course of measurements for valuations.

The comments about the Architect's liability to the Contractor would also apply to the corresponding liability of the Quantity Surveyor.[111]

2.6 Designers

This section considers the generic role of designers whose involvement with the project continues to the construction phase. Where the Architect also carried out the architectural design of the Works, his role as designer described in this section would be additional to his supervisory and administrative responsibilities already discussed in Section 2.4 in this chapter. Except with the simplest of buildings, it is rare for an architect to possess all the skills required to supervise construction to designs by specialists such as structural engineers and mechanical/electrical engineers. For that reason, these specialists are often retained in a limited capacity during the construction phase to assist the architect with supervision. As JCT 05 recognizes only the Architect as supervisor, any instruction by these specialists has to pass through the Architect to the Contractor.

---

108 *Bolam* v. *Friern Hospital Management Committee* [1957] 2 All ER 118; see Chapter 4, Section 4.4.2 for statement of the *Bolam* standard.
109 (1889) *Hudson's*, 11th edn, at para. 2.230.
110 *Hudson's* para. 2.230.
111 See Section 2.4.8 in this chapter.

It is now settled law that a designer who continues as a supervisor is under a duty to review his design to deal with any problems that come to his attention during construction and that ought reasonably to alert him to a serious possibility of errors in his initial design. In *Brickfield Properties* v. *Newton*[112] Sachs LJ said:

> The architect is under a continuing duty to check that his design will work in practice and to correct any errors which may emerge. It savours of the ridiculous for the architect to be able to say, as it was here suggested that he could say: 'True, my design was faulty but, of course, I saw to it that the contractors followed it faithfully' and to be enabled on that ground to succeed in the action.

This is important for reasons of limitation of action under the contract.[113] The duty to review means that the cause of action may not accrue until completion of construction, which may be long after completion of the faulty design.

A designer, as part of the duty to exercise the skill and care of an ordinarily competent designer, may owe a duty to the client to see to it that any assumption upon which the design is based is verified. In *Ove Arup & Partners International* v. *Mirant Asia-Pacific Construction (Hong Kong) Ltd (No. 2)*[114] the foundation of a power station was designed based on an assumption that the insitu bearing capacity of the ground at the relevant location was 3 MPa. It was envisaged in the original design that this assumption would be verified. Although the designer's geologist found, from a surface examination, that the area was generally different from what had been assumed, the assumed strength of the foundation strata was never verified. The Court of Appeal held that, in failing to see to it that the assumption was verified, the designer had acted in breach of the duty under the design agreement with the client.

2.7 The Person-in-Charge (PIC)

Clause 3.2 requires the Contractor always to have on site a competent Person-in-Charge. Traditionally, this person has usually been referred to as the 'Site Agent' although, nowadays, there are other titles such as 'Contract Manager', 'Site Manager', and 'Project Manager'. In practice, he is usually the Contractor's employee with total responsibility for the carrying out of the Works on site. The Contract does not require the Contractor to notify to the employer or the architect the identity of the person acting in that capacity. However, the contractor will usually do so before the commencement of site operations.

The clause states that any AI or direction of the Clerk of Works (COW) given to the PIC is deemed issued to the Contractor. Instructions or directions issued to any other person on site are not binding on the Contractor. It is important to note that the deeming

112 [1971] 1 WLR 862, at p. 873. For application of the duty to review design see also *London Borough of Merton* v. *Lowe* (1981) 18 BLR 130; *Chelmsford District Council* v. *Evers* (1983) 25 BLR 99; *Equitable Debenture Assets Corporation* v. *William Moss* (1984) 1 Const LJ 131; *University of Glasgow* v. *Whitfield* (1988) 42 BLR 66; *New Islington and Hackney Housing Association Ltd* v. *Pollard Thomas and Edwards Ltd* [2001] BLR 74; *Oxford Architects Partnership* v. *Cheltenham Ladies College* [2006] EWHC 3156 (TCC).

113 For discussion of limitation of actions under the Contract see Chapter 1, Sections 1.4.1 and 1.5.5.6.

114 [2005] EWCA Civ 1585; [2006] BLR 187.

provision affects only AI's and COW's directions and that service of notices or other documents required under the Contract is governed by Clause 1.7 as follows.

1. Where there is other specific provision in the Contract for service of any particular document that provision must be complied with.
2. A notice or other document may be served by any effective means.
3. A document is deemed duly served or given if addressed and given by actual delivery or sent by pre-paid post to the Party at an address either agreed or stated in the Contract Particulars.
4. If no applicable address is stated in the Contract Particulars, and there is no other current agreed address, the notice or document is considered effectively served if it is given by actual delivery or sent by pre-paid post, in the case of a registered company, to the registered address or, in other cases, to the last known principal business address.

2.8 The Clerk of Works (COW)

It is explained in Section 2.4.3 in this chapter that normal practice in building contracts does not require the Architect always to be on the site. Clause 3.4 of JCT 05 reflects the normal practice by providing that the Employer has a right to appoint a COW whose function is 'solely as inspector on behalf of the Employer under the directions of the Architect'. He is usually a very experienced foreman appointed on the recommendation of the Architect. On projects of high complexity, he will often be assisted by specialist clerks of works. The RIBA and Institute of Clerks of Works have jointly produced a Clerks of Works Manual, which describes the role and duties of a clerk of works when employed on contracts let on the JCT family of standard forms.

The Conditions are silent on the Contractor's right to raise objections to any person being appointed as a COW. As the Articles expressly provide for such a right in the case of a replacement Architect and Quantity Surveyor, the implication must be that there is no such right. The Contractor must afford the COW every reasonable facility for the performance of his duties. However, the powers of the COW to direct the Contractor in the carrying out of the Works are limited as follows.

- The direction must be given in regard to a matter in respect of which the architect is empowered under the Contract to give instructions.
- The direction is confirmed in writing by the Architect within two working days of the original direction being given.[115]

It is to be noted that the instruction becomes effective from the date of issue of the confirmation and not, as one would expect, the date of its receipt. In practice, many contractors comply with instructions of the COW without waiting for the Architect's confirmation. Although such instructions are often confirmed retrospectively, it should be borne in mind that, technically, the Contractor would be complying with an invalid instruction if he carries out an unconfirmed instruction of the COW. He may therefore lose entitlements to extra payment and extension time if compliance with the instruction entails delay. Even

115 See also Chapter 5, Section 5.2.3.

if the Architect includes payment in respect of such instructions, the Employer would be entitled to challenge the certificate, thereby delaying payment.

It is not uncommon for the COW to carry out some of the express duties of the Architect, some of which may go beyond inspection of matters of detail. This is most common where the Works are very simple in nature. JCT 05 does not provide for such delegation. Where such delegation is anticipated, the Conditions should be amended accordingly, with particular attention being paid to the authority of the COW.

Clause 3.4 makes it clear that the COW, even if appointed on the Architect's recommendation, acts as an agent of the Employer. In any legal action against the Architect for negligent supervision, the Architect can therefore plead any negligence of the COW in the performance of his duties as the Employer's contributory negligence. The effect of such a plea is to reduce the damages recoverable against the Architect by the extent to which the Employer's loss was caused by the COW's negligence.

> *Kensington & Chelsea & Westminster Area Health Authority* v. *Wettern Composites Ltd & Adams Holden & Partners*:[116] pre-cast concrete mullions in the claimant's hospital cracked as a result of the contractor's poor workmanship. The claimant employer's action against the architect for failure to exercise reasonable skill and care in supervising the works succeeded. However, the damages recoverable from the architect were reduced by 20% because of overlooking of the defects by the claimant's COW.

However, the Architect is not entitled to delegate to the COW his obligation to take reasonable steps to ensure that the Works are carried out to his design. This means that the Architect must attend personally to the general scheme of the Contractor's work methods while giving clear directions to the COW on matters of detail to be followed up.

> *Leicester Board of Guardians* v. *Trollope*:[117] this litigation raised the question of an architect's liability for dry rot discovered in a hospital four years after it had been constructed. The cause of the problem was found to be deviations from the work method specified in the contract for constructing ground floors. Although the COW was always on site during construction, he acquiesced to the deviations for corrupt reasons. The architect, who only visited the site from time to time, admitted that he never checked the ground floors of any of the buildings. The architect had therefore left more than matters of detail to the COW. It was held at first instance that the presence of the COW on site did not eliminate the architect's responsibility to see that the work was properly carried out to his design. However, the court suggested that if the architect had seen to it that the work on the first block was properly carried out, and had then told the COW that the rest was to be done in the same way, he might have avoided liability.

The agency relationship between the COW and the Employer suggests that the Employer may be better off not appointing a COW as negligence on his part may reduce the Architect's liability in respect of the matter concerned. Whilst this may be true from a legal standpoint, in practice, prevention of loss through the appointment of a competent COW is better than having legal remedies against An architect. Besides, failure to appoint a

116 (1984) 31 BLR 57.
117 (1911) 75 JP 197.

COW may itself constitute contributory negligence by the employer although there is no direct authority on the issue.

It is not uncommon for the COW to perform some functions of the Architect such as accepting the Contractor's confirmation of oral instructions. The Contract does not provide for this type of delegation. Where such delegation is desirable, appropriate provision must therefore be added as amendments to the Contract or a separate agreement involving the Employer, Contractor and the Architect is drawn up. Another common practice that can raise problems is where the COW uses the official stationery of the Employer or the Architect to communicate with the Contractor. The problem is that the Contractor may be entitled to treat the COW in such circumstances as agent of the Employer or the Architect, as the case may be, and treat the communication accordingly. For example, an instruction to the Contractor on the Architect's notepaper may constitute a Variation without the need for further action from the Architect.

2.9 Suppliers and sub-contractors

The role of a general contractor engaged by a construction client is increasingly becoming one of only management and coordination of actual construction by third parties engaged by the contractor for the purpose. These third parties are referred to generically as 'sub-contractors', of which there are two types. 'Domestic sub-contractors' are selected and appointed by the contractor to carry out specific work for which the general contractor has generally obtained the contract administrator's permission to sub-contract. Some sub-contractors, referred to as 'nominated sub-contractors', are appointed by the general contractor as an obligation under the contract. This type of sub-contractor is usually named in the contract documents or nominated by the contract administrator for appointment by the general contractor. Similarly, nominated suppliers are selected by employers or their contract administrators for appointment by the general contractor. JCT 05 deals expressly with domestic sub-contractors. Previous editions of the Contract contained extensive provisions on nominated sub-contractors and suppliers. These provisions have been dropped completely from JCT 05. The subject of sub-contracting is dealt with in detail in Chapter 8.

2.10 Joint liability

It may be concluded from the contents of this chapter that there are certain types of loss suffered by the Employer for which more than one participant could be liable. The Employer may therefore bring action against any of the parties he considers responsible for the loss. Where two or more parties are found to have contributed to the loss the court has jurisdiction under the Civil Liability (Contribution) Act 1978 (CLCA) to apportion liability between them on whatever basis the court considers just and equitable, taking into account the extent of each person's responsibility for the loss. The CLCA also allows a party who is sued to bring into the proceedings, as a co-defendant, any other party who would also be liable for the same loss. For example, if the Employer sues the Architect for defects arising from negligent supervision, the Architect may add the Contractor to the proceedings for a contribution towards the Employer's loss where the defects represent a breach of the contract by the Contractor.

3

JCT 05 timetable

Certain decisions and actions required under JCT 05 have to be made or taken at specific points in time or within specific time windows. For example, the Architect cannot give certain types of instructions after practical completion. Understanding the general timetable of a contract in JCT 05 Form is therefore necessary for effective administration of the contract. The aim in this chapter is to provide this understanding.

The key events in the JCT 2005 calendar are summarized in Fig. 3.1. This timetable anticipates a project with a single completion date. For such a project, the information required to be stated in the Contract Particulars includes the following:

- Date of Possession of the site, which is the date by when the Contractor is to be given possession of the site in accordance with Clause 2.4 to commence thc carrying out of the Works;
- Date for Completion of the Works, which is the date by when the Contractor must achieve practical completion of the Works;
- deferment of the possession of the site, the maximum delay in granting possession of site by the Employer without being in breach of contract;[1]
- rate of liquidated damages;
- Rectification Period.

The term 'sectional completion' is used to refer to situations where the project has intermediate completion dates. The Contract uses the term 'Section' for each part that has a separate timetable for the performance of the work in it. The Sixth Recital states that 'the division of the Works into Sections is shown in the Contract Bills and/or the Contract Drawings or in such other documents as are identified in the Contract Particulars'. The division of the Works into the relevant sections must therefore be described in the Contract Bills and/or Contract Drawings. Alternatively, the description may be drawn up in a separate document provided it is dated and suitably identified by a reference number or other appropriate identifier. To incorporate the document into the Contract, the reference number and date or other identifier must be completed in the Contract Particulars. A footnote points out the necessity of deleting this Recital if the Works are not required to be completed in Sections.

1 See also Chapter 1, Section 1.5.2 on the Employer's right to defer possession of site.

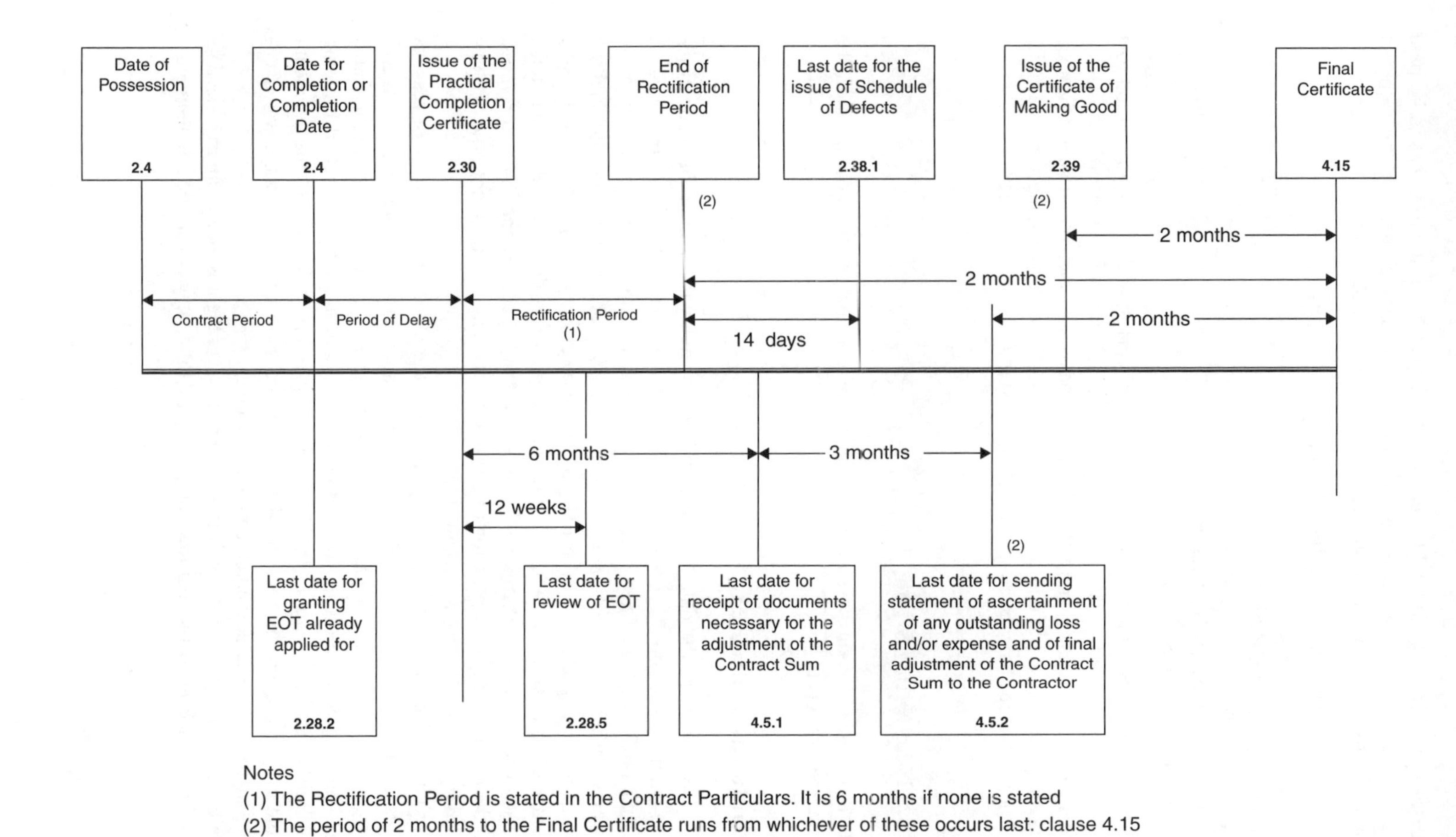

Fig. 3.1 JCT 05 Timetable

For reasons provided in Chapter 11, Section 11.1.2, particular attention has to be paid to the need to define the work content and timetable of each Section in the Contract Particulars. The following information must be stated for each Section:

- Section Sum (which is the total value of work in the Section);
- Date of Possession of Section;
- Date for Completion of Section;
- deferment of possession of Section;
- rate of liquidated damages for Section;
- Rectification Period for Section.

Generally, these terms bear the same meanings as their corresponding use in relation to a project with a single Date for Completion. For reasons of simplicity, most of the discussion in this chapter is primarily in relation to the performance timetable of such a project. However, the principles explained are generally applicable correspondingly to the timetables for performance of any Section properly defined in the Contract Particulars.

3.1 Possession of site

To put JCT 05 provisions on possession of site in context, the general principles on the subject are first explained.

3.1.1 General principles

In construction contracts, the date by which the site for the works should be made available to the contractor to commence construction is referred to as the 'date of possession'. This date should normally be fixed long before there is a need to commence operations on site. However, either because of poor planning or because of circumstances compelling invitation of tenders in advance of availability of the necessary information on the site, contracts are often awarded without any firm idea as to when the site will be available. If, for these or other reasons, the date of possession is not specified, the law will imply a term that the owner will give possession of site to the contractor in sufficient time to allow completion by any agreed date for completion.[2]

It is not uncommon that on the date of possession, the owner, for a variety of reasons beyond his control, is unable to make the site available. For example, in *H.W. Nevill (Sunblest) Ltd* v. *William Press Ltd*[3] a demolition and site clearance contractor failed to get the site ready for the owner to hand over to the contractor for the actual building works. In *Rapid Building Group Ltd* v. *Ealing Family Housing Association*[4] the defendant was unable to give possession of site to the claimant because a part of the site was occupied by squatters whom the defendant could not evict.

Failure to give possession of site by the agreed date or, in the absence of an express date of possession, in sufficient time to allow completion on time, may constitute a fundamental

2 *Freeman & Son* v. *Hensler* (1900) 64 JP 260.
3 (1981) 20 BLR 78 (hereafter *Nevill* v. *Press*).
4 (1985) 1 ConLR 1.

breach of the contract (also referred to as a repudiatory breach). This principle was applied by the High Court of Australia in *Carr* v. *J.A. Berriman Property Ltd.*[5] This means that, in the event of this type of breach, the contractor may, without himself being in breach of contract, refuse to go on with the contract and instead sue for damages. In practice, very few responsible contractors would take such a draconian course of action. A more business-like approach would be to continue with the contract and to claim damages for breach of contract. Even where the contractor chooses to continue with the contract, on principles explained in Chapter 11, the time for completion becomes 'at large' if the contract does not allow for extension of time to cover the delay. Time for completion being at large means that the contractor is only obliged to complete within a reasonable time. Furthermore, the employer loses any rights to liquidated damages. If the contractor fails to complete within a reasonable time, the employer would have to prove any damages that he wishes to recover.

However, mere failure to grant possession in due time will not, in itself, amount to a repudiatory breach. There must be evidence either that the employer has no intention ever to grant possession or that the failure is total or otherwise very serious. In *Wardens and Commonalty of the Mystery of Mercers of the City of London* v. *New Hampshire Insurance Co. Ltd*[6] the claimant (Mercers) entered into a contract with contractors to carry out works to its premises. The contract, which incorporated JCT 80, allowed deferment of possession of site by a maximum of 6 weeks. To avoid forthcoming increases in VAT, the Employer made an advance payment of most of the Contract Sum. The defendant provided a bond guaranteeing that the advance payment would be employed towards the carrying out and completion of the Works. The Employer exceeded the maximum period of allowable deferment of possession of site by a further 4 weeks, a breach of the Contract. The issue before the Court of Appeal was whether the effect of the breach was to discharge the defendant from the obligations of the bond. The answer was in the negative. To arrive at this answer, the court treated the breach as non-repudiatory. Unfortunately the court did not explain their reasoning because both parties were in agreement that the breach was non-repudiatory.

3.1.2 Possession of site under JCT 05

'Date of Possession' is defined in Clause 1.1 of JCT 05 as 'the date stated as such date in the Contract Particulars (against the reference to Clause 2.4) in relation to the Works or a Section'. Clause 2.4 requires the Contractor to be given possession of the site on this date to commence the construction of the Works. In the case of a Section the Contractor is to be given possession of the relevant part of the site on the relevant Date of Possession of Section stated in the Contract Particulars.

To allow for any problems with giving possession of site or the part constituting the relevant Section, JCT 05 contains a mechanism whereby possession of site or Section can be deferred. To avail himself of this opportunity, the Employer, or usually his professional advisers, should indicate in the Contract Particulars that Clause 2.5, which allows deferment of giving possession of site or the relevant Section, will apply. The maximum period of deferment must also be stated in the appropriate place in the Contract Particulars. This

5 (1953) 27 ALJR 273.
6 (1992) 60 BLR 26.

maximum period must not exceed 6 weeks, the default maximum period if that part of the Contract Particulars is not completed. Deferment of possession of site or of any Section for up to the applicable maximum period is the Relevant Event under Clause 2.29.3, which entitles the Contractor to extension of time under Clause 2.28.1. It should be noted that, if the actual period of deferment exceeds the maximum period, the Employer would be acting beyond his contractual rights. Any delay to the completion of the Works caused would amount to impediment, prevention or default by the Employer, the Relevant Event under Clause 2.29.6.

The extent of possession can be a source of disputes (e.g. whether the Employer is entitled to grant possession in stages). Where the extent of possession is expressly stated in the Contract, those express provisions would govern the Contractor's entitlement. In the absence of such provisions, it was said in *Hounslow London Borough* v. *Twickenham Garden Developments Ltd*[7] that the contractor should be given 'such possession, occupation or use as is necessary to enable him to perform the contract'.

Under Clause 2.7, where the Contract Bills indicate that other work on the site is to be carried out by the Employer himself or his other contractors, the Contractor must allow the work to be done. In those circumstances, the Contractor would not be entitled to exclusive possession of the site. Even where it is not so stated in the Contract Bills, the Contractor must not unreasonably withhold consent to the carrying out of such ancillary work (Clause 2.7.2). The Contractor is therefore entitled to exclusive possession only in circumstances where either the Contract provided for exclusive possession or it is reasonably necessary for satisfactory performance of his obligations.

The Contractor is entitled to retain possession of the site up to the date of issue of the Practical Completion Certificate[8] except where the employment of the Contractor is terminated. Also, before the issue of this Certificate, the Employer may, with the consent of the Contractor, use any part of the site or the Works. Such use is referred to as 'Early use by the Employer'.[9] The Employer may, also with the Contractor's consent, take possession of any part of the Works or Section before practical completion.[10] The Contractor must not unreasonably withhold or delay his consent to early use or partial possession by the Employer.

3.2 Completion Date

Clause 2.4 requires the Contractor to complete the Works or Section on or before the 'Completion Date' or 'Section Completion Date'. 'Completion Date' is defined in Clause 1.1 as 'the Date for Completion for the Works or a Section as stated in the Contract Particulars or such other date as fixed either under Clause 2.28 or by a Pre-agreed Adjustment'. Under Clause 2.28, the Architect has powers to revise the applicable Completion Date to take account of specific delays referred to collectively as the 'Relevant Events'. These powers and the procedures governing the granting of extension

7 [1970] 3 All ER 326.
8 For a detailed explanation of the timetable for the issue of this Certificate and its significance see Section 3.3 in this chapter.
9 See Clause 2.6 and Section 3.11 in this chapter.
10 See Clause 2.33 and Section 3.10 in this chapter.

of time are discussed in detail in Chapter 11. A Pre-agreed Adjustment is an extension to the applicable Completion requested by the Contractor in a quotation for a Variation that is accepted by the Employer. The procedure for such an extension is explained in detail in Chapter 6, Section 6.13.3.

From the use of the phrase 'shall complete same on or before the relevant Completion Date' in Clause 2.4, the Contractor is clearly entitled to earlier completion if he can. However, it was pointed out in *Glenlion Construction Ltd* v. *The Guinness Trust*[11] in relation to similarly worded provisions in a contract in the JCT 63 form that there is no corresponding duty on the part of the Employer or the Architect to go out of their way to make such completion possible (e.g. producing drawings or taking other actions required of them earlier than anticipated within the contractual programme).

3.3 Practical Completion Certificate

Clause 2.30 provides that the Architect must forthwith issue a Practical Completion Certificate when practical completion of the Works or Section has been achieved and the following conditions have been met:

- the Contractor has complied sufficiently with his obligation under Clause 3.25.3 to supply information requested by the CDM Co-ordinator for compilation of the Health and Safety File and has ensured similar compliance by his sub-contractors;
- the Contractor has supplied the Employer with all as-built drawings and other information on the Contractor's Designed Portion (Clause 2.40).

3.3.1 Meaning of 'practical completion'

The contract does not specify any factors to be considered by the Architect in deciding whether or not practical completion has been achieved. This is not surprising because practical completion, like the proverbial elephant, is more easily recognized than described. The courts have nevertheless provided some guidance. In the House of Lords case of *Westminster City Council* v. *Jarvis and Sons Ltd*[12] Viscount Dilhorne said:

> One would normally say that a task was practically completed when it was almost, but not entirely finished; but 'practical completion' suggests that that is not the intended meaning and what is meant is the completion of all work that has been done.[13]

However, he added that generally the Architect must not issue the Certificate if there are patent defects. Judge Newey also stated in *Nevill* v. *Press* that practical completion may have been achieved even where minor, *de minimis* (Lat.: trifling) work remains to be done. Continuing on the theme that practical completion does not require absolute 100% completion, he said in *Emson Eastern Ltd (in receivership)* v. *EME Developments*

11 (1987) 39 BLR 89; see also *JF Finnegan Ltd* v. *Sheffield City Council* (1988) 43 BLR 130.
12 (1970) 7 BLR 64.
13 *Supra*, at p. 75.

Ltd[14] that: 'it must be a rare new building in which every screw and every brush of paint is absolutely correct'. The Architect must still be slow to overlook any outstanding work because of possible disagreements with the Employer on whether or not the outstanding work is actually of a trifling nature.

The common understanding of the construction industry is that practical completion is achieved where, except for trifling outstanding work and defects, the Works are so substantially complete that the building can be put to its intended use with safety and convenience. This view is supported by an Australian decision. In *Murphy Corporation Ltd* v. *Acumen Design & Development (Queensland) Pty and Another*[15] the Supreme Court of Queensland stated that the concept of practical completion meant: 'completion for all practical purposes, that is to say, for the purpose of allowing the employer to take possession of the works and use them as intended'. In that case, although minor works remained to be done, the Employer took possession of the building, put it to its intended use and opened it to the public. It was held that the employer's acts were sufficient proof that practical completion had been achieved. The contractor would therefore have been entitled to be put in the position he would have been in if the appropriate completion certificate had been properly issued.

In summary, practical completion is to be considered achieved where the Employer puts the facility being built to its intended use.[16] Also, there is practical completion where the outstanding work or defects, even if apparent, are of a trifling nature. As the Architect has no authority to waive any requirement of the contract, he must still be slow to overlook any outstanding work even if clearly of a trifling nature. An appropriate course of action in such circumstances is to encourage negotiation between the Employer and the Contractor to reach an agreement whereby the Employer agrees to the issue of the certificate in return for written undertakings by the Contractor to carry out the outstanding work without undue delay. This caution is necessary because, under Clause 2.38, the Architect is authorized to issue instructions requiring the Contractor to make good 'defects, shrinkages or other faults in the Works or Section [that] *appear* [authors' emphasis] within the relevant Rectification Period…'. Without such a collateral undertaking, the Contractor may therefore decline to make good defects that existed before the relevant Rectification Period. Although such defects are still a breach of contract for which the Employer would be entitled to damages, exercise by the Architect of his powers to ask the Contractor back to make them good may sometimes be more advantageous.

3.3.2 Effect of the Practical Completion Certificate

The issue of this certificate is one of the most important events in the timetable of the Contract for it triggers off a number of serious consequences:

- the commencement of the Rectification Period (Clauses 1.1 and 2.38 and the Contract Particulars);
- the end of the obligation of the Contractor/Employer to insure against damage to the Works (para. A.1, B.1 or C.2 of Schedule 3 Insurance Options);

---

14 (1991) 55 BLR 114.
15 (1995) 11 BCL 274.
16 See also discussion of *Skanska Construction (Regions) Ltd* v. *Anglo Amsterdam Corp. Ltd* (2002) 84 ConLR 100 in Section 3.10 in this chapter.

- the end of the Contractor's liability for liquidated damages in the event of late completion (Clause 2.32.2);
- half of the Retention Percentage is applied in valuations for subsequent Interim Certificates (Clause 4.20.3);
- start of the period of 6 months within which the Contractor is to supply the Architect and/or Quantity Surveyor with all the necessary documentation for the preparation of final accounts (Clause 4.5.1).

It should be noted that the Works are at the risk of the Employer after the date of issue of the Practical Completion Certificate. The Architect must therefore advise the Employer to ensure that his own building insurance takes effect the following day.

From the consequences listed above, whilst the advantages to the Contractor of practical completion are obvious, the issue of the Practical Completion Certificate is not always in the commercial interest of the Employer. For example, the Employer may not be ready, or is otherwise unable to put the building to its intended use (e.g. occupy, rent, lease or sell). In any such situations, he would be assuming the responsibilities of practical completion with no benefit. For these reasons, the Employer may be tempted either to influence the Architect into withholding the issue of the Certificate or, if it is issued, to dispute its validity. Also, there could be a genuine disagreement between the Architect and the Contractor as to whether the Works have reached practical completion. All such disputes between the Architect or the Employer and the Contractor as to whether or not the Works have reached practical completion must be resolved using the appropriate dispute resolution technique.

Under Clause 8.9.2, interference or obstruction by the Employer of the issue of any certificate under the Contract entitles the Contractor to terminate his employment. Termination by the Contractor and its consequences are discussed in Chapter 16, Section 16.9.3.

3.4 Non-Completion Certificate

If the Works have not reached practical completion on the Completion Date, it is the Architect's duty to issue a certificate to that effect (Clause 2.31). The issue of this certificate, referred to as the 'Non-Completion Certificate', is one of the conditions precedent to recovery of liquidated damages by the Employer. The Employer must also serve a Notice of Liquidated damages[17] and a Withholding Notice.[18]

The total amount of liquidated damages recoverable from the Contractor at any point in time is determined by multiplying the rate of liquidated damages inserted in the Contract Particulars (e.g. £x/week or day of delay) by the total period of delay. If the Employer decides to accept reduced liquidated damages (probably because the actual consequences of the delay are less financially damaging than was anticipated at the time of entering into the contract) he must certify the reduced rate in the Notice of Liquidated Damages. The purpose of this stipulation is probably to avoid challenges to attempted set-off by the Employer on the argument that, because the amount involved is less than the liquidated damages due, it is for something else and therefore not authorized under the Contract.

17 See Chapter 11, Section 11.8.
18 See Chapters 11, Section 11.8 and Chapter 15, Section 15.2.4.

The Architect must issue the Non-Completion Certificate even if there are claims for extension of time still to be assessed. Any subsequent award of extension of time invalidates the Certificate and the Employer must return to the Contractor any liquidated damages recovered for any period before the revised Completion Date (Clause 2.32.3). It is to be noted that the Architect must issue a new Non-Completion Certificate if on the revised Completion Date the Works are still not completed. However, there is no need to serve a repeat Withholding Notice.[19]

3.5 Rectification Period

The Rectification Period is to be stated in the Contract Particulars. If none is stated, it is to be six months and the period begins to run after the day named in the Practical Completion Certificate. Clause 2.38 provides that during this period the Contractor is responsible for making good defects attributable to quality of materials and standards of workmanship being otherwise than specified in the Contract.

The Architect is empowered under Clause 2.38.2 to issue instructions to the Contractor requiring the making good of such defects. He may carry out investigations to identify defects by using his powers to inspect and test under Clause 3.17. The Architect is not limited to any number of instructions. An instruction under this clause is subject to virtually all the provisions in Clauses 3.10 to 3.14. However, there are two significant differences between instructions under Clause 2.38 and other instructions under the Contract. First, whilst with other instructions the Contractor is to comply forthwith (Clause 3.10), an instruction in this case is to be complied with within a reasonable time after its receipt. Second, there is an option, which requires the consent of the Employer, to abandon the instruction, thereby leaving the defect unremedied, and to make an appropriate deduction from the Contract Sum. It is to be noted that there is no requirement that the consent is not to be withheld unreasonably. It would also appear that the Contractor does not have to give his consent to the amount of deduction. However, as the clause expressly states that the amount of deduction must be appropriate, the Contractor may be entitled to challenge any exorbitant deduction on the ground that it is not appropriate.

The last instruction of the Architect requiring the making good of defects is a special one. It should take the form of a Schedule of Defects which the Contractor must make good (Clause 2.38.1). This Schedule is referred to in the construction industry as the 'final snagging list'. The Architect has up to 14 days after the expiry of the Rectification Period to issue this Schedule. After the delivery of this schedule or after the expiry of the 14 days from the end of the Rectification Period, the Architect has no further powers to order the making good of defects under Clause 2.38.2. It follows therefore that:

- the Architect must ensure that all defects for which the Contractor is responsible under Clause 2.38 are included in the Schedule;
- it would be prudent on the part of the Architect to avoid issuing the Schedule too soon, having regard to the time available and the quality of the Contractor's work in general.

---

19 See *Reinwood Ltd* v. *L Brown & Sons Ltd* [2008] UKHL 12.

Where the Architect issues the Schedule outside the specified timetable, the Contractor does not have to make good the defects so notified. However, the Architect would be authorized to take account of such defects in the final adjustment of the Contract Sum, as he is allowed to include in the relevant valuation only work properly carried out. In practice, most contractors would therefore attend to defects notified late.

As stated in Section 3.3.2 in this chapter, the issue of the Practical Completion Certificate triggers off the end of the obligation to maintain insurance against damage to the Works. However, prudence demands availability of alternative cover where the Contractor has to return and carry out extensive remedial works. Such remedial works may even amount to a separate project requiring notification under the CDM Regulations. The Contract is silent on the contractual responsibility for appointing or maintaining a CDM Co-ordinator and principal contractor for the duration of such works. However, as explained in the next section, the Employer would be entitled to recover the cost incurred in such additional compliance with the CDM Regulations from the Contractor as damages for breach of contract.

3.5.1 Liability for defects that appear during the Rectification Period

A defect that appears during the Rectification Period is technically a breach of contract by the Contractor for which, subject to the terms of Clause 2.38, the Employer is entitled to damages in accordance with normal contractual principles. Under Clause 2.38, the Employer has a right to have the defects made good through exercise by the Architect of his powers under that clause. The Architect's duty to allow the Contractor to make good defects that appear during the Rectification Period is worded in mandatory terms in Clause 2.38.1. Failure of the Architect to perform this duty (i.e. produce the schedule of defects) is therefore a breach of the Contract by the Employer for which the Contractor would be entitled to damages. The clause therefore also gives the Contractor a right to be asked back to make good the defects. The value of this right was judicially recognized in *Pearce and High Ltd* v. *Baxter and Baxter*,[20] which arose from a contract let on a version of the JCT Minor Works Form that contained defects rectification provisions similar to those in Clause 2.38 of JCT 05. Comparing the right to return and repair with the situation where the Employer brought in an alternative contractor to make good the defects, Evans LJ stated at p. 104:

> The cost of employing a third party to remedy the works was likely to be higher than the cost of the original contractor remedying the work himself. The right to return to remedy defective works was therefore a valuable one to the contractor. Accordingly, when denied this right, the contractor would not be liable for the full cost of repair … the employer could not recover more than the amount which it would have cost the contractor himself to remedy the defects.

3.5.2 Defects after the Rectification Period

If the defects appeared during the Rectification Period, as explained above, the Architect should have given the Contractor the opportunity of making them good. If he was in fact

20 (1999) BLR 101 (hereafter *Pearce* v. *Baxter*).

given the opportunity but he failed to comply, as a simple matter of contractual analysis, the Employer would be entitled, even after the Rectification Period, to have the work done by a third party and to recover the costs incurred from the Contractor.[21] Such costs may even include costs of the Employer's managerial time expended in getting the work done by the third party (e.g. time used in selecting the alternative contractor and supervising the work).[22] However, as also explained above, where the Contractor was denied the chance of making good the defects, the Employer can recover as damages no more than it would have cost the Contractor if he had been given the chance. In the *Pearce* v. *Baxter* litigation the Court of Appeal rejected emphatically an argument on behalf of the Contractor that the effect of failure to allow the Contractor to return was to absolve the Contractor completely of liability for the defects that should have been notified to him. The court expressly approved of Judge Stannard's statement in *William Tomkinson* v. *Parochial Church Council of St Michael*[23] that there must be clear words to that effect if the Employer is to be denied the normal remedy of damages for breach of contract on account of the Contractor having been refused the chance to effect the repairs himself. There are no such clear words in JCT 05.

The question of liability for defects in respect of which the Contractor was not given the opportunity to return and make good was also considered in *Tombs* v. *Wilson Connolly Ltd*[24] which arose from a contract which provided for the making good of defects in terms similar to JCT 05 Clause 2.38. Expressly applying *William Tomkinson* and *Pearce* v. *Baxter*, HHJ Coulson QC held that the failure to provide such opportunity affected only the measure of damages recoverable for the defects.[25]

A general caveat should now be entered regarding the above discussion on liability for defects after expiry of the Rectification Period. It is that after issue of the Final Certificate, matters of quality expressly left to the Architect's opinion and on which his approval has been received cannot be challenged unless the Employer invoked the appropriate dispute resolution method not later than 28 days after its issue.[26]

3.5.3 Consequential loss from defects

The Employer's right in respect of defects, whether before or after expiry of the Rectification Period, is not limited to having them made good at no extra cost to himself. If the process of carrying out the necessary repairs causes consequential damages, they can be recovered under normal contractual principles. Examples of consequential damages include the cost of getting alternative accommodation or cost of disruption of any

---

21 In *Forsyth v. Ruxley Electronics and Construction* (1995) 73 BLR 1 the House of Lords held that where cost of repair would be unreasonable, only modest damages for loss of amenity could be recovered. For applications of this principle see also: *McLaren Murdoch & Hamilton Ltd v. Abercromby Motor Group Ltd* (2003) SCLR 323, 100 ConLR 63; *McGlinn v. Waltham Contractors Ltd and Others* [2007] EWHC 149 (TCC).

22 For detailed discussion of recovery of loss for managerial time diverted into dealing with a breach of contract, see Chapter 13, Section 13.4.3.

23 (1990) 6 Const LJ 319 (hereafter *William Tomkinson*).

24 [2004] EWHC 2808 (TCC); 98 ConLR 44.

25 See application of this principle by the Singapore Court of Appeal in *Management Corp. Strata Title lan No. 1933* v. *Liang Huat Aluminium Ltd* (formerly Liang Huai Aluminium Pte) [2001] BLR 351; (2001) 17 Const LJ 555.

26 See Section 3.9.3 in this chapter for details on this issue.

business carried on in the building. On the issue of entitlement to consequential damages, in *P. & M. Kaye* v. *Hosier and Dickinson Ltd*[27] Lord Diplock said:

> At common law a party to a contract is entitled to recover from the other party consequential damage of this kind resulting from that other party's breach of contract, unless by the terms of the contract itself he has agreed that such damage shall not be recoverable. In the absence of express words in the contract a court should hesitate to hold that a party had surrendered any of his common law rights to damages for its breach, although it is not impossible for this to be a necessary implication from other provisions of the contract.

He went on to state that no such implication could be drawn from the defects liability provisions, which were very similar to those in JCT 05 Clause 2.38. In *Pearce* v. *Baxter*, although the parties were in agreement on liability for consequential damages, Lord Diplock's statement was quoted by Evans LJ with no indication of any contrary opinion.

3.5.4 Variations after practical completion?[28]

Case law suggests that the Architect has no authority to issue a Variation after the issue of the Practical Completion Certificate. In *TFW Printers Ltd* v. *Interserve Project Services Ltd,*[29] which arose from a contract incorporating the 1993 version of the JCT Standard Form of Agreement for Minor Works, the Employer was under an obligation to take out and maintain a Joint Names Policy against 'loss or damage to the existing structures (together with their contents owned by him or for which he is responsible) and to the Works and all unfixed materials and goods delivered to, placed on or adjacent to the Works and intended therefore…' by listed perils. There was no express provision on the duration of the cover.[30] The Employer argued that the obligation to insure had expired after issue of the Practical Completion Certificate. The Contractor contended that the Employer was required to maintain the cover during the Rectification Period. The Court of Appeal decided that the obligation applied only up to the issue of the Practical Completion Certificate. A major assumption underlying the analysis adopted in arriving at the decision was that the Architect under that contract had no power to issue variations during the Rectification Period. Dyson LJ, who delivered the leading judgment, accepted as correct a statement in paragraph 18.142 of the 7th Edition of *Keating* that the Architect under JCT98 had no such authority.

3.6 Certificate of Making Good

When the Contractor has finished making good the defects contained in the Schedule of Defects to the satisfaction of the Architect, the latter is to issue a certificate to that effect (Clause 2.39). This certificate is referred to in the Contract as the 'Certificate of Making Good'. On a literal reading of Clause 2.39, it appears to have the odd effect that in the rare situation where the Architect does not have to issue a Schedule of Defects because

27 [1972] 1 WLR 146.
28 See also Chapter 6, Section 6.7.2.
29 [2006] BLR 299.
30 JCT 05 provides expressly that the obligation to insure against damage to the Works, existing structures and the contents applies only up to the date of issue of the Practical Completion Certificate (see paras A.1, B.1, C.1 and C.2 of the Schedule 3 Insurance Options).

there is none he cannot issue a Certificate of Completion of Making Good. However, in such a situation, to make commercial sense, it has to be implied that the Architect must still issue the Final Certificate if all other conditions precedent have been satisfied.[31]

3.7 Final adjustment of the Contract Sum

JCT 05 is basically a lump sum contract. The Contractor undertakes to start and complete the Works for the Contract Sum. However, except in the smallest of projects, the Contractor cannot realistically allow for all possible eventualities in the Contract Sum. Examples of such eventualities include variations, and delays and disruptions for which the Employer is responsible. JCT 05 therefore allows for the adjustment of the Contract Sum in defined situations.

Details of the adjustments allowed are provided in Chapter 15, Section 15.8. Not later than 6 months after practical completion, the Contractor is to send to the Architect (or to the Quantity Surveyor if so instructed by the Architect) all documents necessary for the adjustment of the Contract Sum (Clause 4.5.1). Within 3 months after receipt of all the necessary documents, the Architect (or the Quantity Surveyor as the case may be) has to prepare a statement of final adjustment of the Contract Sum (Clause 4.5.2). This final statement is referred to as the 'final accounts' of the Contract. The Contractor will be in breach of contract if he fails to submit all the necessary documentation within the 6 months. Apart from the fact that the Employer would be entitled to damages that he can prove, the period of 3 months within which the Architect and the Quantity Surveyor must prepare statements of outstanding claims and the final account does not begin to run until all the necessary documents have been received. If final accounts are produced on incomplete information, the Contractor would have an exceedingly tough job challenging them, considering that any resulting inaccuracies may be attributable to his own breach of contract. In any such challenge, the Employer would be entitled to set off the cost of responding to it against any increase in the Contractor's entitlement.

On completion of the preparation of the final accounts, the Architect is to send forthwith copies of the statements to the Contractor. Under Clause 4.15.1, the sending off of these documents by the Architect is one of the reference points in the timetable governing the issue of the Final Certificate. On the assumption that the Architect (or the Quantity Surveyor) will prepare the documents promptly if all the necessary documents are made available, the control of the condition is to that extent in the hands of the Contractor. It is therefore in the Contractor's interest to send all the necessary documents and any further documents requested as quickly as possible.

3.8 Final Certificate

Clause 4.15.1 provides that the Final Certificate is to be issued within 2 months of whichever of the following occurs last:

- the end of the Rectification Period;
- the date of issue of the Certificate of Making Good;
- the date on which the Architect sent copies of the final accounts to the Contractor.

---

31 *Penwith District Council* v. *VP Developments Ltd* [1999] EWHC 231 (TCC).

Where there are Sections the timetable for the issue of the Final Certificate in respect of the whole of the Works is to be determined by reference to the last of the Rectification Periods or Certificates of Making Good applicable to the relevant Sections.

3.8.1 Conditions precedent to a valid Final Certificate

This subject is of the greatest importance because of the not infrequent instances where the Contractor or the Employer discovers long after the Final Certificate an overwhelming commercial or other desire to reopen matters on which the Final Certificate is conclusive evidence under Clause 1.10, for example that the Architect undervalued or overvalued the Works. The only way over such conclusiveness is to argue that the Final Certificate was invalidly issued for lack of compliance with conditions precedent to a valid issue of such a Certificate and that, therefore, it does not possess the conclusiveness provided for in the Contract.

Issue of the Certificate of Making Good and the sending of the final account statement to the Contractor before the Architect has to issue the Final Certificate are at the heart of the good practice that Clause 4.15.1 appears to have been designed to encourage. Such practice gives the Architect or the Quantity Surveyor and the Contractor the opportunity to identify areas of differences in relation to defects, valuations and adjustment of the Contract Sum for amicable reconciliation before the issue of the Final Certificate. It also offers the best chances of a Final Certificate acceptable to both the Contractor and the Employer, thus reducing the need for subsequent adjudication, arbitration or litigation on those matters. However, two TCC decisions could be seen by some parties as giving them the green light to ignore this good practice.

Penwith District Council v. *VP Developments Ltd*[32] was a combined appeal on points of law that arose from three arbitrations between the same parties to three contracts in the JCT 80 form. This form provided for the issue of a Final Certificate and its conclusiveness over various matters in terms substantially the same as JCT 05. In the arbitrations, which were commenced some 3 years after the respective Final Certificates, VP (the Contractor) sought to reopen some of the matters on which the Final Certificates were to be conclusive evidence. The arbitrator decided that the Final Certificates had been invalidly issued. This meant that the Contractor could claim additional payment years after the Final Certificates.

One of the points of law before the court was whether provision of the final account statement by the Architect to the Contractor was a condition precedent to the issue of a valid Final Certificate. The arbitrator had decided that it was a condition precedent. HHJ LLoyd QC disagreed.[33] He explained that treating the sending of the statement to the Contractor as a condition precedent would have the effect that, by simply withholding the necessary documentation, the Contractor can hold up the issue of the Final Certificate although such conduct would be in breach of contract. He further pointed out that, apart from the fact that the amount in the Final Certificate may be payable to the Employer, the Employer has an interest in achieving the project closure arising from a Final Certificate and that the provisions should not be interpreted to allow the Contractor

32 [1999] EWHC 231 (TCC) (hereafter *Penwith* v. *VP*).

33 He distinguished the Court of Appeal's decision in *Crestar Ltd* v. *Carr and Carr* (1987) 37 BLR 113 as applicable to only the contract from which it arose (a pre-1980 JCT Minor Works), which had some relevant terms different from their counterparts in JCT 80.

to frustrate realization of such an interest. The same logic works against treating the issue of the Certificate of Making Good as a condition precedent: the Contractor may prevent its issue by refusing to make good the notified defects.

B.R. Cantrell and E.P. Cantrell v. *Wright & Fuller Ltd*[34] also arose from a contract that incorporated JCT 80. The works achieved practical completion on 23 February 1998. The Defects Liability Period[35] expired on 23 August 1998. The document relied upon by the defendant, the Contractor under the contract, was issued on 29 March 1999. Neither a Schedule of Defects nor a Certificate of Completion of Making Good Defects (the JCT 80 equivalent of the Certificate of Making Good under JCT 05) had been issued. The defendant Contractor demanded payment of the unpaid balance of the certified sum. Contending that the Contractor had been overpaid and that they were liable for defects in the Works and liquidated damages for delay, the Employer declined to pay. The resulting dispute was referred to arbitration some 18 months after the dispute had arisen. In the arbitration the Contractor contended that, because of the expiry of 28 days after the issue of the Final Certificate, some of the issues in the Employer's defence to failure to pay the amount certified could not be reopened. The Employer responded by challenging the validity of the Final Certificate. The arbitrator determined that the Final Certificate had been validly issued.

The issues from the arbitration that were appealed to the court raised the question as to what the conditions precedent to the valid issue of the Final Certificate were. Agreeing expressly with Judge LLoyd's analysis in *Penwith* v. *VP*, HHJ Thornton QC stated that:

(i) the issue of the Certificate of Completion of Making Good Defects was not a condition precedent;[36]
(ii) the sending of the ascertainment of loss and/or expense or statement of such ascertainment was not a condition precedent;[37]
(iii) the completion of the final adjustment of the Contract Sum was a condition precedent;
(iv) where the Works were completed later than the Completion Date, the making of the Architect's decision on the Contractor's entitlement to extension of time and the issue of the Non-Completion Certificate would be conditions precedent;
(v) to be the Final Certificate, the document relied upon as such must be, in form, substance and intent, that anticipated under the contract.

The last requirement is that the document must contain the statements expected under the Contract and must be the product of the Architect's own judgment on the matters the Contract expects him to direct his mind to. It must be clear from the document that it is intended as the Final Certificate. A statement on the face of it that it is the Final Certificate is the safest way of showing such intent although, in the absence of such a statement, other factors such as covering letters may be sufficient evidence of such intent. It is the substance of the document rather any specific words that is important. One of the strategies to

34 [2003] EWHC 1545(TCC); [2003] BLR. 412; 91 ConLR 97 (hereafter *Cantrell* v. *Wright & Fuller*). An application to the Court of Appeal for leave to appeal was refused.

35 Now referred to as the 'Rectification Period'.

36 The explanation for this conclusion was that, as the Architect may take account of the Contractor's failure to make good defects by an abatement against the Final Certificate, there is no satisfactory commercial reason for treating the issue of the Certificate as a condition precedent.

37 The rationale for this conclusion was that, as the Contractor knows his entitlement to loss and/or expense and may challenge the Final Certificate within 28 days after its issue, there is no justification for treating this step as a condition precedent.

adopt towards avoidance of doubt is to use the standard templates for the Final Certificate published by the professional institutions, although their use is not mandatory.

The judge determined that the Final Certificate had not been validly issued. Factors that contributed to its invalidity were that: (i) the Architect had delegated to an unauthorized quantity surveyor decisions on the validity of disputed variations and the final adjustment of the Contract Sum; (ii) although practical completion occurred after the Completion Date, the Architect had neither granted extension of time nor issued a Non-Completion Certificate.

3.8.2 Timetable for issue of the Final Certificate

In the light of *Penwith* v. *VP* and *Cantrall* v. *Wright & Fuller*, the timetable for valid issue of the Final Certificate may be summarized as follows:

1. No valid Final Certificate can be issued before expiry of the Rectification Period without the agreement of the parties.
2. The Contractor is entitled to six months after the issue of the Practical Completion Certificate to submit the necessary documentation for preparing the final accounts. No valid Final Certificate can therefore be issued within this period without the agreement of the Contractor.
3. The Architect does not need all the 3 months after receipt of the documentation to issue the final account statement.
4. The Architect has the power to issue the Final Certificate the day after the last of the triggering events under Clause 4.15.1.
5. Provided the conditions precedent are satisfied, a Final Certificate issued more than 2 months after the triggering events may still be valid. However, the power to issue such a late Final Certificate must be exercised 'reasonably and in accordance with any express or implied agreement of, or waiver by, the parties to relax the timetable for its issue'.[38]

As already discussed, the Architect may still issue the Final Certificate where the Contractor fails to submit any documentation for the preparation of the final accounts. However, the Court of Appeal's decision in Tameside *MBC* v. *Barlow Securities Group Services Ltd*[39] suggests that he would not be in breach of his duty to certify under the Contract if he declines to do so. But where either party, mindful of the protection of the Certificate, applies to him to issue it despite the absence of the documentation, he must comply although he would have a duty to the Employer not to include any item of payment not supported with appropriate documentation.

3.8.3 Payment on the Certificate

The final date for payment on the Final Certificate is 28 days from the date of its issue (Clause 4.15.4). A Payment Notice must be served within 5 days after the date of its issue (Clause 4.15.3). The Employer may serve a Withholding Notice later than 5 days before the final date for payment (Clause 4.15.4). If none of these notices is served the Employer

---

38 See *Cantrell* v. *Wright & Fuller*, para. 112.
39 [2001] EWCA Civ 1; [2001] BLR 113; 75 ConLR 112.

must pay the amount stated on the Certificate by the final date (Clause 4.15.5).[40] If only a Withholding Notice is properly served the Employer must pay the certified amount less the set-off notified.[41] Where the Certified amount is in favour of the Employer the Contractor must pay it to the Employer not later than the final date. The Contract does not provide for serving of notices by the Contractor in relation to any amount certified as due to the Employer under the Final Certificate. The relevant provisions of the applicable Scheme for Construction Contracts would therefore apply.[42]

3.9 Effect of Final Certificate

Clause 1.10 deals with the effects of the issue of the Final Certificate. The areas considered are: (i) the Architect's subsequent role under the Contract; (ii) adjustment of the Contract Sum and claims under Clauses 2.28 (extension of time claims) and 4.23 (loss and/or expense claims); (iii) quality; (iv) adjudication, arbitration, other proceedings.

Before they are explained, it is important to note that only a valid Final Certificate can have these effects.[43] To be valid the Certificate must be in form, substance and intent the certificate required by the Contract.[44] It must be clear on the face of it that it is the Final Certificate expected by the terms of the Contract. It must have the expected content. Use of the JCT standard template for the Final Certificate would avoid doubts as to form although its use is not a requirement by the Contract. It must be the outcome of the professional judgment of the Architect and nobody else. A certificate prepared by third parties, for example the Quantity Surveyor or other quantity surveyor engaged by the Architect, but signed by the Architect without independent checks would be invalid. An error on a certificate will invalidate it as the Final Certificate only if, on applying an objective assessment, it is likely to mislead any of the parties as to the contractual document intended. Other documents, such as covering letters, may be accessed in making such an assessment.

3.9.1 Role of the Architect

Upon issue of the Final Certificate the Architect has no further duties under the Contract. For this reason, he is said to become *functus officio* (Lat.: powers ended). This principle was applied in *H. Fairweather Ltd* v. *Asden Securities Ltd,*[45] which arose from a JCT 63 contract. The Architect, after having issued the Final Certificate, realized that he should have issued a Non-Completion Certificate to allow the Employer to deduct liquidated damages. Mr Justice Stabb QC held invalid the Certificate he then purported to issue on the grounds that, after the Final Certificate under the contract, the Architect was *functus officio*.

40 See Chapter 15, Section 15.2.6, for detailed explanation of the principle.

41 The Contractor does not have to accept this deduction; he may immediately challenge it through adjudication to recover any deduction the Employer was not entitled to make.

42 See paragraphs 9 and 10 of Part II.

43 An obvious response of a party wishing to avoid the consequences of a purported Final Certificate is to challenge its validity.

44 See *Cantrell* v. *Wright & Fuller* and Section 3.8.1 in this chapter.

45 (1979) 12 BLR 40; this principle was also applied in *A. Bell (Paddington) & Sons Ltd* v. *CBF Residential Care Housing Association* (1989) 46 BLR 102.

3.9.2 Adjustment of the Contract Sum and claims

With the exception of matters on which dispute resolution proceedings have been commenced before its issue or within 28 days from its issue, the Final Certificate is conclusive evidence that:

- save for errors, all adjustments to the Contract Sum required under the Contract have been correctly made (Clause 1.10.1.2);
- all claims for extensions of time to which the Contractor is entitled have been granted (Clause 1.10.1.3);
- reimbursement made for loss and/or expense is in final settlement of claims in respect of any Relevant Matter regardless of whether the claim is in contract, tort or under statute (Clause 1.10.1.4).

The policy underlying the provisions on the effects of the Final Certificate is to discourage disputes on certain issues, particularly payment disputes, from being started long after final accounts.

Sometimes the amount left to be certified is so small that the Architect ignores it, particularly as the Contractor has little incentive to chase after it. To avoid being surprised and embarrassed by claims many years after completion of the project, the Employer would be well advised to demand issue of the Final Certificate as required by the Contract.

3.9.3 Quality

Clause 30.9.1.1 of JCT 80 stated:

> Except as provided in Clauses 30.9.2 and 30.9.3 (and save in respect of fraud) the Final Certificate shall have effect in any proceedings arising out of or in connection with this Contract (whether by arbitration under Clause 5 or otherwise) as conclusive evidence that where the quality of materials or the standard of workmanship are to be to the reasonable satisfaction of the architect the same are to such satisfaction.

In *Crown Estates Commissioners* v. *John Mowlem & Co. Ltd*[46] the Court of Appeal construed it to the effect that, in the absence of a suitable notice of arbitration, the Final Certificate was conclusive evidence that in respect of all matters of quality the Contractor had complied with the contract (i.e. he was not liable for any defects at all). This had the disastrous implication that, for all projects completed under that and earlier versions of JCT 80, the Contractor's liability for defects ended upon issue of the Final Certificate. Any attempt to avoid this problem by bringing the proceedings in tort had little prospect of success. The decision was received with widespread and very scathing criticism, particularly by Mr Duncan Wallace, QC who opined that the approach adopted by the Court of Appeal to arrive at the decision was unnecessarily legalistic and lacked 'business common-sense'.[47] In *Belcher Food Products Ltd* v. *Miller and Black and Others*[48]

---

46 (1994) 70 BLR 1 (hereafter *Crown Estates*); see also *Colbart* v. *Kumar* (1992) 59 BLR 89 in which His Honour Judge Thayne Forbes, QC reached the same decision with respect to identical provisions in the JCT Intermediate Form of Contract.

47 Duncan Wallace, I. QC, 'Not What the RIBA/JCT Meant: Loose Cannon in the Court of Appeal' (1995) 11 Const LJ 185.

48 (1998) CILL 1415; (1999) SLT 142.

the Scottish Court of Session distinguished *Crown Estates* as limited to only situations where the quality issue in dispute was left to the reasonable satisfaction of the Architect.

That the *Crown Estates* effect was never intended by the JCT is clear from the fact that the clause was very soon after the decision amended through Amendment 15, issued in July 1995. The current provisions on the effect of the Final Certificate on the Contractor's obligations in relation to quality standards are stated in Clause 1.10.1 as:

> Except as provided in Clauses 1.10.2, 1.10.3 and 1.10.4 (and save in respect of fraud), the Final Certificate shall have in any proceedings under or arising out of or in connection with the contract (whether by adjudication, arbitration or legal proceedings) as conclusive evidence that where and to the extent that any of the particular qualities of any materials or goods or any particular standard of an item of workmanship was described expressly in the Contract Drawings or the Contract Bills, or in any instruction issued by the Architect under these conditions or in any drawings or documents issued by the Architect under any of Clauses 2.9 to 2.12, to be for the approval of the Architect, the particular quality or standard was to the reasonable satisfaction of the Architect, but the Final Certificate shall not be conclusive evidence that they or any other materials or goods or workmanship comply with any other requirement of this Contract.

The intended effect is that on all matters of quality not specified objectively (i.e. by reference to a detailed specification clause in the Contract or an appropriate British Standard, but left to the opinion of the Architect), the Final Certificate is conclusive evidence of compliance with the Contract unless properly challenged by the Employer not later than 28 days after the issue of the Final Certificate. The Contractor is therefore not liable for defects relating to matters so left to the Architect's opinion unless the Architect's approval was challenged by commencement of the appropriate dispute resolution procedure at the right time.

Liability for defects relating to matters of quality specified objectively rather than left to the Architect's opinion stays with the Contractor for the appropriate limitation period. It is therefore in the Employer's interest that only minor matters of quality are left to the opinion of the Architect without objective specifications.

Without any remedy from the Contractor for such defects relating to matters left to the approval of the Architect, an obvious alternative open to the Employer is to pursue the Architect who approved the relevant work or the person who drew up such specifications of quality. To succeed in such action, the Employer has to prove that the approval or the relevant approach to specifying quality amounts to negligence. A TCC decision suggests that an Architect sued for such defects would not be entitled to a penny by way of a contribution from the author of the bad work, the Contractor. In *Oxford University Fixed Assets Ltd* v. *Architects Design Partnership and Tarmac Construction Ltd*[49] His Honour Judge Humphrey LLoyd, QC, on a preliminary issue, held that the conclusiveness of the Final Certificate acted as an evidential bar which not only precluded the Employer from establishing the Contractor's liability for the defects, but also prevented proof that the Contractor was liable to make a contribution to the Architect's liability for them. In the words of the Judge, any other conclusion 'would drive the proverbial coach and horses through the structure of the JCT Conditions which has been negotiated over

---

49 (1999) 15 Const LJ 470; 64 ConLR 12; (1999) CILL 1473.

many years and is thus to be taken as representing a fair balance between the competing interests'.

Referring to the practice of contractors bringing contract administrators into proceedings between them and employers for a contribution on account of alleged negligent supervision, Mr Duncan Wallace QC drew an analogy between that conduct and a burglar seeking a contribution from the careless policeman who failed to arrest him and prevent his crime.[50]

3.9.4 Adjudication, arbitration or other proceedings

Clause 30.9.3 of JCT 80 stated:

> If any arbitration or other proceedings have been commenced by either party within 28 days after the Final Certificate has been issued, the Final Certificate shall have effect as conclusive evidence as provided in Clause 30.9.1 save only in respect of all matters to which those proceedings relate.

Another issue raised in the *Crown Estates* litigation was whether, by Clauses 30.9.1.1 and 30.9.3 in JCT 80, arbitration or other proceedings were time-barred unless commenced not later than 28 days after the Final Certificate. If there was such a limitation, arbitration proceedings could be commenced after the 28 days only if the court granted an application to extend time within which to commence them. It was decided that there was no time limit on proceedings other than the applicable limitation period under the general law (i.e. arbitration proceedings could still be commenced after the 28 days without any need for court intervention). However, in any proceedings commenced after the expiry of the 28 days, neither party could question the matters on which the Final Certificate is conclusive evidence.

As Clause 1.10.3 in JCT 05 is in substantially the same wording as JCT80 Clause 30.9.3, an implication of the *Crown Estates* decision is that any matter on which the Final Certificate is conclusive cannot be reopened in adjudication, arbitration or other proceedings commenced after 28 days after the Final Certificate. In such proceedings, there is no bar on reopening matters on which the Final Certificate does not have effect as conclusive evidence.

The issue of what needs to be done for an adjudication to be considered commenced was considered in *Bennett* v. *FMK Construction Ltd*.[51] HHJ Havery QC held that an adjudication was commenced when a Notice of Adjudication was served. It would appear that this would still be the triggering event even if the adjudicator is not appointed and the dispute referred to him within 7 days after the Notice of Adjudication. In that case, which arose from a JCT 98 contract, the Final Certificate was issued on 11th March 2005. The Contractor disputed the certificate and, on 6th April 2005, served a Notice of Adjudication to refer the matter to adjudication. The appointment of the adjudicator and the referral was not achieved within 7 days of the Notice of Adjudication as anticipated under s. 108(2)(b) of the Construction Act and paragraph 7(1) of the Scheme. The adjudicator resigned when his jurisdiction was challenged. On 22nd April 2005, more than 28 days after the issue

50 See note 47.
51 [2005] EWHC 1268 (TCC).

of the Final Certificate, the Contractor served a fresh Notice of Adjudication to refer the same dispute. The Judge declared that the final Certificate was not conclusive evidence on matters covered by the first Notice of Adjudication.

Where an adjudicator's decision is given after the Final Certificate either party has 28 days within which arbitration or legal proceedings may be commenced in relation to the dispute decided by the adjudicator (Clause 1.10.4). Thereafter the decision becomes finally binding if such proceedings have not been commenced.[52]

3.9.5 Dormant Proceedings

It is not uncommon for a Party to take the first step in the commencement of proceedings while still being uncertain whether he will continue them. The reason for such conduct may be to avoid being out of time when good grounds for a claim are found later or simply to create the prospect of being bought off by the other Party who might do so to avoid the nuisance associated with the proceedings. Clause 1.10.2.2 is designed to achieve closure where such proceedings are left dormant for too long. It provides that the occurrence of a period of 12 months after the Final Certificate during which neither Party has taken any step in the proceedings has the effect that the conclusiveness of the Final Certificate applies to the matters to which they relate subject to any settlement.

3.10 Partial possession

With construction projects, it is often possible for parts of the contract works to be usable before completion of the whole project. For example, on a housing contract the completed units may be fit for human habitation before completion of the whole contract. Clause 2.33 of JCT 05 allows the Employer, with the consent of the Contractor, to take possession of such parts for use. The Contractor must not withhold or delay his consent unreasonably. It is submitted that it would not be unreasonable to withhold consent if the ground for the refusal is that use of such parts would seriously impede the carrying out of the uncompleted parts.

Upon the Employer taking over any part, the Architect is to prepare a written statement identifying the part taken over. Any part so identified is referred to as the 'Relevant Part' and the date from which the Employer takes possession is called the 'Relevant Date'. Generally, the effect of partial possession on the Contractor's responsibility for the part taken over is analogous to the effect of practical completion on the whole contract.

Insurance: The Contractor's obligation under Insurance Option A, or the Employer's under Schedule 3 Insurance Option B or paragraph C.2 of Insurance Option C, to insure the Works comes to an end from the Relevant Date in respect of the Relevant Part. Where Insurance Option C (used mainly in contracts involving extension or alteration of existing structures) applies, the Relevant Part becomes part of the existing structures and contents for which the Employer must maintain material damage insurance under paragraph C.1 of Schedule 3 Insurance Option C. This means that the Employer must contact his insurer to

52 For the same construction of similar provisions see *Castle Inns (Stirling) Ltd* v. *Clarke Contracts Ltd* [2005] CSOH 178, which arose from the Scottish Building Contract Jan 2000 revision.

have the policy amended to include the Relevant Part. It is also to be noted that, to avoid the risk of invalidation of the relevant insurance policies already in place and still to be maintained, partial possession must be drawn to the attention of insurers.

Defects: The Contractor's responsibility for defects in the Relevant Part is the same as his responsibility for defects after the issue of the Practical Completion Certificate. The only difference is that the Rectification Period for the Relevant Part runs from the Relevant Date.

Retention: Only half of the full retention applicable to work in the Relevant Part is to be deducted in valuations for Interim Certificates after the Relevant Date (Clause 4.20.2) (i.e. half the retention already deducted is released to the Contractor).

Liquidated damages: There is to be a proportionate reduction in the amount of liquidated damages deductible for delayed completion of the Works or Section. If the value of work in the part is $X\%$ of the total Contract Sum, the liquidated damages are reduced by $X\%$. *Skanska Construction (Regions) Ltd* v. *Anglo Amsterdam Corp. Ltd*[53] suggests that there can be partial possession of the whole of the Works with the consequence that no liquidated damages are recoverable for subsequent delays. In that case the claimant contractor failed to complete an office development by the completion date required by the contract. With the permission of the contractor, the defendant employer allowed a leaseholder to enter the premises to begin fitting out whilst the contractor completed unfinished and defective work. The contractor contended that the arrangement amounted to partial possession of the whole of the works and that the effect of the equivalent of Clause 2.37 of JCT 05 was that no liquidated damages were recoverable. An arbitrator rejected that there could be 'partial' possession of the whole of the Works. On an appeal to the TCC, HHJ Thornton QC agreed with the contractor's contention.

It follows from the discussion of partial possession that if an Employer, with the consent of the Contractor, uses any part of the works, the Architect must issue the statement with all the consequences described above. Sometimes, an Employer may want to make use of part of the works without incurring the attendant responsibility. The procedure for achieving this position is next described.

3.11 Early use/occupation by the Employer

The effect of partial possession is to give the Employer exclusive possession of the Relevant Part. The Contractor is thereafter entitled to access to that part only during its Rectification Period and only for the purpose of making good any defects. Where the Contractor needs access to carry out uncompleted work rather than just making good defects, the partial possession arrangement would not be appropriate. Instead the arrangement under Clause 2.6 is to be used. It allows the Employer to make early use or occupy the site, the Works or any part of them without triggering off the consequences of partial possession. Here, the Contractor retains possession of the part, thus allowing the carrying

53 (2002) 84 ConLR 100 (hereafter *Skanska* v. *Anglo-Amsterdam*).

out of outstanding work, but gives the Employer the right to use or occupy it. It has no effect on the obligation to maintain insurance against damage to the Relevant Part or the Contractor's liability for liquidated damages in relation to that part.

The consent of the Contractor to early use or occupation is required. It is not to be unreasonably withheld or delayed. Before the Contractor gives his consent there must be confirmation from the insurer of the Works that the cover will not be invalidated by such use or occupation. It is the responsibility of the insuring party to sort out the matter of confirmation with the insurer. Where the Contractor is the insuring party (i.e. Insurance Option A applies), any additional premium payable is to be added to the Contract Sum.

The distinction between 'use or occupation' by the Employer and partial possession is not always very clear in physical terms. In view of the huge differences in their effects n the responsibilities and liabilities of the parties, whichever of them is intended must be stated in the clearest possible terms in the parties' agreement if the risk of disputes on such matters is to be minimized. *Impresa Castelli SpA* v. *Cola Holdings Ltd,*[54] which arose from a contract for the design and construction of a hotel, highlights this type of dispute. Completion was delayed and, in order to allow opening of the hotel in line with its advertised opening date, the Employer entered into an agreement with the Contractor to allow use of part of it. The agreement stated that 'no access to the building by the Employer as provided hereunder shall be deemed to amount to Practical Completion for the purposes of the original agreement…'. The Contractor later argued that partial possession had taken place and that the Employer had thereby lost the right to liquidated damages. HHJ Thornton QC held that only use and occupation of the Works, rather than partial possession, had taken place. A major factor he considered, apart from the express terms of their agreement, was the fact that the air-conditioning system, which covered the whole hotel, was still under construction during the relevant period when the Employer was in occupation. However, the decision in *Skanska* v. *Anglo-Amsterdam* suggests that the fact of work continuing does not in itself prevent the Employer being treated as having taken possession. In that case the Contractor was treated as having given up possession and acquired a sub-licence to enter and continue the carrying out of the works.

54 [2002] EWHC 1363; 87 ConLR 123.

4

Contractor designed work

(See also: Section 1.5.2 dealing with Contract Particulars, and 1.5.4 dealing with Optional Documents; also Chapter 6 generally and in particular 6.12.5 dealing with Variations)

4.1 General

The general nature of the JCT standard form of building contract has traditionally been that of a 'build only' contract,[1] in which the Employer specifies the work to be done in exact terms. All the design work is then carried out either by the Architect, or by others in his name; the Contractor is not expressly required to do any.[2] However, there are times when the Employer wishes the Contractor to do all or part of the design. If the majority of the design is to be done by the Contractor, JCT 05 Standard Building Contract is not appropriate, and the form of contract which should be used is the JCT Design and Build Contract (DB).

If the Employer wishes the Contractor's design input to be substantial in an identifiable portion of the Works, but not for the whole of the Works, JCT 05 Contract provides for an optional set of Recitals,[3] which introduce a Contractor's Designed Portion, referred to generally in JCT 05 as 'CDP'. The CDP procedures, in respect of the identified portion, copy the main principles of the JCT Design and Build Contract (DB).

Unfortunately, failure to use the CDP procedures does not bar the Employer, or more likely the Architect, from introducing design to be done by the Contractor informally through the Contract Bills. Work specified by performance is dealt with below in Sections 4.3 and 4.4.

4.2 Contractor's Designed Portion Works (CDP Works)

4.2.1 Introduction

Under JCT 05, the Employer has the option of giving some of the design to the Contractor through the CDP provisions identified in the Seventh to Tenth Recitals. In practice, it will

1 But see Chapter 1, Section 1.4.3, dealing with bills of quantities.

2 The Contractor may have some implied design responsibility arising out of his obligation in respect of workmanship and materials, and his duty to conform to Building Regulations.

3 See Chapter 1, Sections 1.4.1 and 1.5.2, dealing with the Seventh to Tenth Recitals.

usually be the Architect who decides to pass design over, but he should first ensure that he has the Employer's authority. The Architect's engagement with his client is a personal contract, under which the Employer, unless he expresses otherwise, expects the Architect to do the work. It is not for the Architect to decide unilaterally to sub-contract part of the design to the Contractor,[4] or to another design professional, whether it is because the design is complex or specialist, or for any other reason.

The definition of 'CDP Works' in Clause 1.1 is that part of the Works comprised in the Contractor's Designed Portion, which in turn is defined by reference to the Seventh Recital. Simply referring to CDP Works or to Contractor's Designed Portion in the Bills is not sufficient to activate the Contractor's Designed Portion procedures of the Contract; in the absence of the optional Seventh Recital the work described is no more than work specified by performance or design duty introduced by a Bill description. Contractors may view misuse of the contract provisions and the absence of specified procedural rules as being to their advantage, but the reality is that the Contractor's liability is likely to be greater with the protection of JCT 05 provisions removed.

The essence of the CDP procedures is the same as that used by JCT in the Design and Build Contract. The Employer sets out his requirements, and the Contractor responds with his proposals to meet those requirements.

4.2.2 CDP Documents: Employer's Requirements

The Employer's Requirements is a document, or group of documents, identified in the Eighth Recital as setting out what the Employer wants in respect of the portion of work described in the Seventh Recital. No guidance is provided, either in the contract or in the Guide,[5] to assist the Employer in deciding the content and extent of his requirements. So it is likely that owners or their consultants in preparing the requirements will consult the guide relevant to the Design and Build Contract.[6] Nevertheless, it is for the Employer to decide how much information he wishes to provide, and the extent to which he wishes to impose his own ideas on the Contractor. The Employer's Requirements can be a simple explanation of the Employer's problem, which the Contractor is required to solve (e.g. 'Pumping Station and all plant necessary to supply water as required to meet the needs of the Plant') or they can be a definitive design requiring only minor details to be completed. Clearly, a bare performance specification with little detail gives the Contractor wide scope for innovation, leaving him to investigate and decide upon the detailed performance needed before starting his design. It also leaves the Employer not knowing what he is going to get in detail until the design develops. On the other hand, a definitive set of requirements narrows the scope of the Contractor's duty, increasing the Employer's risk in the event of an error.[7] Whichever route the Employer takes, if he is to avoid

---

4 In *Moresk Cleaners* v. *Thos Henwood Hicks* [1966] 2 Lloyd's Rep. 338, it was held that the Architect who had required the Contractor to design a concrete floor had no implied authority from the Employer to allow others to carry out that which he had been employed to do.

5 JCT Standard Building Contract Guide, SBC/G.

6 JCT Design and Build Contract Guide, DB/G.

7 Clause 2.13.2 expressly excludes the Contractor's liability for errors in the Employer's Requirements except with regard to Statutory Requirements.

misunderstanding, he must include sufficient details to spell out what he expects to receive from the Contractor, and for the Contractor to be able to formulate his proposals. Such matters may include:

- Concept or partially completed design, which the Contractor is required to complete;
- Location on site of the CDP Works or any location restrictions – the Employer may give the Contractor an unfettered choice of location, as in the case of a pump house; or the work may be such that location is obvious, as in the case of a specialist installation such as curtain walling;
- A procedure for notifying the Architect if either deep foundations close to the site boundary, designed by the Contractor, or work to a party wall, will require the Employer to give notice to a neighbour under the Party Wall etc. Act 1996;[8]
- Any specific procedure required by the Employer for identifying and auditing the need for removal of waste from the site, to ensure (a) that the Contractor assists the Employer to comply with his obligations as 'Client' under The Site Waste Management Plans Regulations 2008,[9] and (b) that the Contractor complies with his own obligations if he is appointed as principal Contractor[10] under the same regulations;
- The purpose for which the work designed by the Contractor is intended – it is essential that the Contractor is made aware of the detailed purpose. For example, there is a risk of the Contractor providing inadequate lighting if he is told that a church hall is to be used for community meetings, when the intention is to let it to organizations such as sports clubs;[10a]
- Provisional Sums – the CDP provisions deal with adjustment of provisional sums in the Employer's Requirements;
- Planning constraints, if any – under JCT 05 the Contractor takes the risk in complying with Statutory Requirements, so the prudent Contractor will investigate the position regarding planning applications and permissions in any event;
- Information about the site, and the extent to which the Contractor is entitled to rely on it – as a designer the Contractor has a general duty to find out for himself the details he needs to enable him to carry out his design. However, there are circumstances in which the Contractor may find an express or implied warranty[11] as to the accuracy and sufficiency of information provided by the Employer;
- The request for a list of consultants or specialist sub-contractors who the Contractor may wish to engage to carry out design of CDP Works on the Contractor's behalf (to comply with the consent provisions in Clause 3.7.2);
- The amount of detail required by the Employer to be included in the Contractor's Proposals and the Analysis – prudent Employers will describe to tenderers how they wish the Contractor's Proposals to be presented, and how much detail they require in the Analysis. It needs to be borne in mind that the Analysis will form the basis for evaluation of Variations;

---

8 There is no obligation under the contract to give such notice, but the Contractor may be best placed to give early warning in order to avoid delay if a Party Wall Notice is disputed.

9 Statutory Instrument 2008 No. 314; see also Chapter 2, Section 2.1.5 dealing with the Parties' duties.

10 The term 'principal contractor' under the Site Waste Management Plans Regulations 2008, Reg. 4, should not be confused with the 'Principal Contractor' appointed under the CDM Regulations.

10a In *J Murphy & Sons Ltd* v. *Johnston Precast Ltd* [2008] EWHC 3024 (TCC) a sub-contractor who was not told that pipes had to be suitable for an alkaline environment was held not liable for defective pipes.

11 See *Bacal Construction (Midlands) Ltd* v. *Northampton Dev. Corp.* (1975) 8 BLR 88, CA.

- Any amendments required by the Employer to the design submission procedures in the Contract – the JCT Guide SBC/G at paragraph 6 makes clear that the Employer may choose to vary the procedures set out in Clause 2.9.3 and Schedule 1;
- Requirements as to quality – the Employer may wish to refer to Codes of Practice and British Standards as a minimum specification. On high quality projects, there may be a need for a standard of workmanship higher than the minimum, in which case the Employer must find a means of describing what he wants. Superlatives, such as 'first class', 'top quality', 'best', or 'highest international standards' do not assist, for want of a consistent meaning. Such phrases are likely to be construed as meaning 'reasonable in the particular circumstances', which gives neither Party any idea as to what is required. In *Rolls Royce Engineering plc* v. *Ricardo Consulting Engineers Ltd*,[12] it was held that services of 'first class quality' might, practically speaking, be no different from using reasonable skill and care. If an Employer, or his Architect, has a standard in mind, it is likely that there is a sample of it somewhere, and it is not unknown in the design and build sector for other completed projects to be used as a yardstick. Whatever means is used to define quality, for the reasons given under the next bullet point, Employers should avoid terms such as 'work to be to the Architect's (or the Employer's) satisfaction';
- JCT 05 does not require approval of the Contractor's design, but simply the ability to comment if the Architect sees fit. The Employer may wish for the design or installation to be expressly to the Architect's satisfaction, or approval. To the extent that such approval relates to aesthetics or layout, limitation of that nature may be necessary to ensure that the Employer gets what he wants. However, great care needs to be exercised to ensure that the desire to approve does not stray into matters that are strictly the responsibility of the Contractor. The difficulty lies in the decision of the Court of Appeal in *Crown Estates Commissioners* v. *John Mowlem & Co. Ltd*[13] in which it was held that matters required to be to the satisfaction of the Architect become the Employer's risk after the issue of a Final Certificate. I makes sense for the Architect to 'approve' the colour scheme, or (say) the veneer on the doors. However, it would not make sense for 'approval' of the construction details of the fire doors to bind the Employer, unless it is intended that the Employer should take the risk if the design does not comply with the contract. In the example of the fire door, if the design is part of the Contractor's obligations, 'approval' by the Architect will not normally reduce the Contractor's liability but following the *Crown Estates* case, the risk could be the Employer's if work is required to the Architect's satisfaction;
- Alternative specification – if the Employer wishes to invite alternative specifications, they should be set out clearly in the Employer's Requirements.

It is for the Employer to ensure that the information and any design included in the Employer's Requirements are accurate. Clause 2.13.2 expressly excludes the Contractor from responsibility and goes further to remove from the Contractor any implied duty[14] to check the Employer's design (see Section 4.4.3 in this chapter for Contractor's design duty and liability).

---

12 [2003] EWHC 2871 (TCC), 2 December 2003.
13 (1994) 70 BLR 1.
14 See reference to the *Co-operative Insurance* v. *Boot* case in Section 4.4.4 in this chapter – Liability for Employer's design and obligation to check.

4.2.3 CDP Documents: Contractor's Proposals

The Contractor's Proposals are the Contractor's response to the Employer's Requirements. They are a description of how the Contractor intends to solve the problem set by the Employer. The form of the Contractor's Proposals must therefore be reactive to the requirements laid down by the Employer.

It is not strictly necessary under the standard form to list potential sub-contractors in the Contractor's Proposals. However, it is necessary to obtain the Employer's (not the Architect's) written consent before sub-letting any of the CDP design (Clause 3.7.2), and the Contractor's Proposals provide a convenient vehicle for notifying the Employer.

Typically, Contractors will include drawings, specifications, and possibly a list of exclusions and assumptions if there is any risk of misunderstanding later. However, assumptions should be used with care. If the Contractor has made assumptions in preparing his design, he must ensure that he acquires information to verify those assumptions, and to notify the Architect of the need. If he does not, in the event of any resulting design failure, the Contractor would probably be found liable for failing to use reasonable skill and care.[15] If the purpose of an assumption is an intention by the Contractor that his price should be made conditional on the assumption being correct, he would be wise to spell it out with a disclaimer.[16] In an industry where openness is becoming ever more popular, a bare statement of assumption may be construed as no more than interesting information, but at the Contractor's risk.[17]

There is no right or wrong level of information to be provided, except to meet any stipulations in the Employer's Requirements. Again in the interests of avoiding misunderstanding and dispute later, the Contractor's Proposals should be sufficient for the Employer to know what he is getting for his money, and just as importantly for the Contractor to be reminded what he is providing, subject always to their meeting the Employer's Requirements. Often drawings are prepared by consultants to the Contractor, and there is a risk for the Contractor that items or standards not requested by the Employer may be included. If the design is enhanced in this way, it is at the Contractor's expense, leaving him to attempt recovery from his designer.

The Contractor's Proposals should not include prime cost sums or provisional sums. There is no provision for prime cost sums in this contract. However, there is provision in Clause 3.16 for the Architect to expend Provisional Sums, albeit only those in the Contract Bills or in the Employer's Requirements, and in Clause 4.3 for the Contract Sum to be adjusted; but it seems he has no authority to deal with provisional sums elsewhere in the Contract Documents. It is strange that the JCT has continued with the concept of Provisional Sums appearing only in the Employer's Requirements, first introduced in the With Contractor's Design form published in 1981. There are many occasions when a Contractor cannot complete his design, perhaps because of reliance on another as yet

15 See *Ove Arup & Ptnrs International Ltd & Anr* v. *Mirant Asia-Pacific Construction (Hong Kong) Ltd & Anr* [2005] EWCA Civ 1585, paras 91–94.

16 In the *Ove Arup* v. *Mirant* case cited above, the Court of Appeal said (para 91(3)) that without an explicit warning and a disclaimer of an assumption, it would not be sufficient to leave it to the client to acquire and interpret necessary information.

17 In the *Ove Arup* v. *Mirant* case cited above the Court of Appeal held (paras 91–94) that the risk in obtaining information was the designer's, notwithstanding a statement by him that the design was based on assumptions; further notification to the client was required for the designer to divest himself of the risk.

to be designed detail, and the logical solution is to include a provisional amount. If he does, the Contractor must either ensure that he draws it to attention before entering into the contract so that it can be transferred into the Employer's Requirements, or he should seek to have Clause 3.16 amended. If he does not, in the absence of machinery to adjust them, provisional sums in the Contractor's Proposals become an allowance within the Contractor's tender, and at the Contractor's risk.

It is essential that the Contractor ensures that his proposals are compatible with the Employer's Requirements; the position where they are not is dealt with in Chapter 6, Section 6.12.6.5. Sometimes uncertainty or conflict can be created by a request from the Employer for alternative specified materials or specialist systems. He may even ask the Contractor to suggest means of reducing the price, or shortening the programme by alternative design (sometimes termed 'value engineering'). The Contractor should check that any suggestions in his Contractor's Proposals that are not strictly in accordance with the Employer's Requirements are drawn to the Employer's attention, and if accepted by the Employer, are transferred into the Employer's Requirements, before the contract is signed. Likewise, if an option is offered unilaterally, the Contractor's Proposals submitted with the tender should contain both the compliant and the optional designs. Once a choice is made by the Employer, the rejected proposal should be struck out, and if necessary, the Employer's Requirements should be amended to suit. Failure to modify the documents in these circumstances, before the contract is signed, may leave the Proposals ambiguous. In that case, treating the issue as a discrepancy within the Contractor's Proposals, the Employer may take his pick at the Contractor's expense.[18]

4.2.4 CDP Documents: Analysis

The CDP Analysis is simply the breakdown of the Contractor's price in respect of CDP Works. It is not part of the Contractor's Proposals, but is supplied additionally, and is identified separately in the Agreement (see Chapter 1, Sections 1.4.1 and 1.4.4). There is no prescribed form or level of information for the Analysis, except to the extent that form and content may be set out in the Employer's Requirements. The purpose of the Analysis is to provide a pricing basis for interim payments, valuation of Variations and price fluctuations where a formula method applies, so its function is compromised if there is insufficient detail. However, that is a risk that the Parties may wish to take. A Contractor whose price is keen may prefer to rely on a fair valuation of a Variation in the absence of detailed prices. However, it must be remembered that absence of prices in the pricing document relevant to an omission Variation, may produce a credit greater than the amount actually allowed in the price.

The Contract, Ninth Recital, refers to the CDP Analysis as an analysis of the portion of the Contract Sum relating to the Contractor's Designed Portion. This suggests no more than an arithmetic breakdown, although the purpose as a basis for evaluation, including the value of design in Variations, requires more detail.

One element of the price that should be shown, in the interests of both Employer and Contractor, is the amount in respect of design to be added or omitted. The Parties may

---

18 See Chapter 6, Section 6.14.

wish to add or omit simple percentages as being the value of design work, but it is common where a design is changed before its construction starts, for part of the design to have been prepared already. A valuation of design based only on a percentage of the net variation takes no account of the abortive design. With this in mind, the Parties may find it convenient to extend the Analysis beyond a pure price breakdown. Other matters that might be helpful include rates for adding or omitting design value, and Contractor's risk contingencies to be added when providing a Schedule 2 Quotation for a CDP Variation before the design is completed. However, the CDP Analysis should not include Provisional Sums, except as set out in the Employer's Requirements, or prime cost sums (see Section 4.2.3 in this chapter above).

4.2.5 Employer's response to Contractor's Proposals

The JCT deal with the Employer's response to the Contractor's Proposals in the Tenth Recital: 'the Employer has examined the Contractor's Proposals and … is satisfied that they appear to meet the Employer's Requirements'. The Employer does not need to accept the Contractor's Proposals. The Employer is obliged to examine the proposals presented to him, but he is entitled to rely on the fact that he has chosen to pass some of the design to the Contractor. The reason may be that the Employer, or his Architect, has insufficient resources, or that the solution sought requires specialist knowledge, or it may be simply that he wishes to transfer the risk; but the reason is irrelevant. The Employer has set out his problem for the Contractor to solve. The Employer, and his team, may not know how the Employer's Requirements should be met. He need do no more than examine what is proposed, and comment that the proposal seems to be what he wants, but he does not know for sure. A typical example would be the design of a building management system. The Employer and his Architect know which building systems are to be controlled, and that will be set out in the Employer's Requirements, but how the computer software works, or indeed whether it will work at all, may be beyond their expertise. In that case, the Employer is not in a position to accept the proposal, and must rely on the Contractor's, or Sub-contractor's, specialist knowledge.

Disputes arising out of failure to meet the Employer's Requirements are often identified as a conflict between Employer's Requirements and Contractor's Proposals. The issue is dealt with in principle in Chapter 6, Section 6.12.6.5.

Where an Employer has asked for alternative suggestions, and the Contractor has supplied options in his Contractor's Proposals, it is essential that the Employer's requirements are modified to reflect the Employer's choice.

4.2.6 Execution: Contractor's Design Submission Procedure (Schedule 1)

1. Introduction

Clause 2.9 and Schedule 1 deal with the opportunity for the Architect to comment on the Contractor's design. The Contractor is obliged to provide to the Architect when necessary from time to time, two copies of the Contractor's Design Documents. The Architect may also ask for related calculations and information. There then follows a strict procedure of implied approvals and comment, under which the Architect marks the Contractor's

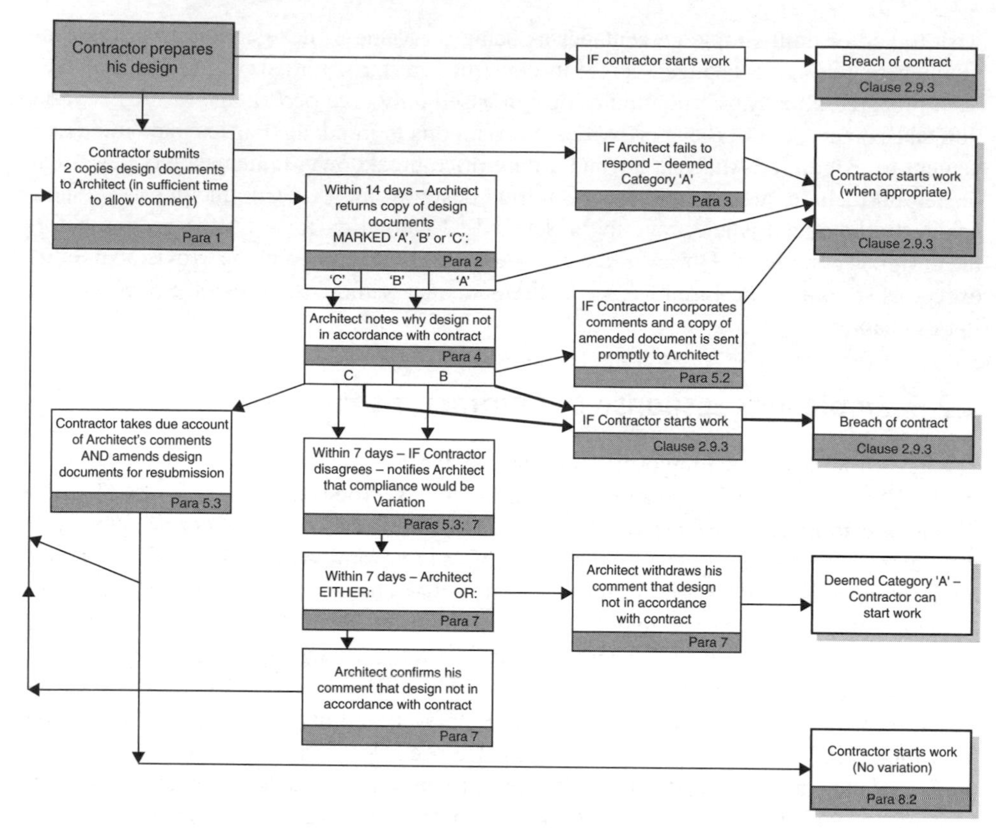

Fig. 4.1 Schedule 1 Contractor's Design Submission Procedures

documents as 'A', 'B', or 'C' category, which either entitles the Contractor to start work or requires him to amend his design.

The submission procedure (see Figure 4.1) is a default system that applies in the event that the Parties have not agreed any other.[19] The JCT suggest[20] that, since the design input by the Contractor does not relate to the whole design, but only to the identified portion, the design submission system should be flexible. This is sensible. The CDP element of the Works may be limited to only small value or simple installations, or it may encompass a large proportion of the whole design. It is for the Employer and Contractor to agree upon a suitable and workable system to suit the need; but if they do not, the Schedule 1 procedures apply.

2. Contractor's Design Documents

The definition of Contractor's Design Documents in Clause 1.1 covers all drawings, details, specifications and other documents prepared by or for the Contractor in relation to the Contractor's Designed Portion. It is worth noting here that, under Clause 3.7.2,

19 See also Section 4.2.2 in this chapter.
20 JCT Guide SBC/G, para 44.

the Contractor must obtain written consent from the Employer (not the Architect) before sub-letting any of the CDP design, whether to an independent designer, or to a specialist sub-contractor.

In many cases, a designed portion will be the province of a specialist sub-contractor, who may consider information such as calculations to be confidential. The task of obtaining such documents clearly falls on the Contractor, and if the procedure is to have any chance of working smoothly, the Contractor needs to be told in the Employer's Requirements (before he enters into sub-contracts) what he will be expected to produce.

Copyright in all Contractors' Design Documents remains with the Contractor (Clause 2.41.1), although the Employer has the right to copy and use them for any purpose relating to the Works (Clause 2.41.2), including completion, maintenance, repair, and advertisement (i.e. he has a licence to use the designs to do work on the project that is the subject of the Contract). However, the Employer cannot reproduce the Contractor's designs to extend the facility constructed under the Contract, although he may copy and use the information in the design documents to facilitate the extension. In this respect the design documents are to be used as no more than a 'survey' of the existing, to enable an extension to be designed and built. In the event that the Employer uses the Contractor's Design Documents for a purpose other than that for which they were prepared, the Contract expressly excludes any Contractor's liability (Clause 2.41.3). Such use would also be a breach of Clause 2.41, rendering the Employer liable to the Contractor for damages, probably amounting to a reasonable charge for a licence.[21]

The Employer's right starts on practical completion, and only exists so long as the Contractor has been paid all monies due and payable to him under the Contract. The amount due and payable does not relate solely to CDP Works, but to the whole of the Works, and refers not to the final value, but to the amount due at any particular time.

3. Submission of Contractor's Design Documents to Architect

Clause 2.9.3 obliges the Contractor to submit his designs in his Contractor's Design Documents, together with any other documents and information that the Architect may request, provided they are reasonably necessary to explain the Contractor's Proposals. Submissions are made from time to time as and when necessary. The Contractor is not required to complete his designs immediately after entering into the contract; he can treat the design as a continuing process to be completed to suit the construction programme. However, the Contractor must take into account the time needed for ordering materials. Paragraph 1 of Schedule 1 reminds the Contractor that submission must be in sufficient time to allow the procedure to be completed, and that the procedure must be completed before the design is used for procurement or carrying out the CDP Works.

The procedure does not prescribe a period for submission, but it is open to the Parties to identify dates or periods in the Contract Documents if they wish. There is no entry in the Contract Particulars for such dates or periods, so any agreement is best placed either in the Employer's Requirements or in the Contractor's Proposals.

---

21 In *Meikle* v. *Maufe* [1941] 3 All ER 144, it was held that a reasonable charge for a licence was the proper measure of damages based on the fact that the design was only of value when repeated in an extension of the building. See also House of Lords decision in *Redrow Homes Ltd* v. *Bett Bros plc (Scotland)*, HL, 22 January 1998, in which it was held that additional damages could be awarded for flagrant infringement.

4. Architect's response to Contractor's design submission

The Architect has the opportunity to comment on Contractor's Design Documents, although he does not have unfettered power to change or veto the Contractor's design. Within 14 days after receiving a Contractor's Design Document the Architect must return it to the Contractor, marked with the classification 'A', 'B' or 'C'. The 'A' category is used where the Contractor can proceed, and the 'B' and 'C' categories, accompanied by reasons, are expressly to be used only where the Architect considers the design is not in accordance with the Contract. Provided the Contractor's design complies with the Contract, it is his choice as to how he achieves it. Such comment is not a vehicle for the Architect to impose his own preferred method on the Contractor. It is simply an opportunity, and an express obligation, to warn the Contractor of deficiency in the design proposed. Provided the Contractor's proposal meets the specified requirement, any insistence by the Architect that the Contractor should use a particular method, or that the design should be 'improved' in some way, would probably amount to a deprivation of choice (i.e. a Variation).[22]

There is difficulty in identifying a difference between categories 'B and 'C' at the point of creation. In either case, the associated comment identifies failure in the Contractor's design to comply with the Contract, but that is all. The difference between categories 'B' and 'C' relates more to the manner in which the Contractor may react than to the comments of the Architect, so in choosing a category the Architect must consider how he wishes the Contractor to respond. The result is that the category chosen is likely to be the one that gives the Contractor the least latitude or opportunity to disagree. The Contractor's response is dealt with below.

If the Architect fails to respond within the timetable, the design document is deemed to be marked 'A' (paragraphs 2 and 3).

Category 'A': If a document is marked 'A', or if the Architect has failed to respond in time to the Contractor's submission, the Contractor must carry out the work in strict accordance with the design represented. To the extent that the Architect's action gives the appearance of accepting or approving the Contractor's design, it does not diminish the Contractor's liability for the accuracy of that design. That is the general position unless the design is specifically required to be to the Architect's approval,[23] and it is also confirmed in the Contract at Schedule 1, paragraph 8.3. However, in this instance, the Architect is not approving or accepting the design at all; he is simply not vetoing it.

Category 'B': If the Architect thinks that the design presented to him does not comply with the Contract, he must mark the Contractor's Design Document 'B', and state why in a written comment (paragraphs 2 and 4). This is not an opportunity for the Architect

---

22 In *English Industrial Estates* v. *Kier Construction* (1991) 56 BLR 93, the Engineer's instruction to use crushed arisings from the excavation in preference to imported fill, when the specification provided for either, was held to be a deprivation of the Contractor's choice and a variation. In *Skanska Construction UK Ltd* v. *Egger (Barony) Ltd*, 3 December 2002, CA, [2002] BLR 236, the Contractor's choice of cabling fulfilled the contractual obligation so the client's instruction for something different was a variation.

23 The Employer should be wary of requiring design or work to be to the Architect's 'approval'. Following the Court of Appeal decision in *Crown Estates Commissioners* v. *John Mowlem & Co. Ltd* (1994) 70 BLR 1, a Final Certificate issued under Clause 4.15 may be construed as conclusive evidence that where the design is required to be to the Architect's 'approval', it complies with the contract.

to reject the design on grounds of preference for another. It is common for Architects, and others in an Employer's design team, to favour particular products, or have prejudices about aesthetic details. However, the detailed design of the CDP Works is the Contractor's choice – that is why the Employer's Requirements need careful drafting if the Employer, or more likely his Architect, wishes to control the design.

Paragraph 4 requires no more from the Architect than a comment as to why he considers that the Contractor's design does not comply with the Contract. Nevertheless, in order for the procedure to operate successfully, he needs to give positive direction if the Contractor is to respond. The significance of category 'B' is in the response available to the Contractor, dealt with below.

Category 'C': Category 'C' is similar to 'B' in that it can be used only when the Architect considers the design does not comply with the Contract. As with 'B', a category 'C' response must identify by written comment why the design is not compliant (paragraphs 2 and 4).

The significance of category 'C' is in the response available to the Contractor, dealt with below.

5. Contractor's response to Architect's comment

The Contractor's ability to get on with procurement and construction relies on the category of response he receives from the Architect.

A category 'A' response, or silence from the Architect, obliges and entitles the Contractor to proceed strictly in accordance with the relevant design documents (paragraph 5.1). The Contractor is able to carry on with procurement and construction, but he is not required to start immediately; his timing is dictated by the construction programme.

A category 'B' response is accompanied by the Architect's written comments indicating why the design is considered not to comply with the Contract. If the Contractor agrees with the comments, he is allowed to proceed with procurement and construction, provided he complies with the comments and promptly resubmits the document to the Architect with the comments incorporated (paragraph 5.2). Clearly, the Contractor can only respond positively if the Architect's comments go further than simple negative criticism. On the other hand, if the Contractor considers that his design complies with the Contract, he is allowed within seven days to notify the Architect that he disagrees with the comments on his design document (paragraph 7). In practice, it is likely that many of the Architect's comments will turn out to be statements of preference rather than non-compliance, and the Contractor will disagree. Disagreement is dealt with below.

A category 'C' response, like category 'B', must be accompanied by written comments identifying why the design is considered not to comply with the Contract. However, unlike category 'B', the Contractor is expressly barred from continuing work based on his design. Paragraph 6 makes clear that the Employer need not pay for any work carried out within the CDP Works, other than work covered by categories 'A' or 'B'. Having received a category 'C' response, the Contractor must take due account of the Architect's comments, and then he has a choice under the Schedule 1 procedure (paragraph 5.3). Either the Contractor can amend his design document in line with the comments, and resubmit it to start the procedure over again, or he may disagree with the comment using the route described in paragraph 7 of the procedure (see below).

6. Contractor's disagreement with Architect's comments

Paragraph 7 describes the procedure when the Contractor disagrees with the Architect's comments.

Contractor's notification of disagreement: If the Contractor disagrees with a category 'B' or 'C' comment that the proposed design does not comply with the Contract, he must, within 7 days, notify the Architect in writing that compliance would give rise to a Variation, giving his reasons. If he does not notify the Architect, the Contractor cannot later claim that the change is a Variation (paragraph 8.2).

Architect's response to Contractor's notification: Within 7 days after receipt of the Contractor's notification, the Architect must either confirm his comment, or withdraw it. It is tempting for the Contractor to view confirmation or withdrawal of a comment by the Architect as a sign that the Architect has in some way approved the Contractor's design; but that is not so. Paragraph 8.1 warns that confirmation or withdrawal of a comment by the Architect does not signify that the design document is accepted by the Architect or the Employer as complying with the Contract; nor does it signify that incorporation of the comment by the Contractor is accepted by the Architect or Employer as giving rise to a Variation. The submission procedure at this stage is no more than a system enabling the Parties to take a stance, while getting on with the project. Any dispute can be resolved straight away or left until later, but either way the actions of the Parties cannot be used as evidence to indicate liability.

Contractor's response: If the Architect has confirmed his comment, the Contractor must amend and resubmit his design document. There is no proscribed timescale, but if the Architect is correct, and the Contractor's proposed design does not comply, it is in the Contractor's own interest to resubmit the document as soon as he can to avoid any delay for which he would not be entitled to extension of time. If the Contractor still thinks that compliance with the comment would give rise to a Variation, his likely course of action would be to refer the matter to an adjudicator.

Comment on Paragraph 7 disagreement: The Architect's comments under the Schedule 1 procedure are expressly required to describe the manner in which the proposed design does not comply with the Contract. However, the Contractor may consider that his design does comply, in which case he can register his disagreement using the procedure set out in paragraph 7, and declaring that the comments, if introduced, would give rise to a Variation. If the Architect's comments are no more than an expression of the Architect's preference, they are not comments given in compliance with paragraph 4, but might nevertheless be used by the Architect as a convenient way of changing the Contractor's design to match his own. In practice, whether a comment complies with paragraph 4 or is simply an attempt to impose the Architect's will on the Contractor, is immaterial; it will be for the Contractor to disagree in either case. The point that both the Architect and the

Contractor must keep to the forefront of their minds is that the Contractor is the designer of CDP Works, at least to the extent set out in the Employer's Requirements. Provided the Contractor meets the Employer's Requirements within any imposed parameters, it is the Contractor's choice as to how he does it.

There is one area of potential difficulty that deserves special mention. The procedure assumes that the Architect's comments, if incorporated either by acceptance or by Variation, would produce a sound design in compliance with the Contract. The Schedule 1 procedure does not deal expressly with a situation where the Architect's comment itself would introduce defective design, perhaps even replacing a Contractor's compliant design. In the event of such a comment the Contractor would be put in an invidious position. If he complies with the comment he may construct a defective building. If he disagrees on the grounds that it is a Variation, and then complies on the instruction of the Architect, again he may construct a defective building. Paragraph 8.3 makes clear that liability rests with the Contractor: 'compliance ... with the Architect's ... comments shall not diminish the Contractor's obligations to ensure that the Contractor's Design Documents and CDP Works are in accordance with this Contract'. This simply reiterates the common law position where the Contractor as the designer follows the design given to him by his client.[24] Unfortunately for the Contractor, his only realistic course if he thinks that the Architect's design is faulty – yet having been warned, the Architect still insists that his comments must be incorporated – would be either to refuse to comply, or to incorporate the comment at his risk, and claim for a Variation. If the Contractor considers that the Architect's comment would lead to a design that was not simply faulty, but also dangerous, his safest route is to refer the matter immediately to an adjudicator. If the adjudicator decides in his favour, the Contractor's reluctance would be vindicated, and he would be entitled to extension of time and loss and expense resulting from any delay. However, if the adjudicator decides that the design is not dangerous, the Contractor would have suffered culpable delay. The Contractor takes a risk, whichever course of action he pursues.

7. Starting work

The Contractor is barred from starting work on his designed portion works on site until the relevant design has passed completely through the design submission procedure. That is not to say that approval must be given; it means simply that the design document must pass through the system to give the Architect an opportunity to comment. To start procurement or construction without completing the procedure would be a breach of Clause 2.9.3.

The circumstances in which the Contractor may start are: (a) when the design document receives an 'A' classification, or (b) when the design document receives a 'B' or 'C' classification and the Contractor chooses to incorporate the Architect's comment, or (c) if the Architect fails to respond to the Contractor's submission of his design document, or (d) when the Architect withdraws his comment that the design does not comply.

---

24 In *McGlinn* v. *Waltham Contractors Ltd* [2007] EWHC 149 (TCC) 21 February 2007, the architect was held liable for his client's design, even though he had warned that it was unsafe.

4.2.7 Execution: Integration into Architect's design

It may be that the Contractor's design proposal is incompatible with the overall design prepared by the Architect, yet still be compliant with the Employer's Requirements. An example would be where an adequate interface detail between CDP Works and surrounding work is designed by the Contractor in a situation where the Architect wishes to create a feature. In those circumstances the Architect must decide how to integrate the CDP Works into the surrounding work, and give instructions accordingly. The Architect may try to use the comments facility in the Contractor's Design Submission Procedure as a convenient vehicle to bring the Contractor's design into line with his own, although the procedure should properly be used only for comments where the Contractor's design does not comply with the Contract. The Contractor may consider the change to be of little consequence and incorporate the Architect's comment, but there may be situations where he does not. If the Contractor considers compliance with such comments would prejudice his design, or be a Variation, he must notify the Architect under the Schedule 1 procedure. However, provided the Contractor's design complies with the Contract, it is for the Architect to work his design around the Contractor's, or instruct a Variation on the CDP Works.

4.2.8 Execution: Divergence and discrepancies

Divergence and discrepancies in and between the CDP Documents and other Contract Documents are dealt with in Chapter 6, Section 6.14.

4.2.9 Execution: Variations

Variations to CDP Works are treated in a similar manner to Variations applying to other work. All Variations, including those relating to CDP Works, are dealt with together in Chapter 6.

4.3 Specification by performance

Sometimes the Employer will wish the Contractor to carry out specified work to meet a defined performance, which necessitates some design input by the Contractor or a Subcontractor. There are many instances in a building project where a proprietary product may be specified which then requires specialist design in adapting to meet the particular circumstances of the surroundings. A typical example would be a composite unit such as a window. Whilst the window may be made from standard extrusions, the choice of extrusions and the manner in which they are put together is determined by wind loadings, security and maintenance needs, or any other factor that may affect the manufacturer's detailed design. These are criteria that determine the performance required of the product. Other examples would include roof trusses, proprietary flooring systems, curtain walling, high bay storage systems, all of which require 'fine tuning' to fit into the project, and all of which require design input by the manufacturer or installer. However, the use of any of these examples in isolation would not normally suggest that the Architect would no longer be designing the Works. He would simply be choosing a product to be built into the Works, albeit specialist design would be required to be completed by someone else.

JCT 98, Part 5 (Performance Specified Work) contained provision for design by the Contractor of simple or easily identifiable types of work. In JCT 05 there are no longer any 'performance specified work' provisions. In their Guide, the JCT explain that the omission results from only rare use and that, where used, the optional Contractor's Design Portion Supplement would have been appropriate. Nevertheless, under JCT 98 it is common for work with a performance specification to be included in Contract Bills without completing the Performance Specified Work entries in the Contract. The result under JCT 98 would be that the Performance Specified Work provisions were not activated, but that the Contractor still had a design obligation.

The position under JCT 05 is similar – if the Bills contain items which the Contractor is required to design, the Contractor has a design duty because that is what he has agreed to do.[25] The absence of express provisions does not remove that obligation or reduce any subsequent liability if the Contractor's design is faulty.

However, in the absence of express provisions in the contract to deal with work specified by performance, other than the defined CDP Works, the Employer takes a risk. There is no obligation on the Contractor to fulfil his related design duties in any particular manner, and the design submission procedure does not apply unless the contract has been suitably amended, or a procedure has been specified elsewhere in the Contract Documents. Likewise, the Contractor also takes a risk. In the absence of provisions dealing with liability, the Contractor who designs and installs work other than CDP Works will probably take onto himself, albeit unintentionally, an obligation to design and install work which is fit for its intended purpose.[26] This is a higher standard than that of reasonable skill and care which applies to the Architect,[27] and indeed to the Contractor in respect of CDP Works, and one which is difficult to cover with professional indemnity insurance.[28]

4.4 Liability for design and installation

There are three standards by which a Contractor may be judged when carrying out design under the same contract as the construction. They are:

- An obligation to provide a design that is fit for its purpose;
- An obligation to provide a design using reasonable skill and care;
- An obligation to provide a design to a standard specified in the contract.

Each is dealt with in turn below.

25 In *Haulfryn Estate Co. Ltd* v. *Leonard J. Multon & Ptnrs and Frontwide Ltd*, ORB; 4 April 1990, Case No 87-H-2794, it was held design obligations placed on the Contractor in Specifications under the JCT Minor Works form were not inconsistent but added to the obligations in the Conditions.

26 See *Independent Broadcasting Authority* v. *EMI Electronics & BICC Construction* (1980) 14 BLR 1, and *Viking Grain Storage* v. *White (T. H.) Installations* (1985) 33 BLR 103.

27 See Section 4.4.2 in this chapter dealing with reasonable skill and care.

28 JCT 05, Clause 6.11 obliges the Contractor to take out Professional Indemnity insurance where there is a Contractor's Designed Portion. However, most insurance companies resist offering to cover fitness for purpose.

4.4.1 Common law standard: fitness for purpose

Where a Contractor designs and installs work under the same contract the common law obligation, in the absence of terms to the contrary, is one of fitness for purpose. The Contractor has the same liability in respect of his design as that in respect of workmanship and materials; it must be reasonably fit for the intended purpose previously made known to him by the Employer. In *Greaves* v. *Baynham Meikle*,[29] Lord Denning, when referring to the relationship between the Contractor and the building owner, said (*obiter*):

> Now, as between the building owner and the Contractors, it is plain that the owners made known … the purpose for which the building was required … It was, therefore, the duty of the Contractors to see that the finished work was reasonably fit for the purpose … It was not merely an obligation to use reasonable care. The Contractors were obliged to ensure that the finished work was reasonably fit for the purpose.

In *IBA* v. *EMI & BICC*,[30] Lord Scarman underlined the position:

> I see no reason why one who … contracts to design, supply and erect a television mast is not under an obligation to ensure that it is reasonably fit for the purpose for which he knows it is intended to be used.

The design must work reasonably; if it does not, the Contractor is liable. That would be the general position if the contract did not limit design liability.

Fitness for purpose is the standard applying to work carried out in respect of performance specification in the Bills (see Section 4.3 in this chapter).

4.4.2 Reasonable skill and care

If a contract for the provision of services is silent as to liability, a duty to use reasonable skill and care is implied by the Supply of Goods and Services Act 1982, Part II, s. 13. The Act does not define the standard to be applied, but the courts have considered the matter in relation to negligence. The extent of liability was described in *Bolam* v. *Friern Hospital Management Committee*:[31]

> Where you get a situation which involves the use of some special skill or competence, then the test as to whether there has been negligence or not … is the standard of the ordinary skilled man exercising and professing to have that special skill. A man need not possess the highest expert skill … it is sufficient if he exercises the ordinary skill of an ordinary competent man exercising that particular art.

Thus where an architect or designer–contractor has an obligation to use reasonable skill and care, he will not have failed if he does what his peers would have done in the same circumstances.[32] A designer is not barred from using new materials or techniques; that would prevent the progress of science in design. However, he must be cautious and make

---

29 *Greaves & Co. Ltd* v. *Baynham Meikle & Ptnrs* [1975] 3 All ER 99, CA.

30 *Independent Broadcasting Authority* v. *EMI Electronics & BICC Construction* (1980) 14 BLR 1.

31 [1957] 1 WLR 582 at 586.

32 For example, in *The London Borough of Merton* v. *Lowe* (1981) 18 BLR 130, the Court of Appeal held the Architect was not negligent when the ceiling was first found to be unsuitable, because it was a new product and he had made the sort of enquiries which a competent architect would be expected to make, although he was negligent for not enquiring further when the ceiling failed.

prudent inquiries. The obligation to use reasonable skill and care is not limited to the substantive design; a designer must also apply the same standard to his choice of what to design, if the choice is his. This would include interpreting his client's requirements in order to decide upon a suitable design.[33]

4.4.3 JCT 05 – Express obligation

Reduction of liability from common law standard: In line with other JCT design and build terms, the liability for Contractor's Designed Portion work is expressly reduced in Clause 2.19.1 to that of an Architect or other appropriate professional designer:

> In sofar as its design is comprised in the Contractor's Proposals and in what the Contractor is to complete in accordance with the Employer's Requirements … the Contractor shall … have the like liability … to the Employer, whether under statute or otherwise, as would an Architect or … other appropriate professional designer holding himself out as competent … (acting independently carrying out design only for a builder to construct).

The design-only architect's liability is the use of reasonable skill and care except to the extent that the design relates to dwellings (see below).

There are two issues that sometimes create misunderstanding amongst contractors and employers who are new to contractors' design duties. The first concerns the liability of an Architect or other professional designer, whose duty is to use reasonable skill and care. Section 4.4.2 above deals with the general position. The duty to use reasonable skill and care is measured by a defendant's peers. So for example, a civil engineer would be judged by the standards of the ordinary competent civil engineer, and a tunnelling engineer by the standards of an ordinary competent tunnelling engineer.[34] It would then follow that a design and build Contractor ought to be judged by the standards of an ordinary competent design and build Contractor. However, instead, the provisions of Clause 2.19.1 apply the standards of an ordinary competent Architect, or other designer doing design only, to a designer–contractor. Under JCT 05, to the extent that he carries out design for CDP Works, the Contractor is not judged by his peers. The Contractor carrying out the design of CDP Works is required to use only the reasonable skill and care of a design-only designer, rather than his higher common law obligation to design work that is reasonably fit for its purpose.

The second issue concerns the term '*or other appropriate designer*'. The Contractor does not need to have actual competence in a particular field of expertise, and nor does he need to claim such competence, to attract liability. A Contractor may be imbued with a skill that he has never expressly claimed to possess, by impliedly holding himself out to have a skill that also could be implied. In the case of *Tharsis Sulphur & Copper Co.* v. *McElroy & Sons*[35] it was said: 'a Contractor who expressly or impliedly undertakes to complete the work … impliedly warrants that he can do so'. In short, the Contractor's act of entering the contract alone impliedly warrants capability.[36]

---

33 See the Court of Appeal in *Platform Funding Ltd* v. *Bank of Scotland plc (formerly Halifax plc)* [2008] EWCA Civ 930, in which a surveyor produced a competent valuation, but of the wrong premises.

34 In *Wimpey Construction UK Ltd* v. *DV Poole* (1984) 2 LLoyd's Rep 499, it was held that the level of skill would be measured against the ordinary level of any special skill claimed.

35 (1878) App Cas 1040.

36 See also *Chaudry* v. *Prabhakah (CA)* [1989] 1 WLR 29, in which a man who assisted his friend to buy a car was held liable when it was found to be seriously defective, on the grounds that he impliedly held himself out to have the skills of a trained mechanic.

The reduction of liability from the designer–builder's common law standard of fitness for purpose applies only to design, and does not reduce the general obligation as regards orkmanship and materials. In summary, materials supplied by the Contractor must be both fit for their general purpose and in accordance with the Contract, and workmanship must be to the standards described in the Contract; design and interpretation of requirement must be carried out using reasonable skill and care of a design-only architect. Clause 2.19.2 expressly applies the liability described in Clause 2.19.1 to work in connection with dwellings. This extension of the limitation on design liability to dwellings seems at first sight to breach the provisions of the Defective Premises Act 1972. Section 1 of the Act provides 'A person taking on work for the provision of a dwelling … owes a duty … to see that the work which he takes on … will be done in a … manner … so that as regards that work the dwelling will be fit for habitation when completed'. However, a professional designer would have a statutory liability to prepare a design fit for the purpose of habitation. Thus, if the contract is for the provision, conversion or enlargement of a dwelling, any design element relating to dwellings in CDP Work will be subject to a fitness for purpose requirement, notwithstanding apparent limitation in Clauses 2.19.1 and 2.19.2.

Express limitation of design liability: Clause 2.19.3 allows the Parties to agree, and insert in the Contract Particulars, a monetary limit to the Contractor's liability for some of the Employer's losses associated with defective design. This provision follows the practice of limited liability applied in many professional appointment contracts. The matters to be limited include only loss of use, loss of profit or other consequential loss. The limit does not affect liability for liquidated damages, and nor does it apply to liability relating to the design of dwellings, which are expressly excluded.

4.4.4 JCT 05 – Liability for Employer's design, and obligation to check

One area of risk often overlooked by Contractors is the tendency to be provided with a partly finished design by the Architect, and then failure to check its integrity. Whilst the specification by the Architect of a performance, or a statement of the employer's requirements, does not necessarily require a concept design to be given to the Contractor, the desire to control the Contractor's contribution is often met by setting parameters within which he must work. This can have the effect of leading the Contractor towards a particular solution, and may result in his preparing a design which has been partly chosen by the Architect. The Contractor faces difficulty in these situations. If the performance given to him is qualified by set parameters, he may be completing an embryo design prepared by others. In the case of *Co-operative Insurance Society Ltd* v. *Henry Boot (Scotland) Ltd and Others*,[37] it was said:

> … someone who undertakes … an obligation to complete a design begun by someone else agrees that the result, however much of the design work was done before the process of completion commenced, will have been prepared with reasonable skill and care. The concept of 'completion' of a design of necessity … involves a need to understand the principles underlying the work done thus far and to form a view as to its sufficiency.

37 1 July 2002, [2002] EWHC 1270 (TCC), [2002] 84 ConLR 164.

The contract in the *Boot* case was JCT 98 with the Contractor's Design Portion, but the principle would also apply if the Contractor were simply finishing off the Architect's design under the guise of work specified by performance, as referred to in Section 4.3 above.

Under JCT 05 provisions for CDP Works, the Contractor has some protection. Clause 2.13.2 relieves the Contractor of both the responsibility for the contents of the Employer's Requirements, and any duty to check the design contained in them.[38] This would include any concept design prepared by or for the Employer for the Contractor to complete, to the extent that it is set out in the Employer's Requirements or in any Variation to the CDP Works.

However, Clause 2.13.2 does not protect the Contractor against errors in Architect's comments made on Contractor's Design Documents. Nor is the Contractor protected to the extent that concept design is given to him through the Bills, in work specified by performance. Following the *Boot* case, the Contractor's duty is then to check what is given to him, and become responsible for its integrity.

---

38 The JCT Guide SBC/G explains that the JCT felt some unease as to the effects of the *Co-operative Insurance* v. *Boot* case so made clear its intention that the Contractor should not be required to check the design given to him.

5

Architect's instructions

5.1 Power to issue instructions, confirmations, etc.

The Architect's actual authority comes from his terms of engagement with his client, which should cover all the duties required of him under a JCT contract. However, irrespective of the Architect's authority under his terms of engagement, the Contractor is entitled to assume that it is the same as that set out in the building contract;[1] if it is not, the Employer will be in breach of his contractual obligation to appoint and maintain a duty-holder to fulfil the duties of the Architect.[2]

The Architect's power or obligation to issue instructions, confirmations giving rise to deemed instructions, directions, consents, requests, and notifications is limited to the type expressly identified in the contract,[3] the most important being as follows.

1. Instructions

- Clause 2.10 (acceptance of errors in levels and setting out);
- Clause 2.12.1 (necessary instructions for the Contractor to complete the Works);
- Clause 2.17.2 (divergences from Statutory Requirements);
- Clause 2.38.1 (schedule of defects delivered to Contractor as an instruction);
- Clause 2.38.2 (making good defects whenever necessary);
- Clause 3.14 (Variations);
- Clause 3.15 (postponement);
- Clause 3.16 (instructions on provisional sums);
- Clause 3.17 and 3.18.4 (opening up and testing);
- Clause 3.18.1 (removal of work not in accordance with contract);
- Clause 3.18.3 (Variations resulting from removal);

1 The use of the RIBA Standard Form of Agreement (SFA) assists compatibility, but it is still possible for the Employer to employ the Architect to carry out only limited duties, even under that form.

2 See *Croudace Ltd* v. *London Borough of* Lambeth (1986) 33 BLR 20, CA, where a Local Authority Employer failed to re-appoint an Architect and attempted to rely on the absence of an Architect to avoid paying the Contractor; see also *Scheldebouw BV* v. *St James Homes (Grosvenor Dock) Ltd*, [2006] BLR 113 in which it was held that an Employer could not appoint himself as certifier.

3 See also Chapter 6, Section 6.4 dealing with Variations.

- Clause 3.19 (carry out Works in proper workmanlike manner);
- Clause 3.21 (exclusion from site of any persons employed thereon);
- Clause 3.23 (antiquities);
- Clause 3.24 (Quantity Surveyor to ascertain loss and expense);
- Clause 4.19.1 (Quantity Surveyor to prepare statement of notional retention);
- Clause 4.23 (Quantity Surveyor to ascertain loss and expense);
- Clause 5.3.1 (Schedule 2 Quotation required);
- Clause 5.9 (deemed instruction for Variation where work affected by Variation);
- Schedule 2, para 4.1 (confirmation that Schedule 2 Quotation not accepted, instruction to proceed with Variation to be valued under Valuation Rules);
- Schedule 2, para 4.2 (confirmation that Schedule 2 Quotation not accepted, instruction that Variation not to be carried out).

2. *Confirmations*

- Clause 3.4 (confirmation of clerk of works' directions);
- Clause 3.12.4 (retrospective confirmation of instruction given otherwise than in writing);
- Clause 3.18.2 (confirmation that defects may remain);
- Schedule 1, para 7 (confirm or withdraw comment on Contractor's Design Document (CDP));
- Schedule 2, para 3.2 (confirmation that Schedule 2 Quotation accepted, instruction to proceed with Variation).

3. *Directions*

- Clause 2.2.2 (directions for integration of CDP design into whole).

4. *Consents*

- Clause 2.3.1 (consent to substitute materials or goods);
- Clause 2.24 (consent to removal of materials);
- Clause 3.7.1 (consent to sub-letting works by Contractor);
- Clause 3.7.2 (consent to sub-letting design by Contractor);
- Clause 3.12.2 (dissent to confirmation of instruction given otherwise than in writing);
- Schedule 5, Part 1, para 2 (authority to use materials other than in accordance with guidelines);
- Schedule 5, Part 2, para 2 (ditto).

5. *Requests*

- Clause 2.3.4 (request for reasonable proof that materials and goods comply);
- Clause 2.20.2 (application requesting dates when information required by Contractor for Contractor's Design Documents);
- Clause 4.23 (request information regarding cause and details of loss and expense);
- Clause 8.7.2.1 (requirement to remove Contractor's plant and buildings, etc.);
- Clause 8.7.2.3 (requirement to assign agreements to supply, to the Employer);
- Schedule 7, para A.7 (requirement for evidence for fluctuations calculation);
- Schedule 7, para B.8 (ditto).

6. Notifications

- Clause 2.16.2 (notification of decision regarding discrepancies in Employer's Requirements);
- Clause 2.28.2 (notification of decision regarding extension of time);
- Clause 2.28.4 (notification of fixing earlier Completion Date);
- Clause 2.28.5 (notification of reviewed fixing of Completion Date);
- Clause 8.4.1 (notice specifying default before Practical Completion);
- Schedule 2, para 2 (comment on Contractor's Design Document).

This diversity of descriptions of communication from the architect creates some difficulty when considering the entitlement (or the obligation) of the Contractor to take action. 'Instructions' are picked out in JCT 05 for special attention, in Clauses 3.10 to 3.21. The issues associated with the giving and receiving of instructions are dealt with separately below (see Sections 5.2 and 5.3 in this chapter), but care needs to be taken also over the status of notifications, requests, consents, confirmations, etc.

Clause 3.12 contains the global requirement that all 'instructions' are to be in writing, and most of the clauses referred to above, dealing with requests or consents, etc., also expressly require communication to be in writing, but a few do not. For example, under Clause 3.4, 'directions' from the clerk of works have to be '*confirmed in writing by the Architect*', but under Schedule 2, paragraph 1, the Architect may simply '*either confirm or withdraw the comment*'. Similarly, under Clause 2.3.1, the Contractor cannot substitute materials without '*the written consent of the Architect*', but under Clause 2.2.2, the Contractor is to '*comply with the directions of the Architect*' regarding integration of Contractor's design into the whole design. It almost goes without saying that, in practice, any communication that is intended to be relied upon should be in writing.

The diverse descriptions of required communications have greater significance when considering the effect of directions or consents given to site personnel. This issue is dealt with below in Section 5.2.3 in this chapter.

5.2 Instructions

5.2.1 Instructions to be in writing

The term 'instruction' is not a defined term, but reference to other types of communication such as confirmations, requests, etc. are distinguished individually where necessary in the Contract, by express qualification to be in writing. Clause 3.12 states all 'instructions' must be in writing. This does not mean that an instruction is only issued validly if it is headed 'Architect's Instruction'. There is no prescribed form for an instruction. Nevertheless, it is in both Parties' interests that wherever possible a form such as the pro forma published by the RIBA should be used, which is immediately identifiable as an instruction. This is particularly important in instances where an instruction gives rise to a Variation, or where it may be unclear as to whether the communication is an instruction, a direction or a consent.[4] However, despite best intentions, instructions are not always reduced to a standard pro

4 For example, permission to sub-let part of the Works under Clause 3.7.1.

forma. The issue of a drawing may be an instruction in writing, and so too may a handwritten note on a drawing. The issue of minutes of a site meeting may contain instructions in writing, providing they are issued by the Architect, although site meeting minutes produced by the Contractor may constitute only confirmation of oral instructions.

An area of uncertainty is in the growing use of e-mail. Clause 1.8 provides that any communications, other than those sent in writing by post or delivered personally,[5] may generally be 'in writing' for the purposes of the contract provided separate specified agreements have been concluded, or to the extent that it is stated in the Contract Particulars. However, certain types of important communication must still be in writing where expressly required in the contract. Thus an Architect's instruction sent by e-mail would be validly given in writing, but a notice of termination under the terms of Clause 8.2 may not.

A note in the Contract Particulars makes clear that in the absence of an entry against Clause 1.8 or a subsequent agreement, all communications must be in writing as required by the Conditions.

5.2.2 Instructions 'other than in writing'

Whilst Clause 3.12 requires all instructions to be in writing, the Contract does not bar the giving of oral instructions. Clause 3.12.2 allows instructions 'otherwise than in writing' to be confirmed in writing by the Contractor, although such instructions are stated to be of no immediate effect. Such instructions may be given under any of the provisions listed above in Section 5.1 in this chapter, but the main source of dissatisfaction amongst contractors tends to be the issuing orally of instructions giving rise to Variations. This is because an oral instruction has an air of immediacy about it, yet it is of no immediate effect so the Contractor acts on it at his risk.

The general procedure for confirming instructions not in writing, which would include oral instructions and perhaps instructions sent electronically, is explained, together with commentary, in the chapter on Variations (see Chapter 6, Section 6.8, Oral Variation instructions).

5.2.3 Instructions given to the Contractor

Throughout JCT 05 power is given to the Architect to issue instructions to the Contractor. Clause 1.7 provides that, subject to specific provisions elsewhere, notices and other documents will be effectively served if given or delivered by pre-paid post to the address stated in the Contract Particulars, or to an agreed address. If no address is agreed service may be to the last known business address, or registered or principal office.

If a communication is sent by email, Clause 1.8 provides that communications may (subject to specific provisions in the contract as to service) be made electronically in accordance with any procedures identified in the Contract Particulars or otherwise agreed in writing. The term 'subject to specific provisions' used here is relevant to important notices where the contract requires a specific form of delivery. Examples are notices of termination and third party rights notices, which under Clauses 8.2.3 and Schedule 5 respectively must be given by actual, special or recorded delivery (see comment in Chapter 1, Section 1.5.2.4 dealing with Contract Particulars entry 1.8).

---

5 This would include facsimile transmission.

Two common difficulties experienced by contractors are the issue of instructions to site personnel, and the issue of instructions direct by the Employer.

Instructions given to site personnel: The Contractor is required by Clause 3.2 to constantly keep on site a person-in-charge, and it is expressly provided that instructions or directions given to that person are deemed to be given to the Contractor. This can create difficulties on some sites, where the contract contains contractor design, or where Schedule 2 Quotation instructions are common, or where the communication given is not an instruction.

Unless the site establishment includes all the necessary disciplines, the person-in-charge at the site may not be the right point of contact. In such cases, the issue of *bona fide* variation instructions can introduce an unnecessary risk of delay in communication. Whilst the Contract does not require instructions to be issued to the site person-in-charge, it does sanction them. Clearly there is a good case for issuing building-related directions direct to site, such as those concerning emergency work; but arguably it is in both Parties' interests to ensure that all variation instructions go direct to the Contractor's project leader, wherever he might be. This is particularly so if the instruction is for a Schedule 2 Quotation, or relates to CDP Works, requiring input from disciplines other than site supervision, within a strict timetable.

Whilst the commonplace receipt of instructions generally on site may create an atmosphere of diligence, it can be a risky practice when communications other than instructions are sent. The Contractor is entitled and obliged to act upon instructions given to the person-in-charge as being given to the Contractor, but JCT 05, deliberately or not, distinguishes instructions from other communications. It is questionable whether consents, authority, directions from the Architect to the person-in-charge are given validly under the Contract, and whether either Party is entitled to rely on the communication. Under Clause 3.2, only instructions from the Architect, or directions from the clerk of works, when given to the person-in-charge, are deemed to be given to the Contractor. That creates the strange position in which a direction from the clerk to the person-in-charge would be given validly, even though under Clause 3.4 a direction from the clerk is of no effect (see below); but a direction given to the person-in-charge by the Architect is not given to the Contractor at all, and arguably, is unenforceable.

Instructions received from Employer (or Employer's representative): There is no provision for the Employer to give instructions personally. Instructing the Contractor is a matter for the Architect. As a party to the Contract the owner may make requests to the Contractor regarding the Works, but he has no right to insist; it is a matter of choice for the Contractor. In any event, compliance with the owner's request would probably result in a separate collateral contract, unless the two Parties agree to vary the contract to incorporate the request.[6] The same applies to the Employer's representative who, if appointed under Clause 3.3, acts as an agent of the Employer. His power is limited to that of the Employer, although it may be further limited by the Employer and notified to the Contractor. However, the Employer cannot, in a notice to the Contractor, give powers to

---

6 See also Chapter 6, Section 6.2 dealing with variations to the contract.

his representative greater than he has himself. For example, the Employer may limit his representative's authority to agreeing Schedule 2 Quotations up to (say) £10,000 in value, but he cannot give his representative authority to instruct a Variation.[7]

Instructions received from clerk of works: The Employer may, if he chooses, engage a clerk of works, whose role is to inspect and to report to the Architect. As a member of the Employer's team resident on site, there is often a temptation for a clerk to give directions to the Contractor. This is often done in an attempt to save time in correcting what he sees as defective work. However, the clerk of works has no power to direct the Contractor. The Contract at Clause 3.4 warns that any directions given by the clerk of works are of no effect, unless it is something that the Architect could instruct and the direction is confirmed by the Architect in writing within 2 days. Clause 3.4 continues to the effect that any such direction confirmed by the Architect is deemed to be an instruction from the Architect. This is a strange provision, and its purpose is obscure, save perhaps to warn the Contractor of an instruction on its way to him. Nevertheless, warning or no, the Contractor is at risk if he acts on a direction before confirmation by the Architect (see also Chapter 2, Section 2.8).

5.3 Compliance and query

5.3.1 Contractor's obligation to comply and entitlement to query

Clause 3.10 requires the Contractor to comply forthwith with any *bona fide* instruction given to him by the Architect (i.e. instructions of the type which the Architect is expressly empowered to issue by the Conditions). There are three express exceptions: two related to Variations and one concerning design co-ordination. The Contractor is not obliged to comply with a Variation regarding the imposition of restrictions, etc. under Clause 5.1.2 if he makes reasonable objection in writing (see Chapter 6, Section 6.5.1); nor is he obliged, or indeed entitled, to start work against a Schedule 2 Quotation until expressly empowered to do so under the Schedule 2 procedures (see Chapter 6, Sections 6.13.2 to 6.13.5). Likewise, he may object if a direction under Clause 2.2.2 regarding integration of his design into the whole, or any instruction, 'injuriously affects' the efficacy of his design.

The Contractor has one other important entitlement to veto an instruction. Clause 3.13 allows the Contractor to request which provision of the Conditions empowers the Architect to issue any specified instruction. If so requested by the Contractor the Architect is obliged to comply 'forthwith'. Whilst there is no express provision for the Contractor to delay compliance with the instruction, the Contractor may put himself at risk if he were to carry out an instruction given by the Architect outside his authority.[8] However, if the Architect delays in notifying the Contractor of the particular provision obliging the

7 Under Schedule 2, para 3, it is the Employer who accepts a quotation and notifies the Contractor; albeit the Architect requests it, and also confirms the acceptance.

8 See Chapter 6, Section 6.5.2, Objection to Purported Variations, regarding Employer's obligation to pay for work ordered by the Architect outside his authority.

Contractor to comply with the instruction, he will put the Employer in breach. Any loss to the Contractor caused by his failure to put the work in hand may then be recovered by the Contractor as damages. One important point here is that the Architect must identify the empowering provision in the 'Conditions'; thus an express power or authority purported to be given under a clause in the Contract Bills (or any other part of the Contract) would not be sufficient, and the Contractor would not be entitled or obliged to comply. However, if the Contractor has any further objection he must invoke one of the dispute resolution procedures. If after receiving a response from the Architect, the Contractor proceeds to comply with the instruction without either Party first commencing one of the dispute resolution procedures, it is deemed under Clause 3.13 to be a valid instruction and enforceable.

The Contract does not deal with the situation where the Contractor queries an instruction because he considers that the design contains an error, or that compliance with an instruction would be dangerous. Where the Contractor perceives, or ought to perceive, that there is an element of danger, he is not only entitled to query an instruction, but he has an obligation to do so.[9] In these circumstances, the Contractor needs to consider carefully whether to proceed if, despite his warning, he is instructed to comply with the queried instruction. A Contractor who carries out work following a design and knowing it to be dangerous to others is likely to be liable in the event of a design failure. It would not be sufficient for the Contractor to claim that he had carried out the design because he had been told to by the Architect. The Contractor's duty to warn generally is dealt with in Chapter 2, Section 2.3.5.

5.3.2 Non-compliance by Contractor

Clause 3.11 provides a remedy for the Employer in the event of the Contractor failing to comply with a valid instruction from the Architect. The Architect may give notice to the Contractor requiring compliance, and if the Contractor fails to comply within 7 days of receiving such a notice, the Employer may employ others to do the work. The Contractor must then give access to the Employer's other contractor, to enable the work to be done. In *Bath and North East Somerset District Council* v. *Mowlem plc*[10] the building owner issued an instruction about remedying defective work, but the contractor did not comply. The contractor then tried to stop a replacement contractor by refusing access to the site, but the building owner was able to obtain an injunction to prevent the contractor's action, and so have the work completed. All the costs associated with engaging others may be deducted from the Contract Sum and thus be recovered through an interim valuation and payment certificate. The use of the word 'costs', limits the Employer's recovery to actual additional cost incurred. This would include fees and the cost of site facilities if the Contractor does not provide them, but excludes losses that might arise as a result of delay. Losses suffered through delay caused by the need to bring in another contractor to do remedial work would normally be recovered from the Contractor through the deduction of liquidated damages.

9 See *Plant Construction plc* v. *Clive Adams Associates and JMH Construction Services Ltd* [2000] BLR 137.
10 20 February 2004, [2004] EWCA 114 (Civ).

6

Variations and provisional sums

6.1 Variations: the general position

In a perfect world a building owner would know exactly what he wanted in advance of detailed design and part construction of his building. Likewise, the designer would produce both the entire conceptual and detailed design before construction work started. There are circumstances when this can be achieved regularly, but they normally relate to projects involving repetition; for example, a standard design fuel filling station, retail store, or fast food restaurant.

However, the nature of construction work and the permanence of the finished product are such that it is rare for the building owner and his professional team to know exactly what the final requirement will be. Even the standard model may need adjustment to meet the peculiarities of the site. Many building owners procuring building work are strangers to the building industry; a new building is often a one-off experience and needs are dictated by circumstances which themselves change with the passage of time. The length of the total construction process enables a building procurer to have second thoughts, to take into account changed fortunes, to fine tune what at the outset were vague ideas; this often happens only as the substantive construction work proceeds, and rightly so if the procurer is to receive the product he needs and wants.

The secret to delivering the required product is flexibility, and whilst the contractor may view changes of mind as disruptive, it is unlikely that building owners as a class would wish to be deprived of the opportunity to develop ideas. Similarly, the designer needs to be able to correct errors in the design, and to translate the procurer's developing requirements into design instructions for the contractor. However, the flexibility expected by the procurer and the designer is not an automatic right.

If the contract for the construction work does not provide for change, flexibility may not be available to the procurer and his designer. In a contract for a specified scope of work, the contractor is bound to provide that scope, no more and no less; he is obliged to produce the work he contracted to do. To do anything else would be a breach of the contract. The contractor is not only obliged to perform the specified work, but he is also entitled to perform it; to prevent him is also a breach of the contract.[1] In short, neither party can change the obligations of the other without the other's agreement. There is no flexibility. In the absence of a Variations clause, there is no term implied in the contract that variations may be ordered.[2] Changes in the work can still be introduced, but only by changing the initial agreement, or by entering into a separate contract; in either case, there is no obligation to enter such an arrangement. This can create difficulties. For example, if the building owner wished to change the colour of the sanitary fittings during the course of the project before they were ordered by the contractor, and the contractor did not want the hassle of changing, the owner would have only two options. He could either accept the colour provided, or he could wait until the work was completed and pay the contractor for the fittings within the contract price. It would then be for the owner to arrange for the new fittings to be replaced, and pay again.

Even if the nature of required changes were such as to enable the parties to enter into a separate contract, the position would be fraught with potential difficulty. Procedures would be duplicated, and work under each contract could materially affect and disrupt performance of the work under each of the other contracts. This would be so, even though the parties were the same.

It is for this reason that most forms of construction contract contain express provisions enabling the procurer to change the obligations of the contractor. Such provisions normally allow wide changes to the scope of work and the circumstances or conditions under which it is to be carried out. However, in setting parameters such provisions also create the limits, albeit wide limits, on the parties' entitlement to introduce change.

6.2 Variations 'to' or 'under' the contract?

Variations to the contract are changes to the contract itself (i.e. changes to the agreement including terms and conditions, or changes to a party's performance which fall outside the types of change provided for in the contract).

Such changes can only be made by consensus between the parties. A typical variation to a contract would be an agreement to accelerate the works where no acceleration provision is made in the contract.[3] Likewise, a late request for the contractor to take on the role of Principal Contractor[4] to comply with waste management legislation, where no such requirement is provided in the contract,[5] would probably need the agreement of the contractor.

Agreements of this type could be formalized, but only by consensus, in an addendum agreement, or, more tidily, in the case of JCT 05, by the use of a Schedule 2 Quotation. The latter has the advantage of acceptance by the Employer, yet it brings the issue under the control of the Architect and gives the Contractor rights to interim payment, extension of time and loss and expense.

---

1 See *Tancred Arrol & Co.* v. *The Steel Company of Scotland Ltd* (1890) 15 App Cas 125.

2 See Waller LJ, *SWI* v. *P&I Data Services*, CA, [2007] BLR 430 at 433.

3 See *John Barker Construction* v. *London Portman Hotels* (1996) 83 BLR 31; (1996) 12 Const LJ 277; [1996] 50 ConLR 43.

4 Under the Site Waste Management Plans Regulations 2008, a Client is required to appoint a contractor as principal contractor to fulfil specified administrative duties.

5 See Chapter 1, Section 1.5.5.3 dealing with amendments to the standard form.

Variations under the contract are changes to the scope of the work or conditions under which the work is carried out to the extent that terms in the contract provide for such changes. The term 'Variation' is often used generically in the construction industry but contracts differ in their definition of the term. Indeed, some contracts use other terms such as 'Change', or 'Alteration'.[6]

Such widespread generic usage of the term 'Variation' sometimes leads to confusion, and often leads both to abuse of the power to vary works, and to unwarranted high expectation of payment amongst contractors. Some of the resulting problems are dealt with later in this chapter.

6.3 Variations under JCT 05:[7] definition

6.3.1 Instructed Variations

Clauses 3.13 and 3.14 are enabling provisions, introducing into the Contract the ability to change the Parties' performance obligations by an instruction. In this Contract such changes, including those in regard to CDP Works, are described as 'Variations' and the extent to which Variations are permitted is expressly set out in Clause 5.1.

Variations are defined in Clause 1.1 by reference to Clause 5.1 as:

a) The alteration, modification of design, quality or quantity of the Works. (This includes the addition, omission or substitution of any work, alteration of materials specifications, and removal of work or materials.)

An important point here is that, although Variations can be wide ranging, Architects may sometimes give instructions that, to the contractor, seem to change the nature of the work; for example, an instruction to build a swimming pool as a Variation under a contract to build a sheltered housing scheme may invite objection from a contractor who only builds houses. In *Blue Circle Industries plc* v. *Holland Dredging Co. (UK) Ltd*[8] an instruction was given under a dredging contract to dump the dredged material to form an island bird sanctuary; it was held to be outside the scope of the original contract and therefore the work had been carried out under a separate arrangement. However, the courts sometimes appear to strain to bring a change within the parameters of the contract. In *McAlpine Humberoak Ltd* v. *McDermott International Inc*,[9] the Court of Appeal were influenced by the purpose of a Variation clause. The contract was for the construction of four pallets (shown on 22 drawings) for an oil rig tension leg. By instruction this was reduced to two pallets (but shown on 161 drawings). The contractor argued that the whole nature of the work had changed, effectively making a new contract. It was held that the contract was still a contract for the construction of pallets, and the purpose of the Variations clause was to deal with such circumstances. Similarly, in *Hallamshire Construction plc* v. *South Holland*

6 See Clause 5.1 of JCT DB 2005, and Clause 51 of ICE *Design and Construct* form 2nd Edition.
7 See also Ndekugri and Rycroft, 'Variations under the JCT Contract' (2002) 18 Const LJ 310.
8 (1987) 37 BLR 40.
9 CA; 5 March 1992; (1992) 58 BLR 1.

District Council,[10] the second phase of refurbishment works were held to be a Variation, albeit separate bills of quantities had been prepared.

b) Imposition, removal or changing by the Employer of obligations or restrictions in regard to site access or use of parts of the site, limitation of working space, working hours, the order in which work is done or finished.

In the interests of flexibility, this entitlement is significant for the Employer. For example, during building work on a factory refurbishment and extension, the Employer may need to change its own working practice to cope with a large order, delays to which could put the Employer in breach of its own supply contract. It is of great benefit to be able to change the sequence of building work on the extension, or to change working hours to free areas of the premises at critical operational times. However, in the absence of terms enabling such changes, it would be a breach of contract by the Employer[11] if he were to prevent the Contractor from working as agreed, or to deprive the Contractor of working areas. In return for the benefit of flexibility, the Employer has to pay. The payment is not compensation; as a Variation it is treated as adjusted sales turnover, thus entitling the Contractor to overhead and profit contributions.

6.3.2 'Deemed' Variations

The majority of Variations result from a change of mind by the Employer, or his team, and are introduced by an instruction from the Architect. In some cases a Variation is not instructed, but nevertheless varied work is sanctioned retrospectively.[12] In both of those situations, the Variation results from the exercise of choice.

A third type of Variation is the 'deemed' Variation. Most 'deemed' Variations result from the operation of the Contract, and are imposed upon the Parties, irrespective of choice. Typical examples are the correction of errors in the Contract Bills, or a change in the construction work brought about by a change in Statutory Regulations. In short, most[13] 'deemed' Variations are changes to the Works or to the Contract Price, that are not chosen by the Architect, but which, for convenience, are treated as though they were instructed as Variations.

The main 'deemed' Variations created by the Contract are:

- Correction of an error in the Contract Bills (Clause 2.14);
- Correction of inadequacy in any design in the Employer's Requirements (Clause 2.14.3);
- Correction of a discrepancy within the Employer's Requirements (Clause 2.16.2);
- Alteration to Contractor's Designed Portion caused by a change in the Statutory Requirements after the Base Date (Clause 2.17.2.1 and 2);

10 [2004] EWHC 8 (TCC), 16 January 2004.

11 In *John Barker Construction* v. *London Portman Hotels*, it was held that the Employer was in breach of an implied non-hindrance clause in a bonus agreement when he instructed Variations under the main contract preventing the Contractor from finishing on time.

12 Sanctioned Variations are dealt with in Section 6.6 in this chapter.

13 The confirmation of the clerk of works' directions is an exception. Since the clerk of works' direction is of no effect, it is difficult to understand why the Architect's confirmation should not simply be an instruction, which may or may not be a Variation.

- Emergency compliance with Statutory Requirements where caused by a divergence between Statutory Requirements and the Contract documents, including instructions and issued drawings (Clause 2.18.3);
- Confirmation by Architect of directions from the clerk of works (Clause 3.4);
- Change in the conditions under which work is carried out, resulting from a Variation to other work (Clause 5.9);
- Emergency Remedial Measures required by the insurer in compliance with the Joint Fire Code, carried out in advance of an instruction (Clause 6.15.2);
- Restoration work including removal of debris following damage, the subject of a claim under a Joint Names Policy covering new or existing buildings Schedule 3 paragraphs B.3.5 and C.4.5.2.

For the purpose of valuation, 'deemed' Variations are subject to the same rules as instructed Variations.

6.4 Architect's power to order Variations

The Architect derives his power from the Contract. If the Contract were silent as to Variations, the Architect would have no power to change the scope of work. In *SWI Ltd* v. *P&I Data Services Ltd*,[14] the Employer required less than the full scope of work to be completed under the lump sum contract. It was held by the Court of Appeal that without some term allowing Variations, the paying party is not allowed to vary the contract by reducing the work to be done.

In JCT 05, the power to vary the Works is in Clause 3.14, which states that the Architect may issue instructions requiring a Variation. The Architect is acting here as an agent of the Employer. The clause gives him authority to change certain of the Parties' obligations under the Contract; but his power is limited to Variations as defined in Clause 5.1. Thus, the Contractor has no obligation to carry out an instruction purporting to be a Variation if the subject matter of the instruction falls outside the definition. An example of an invalid instruction would be one requiring the Contractor to accelerate the works in order to finish before the completion date or the proper extended date; the power to order acceleration is not included in the definition of a Variation.

One area of difficulty concerns the omission of work in order to pass it to another Contractor. However, the Architect cannot unilaterally change the provider of work. The general position is that, whilst work can be omitted (since the Contract provides for it in the definition of a Variation), unless the Contract makes express provision, neither the Employer, nor the Architect on his behalf, can take work away from the Contractor to be given to someone else; to do so is a breach of the Contract. In *Abbey Developments Ltd* v. *PP Brickwork Ltd*,[15] there were complaints about the Contractor's work and the Contractor's employment was terminated. In deciding that the Contractor had been wrongfully dismissed, the Court observed:

> The basic bargain struck between the Employer and Contractor has to be honoured, and an Employer who finds that he has entered into what he might regard as a bad bargain is not

---

14 [2007] BLR 430 [CA].

15 [2003] EWHC 1987 (Technology), 4 July 2003. See also *Carr* v. *J.A. Berriman Pty Ltd* (1953) ConLR 327.

allowed to escape from it by the use of the omissions clause so as to enable it then to try to get a better bargain by having the work done by somebody else at a lower cost once the Contractor is out of the way …

In *Amec Building Ltd* v. *Cadmus Investments Co. Ltd*,[16] a Food Hall was omitted by the Architect, and another Contractor was given the work. The Contractor was awarded damages for loss of profit on the work taken away from him. It made no difference that the value of the Food Hall had been represented by a provisional sum.[17]

The principle applies equally to work covered by a Provisional Sum[18] for defined work.[19]

6.5 Contractor's right of objection

The Contractor's obligation to carry out Variations is not absolute. Clause 3.10.1 provides the Contractor with rights to object to a Variation instruction purportedly given under Clause 5.1.2; those rights are in addition to the right under Clause 3.13 to challenge the Architect's authority to issue a particular instruction.

6.5.1 Objection to *bona fide* Variations

Clause 3.10.1 states the Contractor need not comply with a *bona fide* Variation instruction to the extent that he makes reasonable objection in writing if it adds, omits or varies obligations or restrictions regarding access, working spacc, hours or scqucncc.

The Contract offers no clues as to what a reasonable objection may be. However, it is suggested that the Contractor's personal circumstances may be reasonable grounds. For example, a Variation may affect the sequence on specialist work extending the Contractor's involvement into a period when his specialist resources are already fully committed elsewhere; in other words, stretching his organization beyond its capability. But it may not be a reasonable objection that the Variation would itself be less profitable than the rest of the job, or that the Contractor suspects that the Employer has insufficient funds to meet the extra cost.

Refusal to carry out a *bona fide* Variation instruction without reasonable grounds is a serious breach of contract. It is a risky business for the objecting Contractor; it is a brave Contractor who challenges the Architect's authority, particularly towards the beginning of the job.

6.5.2 Objection to purported Variations

Clause 3.13 gives the Contractor the opportunity to challenge the validity of any purported instruction issued by the Architect (see Chapter 5, Section 5.3). In relation to Variations the Contractor may use the right to challenge an instruction which he considers is outside the definition of a Variation, and thus outside the power and authority of the Architect to instruct.

16 [1997] 51 ConLR 105.
17 See Section 6.9 in this chapter.
18 See Section 6.9 in this chapter.
19 See SMM7, General Rules 10, for distinction between defined and undefined work.

The types of instruction likely to be most commonly challenged are for work that could qualify as a separate contract, and administrative matters outside the apparent authority of the Architect. The latter would include an instruction to provide an invoice prior to an interim valuation, or an instruction to provide a draft final account.

The decision to challenge or accept a Variation instruction when its validity is in doubt should not be taken lightly. If the instruction is carried out, and the Employer subsequently refuses to pay on the grounds that the Architect had exceeded his authority, the Contractor may be left with no remedy. Whilst the Architect, in giving instructions, generally acts as an agent of the Employer, he does so only to the extent that he is authorized. If he acts outside his authority, the Employer is probably not liable. In *Stockport Metropolitan Borough Council* v. *O'Reilly*,[20] Judge Edgar Fay, QC said:

> An Architect's ultra vires acts do not saddle the Employer with liability. The Architect is not the Employer's agent in that respect. He has no authority to vary the Contract … he cannot saddle the Employer with responsibility for them.

That leaves only the Architect in the Contractor's sights. The Architect is not in contract with the Contractor, leaving no direct contractual remedy.[21] The claim could only be in tort, but the Court of Appeal[22] has restricted that route where the parties to the construction contract have made an arbitration agreement. It has been suggested[23] that if the Employer were impecunious the Contractor could get around the Employer to the Architect, but the point is yet to be tested.

6.6 Architect's power to sanction Variations

The Contractor's obligation is to carry out and complete the Works in compliance with the Contract Documents. It is not for the Contractor to provide something different; to do so is a breach of contract.

There is a developing trend in modern bespoke construction contracts for the builder to be brought into the design development process. The concept is carried into some standard form contracts that contain express 'value engineering' provisions[24] to encourage Contractors to propose ideas for change. Surprisingly, the JCT has not taken the opportunity to introduce uninvited Contractor proposals for Variations, so JCT 05 still adopts the basic philosophy that all changes come from the Architect.

However, there may be times when the Contractor is in a position to suggest an alternative to the specification, timing, or method of construction to the benefit of the Employer and Contractor. It may be that the Architect takes the suggestion and converts it into an instruction for a Variation. However, if the Architect is indifferent about the suggestion,

20 [1978] 1 Lloyd's Rep. 595 at 601.

21 The Contract (Rights of Third Parties) Act 1999 may provide rights to the Contractor under the Architect's terms of engagement with the Employer, but identifying such rights is difficult unless the terms are made known to the Contractor.

22 *Pacific Associates Inc.* v. *Baxter* [1989] 2 All ER 159.

23 See Chapter 15, Section 15.9 for futher discussion on this point.

24 See JCT Major Project Construction Contract 2005, Clause 25; GC/WORKS/1 Single stage Design and Build, Condition 40(4).

because it may benefit only the Contractor, it would be understandable if the Architect took the matter no further, requiring the Contractor to comply with the Contract as it stood. That would be the Architect's obligation in the absence of authority from the Employer, since the terms of his own contract with his client usually include administration of the building contract, but not the authority to allow the Contractor to provide something different. The position is dealt with in Clause 3.14.4.

Clause 3.14.4 states that the Architect may sanction in writing any Variation made by the Contractor. There is no requirement here for the Contractor to obtain prior permission, although the prudent Contractor will do so to avoid risk of having to remove work. The main benefit of the clause is to give the Architect the ability to allow retrospectively, a change made without prior permission, without breaching his own terms of engagement, and at the same time avoiding disruptive and perhaps unnecessary correction of the Contractor's work. Clearly, if the Architect sees the change as a financial benefit, such as an omission from the Contract Sum, there is nothing to prevent him from giving an instruction for a Variation in the normal manner; but in so doing he would be transferring risk from the Contractor to the Employer who would be paying less for the Works.[25]

A difficulty arises when the Variation would result in an addition to the Contract Sum. Whilst it may seem strange that the Contractor should be entitled to payment for a Variation which he has introduced without authority, that is the effect of Clause 5.2.1.1, which states 'all Variations required by an instruction of the Architect or subsequently sanctioned by him ... shall, unless otherwise agreed by the Employer and the Contractor, be valued ... (by the Quantity Surveyor)'. If the Architect wishes to sanction a Variation without committing the Employer to payment, he must ensure that he has the authority of the Employer to act as an agent, and then make an agreement in the name of the Employer with the Contractor. The Employer is unlikely, in these circumstances, to withhold authority. However, if the Contractor is unwilling to enter into such an agreement, then the Architect may exercise his discretion to refuse sanction, thereby keeping the Contractor in breach for not complying with the Contract specification.[26]

Occasionally a Contractor will be faced with circumstances, such as site conditions, or gaps in the detailed design, that require work not in the Contract, but for which there is no Architect's instruction. There is a natural temptation to simply get on with the work because it has to be done, in the belief that it is obvious that the Employer should pay for it; but what may seem fair to the Contractor may not be the way that a court views it. The problem was addressed in the Scottish case of *Amec Mining Ltd* v. *The Scottish Coal Company Ltd*[27] in which it was held that if the contractor, under a contract where an instruction was required for a Variation, carried out extra work, he did so at his risk. The Court held that there was no implied obligation to issue an instruction, and suggested that the contractor's option was to not do the work.

---

25 See *Simplex Concrete Piling Ltd* v. *The Mayor and Aldermen and Councillors of the Metropolitan Borough of St Pancras*: (1958) 14 BLR 80, in which a letter assenting to a Variation was held to be an Architect's instruction for a Variation and the Employer was liable to the Contractor.

26 But note the decision in *Howard de Walden Estates Ltd* v. *Costain Management Design Ltd* (1992) 55 BLR 124, in which an architect's instruction letter was qualified to make it at the Contractor's cost; it was held that the variation was at the Contractor's cost because it resulted from the Contractor's defective work; it was said to be the sort of agreement made by sensible parties who wanted to get on with the work.

27 6 August 2003, [2003] ScotCS 223.

6.7 Variations after practical completion

6.7.1 After due date for practical completion

In *Balfour Beatty Building Ltd* v. *Chestermount Properties Ltd*[28] it was held that the Architect is entitled to issue Variation instructions after the date on which the Works should have reached practical completion; this applies whether or not the delay is one for which the Contractor is entitled to an extension of time. In short, so long as the Contractor is still in possession of the site and the Works are incomplete, the Architect can keep him there *ad infinitum*, although the Contractor will be entitled to relevant extensions of time.

6.7.2 After actual practical completion

The process of changing of mind or fine-tuning requirements by the building owner, and correcting defects in the design by the Architect, often continues after the Contractor has achieved practical completion of the Works. The practice of attaching lists of 'snagging items'[29] to the certificate of practical completion often leads to the presence of the Contractor on site after 'actual' practical completion. He is there to finish off minor works, but is often drawn into carrying out extra work, which falls into two broad categories.

The first category is overtly extra work instructed by the Architect taking advantage of the Contractor's continuing presence.

The second category is work done under instructions which have the appearance of requiring the Contractor to put right a defect. For example, a plaster crack could be the result of the Contractor's poor workmanship, or it could be poor design of the building creating excessive vibration when over-specified door closers slam the doors. If it is the latter, the corrective work is an extra.

Whichever of the two categories the additional work falls into, unless the contract expressly provides,[30] the Architect in issuing the instruction at this time acts outside his authority and power. In the first instance, in the case of *Treasure & Son* v. *Dawes*,[31] the judge could find no express time bar in the JCT Contract on the issue of instructions and, surprisingly, concluded that instructions issued after practical completion were enforceable.[32] However, in the Court of Appeal Case of *TFW Printers Ltd* v. *Interserve Project Services Ltd*,[33] Lord Justice Dyson, with detailed reasoning,[34] held that many of the Contractor's obligations, including compliance with Variation instructions, ceased on practical completion. The work is finished (that is, finished within the meaning of the contract), and the final calculations of time and value have started. The Contractor is not then obliged to accept the instruction. He may if he wishes refuse outright, or he may propose a separate contract.

28 (1993) 62 BLR 1.
29 See discussion on Practical Completion in Chapter 3, Section 3.3.
30 See ICE Conditions (7th edn), cl. 51(1) which provides for Variations to be ordered during the Defects Correction Period.
31 [2007] EWHC 2420 (TCC).
32 It is suggested, by the authors, that in the light of the reasoning by Dyson LJ in the Court of Appeal decision in *TFW Printers* (see next), this finding is unsafe.
33 [2006] ABC LR 06/07; [2006] BLR 299, CA.
34 [2006] ABC LR 06/07 at paras 24–34; see judgment of Dyson LJ referring to and agreeing with Keating 7th edition, para 18.142.

Alternatively, he may, by express agreement with the Employer, amend the Contract to incorporate the extras; but in any event he is not obliged to do the work at the Contract rates.

In practice, instructions coming within the second category are often carried out, then disputed later. The danger for the Contractor is that he may carry out such instructions voluntarily, without the protection of a binding agreement to vary the Contract. This may be because the cause is not immediately apparent, or because the list of work is handed to operational supervisors at site who may not be familiar with the detailed provisions of the Contract. It may even be because an agreement to vary the Contract is held to be unenforceable.[35] The Contractor needs to take care, for if an item is later identified as an extra, it will have been the subject of an instruction that the Architect had no power to give, and the Employer may decide not to pay for it.[36]

6.8 Oral Variation instructions

Under Clause 3.12.1 all instructions issued by the Architect must be in writing. Ideally, Clause 3.12 should end there; but unfortunately, it does not. Clauses 3.12.2 to 3.12.4 go on to provide a mandatory system of confirmation, to be initiated by the Contractor in the event that the Architect gives an instruction orally.

Clause 3.12.2 provides:

a) '*(an oral instruction) shall be of no immediate effect*'. The Contractor cannot start to carry out the instruction, even though he could at the time be building work which under the instruction would have to be taken down later. This negates any intended benefit to be derived from an oral (and thus immediate) instruction.
b) '*but the Contractor shall confirm it in writing … within 7 days*'. The Contractor has no choice in the matter, except under provisos in Clause 3.12.4 dealt with below. He must act, although the Contract is silent as to any sanction if he does not. However, the failure to confirm is a breach of the term and could affect the Contractor's entitlement to payment for removal of excessive work caused by late execution of an instruction. Under the provisos, the obligation to confirm is lifted if (i) the Architect himself confirms the instruction within 7 days, or (ii) in the event of the Contractor having complied with the instruction, but not having confirmed it, the Architect himself confirms it before the issue of the Final Certificate. The Architect's confirmation under the latter proviso is not obligatory; the Architect '*may … confirm*' the instruction. The second proviso applies only when the Contractor has complied with the instruction so, if the Architect chooses not to confirm it, the Contractor will have deviated from the scope of work in the Contract. That puts the Contractor in breach unless he corrects his work before practical completion.[37]
c) If the Contractor's confirmation is not dissented from within 7 days after the Architect's receipt, the confirmation is effective; the oral instruction is then as binding from that point as if it had been issued by the Architect.

---

35 See *Treasure & Son Ltd* v. *Dawes* [2007] EWHC 2420 (TCC), where an alleged 'agreement' varying the Contract was held not to have taken place.

36 See Section 6.5.2 in this chapter, Objection to Purported Variations.

37 See Chapter 2, Section 2.4.2.5 concerning the doctrine of Temporary Disconformity.

Unfortunately, the system of confirming instructions other than in writing is open to abuse, and can be treated as an invitation to issue oral instructions. This has the effect of leaving the administration to the Contractor; with it goes the risk of delaying work or carrying out an unauthorized variation potentially in breach of the contract, in the hope that the Architect will confirm it. Clearly, the prudent Contractor will do nothing until the instruction is ratified expressly or by silence; that is both his right and obligation. That being the position, the confirmation provisions in Clause 3.12.2 seem, at best, to be of doubtful benefit, and at worst conducive to poor practice and a spawning ground for dispute. This is particularly so when there are many oral instructions, many of which may impact upon each other.

Some contractors supply their site staff with a pro forma confirmation sheet; it is commonly headed 'Confirmation of Verbal Instruction'. Such contractors should be applauded for their recognition of the procedural requirement, but unfortunately they may suffer from making oral instructions too convenient. A better contractor's pro forma, it is suggested, would be one addressed to himself, headed 'Instruction', for completion by him, but to be signed by the Architect at the time of giving the instruction. The advantages are clear. It avoids the need for retrospective confirmation by the Architect which invites the Contractor's compliance at risk, and it removes the possible 2-week waiting period before the Contractor can be sure of his position.

6.9 Instructions with regard to provisional sums

The Architect is given power and duty in Clause 3.16 to issue instructions with regard to expenditure of provisional sums in the Contract Bills, or in the Employer's Requirements.

There is no machinery for dealing with provisional sums that appear anywhere else in the contract documents. Whilst provisional sums relating to building work are most likely to appear in the Bills, the introduction of provisional sums relating to the Contractor's design in CDP Work (i.e. in the Employer's Requirements) needs care when creating the Contract documentation. The Employer's Requirements set out what the Employer wants, often by way of identifying a problem to be solved, and he may specify an amount to cover an unknown cost. The Contractor responds with his Proposals, but it may be that the design, albeit within his duty, cannot be completed at that stage sufficient to value the work. In those circumstances, the Contractor himself may introduce a provisional sum into his Proposals. An amendment is then required, either to extend Clause 3.16 to include the sums in the Contractor's Proposals, or to transfer the provisional sum into the Employer's Requirements. Failure to make such an amendment creates the risk of dispute over the nature of a purported provisional sum in the Contractor's Proposals, and whether it is no more than a lump sum at the Contractor's risk.

Provisional Sums are sometimes used by Employers to reserve against expenditure by several Contractors (or even the Employer himself) on a project wider than the proposed building Contract. The practice is seen as a simple means of estimating the total cost for the purpose of obtaining funding. The potential difficulty arises when the Provisional Sums are left in the building Contract when it signed. The Architect's power to use Provisional Sums, whether they are in the Bills or in the Employer's Requirements, is limited. As with any other Variation, the Architect may omit work covered by a Provisional Sum if the work is not going to be done or if the risk does not materialize. However, he may not omit such work in order to give the work to someone else, without the Employer running the risk of a claim for breach of contract and damages.[38]

6.10 Variations instruction and Valuation

6.10.1 Introduction

Until recently, the long-standing method in JCT Contracts of instructing and calculating the value of Variations was, in simple terms, for the Architect to instruct, for the Contractor to carry out, and for the Quantity Surveyor then to value what had been done. There was no provision for quotations, and the Architect had no authority to accept a quotation if it were provided. The Contract set out a list of detailed rules, and it was by those rules that a variation had to be valued unless the Employer and Contractor expressly agreed otherwise. The Architect and Quantity Surveyor had no choice in the matter.

In recent years consideration has been given to various growing or apparently habitual practices in the industry; with regard to Variations these include:

- the general preference of many Employers to know the building cost with some certainty, usually by the use of advance prices;
- a tendency for Contractors to measure and value Variations, either for their own purposes or to assist the Quantity Surveyor.

The Employer's need for advance prices is met in Clause 5.3 and Schedule 2; it is an alternative system both of valuation and of administration. Clause 5.3 enables the Architect to require a quotation from the Contractor for acceptance by the Employer. This is carried out under a specific regime of instruction, quotation and acceptance. A quotation, called a 'Schedule 2 Quotation', under this procedure must be accepted before the work described in the instruction is implemented by the Contractor.

As to who values the work, JCT 05 still sets out detailed rules as a basic method of valuation by the Quantity Surveyor; these are to be found in Clauses 5.6 and 5.7. Variations to work where the Contractor designs to meet the requirements of the Employer are valued as a variant of that basic method and are dealt with in Section 6.12 below. The wish of contractors to value variations for themselves was considered, ratified and formalized by the JCT in 1998 by providing alternative evaluation machinery.[39] 'Alternative A – Contractor's Price Statement' allowed the Contractor to value Variations if he wished. 'Alternative B' was the traditional system of valuation by the Quantity Surveyor which applied when 'Alternative A' was not used by the Contractor in respect of any particular variation, or when 'Alternative A' was used but agreement on value could not be reached. However, in drafting JCT 05, the JCT has reverted to the traditional method of valuation by the Quantity Surveyor, and has omitted the Contractor's Price Statement on the grounds of lack of use.

---

38 In *Amec Building Ltd* v. *Cadmus Investments Co. Ltd* [1997] 51 ConLR 105, the Architect omitted a Provisional Sum for a Food Hall and gave the work to another Contractor. The plaintiff Contractor was awarded loss of his anticipated profit as damages.

39 For commentary on 'Alternative A' and 'Alternative B' valuation, see Ndekugri & Rycroft, *The JCT98 Building Contract: Law and Administration*.

The term 'Valuation' wherever used in JCT 05 is defined in Clause 1.1 as a Valuation by the Quantity Surveyor in accordance with the Valuation Rules.

6.10.2 Directions as to choice of method

Clause 5.2.1 directs that the value of all Variations instructed or sanctioned by the Architect including provisional sums, all work treated as though it were a Variation, and work covered by approximate quantities shall be such amount as is agreed by the Employer and the Contractor.

If no amount is agreed by the Employer and Contractor, and unless the Employer and Contractor agree otherwise, valuation is to be made by the Quantity Surveyor. The Quantity Surveyor is required to value in accordance with Clauses 5.6 to 5.10 unless it is dealt with under the Quotation provisions of Clause 5.3 and Schedule 2.

Thus the choice of Variation rules is:

1. an agreement on an amount between Employer and Contractor, or agreement on a method of valuation;
2. a valuation by Quantity Surveyor applying rules in Clauses 5.6 to 5.10;
3. an agreement using Schedule 2 Quotation.

Option 1 is not mandatory, since it only requires failure by either Party to agree, for the default provisions in option 2 to apply. However if option 1 is used it takes precedence. Option 2 applies by default if option 1 is not used and provided a Schedule 2 Quotation has not been accepted. Option 3 is used solely at the discretion and direction of the Architect.

6.11 Variation rules – Agreement by Employer and Contractor

There is no provision in JCT 05 for the Contractor to take it upon himself to supply a quotation for a Variation unilaterally, as a right. However, in Clause 5.2.1 there is provision for the Employer and Contractor to agree an amount. There is no procedure described for reaching that position. The Parties may choose the Schedule 2 Quotation route, but the clause does not specifically limit agreement to that option. It seems that the Employer and Contractor may use whatever route they wish, if they want to agree Variations as lump sums. In order to reach agreement one of the Parties must take the initiative, and that may be a request from the Employer, or a proposal from the Contractor; but neither are duties or rights. It is open to either Party to ignore the other's initiative, and to allow the valuation of the Variation to fall to the Quantity Surveyor.

The Contract does not prescribe what elements should be included in any amount agreed,[40] so to avoid dispute over such matters as loss and expense, and allowance for design, it is essential that the Parties make clear what is included and indeed, if there is any likelihood of confusion, what is not included.

40 Compare with agreement of a Schedule 2 Quotation, which is required to include specified elements including effect on loss and expense.

An important point is that this option is for the Employer and the Contractor, not for the Architect or the Quantity Surveyor, unless they are expressly given powers of agency by the Employer to make valuation agreements with the Contractor, and the Contractor is made aware of their status. Inclusion of the provision allows Employer and Contractor agreements to be effected under the Contract, rather than as ex-contractual arrangements. Clearly, someone must tell the Architect and Quantity Surveyor, although the Contract does not specify who or when.

6.12 Variation rules – Valuation by the Quantity Surveyor

The duty to value the work, in the absence of agreement between the Contractor and Employer to the contrary, lies squarely with the Quantity Surveyor. This is made clear in the Clause 1.1 definition of 'Valuation' and in Clause 5.2 which requires the Quantity Surveyor to make a Valuation in accordance with the Valuation Rules, and to give the Contractor an opportunity to be present if it is necessary to measure work (Clause 5.4). It is not the role of the Contractor to value; albeit the Contractor is entitled to witness measurement on site if he wishes (Clause 5.4), and is obliged to provide information necessary for the calculation of the Final Account (Clause 4.5.1). Too often in the past it has been left to the Contractor to present calculations for checking by the Quantity Surveyor. Contractors probably see the effort as rewarding, driving the timing and content of calculation to their advantage. However, it has often led to disillusionment and dispute when the Quantity Surveyor has ignored the Contractor's submission, or used the Contractor's calculation for interim payment and budgeting purposes, and has then 'done it properly' for the Final Account, reverting to the strict rules.

The Valuation Rules identified in the Clause 1.1 definitions and Clauses 5.6 to 5.10 apply to several types of work set out in Clause 5.2.1:

- variations required or acceded to by the Architect;
- work to be treated as a Variation;
- work covered by Provisional Sums in the Contract Bills, or in the Employer's Requirements;
- work covered by Approximate Quantities in the Contract Bills, or in the Employer's Requirements.

6.12.1 Variations required or acceded to by the Architect including work covered by Approximate Quantities

6.12.1.1 Measured work

Clause 5.6 contains rules for valuation where work, other than CDP Work, can be measured. Clause 5.6.1 sets out a list of criteria to be applied to the description of varied work in order to link Variation prices with the Contract price. Thus, work closely resembling that in the Contract in all respects must be valued at the Bill rates, but the Bill rates become progressively less relevant to any additional work which bears little or no similarity to that described in the Bills. Clause 5.6.2 deals with omitted work. The criteria and consequential rules are:

Clause 5.6.1

1. Work of a similar character, carried out under similar conditions with no significant change of quantity: Bill rates apply.
2. Work of a similar character, but not carried out under similar conditions and/or significant changes in the quantity: Bill rates form the basis of valuation, adjusted to make fair allowance for differences.
3. Work not of a similar character: Bill rates do not apply, and valuation shall be by use of fair rates and prices.
4. Where an Approximate Quantity is a reasonably accurate forecast and provided only the quantity has changed: Bill rates apply.
5. Where an Approximate Quantity is not a reasonably accurate forecast and provided only the quantity has changed: Bill rates form the basis of valuation, adjusted to make fair allowance for difference in quantity.

Clause 5.6.2

6. Omission of work set out in the Bills: Bill rates apply. To the extent that work omitted is not set out in the Bills, the Bills will first need to be corrected in order for the value to be omitted. By this somewhat contrived route the value of the omission is then based on the same rules as for additional or substituted work.

Clause 5.6.3

7. For valuation of measured Variation work, the principles of measurement are the same as those described in Clause 2.13 governing preparation of the Bills.
8. When valuing under Clause 5.6, any percentage or lump sum adjustments in the Bills must be taken into account.[41]
9. With the exception of adjustment of Provisional Sums for defined work,[42] appropriate allowance must be made for adjustment of 'preliminary items'.[43]

6.12.1.2 Work or other matters which cannot be properly measured

Where Variation work cannot be properly valued by measurement, under Clause 5.7 it must be valued by application of daywork rates. Thus, valuation by daywork rates is a conditional method of valuation determined by the Quantity Surveyor's ability to measure; it is not a matter of choice for the Architect. Unlike some contracts,[44] under JCT 05 the Architect does

---

41 Adjustments in the Bills are of particular relevance if valuation requires determination of a fair price.

42 See SMM7, General Rules 10 for distinction between Provisional Sums for defined and undefined work.

43 'Preliminary Items' identified in the relevant Method of Measurement, including for example site establishment costs, standing scaffolding, supervision.

44 See ICE *Conditions Measurement Version* 7th Edition, Clause 52 (6), under which the Engineer can instruct work on a daywork basis.

not have express power to order work to be carried out on a daywork basis, so is deprived of the opportunity to monitor the work as it is being done. Daywork rates are those calculated in accordance with the Royal Institution of Chartered Surveyors' 'Definition of Prime cost of Daywork carried out under a Building Contract' plus the relevant percentage additions stated in the Contract Bills. There is a caveat: detailed work records described as vouchers are required to be delivered for verification to the Architect or his representative not later than the end of the week following that in which the work was carried out. The status of the Contractor's vouchers is dealt with below in Section 6.12.2.6 in this chapter.

Sometimes the change to be valued does not relate to physical work, but to some other change, such as the imposition of working times, or change relating to the site establishment. In those circumstances, the change clearly cannot be measured or recorded on a daywork voucher, and Clause 5.10.1 expressly requires a fair valuation to be made (see Section 6.12.2.5 below).

6.12.1.3 Other work affected

The effect of a Variation on other work, if there is a substantial change in the conditions under which the other work is carried out, is to be treated as though it were itself the subject of a Variation instruction. The term 'substantial change' in Clause 5.9, in the absence of any clues in the Contract, must be viewed objectively by the Quantity Surveyor. In practice, the Contractor is likely to see this as meaning a change in conditions which substantially affects the Contractor's costs, but the measure in the Contract is a substantial change in the conditions, not the costs.[45] There is no formula for deciding what is substantial and the Quantity Surveyor, faced with the unenviable task of forming an objective view, may be unaware of any effect until prompted by the Contractor. Any effect on the other work must be valued under Section 5. This applies to any instruction requiring a Variation, changed Approximate Quantity, or expenditure of a Provisional Sum. However, where such a Provisional Sum is for defined work, it applies only to the extent that the instruction differs from the description for the relevant defined work set out in the Bills.

6.12.2 Some typical problems

Whilst the Variation Rules cover many situations, there are a number of typical recurring problems:

1. Identification of work which can properly be valued by measurement

Contractors will often claim work cannot be measured by reason of a change to conditions under which it is carried out, or that it is of a different character. Clearly, such claims must fail since those are the precise circumstances covered by Clauses 5.6.2 to 5.6.3 for application to measured work. The only logical test can be the application of the measurement rules of the contract (i.e. the Standard Method of Measurement). If the rules can be applied, then the work can properly be measured. In practice, the difficulty often lies in identifiable

45 See Section 6.12.2.2 in this chapter.

parameters. For example, a variation may relate to damage to completed plasterwork. If the boundaries are clear, such as a replacement of a whole wall, then measurement will be possible; but if the damage is in undefined patches, requiring cutting back to firm bonded plaster wherever that might be, the areas may not be measurable in the practical sense.

Quantity Surveyors, on the other hand, will sometimes strain to achieve measurement to avoid resorting to daywork. The result can be measurement by approximation, or the assertion that measurement can be made simply because some form of measurement has been made! It is submitted that the meaning of properly in this context means measurement by the rules, using descriptions categorized in the rules, but applied only where the nature of the work would allow accurate physical measurement.

2. Not of similar character or conditions

Where work is not of a similar character to that described in the Bills, the Bill rate does not apply and a fair valuation must be made. In practice, the concept of similar or dissimilar character seems to cause little difficulty, since detailed specification including performance determines character.

What comprises similar conditions is more likely to create problems. The conditions under which the work is carried out includes such factors as difficulty of access to the work face, distance from stores, difference in height, whether work is carried out in natural summer light or artificial light in winter. In *Wates Construction* v. *Brodero* Fleet,[46] it was held that such conditions are those to be derived from the express provisions of contract documents, and do not include the constructive knowledge and expectations of the parties gained during pre-contract negotiations. The words 'executed under similar conditions' were considered in the case of *Henry Boot Construction Ltd* v. *Alstom Combined Cycles Ltd*[47] in which it was said that they do not refer to economic or financial conditions:

> Intrinsic profitability or otherwise of the rate or price is not … a relevant consideration to be taken into account …

3. Bill Rates apply to measured Variations – even when relevant rates are commercially low or high

A frequent cause of dispute is the rate in the Bill which the Contractor realizes during the contract, was entered in error. Sometimes the rate is patently inadequate; sometimes it is simply a bit high or low.

Contractors frequently argue that they are entitled to have a low rate corrected, and that the Architect is not entitled to order greater quantities than those in the Bills in order to take advantage of a bargain, albeit the error is discovered after conclusion of the contract. Quantity Surveyors will sometimes take pity on the Contractor by valuing increased quantities using a fair rate, at the same time reminding the Contractor that he is obliged to bear the loss for which he contracted (i.e. the quantity in the Bills at the Bill rate).

The problem was considered by the Court of Appeal in *The Mayor Aldermen and Burgesses of the Borough of Dudley* v. *Parsons and Morrin Ltd*.[48] The Contractor had

---

46 (1993) 63 BLR 128.
47 [1999] BLR 123.
48 8 April 1959, CA unreported; *A Building Contract Casebook*, 2nd Edn, 1990, p. 54; see also 'Building and Civil Engineering Claims', 1984, cited by Wood, R.D., at p. 527.

inserted in the Tender Bill a rate of 2 shillings per cubic yard extra-over for excavation in rock. A fair price was £2. The Contractor had been allowed a fair rate by the Architect for all the quantity over the provisional quantity in the Bill. The Contractor sought the whole quantity at a fair rate. Lord Justice Pearce said:

> Naturally one sympathises with the Contractor in the circumstances, but one must assume that he chose to take the risk of greatly under-pricing an item which might not arise, whereby he lowered the tender by £1,425. He may well have thought it worthwhile to take that risk in order to increase his chances of securing the contract.

Clearly, the Court was influenced in this case by the provisional nature of the quantities in the Bill, but the outcome was that the Bill rates were held to apply to the total quantity. The whole principle of using pricing levels in the Contract for Variations would be undermined if it were otherwise. Once the Contract is concluded, and provided the Employer was not aware of, and taking advantage of, a patent error in accepting the erroneous tender, the rate is accepted at risk. The risk is borne by both parties.

Consequently a Bill rate which is erroneously high must also stand. This point was confirmed by the Court of Appeal in the case of *Henry Boot Construction Ltd* v. *Alstom Combined Cycles Ltd*[49] where it was held that the parties had agreed the contract rates, and it was immaterial whether they appeared high or low. It is suggested that the Contractor would have no right of objection if the Architect omitted quantities in order to save money for the Employer, provided the work was not going to be done, and the Architect was not simply omitting high rates in order to get a better bargain elsewhere.[50]

4. Valuation based on Bill Rates, adjusted to make fair allowance for differences

This is often misread by Quantity Surveyors and Contractors alike to mean 'valuation shall be *pro rata* the Bill rates'. It may be that *pro rata* has a part to play in some instances where a factor such as an increase in thickness may be proportionate for the materials part of a rate; but it is rare for a change in thickness, height, colour, etc. to affect the cost of work entirely proportionately. Only when a new description falls squarely between two Bill rates is there any excuse for a simple *pro rata* calculation, and even then care should be taken to ensure that the element of difficulty does not in reality increase or decrease exponentially.

5. Valuation by Fair Rates and Prices

The concept of 'fair rates' can be confusing. Normal market rates can be 'fair rates'; so too, depending on the context, can 'not unreasonable costs excluding profit'.[51] The context in Clause 5.6.3 suggests equal fairness to both Parties. Quantity Surveyors sometimes see this as meaning the best price the Employer could have obtained by approaching local firms, or the price from a pricing book; whereas Contractors sometimes see it as being all their actual costs plus an arbitrary percentage for overheads and profit. However, fairness,

---

49 (1999) BLR 123; [2000] BLR 247, CA.

50 In *Abbey Developments Ltd* v. *PP Brickwork Ltd* [2003] EWHC 1987 (Technology), 4 July 2003, (2003) CILL 2033 it was held that omitting work to obtain a better bargain elsewhere was a breach of Contract.

51 *Semco Salvage & Marine Pte Ltd* v. *Lancer Navigation Co. Ltd*: HL; 6 February 1997.

for the purpose of Clause 5.6.3, is tinged with subjectivity. What is to be considered here is the fair price to be paid to this Contractor (not to some other Contractor) by this Employer under this Contract (not some other Contract). In short, the Contractor's actual net costs must be taken into account, provided he did not waste costs. Whilst costs are no more than a factor to be taken into account, along with many other factors such as efficiency and learning curves, the costs will indicate a maximum. In *Henry Boot Construction Ltd* v. *Alstom Combined Cycles Ltd* it was said:[52]

> A fair valuation … generally means a valuation which will not give the Contractor more than his actual costs reasonably and necessarily incurred plus similar allowance for overheads and profit.

As with costs, the profit margins built into the Tender must be included if they can be identified; but so too must any actual commercial concessions which make up the market level of the Tender, although likely profitability or unprofitability should be ignored. In *Weldon Plant Ltd* v. *The Commission for the New Towns*[53] it was held that a fair price must include something in respect of each element normally found in a contract rate, and that allowance for overheads was a normal inclusion, even though in this case there was no indication of such allowance elsewhere in the tender. It was not necessary to prove that cost was incurred in respect of general overheads since it could be assumed that overheads would be attracted, but specific time related overheads may require substantiation. An example would be the possible need for additional monitoring staff if, say, a Variation were to be instructed requiring the removal of large quantities of dangerous waste, where previously there was anticipated to be little waste to be managed.

Contractors sometimes object to the Quantity Surveyor's request to see the estimator's Tender workings on the grounds that they are confidential; but without them the Quantity Surveyor is unable to fulfil his duty to properly value the Variation. In those circumstances, he is entitled to, and he must, simply do his best.

6. The status of daywork vouchers

The daywork voucher supplied by the Contractor under Clause 5.7 is no more than evidence of the time, materials and plant expended on the described task. The existence of a voucher, whether signed or not, does not give rise to entitlement in principle for work to be valued on a daywork basis. Entitlement follows the test of whether work can be measured, and it is commonly left to the Contractor to guess whether the Quantity Surveyor will be able to properly measure the work. This can lead to a glut of vouchers as the Contractor plays safe and submits a voucher for anything that he considers to be out of the ordinary. The result is often a pile of unsigned vouchers when the Architect resists signing for work he considers is not a Variation, or is measurable, or was done when he was not present.

This last reason is one frequently faced by Contractors, even in circumstances where the work is clearly not measurable. The Contractor should not worry – he is entitled to rely on his vouchers whether signed or not. The purpose of giving the Architect (or, as is often the case in practice, the clerk of works) an opportunity to verify a voucher is

52 [1999] BLR 123 at 137.
53 14 July 2000, TCC, [2000] BLR 494.

to protect the Employer's interest. The Architect who avoids signing vouchers on the grounds of absence from site runs the risk of putting himself in breach of his contract with the Employer.[54] Unsigned sheets were taken as reliable and accurate evidence, in the absence of anything to the contrary, in *JDM Accord Ltd* v. *The Secretary of State for the Environment, Food and Rural Affairs*.[55] A timesheet was supposed to be signed each day by a DEFRA representative, for work in disposing of carcasses during an outbreak of foot and mouth disease. Unfortunately, the outbreak was so severe that there were insufficient DEFRA representatives to discharge the duty. It was held that the failure to sign sheets was a breach of contract by DEFRA. It would then be wrong for DEFRA to benefit from its breach, by being allowed to challenge the sheets as though the representatives had carried out their duties at the proper time. The sheets were accepted as reliable generally, and open to query only in respect of obvious mistakes such as arithmetical errors.

A view sometimes held by architects and quantity surveyors is that, before signing a daywork voucher, there is a right or even a duty to correct the contents of the voucher to what they consider is a fair amount. For example, a voucher may show 10 hours for labour that should, in the Architect's opinion, have been completed easily within 3 hours. However, it is not for the Architect, or later the Quantity Surveyor, to change the hours or any other entry on the voucher, unless the entry is not a correct representation of what actually occurred. This point was considered by the Court of Appeal,[56] in which it was said, after the trial judge had gone behind daywork timesheets:

> [The trial judge was] … wrong to go behind the timesheets. The timesheets were not suggested to be fake. … The most that could be said … was that perhaps the workmen did not work as expeditiously as they might have done. That is the danger of day-work contracts. It is a danger which is often dealt with by the Architect, making sure that the men are on site and working.

Clearly, the Quantity Surveyor's duty in these circumstances (if daywork is the appropriate method of valuation) is to apply the Contract daywork rates to the hours on the voucher, provided it is an accurate record of the time taken. This will be so even if he considers the hours are unreasonably high. If the hours recorded are not a true record of the time expended, then the Contractor may be guilty of a criminal offence under the Theft Act 1968.[57]

6.12.3 Work covered by Provisional Sums in the Contract Bills

A Provisional Sum is simply an amount of money established by the Employer, which he requires the Contractor to include in his Tender to cover the cost of a designated risk. The risk often relates to a section of work, such as a gatehouse or excavation in rock, which could not be measured or described in the Bills for want of design detail or extent. When an item is sufficiently designed as to be able to provide such details as location, construction, quantity and extent it can be included as a 'Provisional Sum for defined work'

54 See *Chartered Quantity Surveyor*, September 1986, 'Questions & Answers', p. 15 for correspondence on this point.

55 16 January 2004, [2004] EWHC 2 (TCC).

56 *Clusky (t/a Damian Construction)* v. *Chamberlain*, CA: 24 November 1994 unreported.

57 S.15(1) – Obtaining property by deception, s.17(1) – False accounting.

in compliance with the Standard Method of Measurement.[58] However, when a risk is unidentifiable – such as provision to cover contingencies, or the unquantifiable risk of finding rock in the excavations – a sum may be included as a reserve in the form of a 'Provisional Sum for undefined work'.[59] Save for affecting allowance in respect of time-related costs, adjustment of the two types of Provisional Sum is the same. When the work or the risk can be properly valued, the sum is omitted and replaced by the proper calculated value. The method of calculation is the same in principle as for any other Variation.

When pre-contract preparation time is short or budgets are tight, there may be a temptation to include a disproportionate number of Provisional Sums in the Contract Bills, although overall neither Party gains. Whilst the practice may not be in the spirit of the 'With Quantities' contract, there is little by way of sanction. The Contractor could argue that where the work could have been measured properly, the Bills are not measured in accordance with the Standard Method of Measurement Rule 10, in breach of Clause 2.13.1. However, the remedy would simply be a proper valuation, which is provided for in any event in the Valuation Rules. For the Employer, the benefits are illusory. The sums included are not maxima, so there is uncertainty over the final price. The work will still need to be measured in detail, at the Employer's expense, and a sparsity of rated items in the Bills, makes it more difficult for the Quantity Surveyor to value the work.

Likewise, Employers need to take care that Provisional Sums are not included in the Bills as a general reserve for doing specified work, with the intention of omitting them in order to fund others to do that work. Even without intention when entering into a Contract, if a sum is omitted in order for another contractor to carry out the work, it will be a breach of contract if done without the contractor's consent (see also Sections 6.4 and 6.9 in this chapter).

6.12.4 Variations in respect of work specified by performance

JCT 05 contains no express provision for 'Performance Specified Work'. However, that does not prevent the Contract Bills including work that the Contractor must design to meet a specified performance. It simply means that there are no rules governing valuation, other than the general rules in the Conditions and the rights and obligations of common law.

Amendment 12 of JCT 80 introduced the concept of 'Performance Specified Work', including provision for valuation and liability, and machinery for dealing with discrepancies. The JCT has omitted the provision in JCT 05 on the grounds that the Performance Specified Work option was rarely used and, where it was used, the Contractor's Design Portion would be more appropriate. Nevertheless, it is common for Bills to include for design by the Contractor, without complying with the provisions for its incorporation.[60] On those grounds, Variations to work specified by performance is worthy of mention here.

Despite the original general 'build only' nature of JCT 05, there are circumstances where a Contractor's specialist building knowledge includes design. Frequently the Contractor supplies a specialist product, often by use of a sub-contractor, which requires detailed design knowledge of the product. A typical example would be a uPVC window

58 See SMM 7, paragraph 10.3, for definition of Provisional Sum for defined work.
59 See SMM 7, paragraph 10.5, for definition of Provisional Sum for undefined work.
60 See Chapter 4.

system designer who decides what profile of extrusion is necessary for the circumstances. If a Variation is required which affects the influences on design (e.g. a requirement for windows to match those on the ground floor, but instead to be fixed at the twentieth floor), then the best person to determine the detailed design and suitability is the system designer. In these circumstances valuation by prescribed rules is inadequate, and open to abuse when the Parties attempt to make it fit.

The concept, in general terms, is that the Employer requires a specified result. It is for the Contractor to achieve that result, and how he achieves it is his choice, limited only by any parameters set out in the Contract Bills. The JCT 98 provisions for Performance Specified Work limited the Contractor's extreme common law obligations and passed risk of the interface with the Architect's design to the Employer. Perhaps that is why the provisions were seldom used.

In so far as an Architect's instruction requires a Variation, identification and valuation is governed by the general rules of Clauses 5.1 to 5.7 and 5.9 to 5.10.

However, where performance specification is incorporated in JCT 05 the Contractor needs to be wary. The general principles are more akin to those of 'design and build', and misunderstanding as to what is, or is not, a Variation is commonplace.

Some examples of typical problems frequently encountered are described in Section 6.12.6 in this chapter below.

6.12.5 Variations in respect of Contractor's Designed Portion

The use of the Contractor's Designed Portion generally is dealt with in Chapter 4.

The rules for valuation are set out in Clause 5.8:

1. The Valuation must include for addition or omission of relevant design related work such as preparation of drawings. There is no right or wrong way to value design, save that valuation must be in accordance with the rules in Clause 5.8. Problems associated with valuation of design are dealt with below (see Section 6.12.6.4 in this chapter).
2. The Valuation 'shall be consistent with values of work of a similar character set out in the CDP Analysis'. The CDP Analysis (see Chapter 4) is used here like a schedule of rates, or if in insufficient detail, to set the market level of the price for this work.
3. Where values of work of a similar character to those in the CDP Analysis are used, due allowance must be made for '*any changes in the conditions under which the work is carried out and/or any significant change in the quantity*'. This is similar to the rules for measured work generally (see Section 6.12.1 to 6.12.3 in this chapter above).
4. Where the CDP Analysis contains no work of a similar character, the valuation must be a fair valuation. The principles applying to fair valuation of CDP Variations are the same as those described in Section 6.12.2.5 in this chapter.
5. Valuation of omission of work shall be in accordance with the values, if identifiable, in the CDP Analysis; otherwise a fair value must be omitted. There is often difficulty in valuing omissions, particularly if the relevant work is not adequately described or priced in detail. The Quantity Surveyor may then be left with the task of determining a fair value for work which will not be done and for which detailed design is not available. Although this may not cause difficulty if the variation is simply one

of quantity, in an instance such as that in the uPVC windows example cited above,[61] there may be need for a new design before the original is prepared. If the new design appears to involve no more than a change in extrusion profile, it is tempting to apply some form of *pro rata* calculation, whereas the changes may in fact be complicated, involving a different manufacturing technique. For this reason, Variations involving specification change in contractor-designed work should always prompt the Quantity Surveyor to ask the Contractor for a price analysis from the system designer.

6. Allowance must be made for any lump sum or percentage adjustments in the Bills which may affect the level of values in the CDP Analysis.
7. The valuation must include allowance for any adjustment necessary in preliminary items. Preliminary items are identified as those referred to in the Standard Method of Measurement.
8. Where the basis of a fair valuation is daywork, the provisions for daywork in Clause 5.7 apply.
9. If a Variation instruction regarding CDP Work, including an instruction in respect of a Provisional Sum in the Employer's Requirements, affects other work by changing the conditions under which that other work is carried out, such other work is to be treated as though it were itself the subject of a Variation. The appropriate valuation rules are those applying to the other work affected, thus other CDP Work is valued by the rules in Clause 5.8.

The rules described above are more or less the same as those for architect-designed work; the only material difference is that the pricing document for CDP Work is the CDP Analysis, rather than the Bills.

Some examples of typical problems frequently encountered are described in Section 6.12.6 below.

6.12.6 Variations in respect of contractor-designed work: a few typical problems

Difficulties in identification and valuation of work designed by the Contractor are often problems normally associated with the principles applying to 'design and build' and lump sum contracts, albeit there may be a unit rate for the work in the Contract Bills in the case of Performance Specified Work, or the CDP Analysis in the case of Contractor's Designed Portion work.

1. Variations introduced by the Contractor

The Contractor's general obligation under JCT 05 is to build what he is told to build by the Architect. When the Architect's instructions relate to contractor-designed work, the Contractor has some discretion in how the performance will be achieved. The amount of discretion will depend on the parameters set out in the description of the work required, either in the Bills, or Employer's Requirements and the Contractor's Proposals describing how he intends to achieve it. Such discretion is not only an entitlement; it is also an obligation.

61 Section 6.12.4 in this chapter.

Many Contractors, particularly those new to providing a specified performance or CDP Work wrongly assume when there is a bill of quantities that they are entitled to be paid for any change to the work. However, there are a number of situations in which the Contractor will be required to carry out work at his expense whether or not it is expressly referred to in any of the Contract or working documents. Such work is implied either by terms in the contract, or by common law.

A common example of necessary work implied by terms is the obligation to comply with statutory requirements, including building regulations. This type of work seems to create few problems; maybe this is because Contractors are generally familiar with regulatory control, and the extent and nature of the work is defined and foreseeable. However, work implied by common law is more vague, although the principle behind the concept is far from vague. In *Tharsis Sulphur and Copper Co.* v. *McElroy & Sons*,[62] it was said:

> A Contractor who expressly or impliedly undertakes to complete the work ... impliedly warrants that he can do so... In consequence any additional work necessary to achieve completion must be carried out by him at his own expense if he is to discharge his liability under the contract.

Contractors will often claim extra for work which, although necessary, is not expressly identified in the CDP Documents. The acid test of whether or not particular work must be done is often no more stringent than the simple question 'Will the installation operate as required, or meet the specified performance without it?' If the answer is 'No', the work is necessary, and must be performed without any adjustment of price.

One of the disadvantages for the Employer of introducing Performance Specified Work into JCT 05, is that there is no machinery for dealing with the Contractor's proposed method for meeting the required performance. Provided the Contractor meets the stated performance within any specified parameters described in the Bills, he discharges his obligation, and it is not for the Employer to dictate further how he does so. In those circumstances, the Contractor has wide choice.

2. Work omitted by the Contractor

The Contractor may wish to leave out work which he considers unnecessary.

The general position is that the Contractor is in breach of the Contract if he provides something other than that which is described in the Contract Documents. Despite the Architect's power in Clause 3.14.4 to sanction a Variation, the Contractor does not have an unfettered right to vary the Works. However, provided the relevant work forms part of the Works within his discretion and is not outside the parameters limiting that discretion, it is submitted he can omit work, on the principles described in (1) above applying to necessary work. The extent to which the Contractor is limited in his choice depends entirely on the amount of detail in the Bill description, or in the case of CDP Work, in the CDP Documents.

3. Performance which the Contractor cannot achieve

The rapid development of technology in recent years has increased the risk of contracting to do the impossible. This applies particularly in the field of electronics, where the impossible today may be commonplace in the near future. Construction contracts often include items such as the supply of computerized components within bespoke equipment

62 (1878) 3 App Cas 1040.

designed by a domestic sub-contractor. The task is sometimes to solve a problem using or developing knowledge at the frontiers of science; but there are times when technological advance does not match expectation. There are occasions when the Contractor cannot achieve the performance required, but unless the Contract provides otherwise,[63] impossibility is no excuse.

When the Contractor enters into the Contract and provides his Contractor's Proposals for CDP Work, he impliedly warrants that he is capable of doing what he has said he will do. If he fails he is in breach of the Contract, and the Employer is entitled to damages.[64]

The difficulty in this situation is in identifying whether the work is merely outside the capability of the Contractor, or really is impossible. If it is the former, the Employer would be entitled to go elsewhere by giving notice under Clauses 3.10 and 3.11 and employing others. However, if it is the latter, the work cannot be done and another solution must be found. This will probably be by way of an instruction varying the Works, but the Contractor is still in breach and the Employer is still entitled to damages. The damages are likely to be the extra cost incurred by the Employer in achieving the required performance by other means, or the extra cost in finding a satisfactory alternative plus some allowance for disappointment.

4. Valuation of design

Clause 5.8.1 provides for 'allowance' to be paid for the 'addition or omission of ... (design work)'. There are several ways of valuing design effort. The principal methods used in practice are by lump sum, a *quantum meruit*, and by percentage of the building cost. It is for the Parties to ensure an adequate means of calculation is incorporated in the Contract, sometimes by setting out the basis in the Contract Bills but normally by insertion in the CDP Analysis.

The difficulty in agreeing a satisfactory basis for 'allowance' lies in the nature of design work. In the case of a Variation requiring innovative design the resource expenditure may be considerable, yet the building cost may not vary. Likewise the building cost may be reduced, by omission or variation, but the cost of achieving that omission may involve positive design costs. The Contractor who agrees a percentage design allowance runs the risk of a shortfall.

However, under Clause 5.8.1 it is clear that the Contractor is entitled to allowance for the varied design work carried out. This is not the same as design allowance on the varied building work. In other words, allowance must be made for greater or less design work irrespective of the effect on the building cost. The variation to building work is no more than a trigger to the Contractor's entitlement. The method of valuation which seems to meet the requirements of the Contract best is some form of *quantum meruit*, either based on proven actual cost, or on an agreed labour rate applied to time records. In either case the Contractor may have difficulty establishing the exact nature of work being carried out, and it is not unusual for Contractors to under-recover on design costs.

To the extent that Performance Specified Work is included in the Bills, there is no express provision to allow for design, and a Quantity Surveyor's valuation will be based solely on the rates in the Bills, or the general rules applying to Architect-designed Variations.

---

63 See ICE *Conditions* 7th Edition, Clause 13.1,

64 The case of *Co-operative Insurance Society Ltd* v. *Henry Boot (Scotland) Ltd* and others, [2002] EWHC 1270 (TCC); 1 July 2002, provides a striking example in which the Contractor was said to have an obligation to complete a design, and had impliedly warranted that he could do so, but was held liable because he failed to check a design which was incapable of being completed without correction.

5. Conflict between Employer's Requirements and Contractor's Proposals

JCT 05 contains only limited provision for dealing with a situation where the Employer has asked for one thing in his Requirements, and the Contractor has offered something different in his Proposals.

If the difference results from an error or an inadequacy in the Employer's Requirements, the position is clear. Clause 2.14.2 requires the Employer's Requirements to be corrected if the issue is not already dealt with in the Contractor's Proposals. In those circumstances, if an error in the Requirements is corrected by the Contractor's Proposals, the Proposals will take precedence. The logic behind this is simply that the Contractor is expressly not responsible for the contents of the Employer's Requirements, or for checking any design contained within them (Clause 2.13.2). Likewise, where the Contractor's Proposals correct a discrepancy within the Employer's Requirements, the Proposals will prevail (Clause 2.16.2). However, that position only applies to the correction of errors, inadequacies and discrepancies.

The situation is more difficult when the Employer's Requirements state what the Employer wants, and the Contractor includes something different in his tender, without identifying the divergence or offering it as an option.

When any Contractor submits a tender containing an element of design, which he has prepared in response to a request from his potential client, commonsense suggests that it ought to reflect what was asked for. If a factory owner commissioning an extension to his premises asks for a new warehouse in which he will store containers of his product, he would not expect to receive an inadequate racking system, even if the racking design supplied were the one specified in the Contractor's tender. Some design-build Contractors do not see it that way – particularly those who incorporate design infrequently – and the question then arises as to which takes precedence, the Employer's Requirements, or the Contractor's Proposals.

The Contract does not deal with the issue, and neither does the Guide. The JCT leave it to the Parties to ensure that the two documents are compatible. It is often suggested by Contractors in this situation that the Proposals should take precedence, since they are the later document, and the Employer has accepted them. Unfortunately, the Employer has done nothing of the sort. The Tenth Recital states that the Employer has examined the Proposals, and is satisfied that they '*appear to meet*' the Requirements. In short, the Employer has not spotted anything wrong, but that does not mean that the Proposals do provide what was asked for in the Requirements. In the example of the racking system, it would probably require extensive checking to discover that the system design was inadequate. At this point it is worth noting that the Employer, whether or not he has design expertise and resources, does not wish for whatever reason to prepare the design himself; that is why he has engaged the Contractor to do the design.

Under several other forms for design and build,[65] the Contract provides that the Employer's Requirements override the Contractor's Proposals, and that is probably the general position in any event. One point in support of that view is in the extent of the Contractor's duty as a designer. Before substantive or detailed design can start, the designer must first interpret what it is that the Employer wants. He must do so using the same skill and care that an independent architect or other designer would use in the same

---

65 See JCT Major Project Construction Contract, clause 10.4; ICE Design and Build Contract clause 5(1)(b).

circumstances. In *Platform Funding Ltd* v. *Bank of Scotland (Formerly Halifax)*,[66] the Court of Appeal was required to consider the liability of a surveyor who had carried out a survey agreed by the Parties to be done competently, but on the wrong house. Sir Anthony Clarke MR said: "...It is surely to be expected that the valuer would owe the same duty in respect of the location of the property as in respect of its inspection...". In the racking example, the Contractor may have failed to use reasonable skill and care in the interpretation of what was needed, resulting in his producing a design for a good system, but unfortunately not the system needed to meet the particular purpose required by the Employer.

There is one situation, other than dealing with errors, discrepancies and inadequacies, in which it could be argued that the Proposals should take precedence. That is where the difference between the Requirements and the Proposals is so obvious that anyone who claimed to have examined the documents ought to have spotted it, even under the most cursory of examinations. For example, if the client asked for a building with a flat roof, and the Contractor offered a building with a steeply pitched roof appearing on all the drawings, it would at least be arguable that the Employer, having examined the Proposals, was happy with what was offered.

In practice, this whole topic is less common than it was after the JCT first published their Design and Build Contract in 1981, mainly because frequent users of the form amend the conditions to make clear which of the Employer's Requirements or Contractor's Proposals prevail.

6.13 Variation rules – quotations

6.13.1 Schedule 2 Quotation – overview

Clause 5.3 provides an alternative variation system for use by the Architect, the aim being to allow the Parties to agree the value of a Variation including effect on time, before the work is executed, and without allowing the Contractor to hold the Employer to ransom over the price. The Architect may, if he wishes, instruct provision of a quotation, applying the rules set out in Schedule 2 of the Contract (see Figure 6.1).

The concept of the Schedule 2 Quotation is admirable. The Employer knows his expenditure with certainty, thus avoiding the need to lock up reserves needlessly; likewise the Contractor knows what his income (turnover) will be, enabling him to plan his financial year. Agreement on Variation value before the work is done is encouraged by The Society of Construction Law:[67]

> Where practicable, the total likely effect of variations should be pre-agreed between the Employer/CA and the Contractor, to arrive at if possible, a fixed price of a variation, to include not only the direct costs ... but also the time-related costs ...

Such an arrangement is common in bespoke contracts, particularly where the Contractor is also the designer but, until its introduction as a supplementary clause in the JCT With Contractor's Design form 1981,[68] provision for quotations was not a feature of many of the

66 [2008] EWCA Civ 930; see para 67.
67 Delay and Disruption Protocol, October 2002, para 1.7.1.
68 See JCT With Contractor's Design Form WCD 81 Clause S6.

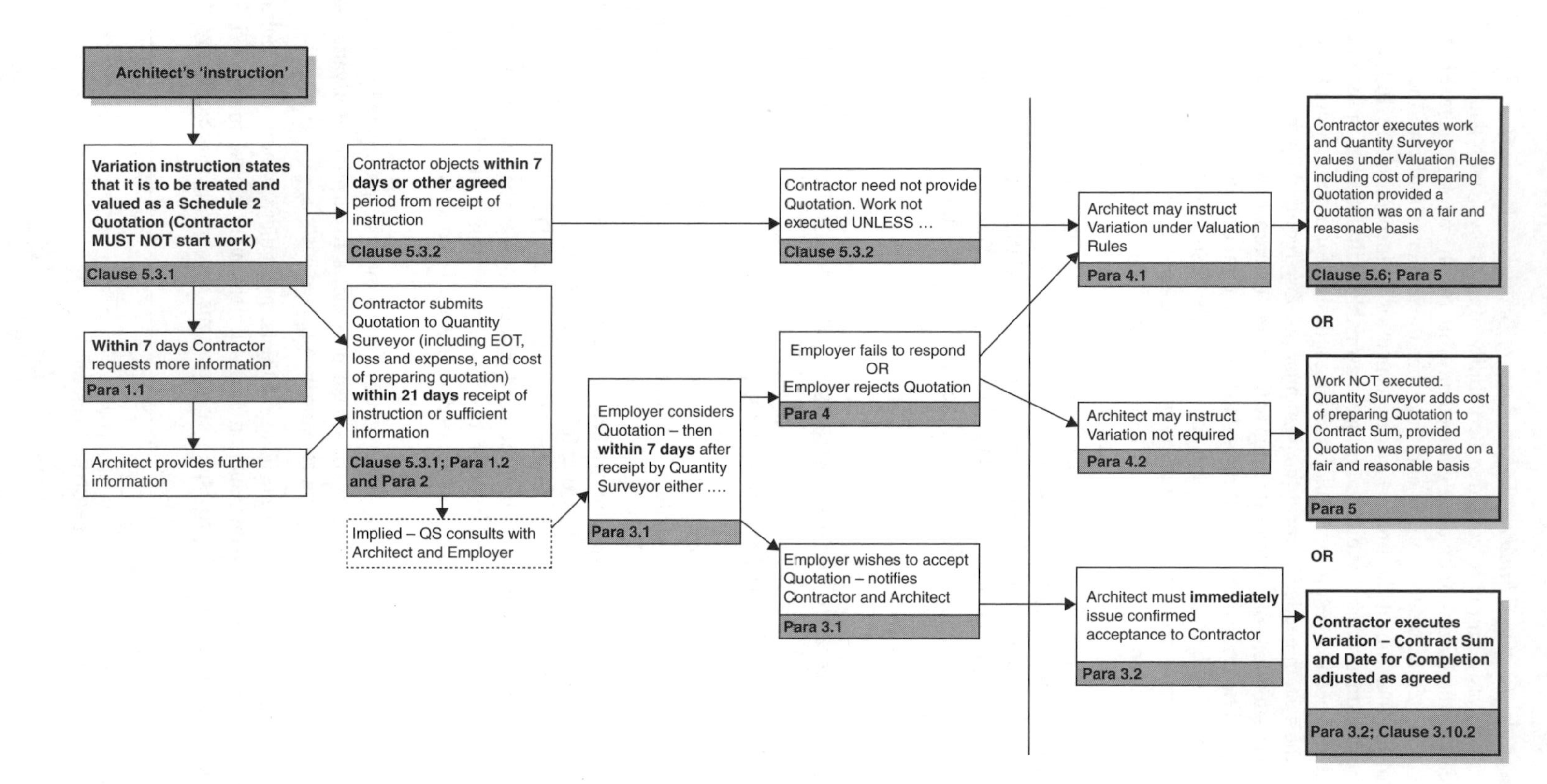

Fig. 6.1 Schedule 2 Quotation Procedures

construction industry standard forms. This often led to amendment of the standard clauses. In *City Inn* v. *Shepherd Construction Ltd*[69] a JCT 80 Contract was amended to oblige the Contractor to start a consultation process by providing an estimate of cost and effect on time, on receipt of any instruction that he considered to be a Variation. The Contractor lost his entitlement to extension of time when he failed to give notice. Under the amended Contract in that case, the onus was on the Contractor to trigger the procedure in order to protect himself. Under JCT 05, the quotation provisions are for the Architect alone to instigate.

JCT 05 Clause 5.3 procedure (Schedule 2 Quotation) is not mandatory; in practice the Architect may use it selectively. The calculation of the Final Account is then eased and accelerated, but possibly not sufficiently to give either Party the full benefit if only a few of the Variations are instructed through the quotation option.

Two important features of the Schedule 2 Quotation are found in the first line of Clause 5.3.1: 'If the Architect …in his instruction for a Variation states that the Contractor is to provide a quotation in accordance with the provisions of Schedule 2 …'. First, it is for the Architect, not the Contractor, to initiate, and second the Architect must use the specified procedure – there is no other machinery in the Contract entitling the Architect to request quotations.

The Architect first instructs the Contractor to supply a 'Schedule 2 Quotation'. The Contractor, unless he disagrees with the principle of a quotation, responds and by submitting a quotation to the Quantity Surveyor starts a somewhat tortuous procedure towards an agreed price. The quotation is then open for acceptance by the Employer. The Employer's 'wish to accept' is sent direct to the Contractor followed by confirmation communicated to the Contractor by the Architect in a 'Confirmed Acceptance'.

The Contractor is barred from starting any work which is the subject of a Schedule 2 Quotation until he has received from the Architect either a Confirmed Acceptance, or a further instruction to proceed on the basis that the Variation will be valued under the Valuation Rules. Any delay in the process by either Party may render that particular application of the Schedule 2 procedure void. Delay by the Employer in accepting a quotation kills the quotation. Delay by the Contractor in waiting beyond the proscribed period for acceptance is at the Contractor's risk; he must carry on with the work ignoring the proposed change.

6.13.2 Schedule 2 Quotation: the instruction

If the Architect wishes to deal with a Variation under the quotation procedure, Clause 5.3.1 states that he must first issue an instruction stating that a Schedule 2 Quotation is to be supplied. Clause 5.3.1 further provides that the Contractor may disagree in writing within 7 days that Schedule 2 should apply to that particular instruction. If the Contractor disagrees, Schedule 2 will not apply.[70] The Architect must then revert to the basic method of instructing work to be valued by a Valuation if he requires the Variation to proceed.

It is significant that the Contractor's right to disagree is unfettered; it is not qualified by an express obligation to be reasonable. The Contractor is not obliged in any way to proceed down the quotation route, although most Contractors would probably see the procedure as a potential benefit in principle. However, one factor that may influence the Contractor is the adequacy of information. Under paragraph 1.1 of Schedule 2 the

69 [2003] BLR 468.

70 see Section 6.13.7.4 in this chapter.

Architect must provide '*sufficient information*' and *Footnote [61]* to Schedule 2 suggests that the information provided should be similar to that provided at the Tender stage. The Contractor then has 7 days under paragraph 1.1 in which to form a reasonable opinion whether the information provided is adequate, and to give notice. This 7-day period seems to be a separate period from that under Clause 5.3.1, albeit running concurrently. If that is so, then an unreasonable opinion formed pursuant to paragraph 1.1 as to adequacy of information may still result in a rejection of the Schedule 2 instruction by application of the unfettered right to reject under Clause 5.3.1.

The 7-day period for the Contractor's statement of disagreement under Clause 5.3.1 may be changed by agreement with the Architect or by extension in the instruction. This is a necessary provision if the procedure is to be seen as credible, particularly in instances where the difficulty of producing a quotation is not immediately apparent, or where the commercial risk of committal to a price cannot be valued readily. However, the means of agreeing an alternative period are not described. It must be presumed unless a global agreement is reached, that the Contractor will give notice of the need for a longer period for disagreement during the first 7 days after receipt of a Schedule 2 instruction, since otherwise the Schedule 2 procedure will apply by default.

6.13.3 Schedule 2 Quotation: the quotation

If the Contractor does not disagree with the application of Schedule 2 within the prescribed notice period, then either:

1. if he reasonably considers the information provided is insufficient to provide a quotation, he must, within 7 days of receiving the instruction, request further information; he must then provide a quotation within 21 days of receiving the further information (paragraphs 1.1 and 1.2.1); or
2. if satisfied with the information, within 21 days after receipt of the instruction he must submit a quotation (paragraph 1.2.1).

The Contractor's Schedule 2 Quotation is submitted to the Quantity Surveyor (not the Architect) and remains open for acceptance by the Employer (not the Quantity Surveyor or the Architect) for 7 days after receipt of the quotation by the Quantity Surveyor.

The periods for disagreeing in principle, requesting further information, submission of the Quotation and acceptance by the Employer referred to in Clauses 5.3.1 and 5.3.2 and in Schedule 2 may be changed by agreement. However, paragraph 7 makes clear that such agreement must be with the Employer (not the Architect or the Quantity Surveyor) who must confirm his agreement in writing to the Contractor. This procedure lacks practicality, and it may assist smooth running if the Employer were to notify the Contractor that the Architect or the Quantity Surveyor had limited powers of agency to agree different timescales to match the circumstances of individual Schedule 2 Instructions.

The content of the quotation is prescribed in paragraph 2. It must be split to show the following elements separately:

1. *The amount of adjustment to the Contract Sum*: This excludes loss and/or expense but includes the effect on preliminary items and on other work. The value must be supported by calculations with reference where relevant to the rates and prices in the

Contract Bills. It is clear that the quotation is not a means for the Contractor to charge what he likes; the pricing level must be the market level of the Contractor's accepted tender which produced the Contract Sum, adjusted only in the manner which the Contract provides.

2. *Any adjustment to time for completion*: The Contractor must state his requirement for any adjustment to the Completion Date including, if relevant, an earlier date than that stated in the Contract Particulars. However, the quotation must not duplicate any allowance for revision of the Completion Date already made by the Architect either under Clause 2.28, or included in any other Schedule 2 Quotation that has been accepted. There is opportunity here for confusion, particularly if the Contractor has previously given notice of delay for other events and the Architect has not yet awarded an extension. The proviso in paragraph 2.2 referred to here does not prevent the Contractor duplicating a request; it simply prevents duplication of a request for an extension where the relevant period has been granted already. Thus the Contractor may be obliged to qualify his request depending not only on extensions awarded, but also on other quotations which may be awaiting acceptance, which if accepted could affect the most recent quotation.
3. *The amount to be paid in lieu of loss and/or expense*: again the Contractor is required to avoid claiming amounts already ascertained under Clause 4.23 or included in other accepted quotations.
4. *A fair and reasonable amount in respect of preparing the Schedule 2 Quotation*: In this provision the JCT have maintained their unique provision, first introduced in JCT 98, to require the cost of preparing a quotation to be identified as a preliminary to reimbursement in the event that the quotation is not accepted. Whilst it would be natural for the Contractor to build the cost of preparation into his price, the cost will almost inevitably be overhead costs, and any allowance will have to be taken into account in ascertaining loss and/or expense if and in so far as it relates to office overheads. The identification of this cost is sometimes ignored by Contractors but is of particular importance for the Contractor when the quotation is not accepted. This aspect is considered below.
5. *A statement of any additional resources required to carry out the Variation, if the Schedule 2 Quotation instruction specifically requires*: The provision of a resources statement is an option for the Architect, but is a useful tool in determining the effect on time as well as cost. It is a pity the JCT did not make the requirement mandatory. The production and subsequent consideration of such a statement forces both Contractor and Employer's teams to address their minds to the real and often disruptive nature of Variations before, rather than after, the event.
6. *A method statement for carrying out the Variation, if the Schedule 2 instruction specifically requires*: The comments made under (5) above apply equally to method statements.

A number of the problems which can arise for the Contractor in producing a Schedule 2 Quotation are considered below at 6.13.7.

6.13.4 Schedule 2 Quotation: acceptance

It is important to remember that the Architect does not have general authority under the Contract to accept quotations from the Contractor or otherwise to make agreements on behalf of the Employer. In order for the Architect to make agreements on the Employer's behalf, the Architect must first be given authority. In addition, that authority must be communicated to the Contractor; otherwise the Contractor may refuse to comply with such an agreement. That position is not altered by the Schedule 2 procedure. It is the Employer who makes it known to the Contractor under paragraph 3.1 that he wishes to accept the Schedule 2 Quotation. This must be done by notification within the 7-day period for acceptance.

Under paragraph 3.2 the Architect then simply confirms the Employer's wish by immediately writing to the Contractor. The notification, referred to as a 'Confirmed Acceptance', must state:

1. that the Contractor is to carry out the Variation.
2. the value of the Variation including any amount in lieu of loss and/or expense and preparation costs.
3. any adjustment of the Completion Date or Section Completion Dates, which can where relevant result in a date earlier than the date in the Contract, being agreed between the two Parties.

Once the Quotation is accepted the agreement is binding; the Contract Price and Date for Completion are adjusted in accordance with that agreement. It is worth noting here that the Architect cannot, at a later date, reduce an extension of time previously granted by agreement under the Schedule 2 Quotation procedure (referred to as a 'Pre-agreed Adjustment'), unless the reduction results from an omission of work (referred to as a 'Relevant Omission'), which itself is the subject of agreement between Employer and Contractor. In a nutshell, this means that prior agreement between the Parties concerning time, can only be changed by further agreement between the two Parties, and that the Architect cannot override it, at any time, simply by omitting work.

6.13.5 Schedule 2 Quotation: not accepted

If the Employer does not accept the Schedule 2 Quotation within the 7-day period for acceptance, it is rejected by default. Under paragraphs 3 and 4 the Employer is not required to state expressly that he does not accept the quotation.

However, the Architect must act to tie up loose ends. The Contractor is neither entitled nor obliged to start work against the Architect's instruction which triggered the Schedule 2 procedure in the absence of acceptance. Consequently, if notwithstanding the Employer's rejection of the quotation, the Architect still requires the work to be done, he would, in the absence of an express procedure, have to regenerate the instruction in some way. Paragraph 4.1 provides the procedure. If the Variation is required to be executed, albeit the Contractor's quotation is not accepted, then the Architect must instruct that the Variation is to be executed and valued under the Valuation Rules in Clauses 5.6 to 5.10.

If the Variation is not required at all, under paragraph 4.2, the Architect must instruct accordingly. Strictly, the JCT could have drafted Schedule 2 to allow the originating instruction and the resulting quotation to die in the absence of acceptance, since there is no obligation on either Party. The general position in common law is that a tenderer incurs costs of preparing a quotation[71] at his own risk in the hope that he will recover in earnings from

71 The terms 'tender', 'quotation', 'bid', 'offer', 'estimate' may each be construed as meaning 'offer' in law if on a true construction it fulfils the requirements for an offer and is capable of being accepted. In *Crowshaw v. Pritchard and Renwick* (1899) 16 TLR 45, it was held that the defendant Contractor's 'estimate' had no special customary meaning and was an offer which had been accepted to create a contract.

successful bids.[72] In the absence of an express term in the enquiry document, an enquirer is only liable for tendering costs if he asks the tenderer to expend resources, either deceitfully,[73] or to obtain something of benefit[74] when the tenderer reasonably expects payment.

The JCT apparently takes the view that the instruction of the Architect to expend the Contractor's resources on preparing a quotation warrants, reimbursement of preparation costs if the quotation is rejected. Thus, under paragraph 5 a fair and reasonable amount in respect of preparation costs must be added to the Contract Sum. It is not clear whether the fair and reasonable amount to be paid to the Contractor for preparation costs is the same fair and reasonable amount which the Contractor included in the Schedule 2 Quotation pursuant to paragraph 2.4. Since there is no reference to paragraph 2.4 in paragraph 5, it may be assumed that the fair and reasonable amount in paragraph 5 must be objective and need not be based entirely on the amount stated in the Contractor's quotation.

There is a caveat to the entitlement to preparation costs; the quotation must have been prepared on a fair and reasonable basis. This is consistent with the often voiced objection that, in giving a quotation for a Variation, a Contractor can hold the Employer to ransom if he does not want to do the work; he can simply price the work too high in the knowledge that it would be impractical for the Employer to get others to carry out the change. Schedule 2 gets around the problem, first by providing the Employer with the right to revert to the basic procedure of valuation by the Valuation Rules, and second with the threat of a sanction on the Contractor. The sanction is deprivation of preparation costs if the Contractor's quotation is not prepared on a fair and reasonable basis. Whether this sanction is sufficient to induce a reluctant Contractor to prepare a price in a reasonable manner is questionable; in any event it is always open to the reluctant Contractor to refuse to provide a quotation on receipt of the originating instruction, by stating his disagreement to the application of the Schedule 2 procedure.

The JCT clearly foresaw reluctance in the industry to pay wasted preparation costs and have dealt with what might have been an excuse used by Architects. Paragraph 5 states that non-acceptance by the Employer is not of itself evidence that the Quotation was not prepared on a fair and reasonable basis. In a situation where an entitlement to preparation costs is triggered expressly by non-acceptance of a Quotation, it is difficult to conceive that the mere fact of non-acceptance could be considered, or even proposed, to be evidence to deprive the Contractor of those preparation costs. The fact that the JCT contemplated, and found it necessary to deal with, the proposition seems a sad reflection on the culture of the modern construction industry.

6.13.6 Schedule 2 Quotation: further variation

There is always the possibility that a Variation instruction, whether issued for valuation under the Valuation Rules or under the Schedule 2 procedure, may affect work which itself forms part of another previously accepted Schedule 2 Quotation.

72 See *William Lacey (Hounslow) Ltd* v. *Davis* [1957] 2 All ER 712.

73 In *Richardson* v. *Sylvester* (1873) LR 9 QB 34, property was advertised for sale by auction with no intention of selling. The costs incurred by a potential bidder in valuing the property were recovered in the tort of deceit.

74 See discussion in Chapter 1, Section 1.8.2.2 and 1.8.2.3.

Clause 5.3.3 requires the Quantity Surveyor to value such later Variation, not under the Valuation Rules, but expressly on a fair and reasonable basis, while having regard to the content of the relevant Schedule 2 Quotation. The Quantity Surveyor must also include any direct loss and/or expense incurred by the Contractor as a result of the Variation.

This clause seems to have been added almost as an afterthought. Why variations on Variations should be treated differently from any other Variation is unclear; arguably it can only lead to confusion and dispute. The rules for Variations in Clause 5 as a whole may be applied just as well to sequential Variations as they do to work which is introduced by variation and is not changed further. This provision can have a far-reaching effect on the value of the Works, for it is common for varied work to be further varied. If on a project, a large number of Variations are instructed under the Schedule 2 Quotation procedure, the result could be a high proportion of the total work in Variations calculated on a fair and reasonable basis, rather than under the Valuation Rules or even under the quotation procedure.

The obligation to include loss and/or expense simply adds more difficulty. Since the further Variation may not be subject to a Schedule 2 instruction, the Contractor is required to comply with the Clause 4.23 to 4.25 procedures if he requires loss and/or expense to be ascertained. The Quantity Surveyor then is obliged to separate out, as best he can, the portion applicable to the Variation. Clearly the task is easier if the further Variation is instructed under the Schedule 2 procedure, since then the Contractor would provide the necessary information related to the individual Variation.

6.13.7 Schedule 2 Quotation: some typical problems

1. Misuse of Schedule 2 procedure

The timetable for dealing with Quotations usually requires an element of estimating rather than historical surveying. The Contractor is obliged to foresee possible risks and to price those risks. The result is sometimes an amount higher than the Employer would like to pay. Architects in that situation may reject the Quotation and proceed under the Valuation Rules. Later when the work is done, if the value of the Variation exceeds the Contractor's Schedule 2 Quotation the Architect may be tempted to hold the Contractor to his earlier price. The Architect is then using the rejected Schedule 2 Quotation as a form of cost advice.

This situation is common whenever a contract provides for quotations against Variations. However, under JCT 05 the Contractor is protected in that paragraph 6 of Schedule 2 provides that neither Party shall use an unaccepted Schedule 2 Quotation for any purpose whatsoever. In short, mimicking general principles of offer and acceptance, the rejected Quotation is dead.

2. An Employer does not express a wish to accept Schedule 2 Quotation

The Architect is not empowered to accept quotations; under paragraphs 1.3, 3.1 and 3.2 the Employer accepts and the Architect confirms acceptance after the Employer has expressed a wish to accept in writing to the Contractor. It sometimes happens that the Architect accepts a quotation without the Employer having first written to the Contractor.

In that situation, the Contractor is not obliged to comply with the instruction since he knows that the Architect is acting outside his authority.

3. Multiple quotation instructions

When a quotation is requested it is not unusual for the Contractor to be instructed to provide prices for alternative schemes. The Contractor is then obliged to provide several quotations in the knowledge that only one (if any) will be accepted. Under paragraph 5 the Contractor is entitled to recover preparation costs of a quotation which is not accepted, so he is protected against the cost of preparing each rejected alternative.

The difficulty for the Contractor is in establishing the base from which the Quotation is calculated. In circumstances where there are few Variations and no other Schedule 2 Quotations pending, the task is comparatively easy. But when the Contractor is aware that the Quotation he is preparing may be affected if another Schedule 2 Quotation previously submitted is accepted, he can do no more than qualify his Quotation accordingly. When there are a number of Quotations awaiting acceptance, the Contractor may need to provide an alternative price for each permutation of accepted Quotations. This is hardly likely to endear the Contractor to the Architect, but it seems there is little choice if confusion is to be avoided.

Strictly under the rules of Schedule 2 the Quotations could be allowed to stack up sequentially, then depending on the order in which they are accepted, could be identified as Variations on earlier Variations. The valuation rule would then be a fair and reasonable basis under Clause 5.3.3. The Contractor's only protection is the short period for acceptance.

There is no simple solution. The important issue is in recognizing that the situation can occur, and drawing it to attention before havoc sets in!

4. Failure to submit quotation

The Contract does not deal with the position where the Contractor fails to disagree with the use of the quotation procedure, fails to require further information, and then fails to provide a quotation within the timetable. The Architect will not be aware of any dilatoriness on the part of the Contractor until the time for submission has passed, by which time at least 21 days will have gone by. If the proposed Variation falls at a critical time in the construction programme, or close to completion, any delay could be significant. It is unrealistic to suggest that the Architect could wring a quotation from the Contractor, pedantically insisting on compliance with the procedure. If the work is required, his only practicable course is to issue an instruction for the Variation, and have the work valued by the Quantity Surveyor in the traditional manner. That suggests that there is no sanction on the Contractor for failing to respond in time, and indeed there is none. Clearly, the Contractor is in breach of paragraph 1 of Schedule 2, but his breach does not occur until the end of the 21 day submission period, since at any time up until then, he has the opportunity to comply. The only delay attributable to the Contractor is a reasonable time taken by the Architect to decide that he needs to act. It is surprising that the JCT have not introduced into this contract the sanction provided in their Design and Build Contract,[75] whereby the Contractor is entitled to payment for the Variation, but is not entitled to interim payment before the final adjustment of the Contract Sum.

---

75 See JCT Design and Build Contract (DB) 2005 Edition, Schedule 2 para 4.6.

6.14 Errors, discrepancy and divergence in and between documents

6.14.1 Generally

A common category of Variation arises from the correction of errors and resolution of inconsistencies in the Contract Documents. If a contract contains an error such as an error in quantity, description of work, or even in a contract term, the general position is that it must stand,[76] unless either the Parties agree when it is discovered that it should be rectified, or the contract expressly sets out what is to be done, or unless a court can be persuaded that the contract does not express what the Parties intended. Under JCT 05 where the calculation of the Contract Sum contains an error, the position is set out clearly in Clause 4.2; any error, whether of arithmetic or otherwise is deemed to be accepted by the Parties.[77]

Sometimes an error appears as an inconsistency, such as a specification on a drawing conflicting with a description in the bills of quantities. In these cases the Parties, or a dispute resolution forum, have to choose which of the specifications describes the Parties' obligations and entitlements. In the absence of provisions in the contract, the general rules of contract construction will apply. So, for example, the written word will normally prevail over the typed word, and the typed word will prevail over the standard printed word although the modern method of contract interpretation is less by use of a set of strict rules, and more by consideration of the factual background, available to, and influencing the two Parties in the use of language, when forming the Contract.[78] If the inconsistency creates ambiguity in the documents as a whole, or if a single document contains conflicting provisions or an ambiguity, then the ambiguity may be construed *contra proferentem*,[79] that is the option least favourable to the party whose document contains the ambiguity.[80]

In some contracts the priority of the various documents is listed,[81] while in others the choice of priority to suit the circumstances is left either to one of the parties,[82] or to a third party such as the administrator of the contract.

JCT 05, Clauses 2.13 to 2.18 deal expressly with specified types of errors and conflict.

6.14.2 Status of Contract Bills and CDP documents

The Contract does not list the priority of the various Contract Documents, although it deals with the status of two project-specific documents, the Contract Bills and CDP Documents, in relation to the Agreement and Conditions.

---

76 *Ewing & Lawson* v. *Hanbury & Co.* (1990) 16 TLR 140.

77 However, there may be an exception where one party deceitfully conceals an error spotted before conclusion of the Contract to his advantage; see for example *MacMaster Univerity* v. *Wilchar Construction Ltd* (1971) 22 DLR (3d) 9, in which the Employer accepted a tender knowing that a page was missing and excluded from the tender price.

78 See Lord Hoffman in *Investor's Compensation Scheme Ltd* v. *West Bromwich Building Society* [1998] 1 WLR 896 at 913.

79 This doctrine normally applies where the Contract is based on one party's unilateral terms; even though the particular terms used have been chosen by the Employer, JCT Contracts terms are not unilateral since they are negotiated by representatives of different sectors of the industry.

80 This doctrine can be seen applied in Clause 2.16, and in JCT Design and Build Design Form DB 2005 Clause 2.14 dealing with errors in the Employer's Requirements and the Contractor's Proposals.

81 For example, see Clause 1.3 of JCT Standard Building Sub-Contract 2005 SBCSub/C.

82 See Clause 2.16 which operates on the *contra proferentem* (Lat. against the one who puts forward) principle.

Clause 1.3 states 'Nothing in the Contract Bills or the CDP Documents shall override or modify the Agreement or these Conditions'. This provision reverses the general rule of construction that the specially prepared conditions prevail where they conflict with the standard printed conditions. It protects the Contractor against terms inserted in the Bills or CDP Documents which may otherwise change the allocation of risk agreed by the members of the JCT. However, the clause, in the similar words of previous editions of the contract, has attracted some criticism and judicial opinion that it should be removed[83] to allow all terms to be read together and not be shut out and left isolated in the project-specific documents (see also Chapter 1, Section 1.4.3, Bills of Quantities, and Section 1.5.5, Amending the Standard Contract Form).

The effect of Clause 1.3 is theoretically wide, but in practice it tends to be limited to dealing with conflict. The difficult area is where an obligation imposed in the project-specific documents adds something new, and it is then a matter of interpretation as to whether the purported obligation modifies the interpretation of the provisions in the printed form. Each case must be taken on its merits, but the courts have demonstrated a degree of robustness. For example, the imposition of design obligations by reference in the specification has been held 'added to but were consistent with obligations imposed by the conditions';[84] similarly, the obligation to submit a programme in a specified form was held not to conflict with the Conditions.[85] If, however, a paragraph in the Bills or CDP Documents purported to place an obligation in head-on conflict with JCT 05 Form, such as, for example, stating a period for payment of an interim certificate different from that in the Contract Particulars, then the Contract Particulars, being part of the Agreement, would take precedence.

6.14.3 Errors and inadequacy in the Contract Bills and CDP documents

1. Contract Bills

The Contractor's obligations include carrying out the works described in the First Recital, i.e. as shown on the drawings and described in the Bills. A difficulty can arise if the Bills contain errors. In the absence of provisions in the Contract, the descriptions in the Bills would be at the Contractor's risk; if work were shown or implied on the drawings, it would form part of the Contract even if it were missing from the Bills. An example can be seen in *Williams* v. *Fitzmaurice*,[86] where it was held that floorboards missed from a schedule of works were necessarily part of the Contract to complete a house. Following this judgment, in *Patman & Fotheringham* v. *Pilditch*[87] it was said that items omitted from the bills of quantities did not necessarily give rise to extra payment; the Employer was not bound to pay for things that everybody must have understood are to be done, but which happen to

83 See dicta of Lord Justice Stephenson in *English Industrial Estates Corp.* v. *Geo. Wimpey and Co. Ltd* (1972) 7 BLR 122.

84 *Haulfryn Estate Co. Ltd* v. *Leonard J. Multon & Ptnrs and Frontwide Ltd:* ORB; 4 April 1990, Case No 87-H-2794. Note that this case dealt with the JCT Minor Works Form, but Clause 4.1 of that form is similar to Clause 1.3 of the JCT Standard Form.

85 *Glenlion Construction Ltd* v. *Guinness Trust Ltd* (1987) 11 ConLR 126.

86 (1858) 157 ER 709.

87 (1904), Hudson's, 4th edn (1914), Vol. 2, p. 368.

be omitted from the quantities. However, it was also held in the *Patman* case that the bills formed part of the Contract, and whilst the Contractor was obliged to do the omitted work, he was entitled to have the price adjusted to account for it.[88] This general position is formalized in JCT 05 where quantities form part of the Contract. It is in the Employer's interests to have the Bills prepared by his professional team,[89] and in return JCT 05 Clauses 2.13.1 and 2.14.1 place the risk in the Bills with the Employer.

Clause 2.13.1 states that the Bills have been prepared in accordance with The Standard Method of Measurement of Building Works,[90] unless the Bills contain specific qualification in respect of each specified item not complying. It is important for the tenderer, and eventually the Contractor, to know what he is pricing, and he is entitled to take the content of the Bills at face value.[91] Thus, to avoid misleading the tenderer, any identification of deviation from the Standard Method of Measurement must be specific. It must be sufficiently specific for the tenderer (and indeed the Employer) to know what the non-compliant description entails; otherwise such a qualification will amount to no more than a denial of warranty as to accuracy. In the case of *Co-operative Insurance Society Ltd* v. *Henry Boot (Scotland) Ltd and others*[92] the bills failed to provide the ground water level, although it was argued that there was sufficient information to calculate the level. It was held that the ability to calculate was not good enough; the Bills had to state the level and when that level was measured. The Bills had to be corrected, and the correction was a Variation.

Clauses 2.14.1 and 2.14.3 provide for errors of description or quantity in the Bills, including omissions and unstated departure from the Standard Method of Measurement, or inadequate information about a Provisional Sum for defined work, to be corrected as a Variation.

The correction of errors, particularly errors in quantity, together with the correction of omissions, is necessary to maintain the concept of a lump sum contract which is adjusted only in respect of Variations and other specified changes, thus avoiding the need to remeasure parts of the Works. Clause 2.14.1 simply provides machinery for rectifying the Contract to what it should have been from the outset.

Unfortunately, the provision is difficult to apply in practice. It is often left to the Contractor to identify errors, which he can do only by comparing records of work done with the Bills, so errors in the Bills often overlap, and become confused with, Variations. Since the Contractor is usually more concerned with being paid for what he has done than identifying the theoretical classification, he is likely to press for remeasurement of suspect quantities. However, the Quantity Surveyor's duty is to apply the rules in the Contract, even if it means identifying his own errors to the Employer.

The requirement to treat Clause 2.14.1 rectification as if it were a Variation enables such adjustment to be considered as a potential 'event' for the purposes of extensions of time, and a potential 'matter' for the purposes of loss and expense.

---

88 See also *Meigh & Green* v. *Stockingford Colliery Co. Ltd* (1922); Hudson's, 11th edn, Vol. 1, p. 975, in which builders were held entitled to be paid for brickwork missed from the bill of quantities.

89 See 1, Section 1.4.3, Bills of Quantities.

90 See Chapter 1, Section 1.4.5, the Standard Method of Measurement.

91 In *C. Bryant & Son Ltd* v. *Birmingham Hospital Saturday Fund* [1938] 1 All ER 503, the equivalent of Rule 1.1 of the SMM7 when read with the conditions stating the Standard Method had been used, was construed as a warranty by the Employer that the information provided to the Contractor was both accurate and sufficient to identify the nature and extent of the Works; see also Chapter 1, Section 1.4.5.

92 [2002] EWHC 1270 (TCC); 1 July 2002.

Notification of such errors by the Contractor is dealt with in Clause 2.15 in the same manner as other discrepancies and divergences (see Section 6.14.4 below).

2 CDP Documents

To the extent that an inadequacy is found in the CDP Documents, Clauses 2.14.2 to 4, following the doctrine of *contra preferentem* referred to above, place the risk depending upon the author of the relevant document. Inadequacy in the Employer's Requirements is the Employer's risk. The Contractor is expressly not responsible for the contents of the Employer's Requirements under Clause 2.13.2,[93] and under Clause 2.14.1 any inadequacy must be corrected, and treated as a Variation, unless the matter has already been dealt with in the Contractor's Proposals.

Any error or omission from the Contractor's Proposals or the CDP Analysis is at the Contractor's risk, and Clause 2.14.4 makes clear that correction of error in description or in quantity, or an omission of an item, including any effect on work other than CDP Work, must be corrected at the Contractor's expense.

6.14.4 Notification and correction of discrepancies, etc.

1. In and between documents generally

Clause 2.15 contains the rules for notifying and correcting errors, omissions and inadequacies including insufficient information about the content of Provisional Sums for defined work,[94] discovered by the Contractor. On becoming aware of errors in or divergence between the Contract Drawings, the Bills, Architect's instructions (other than for Variations), explanatory drawings or documents provided by the Architect and the CDP Documents, the Contractor must immediately give notice to the Architect, who must then issue instructions.

The purpose of Clause 2.15 is to correct errors and ambiguity, yet the action of the Architect having being made aware of a problem is simply to 'issue instructions in that regard'. In so far as the notice relates to divergence between documents, there is no guidance as to the principles to be considered in issuing instructions. In the absence of express direction, the general principles of the *contra proferentem* doctrine are likely to apply. Where two documents conflict, or where there are conflicting provisions within a document, the Architect may choose which provision he wishes to apply. If the Architect's choice is not the same as that of the Contractor, the instruction will give rise in effect to a rectification of the Contract, and to corresponding adjustment of the Contract Price.

The significant feature of Clause 2.15 is the requirement for the Contractor to become aware of the discrepancy or divergence, and then immediately to give written notice to the Architect. This does not impose an obligation on the Contractor to search for discrepancies; he is entitled to stumble across them. It may be that a divergence or ambiguity does not occur to the Contractor until (say during a dispute) an interpretation counter to his own is put to him. In those circumstances, it is suggested, that is the point at which the Contractor becomes aware[95] that there is a discrepancy or divergence. Similarly, it

93 Thus the Contractor is not liable for the Employer's design errors.

94 See SMM7, Rules 10.1 to 10.6.

95 In Revision 1, 1 June 2007, 'If the Contractor finds' was changed to 'If the Contractor becomes aware of'. It is suggested that the meaning was not materially changed, and that the word 'finds' in the unrevised version did not require the Contractor to search, in order to 'find'.

is not for the Architect to search out, or even spot, discrepancies or ambiguities in the documents, but even if he happens upon them, he is not obliged to do anything (save in respect of statutory requirements) until he is notified by the Contractor. Whilst the Contractor is required to give notice '*immediately*', there is no express timescale imposed on the Architect. Nevertheless, it has been held[96] that under a similar provision in JCT 63 a reasonable period in which the Architect should act would be implied if a substantial delay might disrupt the Contract.

2. In and between documents relating to Contractor's Designed Portion Documents

Where the Contractor finds discrepancy in a CDP Document or divergence between different CDP Documents, he is required to give immediate notice to the Architect, as for other discrepancies under Clause 2.15 described above. However, unlike the procedure for other discrepancies, the Contractor cannot simply sit back and wait for the Architect to resolve the matter; he must take the initiative by proposing his preferred amendment. It is then for the Architect to accept, or to decide upon a different solution.

Clause 2.16.1 deals with discrepancy in or between CDP Documents other than the Employer's Requirements. This would include, say, a different specification of flooring shown on a drawing from that described in a specification document, both prepared by the Contractor. The Contractor must send with his notice, or as soon as he can, a statement of his proposed amendment to remove the discrepancy. In the case of the flooring example, he would simply state his choice of the two specifications. On receipt of the statement, the Architect may accept the Contractor's proposal, or he may prefer the other alternative, and issue instructions accordingly.

Whichever of the alternatives the Architect chooses, there will be '*... no addition to the Contract Sum*'. It is significant that the JCT confined adjustment to 'no addition'. Since the error was of the Contractor's making (being in the Documents other than the Employer's Requirements) he should not benefit. If the Contractor chooses the cheaper alternative, and the Architect instructs the more expensive, there will be no addition. If, on the other hand, the Contractor proposes the more expensive alternative, there is nothing to prevent the Architect from instructing the cheaper, but not to reduce the Contract Price. The use of the words '*no addition to*' the Contract Price, instead of '*no adjustment to ...*' invites speculation that the Architect may in that case adjust the price down. However, it is suggested that the definition of Variations in Clause 5.1 together with the list of instructions that may give rise to a Variation in Clause 3.14, is not wide enough to encompass corrective instructions, although a credit could be effected by acceptance of the more expensive followed by a further instruction.

Clause 2.16.2 deals with discrepancy within the Employer's Requirements. If the flooring specification example referred to above relates to a specification and a drawing, both of which were in the Employer's Requirements, the error would be the Employer's. It may be that the discrepancy has been spotted by the Contractor and dealt with in his Contractor's Proposals; if so the Proposals take precedence without any adjustment to the Contract Sum. If the Architect then wants something different he must initiate the normal Variation procedures.

---

96 *R.M. Douglas Construction Ltd* v. *CED Building Services* (1985) 1 Const LJ 232; referring to Clause 1(2) of JCT 63.

If the discrepancy is not dealt with in the Contractor's Proposals, the Contractor must send his notice informing the Architect of his proposed amendment to correct the error. The Architect must then agree the proposal or instruct an alternative. Clause 2.16.3 states that the Architect's agreement, or other decision, shall be treated as a Variation. This procedure is logical where the solution to the discrepancy is not to be found in the Employer's Requirements.

However, the discrepancy often creates an ambiguity, and it is simply a matter of choice between two alternatives. In those cases, it would accord more with commonsense, and the *contra proferentem* doctrine, if the instruction were to be treated as a Variation only if the Architect desired the alternative rejected by the Contractor, or something different.

The position where the Employer's Requirements and the Contractor's Proposals differ is not dealt with by the JCT in this Contract. The issue is discussed in Section 6.12.6, Item 5 in this chapter above.

6.14.5 Divergences, and compliance with Statutory Requirements

1. Generally

Clause 2.17 provides for divergence between the Statutory Requirements and the drawings, Bills, any instruction, the CDP Documents and Variations. In contrast with Clause 2.15, the procedure may be triggered either by the Architect, or by the Contractor: 'If the Contractor or Architect ... becomes aware of any divergence' (Clause 2.17.1). Notice shall be given to the other immediately, although the Contractor's obligation to notify carries with it a liability (see below). To the extent that the divergence relates to CDP Documents, the Contractor must also send the Architect his proposal for dealing with the error. The Contract does not state a period for giving such proposal, but given the nature of the error (breach of Statutory Regulations), the delay should be minimal and certainly no longer than the corresponding period for discrepancies within the CDP Documents (i.e. as soon as is reasonably practicable).

The Architect must then issue instructions within 7 days of his becoming aware of the error or his receiving notice from the Contractor, or in the case of CDP Documents, 14 days after receiving the Contractor's proposal.

The Contract makes clear at Clauses 2.17.2 and 2.17.3 that, except in relation to CDP Documents, the risk lies generally with the Employer, and rectification is treated as a Variation. To the extent that the divergence is between Statutory Requirements and any of the CDP Documents (provided the Statutory Requirements have not changed since the Base Date[97] of the Contract), the risk lies with the Contractor. This is the only area where the Contractor is liable for the errors in the Employer's Requirements, founded on the premise that as a designer the Contractor should check that the Statutory Requirements are being met,[98] and that his own design complies.

If any discrepancy results from change in the Statutory Requirements since the Base Date, the Architect must instruct the modification, which is treated as a Variation.

---

97 See Chapter 1, Section 1.5.2, Contract Particulars entry for Clause 1.1.

98 See reference to *Boot* case in Chapter 4, Section 4.4.4 dealing with Liability for Employer's design and obligation to check; see also JCT Guide SBC/G, paras 47 and 48.

Reference is made above to the Contractor's liability relating to giving notice. Clause 2.17.3 disclaims the Contractor's liability for non-compliance with Statutory Requirements except in relation to CDP Works (dealt with above), and where the Contractor fails to comply with the notice provisions in Clause 2.17.1. It is important to note that Clause 2.17.3 does not suggest that the Contractor is liable, but simply that his exclusion of liability is removed – in short he may be liable. The Architect's duty to give instruction for rectification is triggered by the Contractor's notice or by his own discovery of a discrepancy. The position is straightforward if the Contractor alone actually finds a discrepancy but then fails to notify, and carries on with the work unamended – the Contractor is likely to be liable for the discrepant work. The difficulty for the Architect is in identifying that the Contractor had found the discrepancy at all. The position is more difficult where the Contractor does not find the discrepancy – but should have done. To the extent that the work is not CDP Work the Contractor has no general obligation to check for discrepancy; but if the discrepancy involves some element of danger, it is likely that the Contractor would be expected to spot the danger and give notice.[99] The Contract does not deal with the position where the Contractor has not given notice and neither has the Architect. If the discrepancy was obvious and had been spotted by both the Architect and the Contractor, then both are equally in breach of Clause 2.17.1, and neither are protected against liability.

2. Emergency compliance with Statutory Requirements[100]

Historically the Contractor has been at risk under JCT Contracts when acting without formal instruction. That has led to Contractors being forced to take risks in emergency situations, where either the Contractor could sit back and await instruction, or instigate immediate action, in the hope that the action will be endorsed by the Architect after the event.

Clauses 2.18.1 and 2.18.2 give the Contractor limited authority to act immediately. In responding in an emergency, if the Contractor needs to supply work or materials before receiving an instruction regarding correction of a divergence from Statutory Requirements, he must carry on, and he must give notice to the Architect of the action he is taking. The Contractor cannot wait on the Architect, and cannot use a delayed Clause 2.17.2 instruction as grounds for extension of time. Where the emergency arises because of divergence between the Statutory Requirements and the drawings, Bills and instructions (other than CDP Documents) the Contractor's action is treated as a Variation, provided he has given notice of what he is doing. Nothing in the Contract provides any clues as to what the JCT intend by the term 'emergency'. Clearly, making work safe after an accident, or dealing with an unexpected ground condition, could fall into the category of emergency; but Clause 2.18 relates only to immediate action in complying with Statutory Requirements. It is unlikely that urgent action to maintain progress would be considered an emergency, if taking no action would do no more than cause delay.

---

99 See Chapter 2, Section 2.3.5 for commentary on a Contractor's duty to warn generally.

100 See also Chapter 2, Section 2.3.8.2.

7

Risks to health and safety: allocation and insurance

Construction work has always involved three main categories of risks relating to health and safety. First, there is the risk of injury or death to people on or passing by the site. Second, the carrying out of the Works may cause damage to property other than the Works themselves or unlawful interference with personal rights associated with ownership or occupation of such property (e.g. noise, dust, vibrations and smells). Third, there is the possibility of damage or loss to the Works themselves or to Site Materials. The best strategy against these risks is, of course, one of prevention through the appointment of competent designers, contractors, sub-contractors, and suppliers with good health and safety records. Even where this is done, prudence still demands that construction contracts contain clear provisions on these risks because it is impossible to eliminate them completely. The aim in this chapter is to examine how JCT 05 deals with these risks. Adequate appreciation of the nature of the problem that they present is a prerequisite to developing sufficient understanding of the provisions. For this reason, this chapter begins with an explanation of the general nature of the problem.

7.1 General nature of the problem

A particular feature of the risks is that they affect a large number of people. The usual reaction of anybody injured or who suffers damage to his property is to sue whoever he thinks responsible, either in part or wholly, for the injury or damage. This could be the employer, contractor, a sub-contractor, the architect or other designer. The ways in which liability may attach to any of these potential defendants include:

- negligence;
- nuisance;
- trespass;
- strict liability;
- liability under *Rylands* v. *Fletcher* principle;
- liability in contract;
- breach of statutory duty.

In deciding whom to sue for his injury or loss, the strategy of the typical claimant is often that of the proverbial 'duck shoot' or the so-called 'scatter gun' approach (i.e. he sues everybody involved on the site). Under the Civil Liability (Contribution) Act 1978, a defendant may also bring in as joint defendants others whom he considers to have contributed to the injury, damage or loss. With more than one defendant, the tendency is for each of them to blame the others. It is then the court's duty to find out who is responsible for what and to what extent. This can be a long drawn out and very expensive process. Most construction contracts seek to avoid this situation by spelling out clearly in the contract itself who, as between the employer and the contractor, should be responsible for specific categories of the risks. Such statements of liability are referred to as 'risk allocation clauses' or, simply, 'risk clauses'.

It is to be noted that the mere fact that a particular risk is allocated to a party does not mean that that party caused or is otherwise to be blamed for its occurrence. In some cases, the employer and the contractor may both be blameless but nonetheless one of them must bear the financial consequences, which he may or may not be able to pass on to the third party ultimately responsible for the cause of the loss. For example, a member of the public who is injured by structural collapse due to faulty design by a structural engineer, may sue the contractor because of lack of knowledge of who is really responsible. Although the contractor should usually win in such an action, there is no guarantee that he will get all the costs of defending the action from the victim. In a construction contract, as between the employer and the contractor, this risk will usually be allocated to the employer (i.e. the contractor can claim his unrecovered costs from the employer who ought to bear the consequences of mistakes of any designers that he appoints). The employer may, in turn, pursue the engineer for compensation.

In a contract between parties A and B, allocation of a risk to party A carries the implication that if party B incurs liability as a consequence of occurrence of that risk, A must compensate B. However, many construction contracts contain express statements to the effect that 'A shall indemnify B if …'. Such statements are referred to as 'indemnity clauses' or, simply, 'indemnities'. As explained in Chapter 2, Section 2.3.9, indemnities are very onerous obligations because they can still be enforceable many years after expiry of the limitation period applicable to the contract containing them.

Risk and indemnity clauses in themselves are of limited value because it would be a fruitless exercise to determine who should shoulder the consequences of an accident if the responsible party has no funds to meet the liability. Although knowledge of potential liability will often spur a prudent party to obtain appropriate insurance cover, it would be ill advised to leave the availability of such insurance to the party's discretion. To guarantee availability of funds, construction contracts complement risk allocation clauses with additional provisions requiring a party to take out insurances against identified risks for which he is responsible. Referred to as 'insurance clauses', such provisions often state that if a party fails to maintain the specified insurance cover, the other party may take it out or pay the premiums and recover the cost of so doing from the defaulting party.

In summary, the contractual approach to dealing with the problem of risks from construction work involves clear risk allocation clauses, requirements for certain risks to be covered by appropriate insurance policies and clauses designed to allow easy and effective enforcement of the insurance obligations.

7.2 Allocation of risks under JCT 05

For the purposes of risk allocation, the following categories of risk are identified in the Contract:

- risk of personal injury/death;
- risk of damage to property other than the Works and Site Materials;
- risk of damage to the Works and Site Materials;
- damage from the Excepted Risks.

Some of the risk allocation clauses identify a group of risks referred to collectively as the 'Specified Perils'. The risk allocation cannot therefore be properly understood without prior development of understanding of the nature of this group of risks. It is for this reason that they are explained next.

7.2.1 The Specified Perils

The Specified Perils are defined by Clauses 1.1 and 6.8 as consisting of:

> fire, lightning, explosion, storm, flood, escape of water from any water tank, apparatus or pipes, earthquake, aircraft and other aerial devices or articles dropped therefrom, riot and civil commotion, but excluding Excepted Risks.[1]

Damage by fire, flooding and other forms of water damage have been the subject of considerable litigation. The scope of those perils must therefore be seen in the light of the interpretation given in the case law to the use of those words in similar contexts. In *Computer & Systems Engineering plc* v. *John Lelliott (Ilford) Ltd and Another*,[2] which arose from a contract let on a revision of JCT 80, a pipe in a sprinkler system was sheared off through the negligence of a sub-contractor. This caused discharge of 1600 gallons of water with resultant damage to the claimant Employer's property. The litigation concerned whether the damage arose from the equivalent of the Specified Perils. The Court of Appeal held that the word 'flood' used in a similar context suggested rapid accumulation of larger volumes of water from an external source and that it did not cover discharge from the sprinkler system.

In *The Board of Trustees of the Tate Gallery* v. *Duffy Construction Ltd*[3] water was being conveyed round a construction site by a flexible polymer pipe. Damage to the water main was repaired by reconnecting it to the flexible polymer pipe with a series of valves and reducers which allowed water to be turned on and off. Over an extended holiday period the connection failed and water escaped and entered the basement of the Tate Gallery to a depth of 1.4 metres, causing damage in excess of £5 million. One of the issues before the court was whether the inundation amounted to a 'flood'. After a comprehensive review of the authorities on the meaning of 'flood' used in similar contexts in construction contracts and insurance policies, Jackson J stated that deciding whether an unwelcome arrival of

---

1 The change in the wording of this definition is to be noted.
2 (1990) 54 BLR 1 (hereafter *CASE* v. *Lelliott*).
3 [2007] BLR 216.

water upon property amounts to a flood requires consideration of: (a) whether the source of the water was natural; (b) whether the source of the water was external or internal; (c) the quantity of the water; (d) the manner of its arrival; (e) the area and character of the property upon which the water was deposited; (f) whether the arrival of the water was an abnormal event. He concluded that whether the occurrence amounts to a flood is a question of degree of these factors. Applying the principles to the facts, he determined that the escape of water in that case had resulted in a flood.

The court in *CASE* v. *Lelliott* also confined 'bursting of pipes' (one of the Specified Perils under JCT 98) to damage from internal stresses. Two changes have been made to the definition of Specified Perils in JCT 98: (i) tempest has been omitted from the list; (ii) 'bursting or overflowing of water tanks, apparatus or pipes' has been replaced with 'escape of water from any water tank, apparatus or pipes'. As this new peril is wider, *CASE* v. *Lelliott* would probably be decided differently under JCT 05.

7.2.2 Risk of personal injury/death

Clause 6.1 deals with liability for personal injury and death in the following terms:

> The Contractor shall be liable for, and shall indemnify the Employer against, any expense, liability, loss, claim or proceedings whatsoever in respect of personal injury to or the death of any person arising out of or in the course of or caused by the carrying out of the Works, except to the extent that the same is due to any act or neglect of the Employer or of any of the Employer's Persons.

The above wording shows very clearly that Clause 6.1 is a combined risk and indemnity clause and that the Contractor is not responsible for every incident of personal injury or death on or near the site. For the liability and indemnity to apply, two conditions must be met. First, it is limited to injury or death "arising out of or in the course of or caused by the carrying out of the Works". *Richardson* v. *Buckinghamshire County Council and Others*,[4] in which the claimant was injured when riding a scooter by a construction site, illustrates this qualification. His claim against the Employer, incidentally the Council, failed. However, the Council had to pay the claimant's legal costs because local authorities were then responsible for the legal aid of their residents. The Council sought to recover its expenses from the Contractor under a similarly worded indemnity clause. The court ruled that the loss did not fall under the indemnity because the Council suffered the loss not by reason of the carrying out of the Works but because it was responsible for paying legal aid for its residents. Second, the liability and indemnity will be reduced to the extent that the death or injury is caused by any act or neglect of the Employer or any of the Employer's Persons, which is defined under Clause 1.1 as:

> all persons employed, engaged or authorized by the Employer, excluding the Contractor, Contractor's Persons, the Architect, the Quantity Surveyor and any Statutory Undertaker and including any such third party as is referred to in Clause 3.23...

Such persons include:

- the CDM Co-ordinator, if different from the Architect (Article 5);
- the Principal Contractor if different from the Contractor (Article 6);

---

4 (1971) 69 LGR 527 (hereafter *Richardson* v. *Buckinghamshire*).

- the Employer's Representative (Clause 3.3);
- the Clerk of Works (Clause 3.4);
- other consultants of the Employer;
- employees of the Employer;
- other contractors employed by the Employer under Clause 2.7;
- third parties invited by the Architect to examine fossils, antiquities, etc. found on the site (Clause 3.23);
- statutory undertakers carrying out work outside their statutory duties.[5]

The burden of proof that death or personal injury is caused by the act or neglect of the Employer or of the Employer's Persons rests on the Contractor.

7.2.3 Damage to property

Clause 6.2 states the extent of the Contractor's liability for loss or damage to property in the following terms:

> The Contractor shall be liable for, and shall indemnify the Employer against, any expense, liability, loss, claim or proceedings in respect of any loss, injury or damage whatsoever to any property real or personal in so far as such loss, injury or damage arises out of or in the course of or by reason of the carrying out of the Works and to the extent that the same is due to any negligence, breach of statutory duty, omission or default of the Contractor or any of the Contractor's Persons. This liability and indemnity is subject to Clause 6.3 and, where Insurance Option C (Schedule 3, paragraph C.1) applies, excludes loss or damage to any property required to be insured thereunder caused by a Specified Peril.

The use of the term 'property real or personal' is significant because, in law, property has a wider meaning than physical property. In addition to damage to physical (real) property, the wording of the clause appears wide enough to include infringements of intangible property rights such as easements, rights to light and air and the rights of people not to have enjoyment of their property detracted from by unreasonable levels of smells, noise or vibrations.

Examination of the terms of Clause 6.2 identifies four qualifications to the Contractor's liability for damage to property that need to be examined in some detail. The first three are that liability applies: (i) 'so far as such loss, injury or damage arises out of or in the course of or by reason of the carrying out of the Works'; (ii) 'to the extent that the same is due to any negligence…'; (iii) subject to Clause 6.3. There is a fourth qualification that, where paragraph C.1 of Insurance Option C applies, the Contractor is not answerable for damage against which the Employer is required to insure under that clause. As paragraph C.1 requires the existing structures and their contents to be insured against damage by the Specified Perils, the general intention appears to be that the Contractor is not liable to the Employer for this type of damage.

First Qualification (causation)

The effect of the first qualification is illustrated by the facts in *Richardson* v. *Buckinghamshire.*

---

5 See *Henry Boot Construction Ltd* v. *Central Lancashire New Town Development Corporation* (1980) 15 BLR 1.

Second Qualification (negligence or other fault)

It is illustrated by *Gold* v. *Patman & Fotheringham Ltd*[6] in which piling work damaged adjoining property and the Employer was found liable for the damage in nuisance. His claim under an indemnity clause similar to Clause 6.2 failed because, as the piling had been carried out according to drawings and instructions of the Architect, there had been no 'negligence, omission or default' by the Contractor. The use of the terms 'negligence, breach of statutory duty, omission or default' in this qualification gives rise to the problem that whilst the terms 'negligence' and 'breach of statutory duty' have precise legal meanings, different meanings have been attached to 'default' in the case law. According to Parker J in *Re Bayley Worthington & Cohen's Contract*,[7] it refers to personal conduct raising questions of either not having done what one ought to have done or having done what one ought not to have done. Breach of contract is therefore just an instance of such conduct. This principle was followed in *City of Manchester* v. *Fram Gerard*,[8] another case involving an indemnity, and *Northwood Development Co Ltd* v. *Aegon Insurance Co (UK)*,[9] which concerned the construction of a performance bond. However, the Court of Appeal in *Perar BV* v. *General Surety and Guarantee Co. Ltd*[10] rejected the reasoning in the earlier cases and equated 'default', as a precondition for the enforcement of a performance bond, with breach of contract. Considering the inherent uncertainty of 'not having done what one ought to have done or having done what one ought not to have done' the Court of Appeal's decision is much to be preferred in the context of indemnities.

Third Qualification (the Works are excluded as property)

The 'subject to Clause 6.3' qualification has the general effect that before practical completion or termination of the Contractor's employment under the Contract, the Works and Site Materials (defined under Clause 1.1) are not part of 'property' for the purpose of Clause 6.2. The House of Lords in *Co-operative Retail Services Ltd and Others* v. *Taylor Young Partnership and Others*[11] explained that the effect of the reference in JCT 98 to the equivalent of Clause 6.3 is also that, prior to practical completion, the Contractor is not liable to pay compensation to the Employer for loss or damage to the Works caused by the Contractor's negligence, breach of statutory duty or other default.

As the Contractor promises in Article 1 and Clause 2.1 to carry out and complete the Works, he is under an obligation to reinstate any loss or damage to the Works and Site Materials at his own expense. This is an obligation for which insurance under Insurance Clauses Options A, B or C is required. As explained later, the Contractor is required under Clause 6.4.1 also to insure against his liability for damage to property. The reason for this qualification must therefore be a need to avoid the duplication in insurance cover that would otherwise result.

Any Section for which a Section Completion Certificate has been issued becomes part of property to which Clause 6.2 applies after the date of issue of the Certificate

---

6 [1958] 1 WLR 697; [1958] 2 All ER 497 (hereafter *Gold* v. *Patman*).
7 [1909] 1 Ch. 648.
8 (1974) 6 BLR 70.
9 (1994) 10 Const LJ 157.
10 (1994) 66 BLR 72.
11 [2002] BLR 272; see Section 7.3.4.2 in this chapter for its outline.

(see Clause 6.3.2). Also, where under Clause 2.33, the Employer takes possession of parts of the Works before practical completion, the indemnity applies to the Relevant Part after the Relevant Date.

Fourth Qualification

Qualifications of this kind in the JCT family of contracts have been the subject of much litigation. At the centre of the controversy is the question whether the effect of the qualification is to exonerate the Contractor and his sub-contractors completely from liability for loss or damage to existing structures and their contents caused by a Specified Peril even if that risk, usually fire, was created by the negligence of the Contractor or of his sub-contractor. The answer to this question is very important for two reasons. First, it is possible that even if the Employer effects the required insurance, the insurance proceeds may prove inadequate for the cost of their repair or replacement (e.g. price increases between payment on the claim and time of repair/replacement). Second, the insurance required is of the material damage variety (i.e. it only covers the costs of repair or replacement of the thing insured). There is no requirement for cover against consequential losses arising from the physical damage (e.g. loss of business for the period within which the building is unavailable or is used inefficiently).

The court decisions on the issue have arisen from previous editions or versions of JCT 80 or other contracts in the JCT family. They therefore have to be treated with considerable caution because minor differences in wording or even the use of commas at different places in contracts may give rise to major differences of interpretation. However, most of them are very useful in that they throw some light on the approach followed by the courts in resolving disputes of that kind.

Before examining the cases, it is important to re-emphasize a fundamental conceptual difference between risk and insurance clauses. A risk clause specifies who shall be liable for what risk, thus avoiding disagreements after occurrence of the risk as to who shall shoulder its burden. An insurance clause states who is to take out what insurance against what risk. Its function is to ensure availability of funds to deal with the occurrence of the relevant risk. It is therefore possible that an insurance clause requires one party to take out insurance against risk for which the other party is responsible. The courts have pointed out that it is a misconception to conclude that a party is responsible for a risk simply because the contract requires him to take out insurance against it. For example, on such arguments advanced by sub-contractors in *The National Trust* v. *Haden Young Ltd*,[12] Nourse LJ said at page 11A-B:

> The essential fallacy in them is the assumption that an obligation to insure against loss can, without more, throw that loss on the insured when, by another provision of the contract, it is to be borne by the other party. It is misleading to speak of the apportionment or allocation of risk in this context.

The main source of the problem has been that some risk clauses, epitomized by JCT 05 Clause 6.2, make references to insurance clauses, thus raising the question of whether the intention is to modify the risk allocation structure that would apply without the reference. This issue is essentially a question of the construction of the particular risk clause

---

12 (1994) 72 BLR 1.

together with the insurance clause referred to. In *Surrey Heath Borough Council* v. *Lovell Construction Ltd and Another*[13] Dillon LJ said at p. 121:

> The effect of the contract agreement must always be a matter of construction. People are free to contract as they like. It may be the true construction that a provision for insurance is to be taken as satisfying or curtailing a contractual obligation, or it may be the true construction that a contractual obligation is to be backed by insurance with the result that the contractual obligation stands or is enforceable even if for some reason the insurance fails or proves inadequate.

In a line of cases involving contractors and employers, the contracts, in addition to requiring insurance by the employer, stated that the risk was solely the employer's. In such cases, the decisions were that the contractor was not liable for the risk even where the risk was created by the negligence of the contractor or of his sub-contractors.

> *James Archdale & Co. Ltd* v. *Comservices Ltd*:[14] this arose from a contract let on the predecessor of JCT 05 current in 1952. The Employer was required by the equivalent of today's Insurance Option B to take out All Risk Insurance against damage to the Works. The contract also provided that damage to the Works by fire was the sole risk of the Employer. The Court of Appeal held that the Contractor was not liable for destruction of the Works by fire caused by his own negligence.

In another line of cases, the contracts required the employer to take out insurance against the risk concerned but it was not stated that the risk was the employer's. In those cases, the contractor was held liable for fire damage caused by his own negligence or that of his sub-contractors.

> *The National Trust* v. *Haden Young Ltd*:[15] this arose from a contract for works of repair to Uppark House let on the January 1987 revision of the JCT Minor Works form. Lead work was sub-contracted to the defendants. Clause 6.2 provided that the Contractor should be liable for loss to any property in so far as it arose out of or in the course of the works and to the extent that it was caused by the negligence of the Contractor or of people for whom the Contractor was responsible in law. Clause 6.3(B) required the Employer to insure, in the joint names of the Contractor and the Employer, against loss or damage to the existing structures and their contents. During the course of the sub-contract works, two employees of the sub-contractor went for a tea break leaving a lit oxyacetylene torch. This caused severe destruction of the building and its priceless contents. The sub-contractors argued that the main contractor was not liable and that therefore they were also not liable. The Court of Appeal decided that Clause 6.2 imposed unlimited liability on the main contractor for the loss that had occurred and that the failure of the claimant to insure was an entirely different matter that did not affect the Contractor's responsibility for the risk. *James Archdale* and *Scottish Special Housing* were distinguished on the grounds that, unlike in the present case, the contracts in those cases expressly stated that the risk of fire damage was solely the employer's.

---

13 (1990) 48 BLR 108.

14 [1954] 1 WLR 459 (hereafter *James Archdale*); the House of Lords reached the same decision in *Scottish Special Housing Association* v. *Wimpey* (1986) 34 BLR 1 (hereafter *Scottish Special Housing*) which arose from a contract that contained similar provisions on fire damage to existing structures of the Employer.

15 (1994) 72 BLR 1; see also: *Dorset County Council* v. *Southern Felt Roofing Co. Ltd* (1989) 48 BLR 96; *London Borough of Barking and Dagenham* v. *Stamford Asphalt Co. Ltd and Others* (1997) 82 BLR 25 (hereafter *Barking & Dagenham*).

In *Barking & Dagenham*, which arose from a contract in the terms of the *JCT Agreement for Minor Works*, 1988 revision, the Employer was required to take out and maintain a joint names policy against damage to existing structures and their contents from risks including fire and most of the other Specified Perils under JCT 05. The indemnity clause in favour of the Employer made reference to neither the existing structures nor the Employer's obligation to insure them. The Employer had not taken out the required insurance when fire broke out and caused damage to the building on which the works were being carried out. The Court of Appeal determined that the terms of the Contract did not require the Employer to insure against fire caused by the negligence of the Contractor and that, therefore, the Employer was entitled to compensation under the indemnity clause if the fire was caused by the Contractor's negligence.

It is to be noted that, although *Barking & Dagenham* is generally considered correctly decided on its own facts and contractual terms,[16] it has little relevance to JCT 05 because of the indemnity's expressly existing structures and their contents where the Employer is required to insure them. *Scottish & Newcastle plc* v. *GD Construction (St Albans) Ltd*[17] arose from a contract incorporating the 1984 edition of the *JCT Intermediate Form of Contract* for the refurbishment of the claimant Employer's public house. The contract, in terms almost identical to paragraph C.1 of Insurance Option C in JCT 05, required the Employer to take out insurance in the joint names of the Employer and the Contractor against damage to the existing structures and the contents from listed perils including fire. The Employer failed to do this. A fire broke out and caused considerable loss to the Employer. Relying upon the equivalent of the indemnity under Clause 6.2 of JCT 05, the Employer sought to recover this loss from the Contractor. The Court of Appeal, on a preliminary issue, decided that the indemnity did not apply to loss required to be insured by the Employer. *Barking & Dagenham* was distinguished as limited to the specific wording of the indemnity clause in the contract from which it arose.

The scope for this type of controversy in relation to JCT 05 has been reduced considerably because Clause 6.2 now states expressly that the indemnity of the Contractor under that clause, 'where Insurance Option C (Schedule 3, paragraph C.1) applies, excludes loss or damage to any property required to be insured thereunder caused by a Specified Peril'. The property required to be insured under paragraph C.1 of Insurance Option C is existing structures and their contents either owned by the Employer or for which he is responsible. The loss to be insured against is 'the full cost of reinstatement, repair or replacement'. The Contractor is therefore not liable for these costs arising from damage or loss to existing structures and their contents from a Specified Peril. This was the conclusion as to the effect of the same terms in JCT 80 in *Ossory Road (Skelmersdale) Ltd* v. *Balfour Beatty Building Ltd and Others*.[18] In that case, Judge Fox-Andrews QC held that the Contractor under JCT 80 was not liable to the Employer for the costs of reinstatement of damage to existing structures and their contents by fire caused by the Contractor's own negligence.

The insurance required under paragraph C.1 of Insurance Option C covers only material damage. Consequential losses (e.g. cost of disruption and other additional costs incurred in the carrying on of any business by the Employer in the premises) are not required to be

---

16 See, for example, the judgment of Lord Justice Rix in *Tyco Fire and Integrated Solutions (UK) Ltd* v. *Rolls Royce Motor Cars Ltd* [2008] EWCA Civ 286.
17 [2003] EWCA Civ 16; BLR 131; 86 ConLR 1.
18 (1993) CILL 882 (hereafter *Ossory Road*).

insured against. The Contractor's liability for this associated type of damage was raised in *Kruger Tissue (Industrial) Ltd* v. *Frank Galliers Ltd and Others*[19] which arose from a contract let on JCT 80. Judge Hicks QC started off by construing the equivalent of JCT 05 Clause 6.2 as imposing liability on the Contractor for loss (including consequential loss) arising from damage to property. He then decided that the equivalent of paragraph C.1 of Insurance Option C was an insurance clause and that the Contractor's liability under the equivalent of Clause 6.2 was curtailed only to the extent that the Employer was required to insure against the risk concerned. As the insurance required under the equivalent of paragraph C.1 was of the material damage variety, there was no contractual obligation on the Employer to insure against his consequential loss arising from fire damage. *Ossory Road* was therefore distinguished as limited to only material damage claims.

7.2.4 Personal injury and property damage from Excepted Risks

The effect of Clause 6.6 is that the Employer is not entitled to indemnification against the risks of personal injury or damage to property caused by any of the Excepted Risks, which are defined in Clause 6.8 as consisting of:

> ionising radiations or contamination by radioactivity from any nuclear fuel or from any nuclear waste from the combustion of nuclear fuel, radioactive toxic explosive or other hazardous properties of any explosive nuclear assembly or nuclear component thereof, pressure waves caused by aircraft or other aerial devices travelling at sonic or supersonic speeds.

7.2.5 Risk of damage or loss to Works and Site Materials

Under Article 1 and Clause 2.1, the Contractor undertakes to carry out and complete the Works in accordance with the Contract. Such an undertaking carries with it an obligation to reinstate any damage to work executed or unfixed materials at no additional cost to the Employer unless the Contract is terminated by frustration or, in respect of any particular risk, the Contract itself states expressly that the Contractor is not to have such responsibility. The responsibility of the Contractor to reinstate any damage was asserted in *Charon (Finchley) Ltd* v. *Singer Sewing Machine Co. Ltd*[20] as a general principle applicable to construction contracts in these terms:

> Indeed, by virtue of the express undertaking to complete (and in many contracts to maintain for a fixed period after completion) the contractor would be liable to carry out his work again free of charge in the event of some accidental damage occurring before completion even in the absence of any express provision for protection of the risk.

It is provided in Clause 6.6 that the Contractor is not to indemnify the Employer in respect of loss or damage to the Works and Site Materials from the Excepted Risks. It is submitted that, subject to frustration at common law, this provision does not in any way exempt the Contractor from the responsibility to reinstate the Works. The effect of Clause 6.6 is probably that the Contractor is not responsible for the cost of dealing with contamination from the Excepted Risks.

---

19 (1998) 57 ConLR 1.
20 (1968) 112 SJ 536.

7.3 JCT 05 insurance requirements

As explained earlier, the Contractor's obligations to indemnify the Employer or to reinstate any damage to the Works and Site Materials are useless unless he has the necessary funds. It was also explained that insurance is the customary way of ensuring availability of the necessary funds. It is for these and other reasons explained later that JCT 05 provides for insurance against the following:

1. the Contractor's liability under Clause 6.1 – liability for death and personal injury (Clause 6.4.1);
2. the Contractor's liability under Clause 6.2 – liability for injury and damage to real and personal property (Clause 6.4.1);
3. injury or damage to property due to collapse, subsidence, vibrations, weakening or removal of support or lowering of groundwater attributable to the carrying out of the Works (Clause 6.5.1);
4. damage to the Works and Site Materials (paragraph A.1 of Insurance Option A, or paragraph B.2 of Insurance Option B or paragraph C.2 of Insurance Option C);
5. damage to existing structures and their contents arising from the Specified Perils (applies to contracts involving extensions or refurbishment of existing facilities) (paragraph C.1 of Insurance Option C);
6. loss or damage to off-site materials or goods from the Specified Perils (Clause 4.17.2.2);
7. the Contractor's liability for design of any Contractor's Designed Portion (Clause 6.11.1).

With most of these items of insurance, the party to insure (the Insuring Party) is to obtain a type of cover referred to in the Contract as a 'Joint Names Policy'. JCT 80 defined this term under Clause 1.3 as 'a policy of insurance which includes the Contractor and the Employer as the insured'. The intended effect of this requirement was to prevent an insurer of those risks from exercising its subrogation rights against the Contractor or the Employer. An insurer's subrogation right includes that of being able to stand in the shoes of the insured and to seek remedies against whoever actually caused the relevant damage or loss. In *Petrofina (UK) and Others* v. *Magnaload Ltd and Others*[21] it was held that where a policy names more than one party as insured under it, the insurer could not exercise its subrogation rights against a co-insured after meeting a claim from another co-insured. However, the Court of Appeal's decision in *Barking & Dagenham*[22] cast some doubts as to whether a Joint Names Policy as defined above was effective in limiting the insurer's subrogation rights in the intended way. In that case, it was argued that where the insurance covers damage to property, the insurer may still have subrogation rights against any insured without any insurable interest in the property. Accepting this argument, the trial judge held that if the Employer takes out a Joint Names Policy against damage to the Employer's existing buildings and their contents, the insurer would still be entitled to look

21 (1983) 25 BLR 37.

22 (1997) 82 BLR 25. See also *AMEC Civil Engineering Ltd* v. *Cheshire County Council* [1999] BLR 303 in which it was held by Judge Gilliland QC that a Contractor under the 5th edn of the Institution of Civil Engineers Conditions was entitled to payment by the Employer under the Contract for dealing with a risk insured under a joint names policy without any allowance for receipts under the policy.

to the Contractor for compensation if the Contractor caused the damage, because of his lack of insurable interest.

The trial judge's decision was upheld on appeal but on different grounds. Although the Court of Appeal expressly declined to consider that argument,[23] the JCT still responded by re-defining the Joint Names Policy which has been continued in JCT 05 as:

> A policy of insurance which includes the Employer and the Contractor as the composite insured and under which the insurers have no right of recourse against any person named as an insured, or, pursuant to Clause 6.9, recognized as an insured thereunder.

This requires that the policy expressly waives the insurer's rights of subrogation against every party named in the policy as an insured.

It is to be noted that, under Clause 6.6, there is no obligation on the Contractor to insure against any personal injury to or death of any person or any damage to any property (the Works and Site Materials included) by the effect of the Excepted Risks. As such insurance is not available in the insurance market at affordable rates, this provision reflects commercial reality.

It is explained in Section 7.3.4.2 in this chapter that a third party liable to the Employer for loss or damage covered by a Joint Names Policy is not entitled to a contribution from a co-insured whose default also caused the loss or damage.

7.3.1 Insurance against personal injury or death

This category of insurance is to cover the Clause 6.1 liability of the Contractor for death and personal injury. By Clause 6.4.1.1 the Contractor's insurance against this risk must comply with all relevant legislation. For insurance purposes, it is important to distinguish between the Contractor's own direct employees and others. The latter group includes the public at large and other parties working on the site. The minimum cover against injury to the Contractor's employees is as stated in the Employer's Liability (Compulsory Insurance) Act 1969. The Contractor's insurance against liability for death and injury of people other than his own employees should comply with the minimum amount of cover stated in the Contract Particulars.

The cover is for any occurrence or series of occurrences of death or injury arising from the same event. If the Contractor's actual liability exceeds the amount of cover he must make up the shortfall from his own resources. It follows therefore that the minimum amount of cover should be decided bearing in mind the magnitude of the risk and the Contractor's financial capabilities. The Contractor's potential liability under Clause 6.1 may be substantially more than the minimum cover required under the Contract. This is because the amounts of damages for personal injuries being awarded by the courts have increased tremendously over the years. For this reason, the equivalent of Clause 6.4.1 in previous editions of the Contract was often amended to require unlimited cover. This strategy must be reconsidered because, in recent years, unlimited cover is rarely available in the insurance market.

---

23 Decisions after *Co-operative Retail Services Ltd and Others* v. *Taylor Young Partnership and Others* [2002] BLR 272 have tended to distinguish *Barking & Dagenham* as limited to the particular wording of the contract from which it arose.

By law the Contractor must maintain Employer's Liability Insurance for as long as he is an employer. JCT 05 is silent on the question of the period for which the Contractor is to maintain his insurance against death of and injury to people other than his employees. The common practice is to maintain the cover, at least, up to practical completion. Without such cover, the Contractor would be exposed to this risk if he has to return to remedy defects during the Rectification Period.

7.3.2 Insurance against damage to property

The Contractor is to take out and maintain insurance against his liability under Clause 6.2 (i.e. liability in respect of damage or injury to real and personal property). The amount of cover is subject to a minimum stated in the Contract Particulars.

7.3.3 Insurance against Employer's Clause 6.5.1 Risks

The Contractor is only responsible for damage to property due to negligence, breach of statutory duty, omissions or default on the part of the Contractor or of parties for whom the Contractor is responsible in law. Sometimes, damage can occur without any fault on the part of the Contractor (e.g. the damage in the case of *Gold* v. *Patman*[24]). Against this type of damage, the Employer may have some protection through the professional liability insurance of his professional team if the damage is due to their professional negligence. Where the damage is due to neither the fault of the Contractor nor professional negligence, the Employer will have to foot the bill. Clause 6.5.1 envisages that a prudent Employer may wish to insure against this type of liability. If that is the case, it should be indicated in the Contract Particulars that insurance under Clause 6.5.1 may be required. The amount of cover per occurrence or series of occurrences arising from the same event must also be stated. Furthermore, the Architect must instruct the Contractor to take out the insurance. The cost of effecting and maintaining this category of insurance is to be added to the Contract Sum.

Clause 6.5.1 insurance covers injury and damage to any property caused by collapse, subsidence, vibrations, weakening or removal of support or lowering of groundwater attributable to the carrying out of the Works. The clause expressly excludes, from the required cover, damage and injury:

- for which the Contractor is responsible under Clause 6.2;
- attributable to defective design (which would normally be covered by the professional indemnity insurances of the designers and the Contractor's Designed Portion Professional Indemnity Insurance);
- which are reasonably foreseeable as the inevitable consequence of the carrying out of the Works;
- to existing structures and their contents if Insurance Option C applies;
- to the Works and Site Materials;
- covered by any other insurance already arranged by the Employer;
- arising from the consequences of war, invasion, etc.;

24 [1958] 1 WLR 697; [1958] 2 All ER 497.

- arising from or caused by the Excepted Risks;
- caused by or arising from pollution or contamination of buildings, land, water, or the atmosphere;[25]
- that results in the Employer's liability for his breach of contract.

The reason for their exclusion is that either they are covered by insurances required under other clauses of this Contract or other contracts or cover is not available in the insurance market at commercially reasonable rates.

The insurance must be in the joint names of the Contractor and the Employer and placed with insurers approved by the Employer. The Contractor must send the policy and all premium receipts to the Employer through the Architect. If the Contractor fails to take out or maintain the insurance, the Employer can arrange remedial insurance or pay the premiums himself. Apart from seeking damages under the general law, the Contract does not provide the Employer with any sanction against this type of default. In practice, the cover required by Clause 6.5.1 is commonly provided as an extension to the Contractor's public liability policy (explained later). Where this practice is not followed, it is underwritten as a separate policy.

7.3.4 Insurance of Works and Site Materials

There are three alternative clauses for the insurance of the Works and Site Materials: Insurance Options A, B, or C of the Schedule 3. Whichever of these is to apply to the particular project should be completed in the Contract Particulars.[26] Each of these clauses requires the same insurance cover described as 'All Risk Insurance'. However, there are differences between them regarding who is responsible for taking out and maintaining the cover and the administrative procedures to be complied with concerning verification of compliance and management of occurrence of the risk.

Insurance Option A is intended for projects involving new work where the intention of the parties is that the Contractor is the Insuring Party. Insurance Option B is also intended for new work but where it is the Employer who is responsible for obtaining the cover. At first sight, it appears strange that an Employer should take on such responsibility when it is the Contractor who carries the relevant risk. This arrangement is clearly not suitable for inexperienced employers. However, where the Employer has repetitive programmes of procurement (e.g. developers, local authorities and retail chains) he may be in a position to obtain cheaper insurance on account of bulk business and the economies of scale. Insurance Option C is to be used where the Works consist of refurbishment of, or modifications to, or extension of existing structures. Here the Employer is the Insuring Party. The rationale for putting this responsibility on the Employer is that, as an owner of the existing buildings, the likelihood is that he will already have building insurance cover. It would therefore be more convenient and cheaper for the Employer to negotiate the insurance required under Insurance Option C as an extension to any such cover.

---

25 The insurance requirement covers pollution or contamination caused by a sudden identifiable, unintended and unexpected incident which occurs completely at a specific moment during the period of the insurance.

26 See Chapter 1, Section 1.5.2 dealing with the completing the Contract Particulars entries for Clause 6.7 and Schedule 3 Insurance Options A, B or C.

7.3.4.1 All Risk Insurance

'All Risk Insurance' is defined in Clause 6.8 as 'insurance which provides cover against any physical loss or damage to work executed and Site Materials and against the reasonable cost of the removal and disposal of debris and of any shoring and propping of the Works which results from such physical loss or damage'. However, costs of repair, replacement or rectification of the following are expressly excluded:

1. defects in property due to wear and tear, obsolescence, deterioration, rust or mildew;
2. defective work and Site Materials attributable to design, materials or workmanship of poor quality;
3. loss/damage caused by war, invasion, revolution, etc., ordered by any government or public authority;
4. loss/damage from disappearance or shortage revealed only by an inventory and not attributable to any specific event;
5. loss or damage from the Excepted Risks.

From the list of exceptions, it should be apparent that All Risk Insurance does not in any way cover all risks of damage to the Works and Site Materials and that the term is therefore a misnomer. For example, as illustrated by the facts of *AMEC Civil Engineering Ltd* v. *Norwich Union Fire Insurance Society Ltd*,[27] it is unusual for a contractor's All Risk Insurance policy to cover the risk of defective materials or work. In that case it was discovered that concrete blocks used in the construction of a sea wall were defective. It was held in the Technology and Construction Court that the contractor was not entitled to claim the cost of replacing the defective items under the contractor's all risk insurance policy.

7.3.4.2 Insurance under Insurance Option A (by Contractor)

By paragraph A.1 the Contractor is responsible for taking out and maintaining this insurance which must be a Joint Names Policy against the risks covered by the Clause 6.8 definition of 'All Risk Insurance'. As explained earlier, the intention behind the requirement for a Joint Names Policy is to prevent the possibility of the insurer pursuing the Employer for damages or a contribution through exercise of its rights of subrogation.

The minimum amount of cover should be the full reinstatement value of the Works plus a percentage for professional fees. This percentage is to be stated in the Contract Particulars. Failing such statement, a default figure of 15% applies.

As the reinstatement value of the works will increase over the contract period, ideally the policy must allow a gradual increase of the amount of maximum cover but this is rarely done in practice. The worst possible loss is where the Works are destroyed just before practical completion. At this stage their value may be more than the Contract Sum because of inflation and Variations. Whilst conservatism would call for an appropriate allowance for these increases, it is often argued that, as a complete loss of the Works is uncommon, such an approach would be wasteful of premiums. The strategy to adopt would depend upon the risks inherent in the particular project and the parties' preferred ways of dealing with them. However, it has to be borne in mind that where the preferred strategy demands a departure from the requirements of Insurance Option A, the clause should be amended accordingly.

The insurance is to be maintained up to and including the date of whichever of these occurs first: (i) the issue of the Practical Completion Certificate; or (ii) termination of

---

27 [2003] EWHC 1341(TCC).

the employment of the Contractor. There is therefore a need on the part of the Architect to advise the Employer to ensure that the right insurance is in place by the time the Contractor's insurance comes to an end. The obligation to maintain All Risk Insurance ceases in relation to any Section for which a Section Practical Completion Certificate has been issued. It also terminates in respect of the Relevant Part from the Relevant Date where the Employer enters into partial possession.[28]

Paragraph A.2 makes elaborate provisions for the enforcement of the obligation of the Contractor to take out and maintain this element of insurance. First, the insurers must be approved of by the Employer. Second, the Contractor must send to the Employer through the Architect, the policy, premium receipts, and any endorsements for safekeeping. Finally, if the Contractor fails to comply with any of the insurance obligations, the Employer may take out any remedial cover. In such an event, the premiums paid or payable are recoverable from the Contractor against payments due or as debt. It is submitted that the Employer may also be entitled to recover other sums as damages for breach of contract (e.g. cost of managerial time expended).[29]

Paragraph A.3 recognizes the common practice whereby contractors take out, on an annual basis, a single policy, the so-called Contractor's All Risk Insurance, for all their contracts undertaken in the year covered by the insurance. Provided the annual policy meets the Joint Names Policy requirements and covers the Works, the Contractor will not be required to take out a policy specific to the present Contract. Also, although there is no requirement that the annual policy is deposited with the Employer, the Contractor must provide him with evidence that the policy is being maintained (e.g. premium receipts[30]). If the Contractor fails to comply with these requirements the Employer may take out and maintain the relevant insurance with consequences similar to those explained already.

Upon discovery of any occurrence of loss or damage covered by the policy, the Contractor must forthwith give notice in writing to both the Architect and the Employer (paragraph A.4.1), specifying the nature, extent, and location of the loss or damage. He must also comply with the requirements of the policy. This will usually mean informing the insurer and submitting a claim on the insurer's standard form. It is common practice for the insurer to instruct loss adjusters to visit the Works and assess the amount of damage before paying on the policy. After inspection by the insurer or his agents the Contractor must carry out all the necessary work of repair, restoration, and clearance and disposal of wastes. He is also to proceed with the carrying out of the Works to completion.

The Contractor, for himself and all his sub-contractors insured in respect of the damage, must instruct the insurer to pay the insurance proceeds to the Employer (paragraph A.4.4 of Insurance Option A). Without this provision, the insurer would have to make payment in the joint names of all the insured. After deducting the professional fee element, the rest is to be paid to the Contractor by instalments under payment certificates. Eaglestone[31] notes that, in practice, in getting paid for the remedial work, the Contractor deals directly with the insurer and that the Architect and the Employer are hardly ever involved.

The damage is to be ignored when carrying out valuation for certification (paragraph A.4.2). This means that, in effect, the Contractor is paid twice for the same work. It is to

---

28 See Clause 2.36.

29 For detailed explanation of the recoverability of the cost of management time spent in dealing with a breach of contract, see Chapter 13, Section 13.4.3.

30 Where the annual policy is to cover the Works the date of its renewal is to be completed in the Contract Particulars.

31 Eaglestone, F., *Insurance under the JCT Forms*, 2nd edn (Blackwell Science, 1996), at p. 175.

be noted that for the remedial work the Contractor is paid no more than the insurance proceeds less the professional fees. Any shortfall between actual costs and the residue of the proceeds is therefore to be borne by the Contractor.

Case law suggests that the effect of the Joint Names Policy is that any party not covered by the policy (e.g. the Architect, and whoever caused insured loss or damage) would be answerable to the insurers for the whole loss even if negligence of the Contractor or his sub-contractors contributed to it. This was the unanimous conclusion of the House of Lords in *Co-operative Retail Services Ltd and Others* v. *Taylor Young Partnership and Others*.[32] The claimant (CRS) engaged Wimpey as the Contractor for the construction of a new office block under JCT 80, Private with Quantities. Taylor Young Partnership (TYP) was the firm of the Architect. Hoare Lea and Partners (HLP) were consulting engineers to the Employer. Clause 22A (the equivalent JCT 05 Insurance Option A) applied. Just before practical completion a fire broke out on the site, causing extensive damage to the works. In exercise of their subrogation rights through CRS, the insurers brought action against TYP and HLP, who brought in Wimpey and its sub-contractor (Hall) for a contribution on the grounds that the fire was caused by Wimpey's breach of the contract and Hall's breach of a collateral warranty provided to CRS. The insurers relied on s. 1(1) of the Civil Liability (Contribution) Act 1978, which provides that '... any person liable in respect of any damage suffered by another person may recover contribution from any other person liable in respect of the same damage (whether jointly with him or otherwise)'. The insurers' success therefore depended on Wimpey and Hall being liable to CRS for the loss. The House of Lords held, on a preliminary issue, that the effect of Wimpey and Hall being co-insured under the policy covering damage to the Works from fire was that any claim by CRS against them in respect of the damage was barred.

This outcome was extremely harsh on the designers as they carried sole responsibility for the damage although they were not the sole cause of the fire. In the interest of maintaining team spirit within the supply chain this type of insurance arrangement needs to be rethought in the context of collaborative working arrangements.

7.3.4.3 Insurance of Works under Insurance Option B (by Employer)

The provisions in this clause are a mirror image of Insurance Option A. However, there are two key differences. First, the positions of the Contractor and the Employer are interchanged regarding the insurance obligations. It is the Employer who must take out the Joint Names Policy. Second, the remedial works are to be valued and paid for as a Variation in accordance with Section 5. The Employer must therefore meet any shortfall between the insurance proceeds and the total valuation of the remedial work.

7.3.4.4 Insurance of Works, Insurance Option C, Paragraph C.2 (by Employer)

These provisions are virtually the equivalent of those under Insurance Option B. However, there is one major difference. Either party may terminate the employment of the Contractor within 28 days of the occurrence of the loss or damage provided it is just and

32 [2002] BLR 272.

equitable to do so. The procedure for any such termination and the financial settlement are discussed in Chapter 16, Section 16.13.

7.3.4.5 Insurance of existing structures under Insurance Option C Paragraph C.1 (by Employer)

The cover required here is referred to in some circles as 'Fire and Special Perils' policy. The Employer is to take out and maintain a Joint Names Policy against damage to the existing structures and their contents from the Specified Perils. The amount of cover is the full cost of reinstatement, repair or replacement arising from the damage. The existing structures are to include from the Relevant Date any Relevant Part handed over under Clause 2.33. Surprisingly, there is no requirement to provide the same protection for any Section for which a Section Practical Completion Certificate has been issued. The insurance is to be maintained up to and including the date of whichever of these occurs first: (i) the issue of the Practical Completion Certificate; or (ii) termination of the employment of the Contractor under the Contract whether or not the termination is contested.

When a claim is submitted for compensation for loss or damage arising from the insured risks, the Contractor must authorize the insurer to pay the insurance proceeds to the Employer.

The *Co-operative Retail Services* and *Scottish Special Housing Association* cases discussed in Sections 7.2.3 and 7.3.4.2 in this chapter highlight the principle that a party co-insured under a Joint Names Policy is not liable for loss or damage covered by the policy even where that party's negligence or other breach of duty caused the loss or damage. As demonstrated by *Scottish & Newcastle plc* v. *GD Construction (St Albans) Ltd*[33] this principle can have the most drastic of consequences for the Employer if the Employer fails to take out and maintain the required insurance. The Court of Appeal, on a preliminary issue, decided that the Employer was not entitled to be indemnified by the Contractor for loss required to be insured by the Employer.

7.3.4.6 Terrorism cover

The definition of 'All Risk Insurance' under Clause 6.8 is so wide that the obligation under paragraphs A.1, B.1, and C.2 of Schedule 3 to take out and maintain insurance against loss or damage to executed work and Site Materials must cover damage or loss from terrorism. Also, because there is nothing in the Clause 6.8 definition of Special Perils to exclude damage from terrorism, insurance of existing structures and their contents under paragraph C.1 of Schedule 3 must also include damage to existing structures and their contents from a Specified Peril. This width of the scope of insurance has been a common feature of the preceding editions of JCT 05 contract.

In the 1990s, increased terrorist activity in England (e.g. bomb explosions in Canary Wharf in London and in the commercial centre of Manchester) resulted in withdrawal by insurers of full cover in respect of terrorism from their standard commercial policies on account of astronomical rises in levels of claims. The Government stepped in to make full cover available by establishing Pool Reinsurance Company Ltd, the function of which

---

33 [2003] EWCA Civ 16; BLR 131; 86 ConLR 1; see Section 7.2.3 in this chapter under 'Fourth Qualification' of the present Chapter for commentary on this case.

is to take the risk away from insurers. This enables terrorism cover withdrawn from the standard policies to be bought back with premiums graded by the risk associated with the zone of the location of the works. It is these premiums that go to fund the Government's scheme.

It is assumed in the drafting of the insurance clauses in JCT 05 that the Pool Reinsurance arrangement is in force and that whoever is responsible for taking out the All Risk Insurance will purchase any shortfall between the wide requirements of the Contract and the standard policies available in the insurance market. However, the Government has retained the right to cancel the scheme whenever it proves unworkable, giving rise to the possibility that terrorism risk insurance effected at commencement may cease to be available prior to practical completion. The strategy adopted in the drafting of the insurance clauses appears to have been one of demarcating the insurance requirements associated with the terrorism risks and addressing the uncertainty about future availability or affordability of cover through appropriate provisions. At the heart of the demarcation is the definition of 'Terrorism Cover' in Clause 6.8 as:

> Insurance provided by a Joint Names Policy under Insurance Option A, B, or C for physical loss or damage to work executed and Site Materials or to an existing structure and/or its contents caused by terrorism.

The widening of the scope of the cover is to be noted. Whilst the definition under JCT98 was limited to physical damage caused by fire or explosion, the current definition covers other risks such as physical damage by chemical and other types of contamination.

Clause 6.10 provides for how the parties are to deal with non-availability of Terrorism Cover before the Works reach practical completion. The provisions anticipate the insurer of the Joint Names Policy notifying the Insuring Party that from a stated date, referred to in the contract as the 'cessation date', Terrorism Cover will cease to be available. This notice is referred to hereafter in this work as the 'Insurer's Notification'. The Insuring Party must then inform the other party in writing of this notice.

It is for the Employer to decide whether the Contractor is to continue with the carrying out of the Works or the Contractor's employment is to be terminated. The Employer is to serve written notice on the Contractor specifying which of these courses of action he has decided to take. If the Employer decides on termination of the Contractor's employment he must state in the notice the date from when the termination is to take effect. This date must be later than the date of the Insurer's Notification but not later than the cessation date. The effect of termination is explained in Chapter 16, Sections 16.12 and 16.15.

If the Employer elects to continue with the project, he thereby accepts complete responsibility for damage or loss to executed work or Site Material from terrorism. Any work of restoration and related work arising from such damage or loss is therefore to be paid for as a Variation.

Where Insurance Option A applies, the Contractor is entitled to recover, at renewal, any additional premium on the Terrorism Cover and the Contract Sum is to be adjusted accordingly (Para A.5.1). A Local Authority Employer may instruct the Contractor not to renew the cover (Para A.5.2). The effect of such an instruction is that from the date when Terrorism Cover ceases the Employer assumes complete responsibility for damage or loss to executed work or Site Materials from terrorism as already explained. As there is no requirement for advance notice of such premium changes, the Employer may discover the increase in premium when it is too late to terminate the cover.

What amounts to terrorism is not defined in JCT 05. Section 2(2) of the Reinsurance (Acts of Terrorism) Act 1993, the legislation from which the Pool Re arrangement came into existence, defines 'acts of terrorism' as:

> acts of persons acting on behalf of, or in connection with, any organization which carries out activities directed towards the overthrowing or influencing, by force or violence, of Her Majesty's government in the United Kingdom or any other government *de jure* or *de facto*.

Most of the policies available in the insurance market have therefore adopted this definition. It is to be noted that an important requirement of the definition is that the act in question must be linked to an organization that carries out activities aimed at overthrowing or influencing a government by force or violence. The activities of the organization itself need not be violent. For example, an organization that does not engage in violent activities but funds the overthrow or influencing of governments by violent means would be within the definition.

Section 2(3) provides that the term 'organization' includes 'any association or combination of persons'. There is no upper or lower limit to the number of members required to amount to an 'organization'. Neither is there a minimum duration of existence required for the purposes of the definition. However, the ordinary meaning of the word suggests that a fleeting combination of people is unlikely to constitute an organization.

The Reinsurance (Acts of Terrorism) Act 1993 was passed to address problems of reinsurance in relation to terrorism mainly connected with Northern Ireland. The sources and underlying motives of terrorism have been on the increase. The Terrorism Act 2000 therefore defines 'terrorism' in much wider terms. It now includes acts or threats of acts of serious violence against persons or serious damage to property where such action or the threat of it has the purpose of advancing a political, religious or ideological cause.

7.3.4.7 Insurance and sub-contractors

Under Clause 6.9, all sub-contractors are to be protected against liability for damage/loss to the Works from the Specified Perils. Two methods to this end are provided: (i) either it is provided in the Joint Names Policy that, in respect of those risks, each sub-contractor is insured; (ii) or it is stated that the insurer's rights of subrogation against the sub-contractor are waived.[34]

It is to be noted that where Insurance Option C applies, there is no requirement to protect sub-contractors against liability for damage/loss to existing structures and their contents from the Specified Perils. This omission has two serious consequences for the liability of a sub-contractor for damage or loss to the existing structures and their contents from a Specified Peril caused by his negligence. First, as evidenced in *British Telecommunications plc* v. *James Thomson and Sons (Engineers) Ltd*,[35] the Employer may successfully sue the sub-contractor in tort. In that case, which arose from a contract in the terms of JCT 80, nominated sub-contractors, but not domestic sub-contractors, were required to be protected by the Employer's insurance of the existing structures and their contents against damage/loss

34 The *Barking and Dagenham* litigation suggests that merely stating that the sub-contractor is an insured may not be enough and that it is safer to require the insurer to waive its subrogation rights.
35 (1999) BLR 35.

caused by the Specified Perils. A domestic sub-contractor, invoking the so-called 'contract structure' defence to liability in tort, argued that, in view of the Employer's obligation to insure against the risk, it would not be fair, just and reasonable to impose on it a duty of care in tort to avoid causing the damage or loss. Rejecting the argument, the House of Lords pointed out that it was instructive that the main contract provided expressly that nominated sub-contractors, but not domestic sub-contractors, were to be protected against liability. Second, the Employer's insurer would be able to exercise its subrogation rights against the sub-contractor. As noted by their Lordships, it would therefore appear that the exclusion of domestic sub-contractors from the protection of the Employer's policy was intended to reduce the premiums for such cover.

The contract structure bar was successfully applied in *John F Hunt Demolitions Ltd* v. *ASME Engineering Ltd*[36] to negative the existence of a duty of care on the part of a domestic sub-contractor in respect of damage to existing structures by fire caused by the sub-contractor's negligence. That case arose from a contract in the form of the *JCT Standard Form of Building Contract, With Contractor's Design, 1998 edition*, which provided for liabilities, indemnities and insurance in respect of existing structures and their contents in terms substantially the same terms as JCT 05. It was an express requirement in the main contract that the Employer's Joint Names Policy against damage to the Works and Site Materials should cover sub-contractors by way of naming them as co-insureds or express waiver of the insurer's subrogation rights against them. There was no requirement to give sub-contractors the corresponding benefit of the Employer's Joint Names Policy against damage to existing structures and their contents. It was therefore arguable that, by analogy with *BT* v. *Thomson*, the Employer's insurer of the existing structures and their contents should have been entitled to exercise its subrogation rights against the sub-contractor by a tort action in the Employer's name.

However, HHJ Coulson QC distinguished *BT* v. *Thomson*[37] as having been decided on the precise terms of the contract in that case. The analysis adopted by the judge drew support from his perception of the combined effect of the main and sub-contract conditions in the instant case, particularly two provisions in the main contract. First, the main contract provided for an extensive Contractor's indemnity to the Employer in respect of the negligence of the Contractor and his sub-contractors and agents in relation to damage to property but with express exclusion of damage to the existing structures and their contents. Second, the contract required the Employer to take out insurance in the joint names of the Employer and the Contractor against damage to the existing structures and their contents by the Specified Perils. He concluded from these provisions that the sub-contractor would have tendered with the understanding that the Employer was to look to the insurer to pay for the cost of reinstatement of any damage to the existing structures by the Specified Perils and that, therefore, the sub-contractor did not owe the Employer a duty of care in respect of the damage to the existing structures by fire.

The position under JCT 05 therefore appears to be that sub-contractors are to have the full benefit of the Employer's Joint Names Policy in respect of the existing structures and their contents. Such wrap-up cover would be in the spirit of collaborative working arrangements although premiums will reflect the fact that the loss cannot be passed by the insurer to anybody else on the project.

---

36 [2007] EWHC 1507 (TCC).
37 (1999) BLR 35.

7.3.5 Insurance of off-site materials

Under Clause 4.16, the value of off-site materials intended for the Contract must be included in Interim Certificates if they belong to the Listed Items, which is defined in Clause 1.1 as 'materials, goods and/or items prefabricated for inclusion in the Works which are listed as such items by the Employer in a list supplied to the Contractor and annexed to the Contract Bills'. A precondition for such inclusion is that such materials are insured for their full value against damage or loss from the Specified Perils: Clause 4.17.2.2. The cover is required for the period commencing with transfer of ownership of the materials to the Contractor until their delivery to the site. It is further required that the policy protects the interests of the Employer and the Contractor in respect of the insured risks. The Employer's interest is as an owner of the materials where, in accordance with Clause 2.2.5, they have been included in Interim Certificates and paid for by the Employer. To protect this interest, the policy must, at the very least, provide that the Employer is an insured and that he is to be indemnified against loss or damage to the materials from the Specified Perils after they have been so paid for.

7.3.6 Contractor's Designed Portion Professional Indemnity Insurance

It is explained in Chapter 4, Section 4.2 that, in carrying out design of any Contractor's Designed Portion, the Contractor is under a duty to exercise the standard of skill and care expected of a reasonably competent architect or other appropriate designer. He would be liable for any loss or damage caused to the Employer by failure to exercise such skill and care. Depending upon how the Contract Particulars are completed against Clause 6.11 and commercial availability, the Contractor would be under a duty to take out insurance against this liability. This type of insurance is referred to in the Contract as 'Professional Indemnity Insurance'(PII). The entries called for against Clause 6.11 are designed to specify the level of cover and the period for which the cover must be maintained.

7.3.6.1 Level of cover

The Contract Particulars allow for the insertion of separate amounts for the minimum cover for PII. There is provision for insertion of a separate amount of minimum insurance cover against pollution/contamination. The consequence of stating the amount of cover for PII but omitting to do so for the pollution/contamination insurance is that the same amount of cover must be maintained for the latter. If no amount is inserted for PII the Contractor is under no obligation to take out such insurance. It is to be further stated whether the amount of cover applies to claims or a series of claims arising from one event or it is an aggregate amount of cover for all claims made within any one period of insurance. The default position is that the cover required is for the aggregate amount for any one period. The period of insurance is one year unless a different period is stated.

7.3.6.2 Period of cover

Professional indemnity insurance in general is available on a 'claims made' basis (i.e. the insurer is obliged to indemnify the insured only if the claim is made within a period during which the cover is in force). As liability for design error could arise many years after the making of the error or even after practical completion, the cover must be maintained well beyond practical completion. The period after practical completion during which cover must be maintained should be specified in the Contract Particulars. It is either 6 or 12 years. If none is specified a default period of 6 years applies.

7.3.6.3 Commercial availability

The Contractor's obligation to carry PII is qualified by the proviso that 'it remains available at commercially reasonable rates'. The Contractor is required by Clause 6.12 to serve notice upon the Employer if the cover ceases to be available. The parties are then to reach agreement as to how best to protect their respective positions in the absence of such insurance. It is important to note that non-availability does not affect the Contractor's design liability.

7.4 Enforcement of insurance requirements

Reflecting the importance of having appropriate insurance cover, the Contract, in addition to stating the insurance requirements, addresses how they may be enforced. Generally, a party with the responsibility to arrange the cover must, upon request by the other party, provide evidence of compliance. Furthermore, in all cases of failure of a party to take out and maintain a policy as required by the Contract, the other party may do so at the expense of the defaulting party. In practice, this is often almost impossible to do because the party to arrange in default will not usually have the information required by insurers as part of their underwriting procedures. For example, the Employer is unlikely to have details of the Contractor's safety records.

7.5 Effect of Joint Fire Code

As a result of grave concerns over high levels of claims on account of fires on construction sites, a *Joint Fire Code of Practice on the Protection from Fire of Construction Sites and Buildings Undergoing Renovation* has been published by the Construction Confederation and the Fire Protection Association with the support of the Association of British Insurers, the Chief and Assistant Chief Fire Officers Association and the London Fire Brigade. As one of the definitions under Clause 1.1, JCT 05 refers to the document as the 'Joint Fire Code'. Designed to ensure that adequate fire detection and protection systems are incorporated during the design and planning stages of projects and that work on site is carried out to the highest standards of fire safety, it imposes certain obligations on employers, contractors, and construction professionals involved on construction projects. For example, the Employer must require the parties he appoints to the project to discharge their duties under the CDM Regulations to ensure that the risk of fire is properly assessed and kept to a minimum.

The code applies only if it is completed against Clause 6.13 in the Contract Particulars that it is the case. Although the application of the code is not mandatory, whether or not it applies affects the level of premiums or even the availability of cover from some insurers. It is stated in the code that its scope should normally be appropriate to projects of original contract values of £2.5 million or more although it should also be required on smaller projects with serious risks of fire. Where it applies, it should be further completed in the

Contract Particulars whether or not the insurer of the Works has specified the Works as a 'Large Project'. According to the Joint Fire Code current at the time of writing, 'large projects' are those with an original value of £20 million or more although a lower value contract may be so designated, particularly where it involves considerable 'hot works' (i.e. welding, grinding, the use of open flames or the application of heat).[38] Where Insurance Option A applies, the Contractor, as part of his tendering procedures, must have contacted insurers and collected this information. Similarly, where Insurance Option B or C applies, the Employer must consult his insurers to collect this data.

Clause 6.14 requires the Employer and the Contractor to comply with their obligations under the Code. They are also under a duty to ensure compliance by the Employer's Persons and the Contractor's Persons, respectively. A party suffering loss from a breach of the Code for which the other is responsible under the Contract is entitled to damages for breach of contract by the other.

Clause 6.15.1 contemplates a situation where an insurer of the Works, in response to a breach of the Code, specifies Remedial Measures to be carried out by a specified date. The party who receives this communication is required to copy it to the other party and the Architect. It is the Contractor's responsibility, in compliance with any relevant Architect's instructions, to carry out the measures by the specified date if they relate to the Contractor's obligation to carry out and complete the Works.

Clause 6.15.1.2 provides that where Remedial Measures ordered by the insurers require a Variation or other Architect's instruction the Architect must issue the appropriate instruction to the Contractor. Where the Remedial Measures are urgent the Contractor may supply the necessary material and carry out the necessary work without waiting for the instruction. Such materials and work are to be treated as a Variation unless the work or materials relate to the Contractor's Designed Portion.

Clause 6.15.2 authorizes the Employer to arrange the carrying out of the Remedial Measures at the cost of the Contractor if the Contractor, without reasonable cause, fails to start them within 7 days of receipt of the insurer's notice or to proceed regularly and diligently[39] with them. It is important to note that, in contrast to the provision in Clause 3.11 on the Contractor's failure to comply with an Architect's instruction, there is no requirement for the Contractor to be reminded in writing of this default before the Employer is entitled to intervene in this way. It is also to be noted that the Contractor is responsible in this regard for 'all additional costs incurred by the Employer in connection with such employment'. This would include the Employer's internal costs and not just monies paid to others external to the Employer's organization.

The parties need to have considered the possibility of the Joint Fire Code current at the Base Date being revised or amended before practical completion. Whether such a new edition applies to the Contract depends on the terms of the Joint Names Policy. Where the policy states that the new code becomes applicable, the parties need to have completed in the Contract Particulars against Clause 6.16 the party responsible for any additional cost arising from complying with the new edition. The default is that the Contractor bears the additional cost. Clause 6.16 provides that where the Employer is stated to be responsible, the additional cost is to be added to the Contract Sum.

---

38 This calls for the Contractor, as part of his tendering procedures, to have contacted insurers and confirmed this information.

39 For explanation of the meaning of this term see Chapter 16, Section 16.6.2.

7.6 Commercial insurance products

There is some disparity between the terminology of JCT 05 and the jargon of the insurance market. This disparity calls for awareness of the terms used by the insurance market to describe their standard products intended for the construction industry. It has to be noted that, even where the name of a product is similar to the wording of JCT 05, it may not necessarily provide the cover required by the Contract.

There are three main types of insurance products available to the construction industry to meet the risks discussed in this chapter: (i) Employer's Liability Insurance; (ii) Public Liability Insurance; (iii) Contractors' All Risk Insurance. These have not been designed to meet the insurance requirements of any particular standard form of contract. Also, for the same type of product, terms may vary from insurer to insurer. It is therefore important to examine the products to ensure that, together, they meet the requirements of the Contract.

As already explained, Employer's Liability Insurance protects the Contractor as an employer of people against his legal liability for accidents to or diseases sustained by his employees in the course of their employment. The reason why it is compulsory under the Employer's Liability (Compulsory Insurance) Act 1969 is to ensure that an injured employee, or his dependants, obtains compensation regardless of whether or not the Contractor has the financial resources to meet the employee's claim.

The Contractor's Public Liability Insurance, also called 'Third Party Policy', covers the Contractor's liability for injury to or death of people other than the Contractor's own employees and damage to property other than the construction works. It therefore covers only part of the Contractor liabilities under clauses 6.1 and 6.2. Sometimes, it is extended to provide the cover required by Clause 6.5.1.

The Contractor's All Risk Policy covers loss/damage to the Contractor's construction works and site materials. As already explained, the name is a misnomer as no policy covers all possible forms and causes of loss or damage. Although JCT 05 does not require it, for reasons of prudence, this policy is often extended to cover the Contractor's temporary accommodation, plant, tools and equipment.

7.7 Amendments to the insurance provisions

There may be a need to amend the provisions either to remedy perceived shortcomings or to meet the special requirements of the particular project or parties or commercially available insurance products. Where this is the case, care has to be exercised to avoid the amendments being invalidated by Clause 1.3 of the Contract.[40]

7.8 Role of the Architect

The Architect is under a duty to check that the insurance arrangements of the Contractor and the sub-contractors are as required under the Contract. For details on this duty, see Chapter 2, Section 2.4.7.

---

40 See Chapter 1, Section 1.5.5.

8

Novation, assignment and sub-contracting

It is common practice for a party to a building contract to involve third parties in its performance either as recipients of some of his benefits, or undertakers of his obligations under the contract. The common legal mechanisms whereby such third party involvement may be allowed include novation, assignment, sub-contracting, conferment of rights under the Contracts (Rights of Third Parties) Act 1999 and collateral warranties.

This chapter first explains the general principles governing novation, assignment and sub-contracting. Their intended operation under JCT 05 is then examined. Rights of third parties and collateral warranties are covered in Chapter 9.

8.1 Novation

This is a process where a contract between A and B is converted into a contract between A and C. The contract is then said to be novated to C. This can be achieved only with the agreement of all the three parties concerned. There are two common situations in which novation of building contracts occurs. First, where an employer is in serious financial difficulties, the contractor, a financier of the development and the employer may enter into an agreement whereby the financier steps into the shoes of the employer to complete the project by acting as the employer for the outstanding work. The same principle applies to the contracts of engagement with the insolvent employer's consultants. Second, a receiver or liquidator of an insolvent building contractor may, with the agreement of the employer, novate a building contract entered into by the contractor to another contractor who is prepared to take it over at an agreed price.[1]

8.2 Assignment

To understand correctly the concept of assignment as used in relation to building contracts, it is essential to be aware of the distinction between the burdens of a contract and its

---

1 See Section 16.7 for discussion on novation for the Contractor's insolvency.

benefits. In a construction contract, the benefit to the contractor is the right to be paid in accordance with its terms whilst his burden is the obligation to carry out and complete the work. The employer's benefit is his right to have the work completed whilst his burden is to make payment.

Assignment, in the context of a construction contract, refers to the legally recognized transfer by a party to a contract of his rights or obligations under it to a third party to the contract, the assignee. After the transfer, the party effecting the transmission (the assignor), no longer has the assigned rights or obligations whilst the assignee can himself sue and be sued under the contract to enforce them. Because of the doctrine of privity of contract, a party to a contract may not assign its burdens at common law without the consent of the other party. However, the benefits of a contract can be validly assigned under statute or in equity without the consent of the other contractual party unless the contract itself expressly forbids such assignment.[2] The rationale for this is that the benefit of a contract (i.e. the right to call upon the other party to perform his obligations) has some economic value and is considered as belonging to a class of personal property called 'choses in action' or 'things in action'. As such, it is capable of being assigned under the Law of Property Act 1925 (LPA 25), which allows assignment of this class of property. For assignment of choses in action to be effective, s. 136 of the Act requires that it is:

- in writing;
- signed by the assignor;
- absolute (i.e. the entire right must be assigned and not just a part of it – for example, a contractor assigning his right to payment due from the employer must assign his entitlement to all the payment);
- not purported to be by way of a charge only (e.g. payment on interim certificates to a stakeholder would not be assignment);
- accompanied by express notice in writing to the debtor (the party against whom the assignor had the contractual right). The consent of the debtor to the assignment is not necessary unless the contract giving rise to the right contains a specific requirement for his approval.

Where there was an intention in a transaction to assign a contractual right but s. 136 of the LPA 25 has not been satisfied completely, there may be valid assignment at the discretion of the court. This type of assignment is referred to as 'equitable assignment' or 'assignment in equity'.

An example of this type of third party involvement is assignment by a contractor of his right to payment from the employer to a bank that is financing his performance of the contract. On the employer's side, the right to require the builder to make good defects after completion may be assigned to persons occupying the buildings, such as purchasers from the employer or his tenants.

In most situations, owners and contractors choose whom to contract with on the basis of mutual expectations of good performance and a co-operative attitude in the event of difficulties. For example, the contractor would be expected to apply interim payment towards the carrying out of the works and to overlook minor breaches of the contract by the employer (e.g. very short delays in payment) in the interests of future good relations.

---

2 This principle was expressly affirmed by the House of Lords in *Linden Gardens Trust Ltd* v. *Lenesta Sludge Disposal & Others* (1993) 63 BLR 1, the facts of which are examined in the next section.

Proper performance of the work and maintenance of good relations with the employer may not be very high on an assignee bank's list of priorities. It is for this reason that many standard forms of contract contain provisions limiting the parties' rights of assignment.

8.2.1 Assignment under JCT 05

Clause 7.1 of JCT 05 states that 'Subject to clause 7.2, neither the Employer nor the Contractor shall without the written consent of the other *assign this Contract* [our emphasis] or any rights thereunder'. Whilst statute and equity allow the assignment of the benefits of a contract without the consent of the other contracting party, there is no valid way of doing the same for the burdens of a contract. Thus, it is impossible, in the strict sense, to 'assign a contract' (i.e. both burdens and benefits) without the consent of the other party.

For a long time, the meaning of contractual provisions of the kind in Clause 7.1 was therefore a matter of considerable controversy in some legal circles. It was argued in the 10th edition of *Hudson's Building and Engineering Contracts* that, since it is impossible to assign a contract in the strict sense, the intention behind clauses which prohibit 'assignment of a contract' is not to prevent assignment of the benefits of the contract but to prohibit the parties from obtaining vicarious performance of contractual obligations without consent. In *Helstan Securities Ltd* v. *Hertfordshire County Council*[3] it was held that the effect of a similarly worded prohibition in the contract was to make the purported assignment of the benefits ineffective. The issue finally came before the House of Lords in *Linden Gardens Trust Ltd* v. *Lenesta Sludge Disposal & Others*.[4]

> Stock Conversions engaged McLaughlin & Harvey as main contractors on a JCT 63 contract to carry out work on their premises including removal of blue asbestos. Clause 17 of JCT 63 prohibited assignment in terms similar to those of JCT 05 Clause 7.1. Lenesta were in effect nominated sub-contractors for the removal of the asbestos. On 3 July 1985 Stock Conversions issued a writ against Lenesta claiming damages for breach of a direct warranty provided by them. In August 1985 Stock Conversions transferred their leasehold interest in the building to Linden Gardens. In January 1987 Stock Conversions assigned all rights of action under the contracts to Linden Gardens who then became second claimant in the action already commenced. The required consent was not obtained. In March 1989 McLaughlin & Harvey and a third party were brought into the action as defendants. Linden Gardens were seeking to recover the cost of remedial works incurred by Stock Conversion and costs of additional works incurred by themselves after the assignment. The House of Lords held that the effect of Clause 17 was to make the purported assignment by Stock Conversions without consent invalid. Linden Gardens could not therefore enforce the benefit of the contract.

The debate surrounding 'assigning a contract' attracted widespread incredulity in the construction industry. To many in the industry, it is simple logic that if one is forbidden from two courses of action one of which is impossible to do anyway, it simply means the possible course is forbidden. It is ironical that a visit to the House of Lords was necessary to confirm what most people in the construction industry have always known. The layman's

3 [1978] 3 All ER 262.
4 (1993) 63 BLR 1; [1994] 1 AC 85; 36 ConLR 1 (hereafter *Linden Gardens*).

approach to the debate is hardly different from Lord Browne-Wilkinson's statement in his leading judgment that:[5]

> although it is true that the phrase 'assign this contract' is not strictly accurate, lawyers frequently use those words inaccurately to describe an assignment of the benefit of a contract since every lawyer knows that the burden of a contract cannot be assigned.

It should be noted that assignment by the Contractor is subject to the consent of the Employer himself and not the Architect's and that there is no requirement for the withholding of this consent to be reasonable. It follows therefore that any refusal of consent by either party cannot be a dispute.

However, statute may permit valid assignment without the consent required by Clause 7.1. In *Stansell Ltd* v. *Co-operative Group (CWS) Ltd*[6] the JCT standard form contract was between Stansell (the contractor) and X, an industrial and provident society, as employer. The contract prohibited assignment without consent in the same terms as JCT 05 Clause 7.1. X transferred the whole of its property, assets and engagements to Co-operative Group under the Industrial and Provident Societies Act 1965 and ceased to exist thereafter. Neither X nor Co-operative Group had informed the contractor of the transfer. In an action commenced by Co-operative Group, the contractor disputed its ability to do so in the face of the prohibition. The Court of Appeal held that the legislation vested X's interests in the building contract in Co-operative Group notwithstanding the prohibition against assignment without consent.

Under Clause 8.4.1.4 the Employer may terminate the employment of the Contractor for assignment without consent. Similarly, under Clause 8.9.1.3, the Contractor may terminate his own employment if the Employer assigns without consent.

8.2.2 Assignment after Practical Completion

On completion the Employer may wish to transfer some interest in the completed facility (e.g. leasing or selling the building) to other parties. With outright sale the purchaser would not normally have remedies for defects against the Employer because of the *caveat emptor* ('let the buyer beware') rule. In leases containing fully repairing covenants, the Employer may even be entitled to insist on the lessees themselves making good the defects. Thus, although the Employer would be entitled to bring proceedings against the Contractor to enforce the terms of their contract, there will usually be virtually no incentive on his part to do so after the sale or lease of the building. It is therefore only prudent for potential purchasers and lessees to insist on the Employer assigning to them, as part of the sale or lease transactions, the right to bring such proceedings in the name of the Employer to enforce the terms of the contract. The ability to effect such an assignment would therefore make the property more marketable.

Even where the Employer is ready to institute the proceedings, such a course of action suffers from two principal drawbacks. First, it is far more convenient for the purchasers and lessees to do so themselves. Second, where the Employer transferred the building for

5 *Linden Gardens* (1993) 63 BLR1, at p. 27.
6 [2006] EWCA Civ 538; [2006] BLR 233; [2006] TCLR 5.

its full market value, he may be entitled to only nominal damages because he suffers no loss.[7] This flows from the general common law principle that, in an action to recover damages for breach of contract, the claimant can recover only his own losses even if the contract was intended to benefit the third party who suffers the loss.

One of the ways in which the Contract allows a third party to stand in the shoes of the Employer and to enforce directly the Employer's rights under the Contract is to assign the rights to the third party. As already explained, assignment by the Employer generally requires the consent of the Contractor. As an exception to this general requirement, Clause 7.2 allows assignment by the Employer without the Contractor's consent but such assignment can be done only after practical completion of the Works or the relevant Section. Also, it is important to note that the clause applies only if it is indicated in the Contract Particulars to be the case. The terms of any such assignment must be such that the assignee is bound by any prior agreement reached between the Employer and Contractor regarding the rights assigned. For example, if the Employer reached a compromise with the Contractor regarding errors in the setting out of the Works or departures from the specification, the assignee should not be able to go back on the compromise.

The efficacy of the 'no damage' argument against recovery by the Employer of damages suffered by third parties has been reduced by the House of Lord's decision in *St Martins Corporation Ltd* v. *Sir Robert McAlpine & Sons* Ltd,[8] which was heard together with the *Linden Gardens* case:

> St Martin's Property Corporation (Corporation) engaged Sir Robert McAlpine & Sons on a development contract which incorporated a version of JCT63 Clause 17 of which was exactly the same as in the *Linden Gardens* case. Corporation later transferred the property and assigned the benefit of the contract with McAlpines to St Martin's Property Investments Ltd (Investments) for its full market value. After practical completion defects were discovered and were made good at great expense to Investments. Corporation and Investments were joint claimants in a claim against McAlpines for damages for breach of contract. The House of Lords held that Investment's claim should fail because the purported assignment underlying their claim was invalid without the consent required under Clause 17. On Corporation's claim, McAlpine argued that they were entitled to only nominal damages because they had suffered no loss from their breach. However, the House of Lords decided that, as an exception to the general principle in McAlpine's argument, Corporation were entitled to recover Investment's loss on their behalf. It was explained that the exception applies to large developments which, to the knowledge of the employer and the contractor, were going to be occupied or purchased by third parties. In such cases an employer may recover for the benefit of third parties their losses arising from the contractor's breach.

The potential impact of this decision was considered very significant at the time, particularly for construction contracts between developers and contractors and other developments using PFI/PPP type procurement arrangements. It was applied almost immediately

---

7 However, as explained later, in certain circumstances the Employer may claim damages suffered by third parties as their trustee.

8 (1993) 63 BLR 1 (hereafter *St Martins*).

in *Darlington B. C.* v. *Wiltshier Northern Ltd*[9] by the Court of Appeal and in several other cases thereafter.[10]

However, as JCT 05 now allows benefits to be conferred prospectively on third parties under the third party rights legislation or collateral warranties,[11] its significance is likely to decline. In *Alfred McAlpine Construction Ltd* v. *Panatown Ltd (No.1)*[12] the House of Lords held that, where the third party has a remedy for defects, the employer cannot claim the loss of the third party on the third party's behalf. In that case the third party had a collateral warranty from the contractor.[13] The residual effects of *St Martins* are therefore likely to be felt mainly in situations where, for whatever reason, the Contract Particulars were not completed to allow conferment of benefits on third parties.

8.3 Sub-contracting

The term 'sub-contracting' or 'sub-letting' is used to refer to the situation whereby a party to a contract obtains vicarious performance of his duties by a third party to the contract, whilst still remaining primarily responsible to the other party to the contract for the adequacy of the performance by the third party. This is in contrast to both assignment and novation where the assignor and the novator cease to have rights and/or obligations, as the case may be, under the contract.

Whether or not a party to a contract is entitled to sub-let the performance of any of his duties under it depends on the nature of the duties and any express contractual provisions on the point. As a general rule, under the general common law, where a duty is of a personal nature, the obligation to perform it cannot be discharged by sub-letting. Indeed, subletting in such circumstances would be a breach of the contract.

Davies v. *Collins:*[14] the appellants operated a dyeing and cleaning business to which the respondent US army officer entrusted his uniform for cleaning and minor repairs. The appellants sub-contracted the cleaning to sub-contractors who failed to return the uniform. The Court of Appeal decided that, on the facts of the case, the sub-letting was a breach of contract.

However, if by the nature of the duty, it does not matter who does the work, sub-letting without consent may be allowed.

British Waggon Co. v. *Lea & Co:*[15] Parkgate hired out their railway waggons to Lea & Co. It was a term of the hire contract that Parkgate would keep the waggons in good repair. Subsequently Parkgate assigned the benefits of the contract, i.e. the right to rents, to British

9 (1994) 69 BLR 59 (hereafter *Darlington*).

10 *Catlin Estates Ltd* v. *Carter Jonas (A Firm)* [2005] EWHC 2315 (TCC); *Mirant Asia-Pacific Construction (Hong Kong) Ltd* v. *Ove Arup & Partners International Ltd* [2007] EWHC 918 (TCC); *Biffa Waste Services Ltd* v. *Maschinenfabrik Ernst Hese GmbH* [2008] EWHC 6 (TCC), [2008] BLR 155.

11 Although in relation to sub-contractors JCT 05 provides only for collateral warranties. For commentary on these provisions see Chapter 9.

12 [2001] 1 AC 518; [2000] BLR 331; 71 ConLR 1.

13 For the similar treatment of a collateral warranty on third party rights see *Biffa Waste Services Ltd* v. *Maschinenfabrik Ernst Hese GmbH* [2008] EWHC 6 (TCC); [2008] BLR 155.

14 [1945] 1 All 247; see also Court of Appeal's application of this principle to construction work in *Southway Group Ltd* v. *Wolff and Wolff* (1991) 57 BLR 33.

15 [1880] 5 QB 149.

Waggon Co. and arranged with them to carry out the repair obligations. After taking over the workmen of Parkgate, British Waggon was ready and willing to perform these obligations. One of the issues in dispute was whether or not Lea was bound to accept performance of the repair duties by a third party. It was held that, where a party contracts to carry out work of so rough a nature that ordinary workmen conversant with the business would be perfectly able to execute the work, the party may be entitled to sub-let it without the consent of the other contracting party.

There can be little doubt that the service rendered by the contractor in a construction contract is of a very personal nature. This is the only conclusion that can be reached considering the elaborate vetting procedures usually followed in the selection of contractors. It follows, therefore, that even in the absence of express provisions disallowing sub-letting of any part of the works, such as contained in JCT 05 Clause 3.7.1, the Contractor may not be entitled to sub-let in any case.

8.3.1 Sub-contracting under JCT 05

Sub-contractors are often classified into domestic sub-contractors and nominated sub-contractors. The former type is one selected and appointed by the main contractor. The only involvement of the contract administrator, if the main contract specifies any, is to give or withhold consent to the sub-letting of the relevant work package. A nominated sub-contractor is selected either in the main contract or by the contract administrator but he is appointed by the main contractor.

A main contractor carries all the risks in relation to the performance of a domestic sub-contractor. The employer, however, shares the performance risks of nominated sub-contractors. For example, if a nominated sub-contractor fails to complete the sub-contract works, the employer (or the contract administrator on his behalf) must nominate another sub-contractor to complete the outstanding work. Any delay suffered or additional costs incurred as a consequence of the original sub-contractor's failure to complete are the employer's risks.[16] Previous editions of this standard form had extensive provisions on nominated sub-contracting. However, on account of bad experience from the inherent pitfalls of this type of sub-contracting and the complexity of the provisions, nomination was hardly used. The JCT have therefore omitted to allow for nomination under JCT 05. However, parties may still amend the contract to allow for it but they have to bear in mind that the position would be governed by common law in relation to matters not expressly provided for through the amendments.

8.3.2 Sub-contractors under JCT 05

Most of the provisions on sub-contractors are contained in Clauses 3.7 to 3.9. They allow for three methods of appointing sub-contractors. With the first method, the choice of sub-contractor is solely the Contractor's, subject only to the written consent under Clause 3.7.1 of the Architect to the sub-letting of that portion of the Works. The Architect must not withhold or delay his consent unreasonably. Reasonableness of objection to sub-letting will usually relate to the integrity and the technical, financial and organizational capability

16 *Bickerton & Sons Ltd* v. *North West Metropolitan Regional Hospital Board* [1970] 1 WLR 607.

and capacity of the sub-contractor concerned. This restriction, in the widest sense, applies to sub-letting to a 'labour only sub-contractor (LOSC)' even though, in practice, most contractors engage them without the consent of the Architect. Sub-letting without consent entitles the Employer to determine the Contractor's employment (Clause 8.4.1.4). A more prudent course of action is therefore to obtain a blanket consent to engagement of LOSCs.

The Conditions do not expressly empower the Architect to vet sub-contractors. This is because the Contractor is only required to obtain consent to sub-letting and not to his choice of sub-contractors. In principle, the Contractor can therefore obtain consent to sub-let before he even considers the question of whom to appoint. However, in practice, the Architect may vet sub-contractors indirectly by simply withholding his consent until he is satisfied with the identity of the organization or person to whom the work will be sub-let.

There is anecdotal evidence of Architects under previous editions of the contract demanding, as a condition of giving consent to proposed sub-letting, the provision of assignable collateral warranties in favour of employers from the sub-contractors. Such practice is open to attack on the basis that withholding consent on grounds of failure to provide the desired warranties is unreasonable. This practice should no longer be necessary as JCT 05 has a formal procedure for imposing on the Contractor an obligation to procure collateral warranties from sub-contractors.[17]

The second method of appointing a sub-contractor is provided for by Clause 3.8. It requires the annexation of lists of sub-contractors to work packages defined in the Contract Bills. Although each work package is priced by the Contractor, it is to be carried out by the Contractor's choice from the list annexed to the particular work package. The list must contain a minimum of three sub-contractors but, subject to agreement between them, the Employer (or the Architect on his behalf) or the Contractor may add other names to the list before the final choice is made. If there are withdrawals from the list resulting in less than three available sub-contractors before the sub-contract is awarded, there are two alternative courses of action:

- The list is expanded to the minimum of three by the Employer or the Contractor subject to their agreement (not to be withheld unreasonably). It is to be noted that here the Architect cannot himself add to the list.
- The work is given to the Contractor who may sub-contract it under Clause 3.7 if he so wishes.

For the provisions for specified sub-contractors to work smoothly, they must be ready and willing to submit tenders for the sub-contract work packages before the Contractor returns the tender for the main works. Without such sub-contract tenders, the Contractor would be in the untenable position of pricing work without any knowledge of how much he will have to pay the sub-contractor selected at the end of the day. Failing such tenders, a prudent contractor would make allowance for this uncertainty in his tender.

It is suggested that the Contractor may be able to get round these problems by pricing the work packages on the basis of *bona fide* quotations from competent sub-contractors irrespective of whether or not they are specified in the Contract Bills. He can then seek the agreement of the Employer or the Architect to the inclusion of those sub-contractors not already listed in the Contract Bills. In the event of consent being refused, the likelihood is that the

---

17 See Chapter 9, Section 9.4.3.

Contractor will succeed in adjudication or arbitration if the reason for seeking their inclusion is failure of the listed sub-contractors to submit quotations or withdrawal from the lists.

Unreasonable delay in giving consent to the Contractor's proposals of sub-contractors to complete the list may amount to the 'impediment/prevention/default' Relevant Event under Clause 2.29.6 for which the Contractor would be entitled to extension of time if the works are delayed.

The third method applies to sub-letting of the design for any CDP. It requires the written consent of the Employer, which must not be unreasonably delayed or withheld.

A sub-contractor would carry the statutory duties of a contractor and/or a designer under the CDM Regulations.[18] Regardless of the method of sub-letting, the Contractor must, therefore, take reasonable steps to ensure that sub-contractors are competent to discharge their statutory duties before they are appointed.[19] Under Clause 3.25.3, where the Contractor is not the Principal Contractor, he must promptly inform the latter of the identity of every sub-contractor appointment.[20] The equivalent obligation applies to sub-sub-contractors notified to the Contractor.

8.3.3 Sub-contract conditions

JCT 05 does not require the use of any particular standard form of sub-contract conditions. However, it specifies certain terms which must be included in all such sub-contracts irrespective of whether or not the sub-contractor is appointed from a list annexed to the Contract Bills. If the Contractor fails to incorporate these terms he would be in breach of contract, thus entitling the Employer to damages, which would be only nominal if the sub-contract works are carried out to the specified standard.

8.3.3.1 Effect of termination of the Contractor's employment

The sub-contract must provide that the employment of the sub-contractor under the sub-contract is automatically terminated upon the termination of the employment of the Contractor under the main contract (Clause 3.9.1). This requirement is calculated to avoid any problems with removing the sub-contractors from the site after termination of the main contract.

8.3.3.2 Unfixed materials

Clause 3.9.2 requires that, in respect of unfixed materials and goods brought onto the site by a sub-contractor for the purpose of carrying out the sub-contract works, the sub-contract must provide that:

- such materials and goods must not be removed from the site for any purpose other than the carrying out of the Works without the written consent of the Contractor, who must not consent unless he obtains the written consent of the Architect as required under Clause 2.24;
- where such materials and goods are included in any Interim Certificate they become the property of the Employer after the amount properly due to the Contractor is discharged

18 See Chapter 10, Sections 10.6 and 10.8, respectively, for outlines of these duties.

19 See Regulation 4(1)(a) of the CDM Regulations. For a commentary on ensuring competence, see Chapter 10, Section 10.2.

20 This obligation was introduced in April 2007 by Amendment 1.

by the Employer to the Contractor and the sub-contractor must not dispute this transfer of title;
- such materials and goods become the property of the Contractor after the sub-contractor is paid for them by the Contractor;
- regardless of the terms of the sub-contract, after payment for off-site materials through Interim Certificates, they become the property of the Employer.

Under Clauses 4.16.1.2 and 4.16.1.3 the Architect must include the value of unfixed materials, including materials belonging to sub-contractors, in Interim Certificates. Payment for unfixed materials presents certain risks to the Employer, which are explained in Chapter 15, Section 15.5. Certain provisions in JCT 05 are designed to minimize these risks. Basically, they are calculated to ensure that materials paid for by the Employer are actually used for the Works. The purpose of 3.9.2 is to step these provisions down into sub-contracts so as to be binding on sub-contractors.

However, it has to be borne in mind that even complete compliance with Clause 3.9.2 would be ineffective to transfer title to goods and materials to the Contractor or Employer unless the sub-contractor concerned acquired title in them in the first place. For example, where a supply contract reserves title to the supplier until materials are paid for in full, title in the materials cannot pass to the Contractor or Employer until this condition is complied with.[21] The maxim *nemo dat quod non habet* (Lat. 'you cannot give away what you do not own') catches the essence of the problem. The facts of *Dawber Williamson Roofing Ltd* v. *Humberside Council*,[22] on which there is commentary in Chapter 15, Section 15.5.1, illustrate this problem.

8.3.3.3 Access for the Architect

Under Clause 3.1 the Architect is entitled to access at all reasonable times to the Works and workshops and other locations where work is being done for the Contract. The Contractor must ensure that the requirements of Clause 3.1 are stepped down into sub-contracts. Without such provisions, the Architect would have no rights of access to workshops and the like of domestic sub-contractors.

8.3.3.4 Interest on overdue payment

Clause 3.9.2.3 requires the sub-contract to provide that the sub-contractor shall be entitled to simple interest if the Contractor fails to meet any payment due under the sub-contract by its final date for payment. The rate of interest is to be 5% over the Base Rate of the Bank of England that was current when the payment became overdue. The sub-contract must further provide that the right to interest is without prejudice to the sub-contractor's statutory right to suspend performance of the sub-contract for non-payment or the contractual right to terminate his own employment. This requirement is designed to step down into sub-contracts the corresponding right of the Contractor under Clauses 4.13.6 and 4.15.6.[23]

---

21 Such a clause in a supply contract is referred to as a 'Retention of Title' (ROT) or Romalpa clause after *Aluminium Industrie Vaasen BV* v. *Romalpa Aluminium* [1976] 2 All ER 552, a famous case in which the court upheld the effectiveness of clauses of that kind.

22 (1979) 14 BLR 70 (hereafter *Dawber Williamson*).

23 See Chapter 15, Section 15.6.1 for commentary on the quantification of the amount of interest claimable for delayed payment.

8.3.3.5 Fluctuations

Finally, where Fluctuations Option A or B applies to the Contract, the terms of the applicable fluctuation clause must be incorporated into all sub-contracts (Paragraph A.3 and B.4 of Fluctuations Options A and B, respectively).

8.3.3.6 Collateral Warranties

Sub-contracting requires the use of at least two contracts. First, there is the contract between the employer and the contractor: the main contract. Second, there is a separate contract between the contractor and the sub-contractor in question: the sub-contract. A particular problem posed by sub-contracting is that, very often the sub-contractor undertakes some special duties, such as design, that are not part of the main contract. It follows from the doctrine of privity of contract that within the framework of the two contracts, the employer has no remedy if the sub-contractor defaults in the performance of these special duties. However, in cases where the sub-contractor is recommended to the contractor by the employer on the strength of promises made to the employer by the sub-contractor, the law may treat the promise as a separate contract between the employer and the sub-contractor. This type of contract existing alongside a main contract is referred to as a 'collateral warranty'. A classic case which illustrates this principle is that of *Shanklin Pier* v. *Detel Products Ltd.*[24] A supplier of paint made a statement that his paint would last between 7 and 10 years to the claimant employer. On the strength of this statement, the employer specified the defendant's paint to his contractor. The paint did not last anywhere near even a year. It was held that the statement constituted a warranty and the employer was therefore entitled to damages for its breach.

It is now common practice for employers to demand collateral warranties from sub-contractors. In such warranties duties which are not part of the main contract are sometimes imposed on sub-contractors. This is often the case where the sub-contractor provides free design services.

It is explained in Chapter 9 that JCT 05 Contract Particulars may be completed to impose on the Contractor not only obligations to enter into collateral warranties with Purchasers, Tenants and Funders[25] but also obligations to procure from sub-contractors collateral warranties in favour of the Employer and/or such third parties. To cascade this obligation down to the relevant sub-contractors, Clause 3.9.2 requires every sub-contract to contain a term to the effect that, where applicable, within 14 days after receipt of the Contractor's written request to do so, the sub-contractor must execute and deliver collateral warranties in accordance with details in the main contract's Contract Documents.

8.3.3.7 Incorporation of the terms into sub-contracts

Unless the terms required to be incorporated are actually incorporated into a sub-contract they cannot be enforced against the sub-contractor concerned because of lack of privity of contract between the Employer and the sub-contractor. In such an event, the Employer's only remedy is to recover any damages from the Contractor for failure to comply with the requirement. However, this remedy is likely to be only academic because, in the

---

24 [1951] 2 All ER 471.
25 See Clause 1.1 for definitions of these types of third parties.

circumstances in which the Employer is likely to need it, the Contractor may be insolvent. One way of avoiding this problem is for the Architect to make his consent to sub-contracting conditional upon an appropriate sub-contract being entered into and to require the Contractor to terminate the sub-contract if it does not comply.

The JCT Standard Building Sub-Contracts[26] comply with the requirements of Clause 3.9. The easiest way of complying with them is therefore to use this form. However, as explained above, such compliance does not guarantee the Employer complete protection against the risk of *de facto* owners of goods and materials claiming them after they have been paid for through Interim Certificates.

Where an *ad hoc* (special purpose) form is used, it is not uncommon for contractors to seek to comply with the clause by sub-contract stipulations such as 'the sub-contractors shall carry out the sub-contract works in accordance with the terms of the main contract' or 'the sub-contractor shall be deemed to have notice of all the provisions of the main contract'. Whether such stipulations are effective to incorporate the relevant terms of the main contract into the sub-contract is a matter of the interpretation of the particular sub-contract. The following two cases illustrate that different outcomes are possible depending upon the interpretation of the contracts in the context of all their surrounding circumstances.

> *Chandler Bros Ltd* v. *Boswell*:[27] the sub-contract terms were generally similar to those of the main contract. Although the engineer's power under the main contract to instruct the main contractor to remove a sub-contractor was not a term of the sub-contract, one of the sub-contract recitals required the sub-contractor to carry out the sub-contract works in accordance with the terms of the main contract. The Court of Appeal refused to imply into the sub-contract a term that the contractor was entitled to remove the sub-contractor if so instructed by the Engineer. The fact that the sub-contract, in addition to the general reference to the main contract, repeated provisions on most other matters but omitted to do the same for the term in dispute suggested that it was never intended to be part of the sub-contract.

> *Sauter Automation Ltd* v. *Goodman (Mechanical Services) Ltd*:[28] this concerned ownership of materials and goods brought onto a construction site by a sub-contractor. The sub-contractor's quotation contained a *Romalpa* clause. The Contractor's order in response to the quotation stated that the main contract was in the form of the then current version of the GC/Works/1. It was a term of this contract that every sub-contract was to contain a term vesting in the Contractor ownership of materials and goods brought onto the site. It was held that the terms of the contractor's order governed the sub-contract and not those of the sub-contractor's quotation and that the reference to the main contract conditions was effective to vest ownership in the Contractor.

---

26 These are *The Standard Building Sub-Contract*, 2005 Edition, which is for use where the sub-contractor does not carry out any design of the sub-contract works, and *The Building Sub-Contract with Sub-Contractor's Design*, 2005 Edition, where the sub-contractor takes responsibility for design of part or all of the sub-contract works.

27 [1936] 3 All ER 179; in *Dawber Williamson Roofing Ltd* v. *Humberside Council* (1979) 14 BLR 70 an attempt to incorporate these provisions into the sub-contract by a stipulation in the sub-contract that the sub-contractor was deemed to know of the main contract failed to achieve the intended effect.

28 (1986) 34 BLR 84.

9

Third Party Rights and Collateral Warranties

Note: throughout this chapter, unless stated otherwise:

- 'CRTPA' refers to the Contracts (Rights of Third Parties) Act 1999.
- 'Clause x' refers to the relevant clause of JCT 05.
- 'Paragraph x' refers to the relevant paragraph of JCT 05, Schedule 5.
- 's. x' and 'ss. x' refers to the section(s) and sub-section(s) of CRTPA.

9.1 Introduction – the privity rule

Clause 1.6 of JCT 05 represents innovative thinking on the part of the JCT, at least in construction industry terms. It reads as an exclusion clause, denying rights to persons who are not a party to the Contract, but with an important caveat. The caveat is where the JCT's approach differs[1] from most of the other standard contract publishing bodies in the construction industry. The denial of rights does not apply to any Purchasers, Tenants, or Funder identified in the Contract. In Clause 1.6 the JCT make use of a common law rule (the doctrine of privity of contract) in respect of outside persons generally, and make use of statute (The Contracts (Rights of Third Parties) Act 1999) in respect of identified parties in the caveat. Before 1999, such a caveat would have been unenforceable, and the remainder of the provision would have been unnecessary, due to the 'doctrine of privity'.

The privity rule is summed up by the House of Lords in *Dunlop Pneumatic Tyre Co. Ltd* v. *Selfridge & Co. Ltd*:[2]

> In the law of England certain principles are fundamental. One is that only a person who is a party to a contract can sue on it.

Whilst the doctrine protects the parties against claims from strangers, there are times when the parties expressly agree that someone else is to benefit, and then the privity rule

---

1 See, for example, ICE *Measurement Version*, 7th Edition 2004, clause 3(2), which states that there is no intention that any third party should have rights to enforce the terms of contract.

2 [1915] AC 847, per Viscount Haldane LC at p. 853.

thwarts their intentions. An example of the problem, and the way in which courts have allowed ways around the doctrine, can be seen in the House of Lords decision in *Beswick* v. *Beswick*.[3] A coal merchant, Peter Beswick, sold his business to his nephew. The price was £6 per week until the uncle's death, then £5 per week to his widow. The uncle died and his widow became administratrix of his estate. When the nephew failed to continue payment, the widow sued, both personally and as administratrix. She won her case as administratrix, but it was said by the House of Lords that she had no rights in her own name to enforce the contract, because privity prevented it.

Cases such as *Beswick* eventually led to consideration by the Law Commission, and a report[4] was published in 1996. The report contains arguments against the privity rule, including amongst others that the rule frustrated the intentions of the parties, and caused difficulty in commercial life, particularly in construction contracts and insurance law.[5]

The effect of privity on construction contracts relates mainly to the risk of defective design or work by someone not in a direct contractual relationship with the party suffering loss as a result of the fault. For example, a purchaser of a building from the developer may suffer if the designer's work or a sub-contractor's work is defective. Whilst the parties are linked by a chain of contracts, there is no direct relationship. The purchaser would be a third party so far as the contract under which the work was executed is concerned (i.e. respectively, the designer's terms of engagement with the developer, or the sub-contract with the contractor).

The developer would be in a similar position in relation to a sub-contractor's work, and with design increasingly being carried out by specialist sub-contractors, the developer's recourse in the event of defects became ever more remote. The position was exacerbated in the early 1990s by developments in the law of negligence. In 1990 the House of Lords[6] withdrew the concept that defective work could be considered to be physical damage, which was necessary to give rise to a claim in negligence.[7]

Without the right to claim against third parties under the construction contract, and having lost the right to claim in negligence, developers were obliged to protect themselves by creating a direct contractual link with the actual providers of the design and work. The resultant link is known as a collateral contract.[8] Since the main purpose of the collateral contract is for the supplier of goods or services to make binding assurances (i.e. warranties) the documents used in the construction industry are known as collateral warranties.

Each collateral warranty needs to be signed by the parties, which, when there are many warranties related to a project, can be a daunting task. This is often made particularly onerous when suppliers are being urged to commit themselves to gratuitous promises other than those made in the contract under which they are being paid. Unfortunately, it is common for collateral warranties to be unilateral terms drafted on behalf of the developer, designed

---

3 [1968] AC 58.

4 Law Commission No. 242 'Privity of Contract: Contracts for the Benefit of Third Parties', CM 3329, HMSO July 1996.

5 Law Commission Report No. 242, pp. 39–52.

6 See *Murphy* v. *Brentwood District Council* (1990) 50 BLR 1.

7 In *Junior Books Ltd* v. *Veitchi Co Ltd* (1982), HL 21 BLR 66 a sub-contractor was held liable to the owner for defects in the floor topping. This case is considered to be a high point in the development of the meaning of 'physical damage' in negligence cases and was followed by a series of cases up to 1990 reducing such defects to simply 'economic loss' resulting from an act by the sub-contractor, and therefore not recoverable. Economic loss is still recoverable in special cases where there is reliance, such as loss resulting from negligent statements, but not from acts, where physical damage is still required.

8 See Chapter 8, Section 8.3.3.6, and reference to the case of *Shanklin Pier* v. *Detel Products Ltd.*

for self-protection and protection of his purchasers, tenants or funder, without any consideration for the contract under which the services or goods are provided. This has led to a general notoriety amongst warranty providers; but there is no getting away from the fact that without a direct relationship, and in the absence of assignment,[9] privity prevents an injured party from recovery of his loss, when he has to rectify the defective work of others.

One answer to part of the problem is to use only standard form warranties, such as those published by the JCT, but that does not get around the administrative difficulty of collecting together the numerous warranties that may be sought on a large or technically complex project.

An alternative answer was presented by Parliament. The Law Commission recommended abolishing the privity rule generally, but with some flexibility. The Contracts (Rights of Third Parties) Act 1999 came into force in November 1999.

9.2 The Contracts (Rights of Third Parties) Act 1999

9.2.1 Generally

Section 9.2 is a brief overview of the main provisions of CRTPA insofar as it is relevant in the context of the Standard Building Contract. It is intended as no more than a backcloth to assist understanding of the provisions in JCT 05. It is not a detailed exposition of CRTPA.

The purpose of The Contracts (Rights of Third Parties) Act 1999 (CRTPA) is to alter the doctrine of privity, so that where the parties to a contract intend that a third party should benefit, the law will not prevent it. CRTPA does not abolish the privity rule; it contains flexibility suggested by the Law Commission to apply only where the parties choose. Where the parties do not wish others to enforce a benefit from their contract, CRTPA does not apply, and the common law privity rule survives.

CRTPA is triggered by express or construed intention. If a contract states the parties' intention that an identified third party should have some specified enforceable right, CRTPA will support the third party. Alternatively, if the parties to the contract state expressly that no other person shall enforce its terms, the privity rule applies, and third parties cannot rely on any apparent benefits in the contract.

9.2.2 Third parties

Third parties may achieve entitlement to an enforceable benefit through either of two routes provided by s. 1(1) and s. 1(2) of CRTPA. The two routes are:

- where the contract expressly states that they may enforce a term in their own right, or
- where terms of the contract purport to confer benefits on them, provided, under s. 1(2), that it cannot be construed, on a proper construction, that the parties to the contract did not intend the term to be enforceable by others.

Section 1(3) provides that third parties may be identified (a) by name, (b) as a member of a class (e.g. purchasers, tenants, sub-contractors), or (c) by description (e.g. the building owner's funder). Under this section of CRTPA a third party need not be in existence when

9 See Chapter 8 generally.

the contract is entered into, thus allowing rights to be granted prospectively to future class members, such as tenants.

The first route is clear – if an identified party is expressly stated to be able to enforce a term, then he may.

Under the second route it is not sufficient simply to identify a benefit; there has also to be an intention that the benefit should be enforceable, or to be more precise, that there is no intention that it should not. In short, CRTPA appears to favour the third party. The effect of s. 1(2) seems to be that intention under the second route is presumed, and it is for a defending party to find, in the contract as a whole, an intention that the third party should not enforce its benefit. There are no guidelines in CRTPA[10] as to how intention may be rebutted and what factors are relevant, so interpretation in this area is a matter for development by the courts. Unfortunately, the Court of Appeal, in *Themis Avraamides* v. *Bathroom Trading Company*,[11] took a strict view, leaning towards the traditional privity rule. A clause in an agreement to transfer assets and goodwill in a company contained an undertaking to 'complete outstanding orders to customers … as at 31 March 2003, and to pay … any liabilities properly incurred by the company as at 31 March 2003'. The claimant was a customer of the former company, having had refurbishment work done to two bathrooms, which he considered to be defective. The Court found no difficulty deciding, in relation to the first part of the paragraph, that the claimant was 'a customer', but could not identify a benefit. In the second part of the paragraph the reference to paying liabilities was not linked to customers, or any other third party, either by name or by class. It was held that the claimant was unable to rely on the undertaking in the transfer agreement.

Whichever of the two routes applies, the important issue is identification of a third party and a clear link to a specified benefit. In the *Themis Avraamides* case, Waller LJ concluded his judgment, saying: 'I am actually doubtful whether it can be said that this agreement, on a true construction, was one under which it was intended that any persons with rights against the company were able to enforce them directly …'.[12]

Many standard form contracts make the parties' intention clear by a term expressly denying any intention that anyone, other than the parties, may enforce its terms.[13]

9.2.3 Enforcing a term of the contract

Section 1(5) of CRTPA gives a qualifying third party the same remedies that he would have if he were a party to the contract. This includes damages, injunctions, and specific performance. However, it excludes remedies such as rescission which impact on the contract itself. So, a third party may claim damages if (say) work promised to be of good standard in the contract, is defective; or he may enforce payment, if payment by one of the parties is the subject of a term. However, as a third party, he cannot terminate the contract; his entitlement is to enforce particular terms in the contract, not to do away with it.

If the parties to the contract are bound to use arbitration as the forum for dispute resolution, the third party must use arbitration;[14] if the parties to the contract may use adjudication, so too

10 But see Law Commission Report No. 242, pp. 85–89 for examples.

11 [2007] BLR 76.

12 [2007] BLR 76 at p. 80.

13 See ICE *Measurement Version*, 7th Edition 2004, clause 3(2); also JCT Minor Works Building Contract (MW) 2005 Edition, clause 1.5.

14 In the case of *Nisshin Shipping Co. Ltd* v. *Cleaves & Co. Ltd* [2003] EWHC 2602 (Comm), a claimant third party who sued was obliged to recommence his action in arbitration under the arbitration rules in the contract.

may the third party. Likewise, if the contract is a specialty,[15] the third party has 12 years after a breach in which to pursue his rights, and he has 6 years if the contract is a simple contract.

Third parties are not put in a better position than they would be as a party to the contract. Any exclusion or limitation[16] affecting the rights of the parties apply equally to third parties.[17] Reciprocally, any defence or right to counter-claim, available to the parties under the contract, are available against a claim from a third party.[18] A third party has the same rights and limitations as the parties, no more and no less. Nevertheless, a third party is still a third party, an outsider; he does not become a party to the contract.[19]

9.2.4 Restriction on the parties

The grant of rights to third parties in a contract may bring with it restrictions on the parties. Unless the contract states otherwise,[20] a contract promising an enforceable right to another cannot be rescinded, nor can a term be varied, without the consent of the third party,[21] but that is provided: (a) the third party has assented by words or conduct to the term, and (b) the promiser knows, or could reasonably be expected to foresee, that the third party would rely on it, and the third party has in fact relied on it.[22]

CRTPA is a one-way deal. Whilst rights may be given to third parties, CRTPA does not allow the parties to impose burdens. Thus a provision in a contract that requires payment in return for a right, such as payment to a sub-contractor in return for exploitation of copyright, would be unenforceable without a separate direct agreement.

Clearly, a third party's assent to a term, or indeed his ability to enforce it, relies on his knowledge that such a term exists. Therein lies a major stumbling block, for CRTPA does not place any obligation on the parties to notify the possible beneficiaries; it is for a beneficiary to find out for himself as best he can. To a great extent, intention for an outsider to enforce a right may be measured by the eagerness and thoroughness with which the parties notify those concerned.

9.3 JCT 05 – rights of third parties

(See also chapter 1, Section 1.5.2.5).

9.3.1 JCT 05 – Third Party Rights Generally

When CRTPA was introduced, the drafting bodies in the construction industry all wrote CRTPA out of their contracts, by inserting a clause denying intention, or to be more accurate,

---

15 See Chapter 1, Section 1.4.1.2 at 'Main differences between "Simple" and "Specialty" contracts', and at 'Attestation'.

16 See *Prudential Assurance Co. Ltd* v. *David Monroe* [2007] EWHC 775 (ch): a previous tenant was able to benefit as a third party from limited liability under a contract.

17 See s. 1(6) and s. 3(6).

18 See s. 3(4).

19 See s. 7(4); also ss. 1(4), 1(5), 3(4), 3(6).

20 See s. 2(3) which provides for the parties to agree that they may rescind or vary the contract without consent from third parties.

21 See s. 2(1).

22 See ss. 2(1), 2(2).

by expressing a negative intention. This was somewhat bizarre coming from the industry that the Law Commission had picked out as needing assistance. The JCT moved from that position in 2003 with the publication of their Major Projects Form (MPF 03), in which they utilized CRTPA to give limited rights from the contractor to funders, purchasers and tenants. In the Guidance Note to MPF 03 the JCT explain that their action was a means of avoiding the proliferation of separate warranties and other collateral agreements. The concept has been carried through to the Standard Building Contract. However, it seems that, somewhere in the interim, the JCT lost faith in their customers' resolve to reduce collateral warranties. In JCT 05, a further option has been incorporated for the use of collateral warranties instead of third party rights.

Whilst the JCT have, admittedly, made use of CRTPA, it is perhaps regrettable that they have seen third party rights as applying only to a very limited group of beneficiaries. Under CRTPA, everyone may be a potential third party to someone else's contract, and the promise of benefits may be given by either party to that contract. Yet the JCT have limited the beneficiaries to recipients of the Works (purchasers, tenants, funders), and only the contractor is cited as a promisor. Perhaps, in time, when CRTPA is well established, its use may be more even-handed. Examples would include promises by the employer to (say) specified sub-contractors regarding wider circumstances for direct payment, or granting sub-contractors an entitlement to enforce any main contract provisions, such as those relating to the issue of the Final Certificate, or that require sub-contracts to contain specified clauses in their favour. But that is a matter for the future.

9.3.2 JCT 05 – Third parties

JCT 05 utilizes ss. 1(2) and 1(3)[23] of CRTPA by expressing the Parties' intention in the Contract. Clause 1.6 makes clear that no person (which under Clause 1.4 includes any individual or organization) is given any right to enforce any term of the Contract, other than the Parties themselves, or persons identified in the Contract. Clauses 7A and 7B then go on to name Purchasers, Tenants and the Funder as third parties who may enforce specified rights,[24] and their identities, in sofar as accurate identity is required, are set out in Part 2, in the Contract Particulars (see Chapter 1, Section 1.5.2 commentary regarding entries in Part 2 of the Contract Particulars). As stated in Section 9.2.2 of this chapter, CRTPA allows a qualifying third party to be identified by name, or by class or description, so it is adequate if, in the Contract Particulars, reference is made simply to (say) 'all tenants'. Any person or organization wishing to pursue rights granted under the Contract must be able to identify themselves as one of the third parties specified in the Contract Particulars. If they cannot, they will be barred under the express exclusion provisions in Clause 1.6.

9.3.3 JCT 05 – Enforcing a term of the contract

The rights (and restriction) of a third party generally to enforce a term of the Contract under CRTPA are dealt with in Sections 9.2.3 and 9.2.4 of this chapter.

23 See Section 9.2.2 of this chapter.
24 See Section 9.3.3 of this chapter.

In JCT 05, it is not left to the third parties to search for their rights; they are set out in Clauses 7.3, 7.4, 7A, 7B, and Schedule 5 of the Contract. Provided a person is identifiable by name or class, or by description, in the Contract Particulars, he may enforce the rights ascribed to him in Schedule 5, when they become active. However, third parties will only be able to enforce those rights if they know of their existence, and it is impliedly for the Employer to inform them.[25] There is no express obligation on the Contractor to inform third parties of their rights.

Whilst third parties may be named, and rights from the Contractor may be described (see Sections 9.3.4, and 9.3.5 of this chapter below), the entitlement to enforce those rights is restricted. Clauses 7A.1, 7B.1, and 7.4 provide that rights are not triggered until the Contractor receives a notice from the Employer by actual, special or recorded delivery. Until then, the rights are simply latent. The notice must identify the relevant third party, who in turn must be identifiable from the list in Part 2 of the Contract Particulars. For example, a Purchaser's name may be known at this point who was not known to be a purchaser when the Contract was entered into. This is entirely consistent with s. 1(3) of CRTPA.

JCT 05 provides for access to adjudication of disputes, so likewise, third parties have the right to use adjudication. The Contract gives the Parties the option of using the courts or arbitration for final resolution of disputes; a Funder will be expressly bound by whichever option is chosen in the Contract,[26] together with any procedural requirements that go with it. Purchasers and Tenants, on the other hand, are expressly obliged to use the English courts.[27] The period during which a third party may pursue his rights are the same as those under the Limitation Act 1980 (6 or 12 years),[28] and will depend on whether the Contract is signed under hand as a simple contract, or as a deed. Whichever period applies, a legal action cannot be started after the end of that period taken from the date of practical completion of the work (Part 1, para. 8, and Part 2, para. 12).[29]

A third party's right to veto attempts by the contracting parties to terminate or vary the contract under s.3(2)(b) of CRTPA, is dealt with in different ways between Purchasers and Tenants, and a Funder. Under Clause 7A.2 and 7B.2, the parties to the Contract retain the right to amend, rescind, or terminate their agreement, without the consent of any third party, until the rights described in Schedule 5 are triggered. At that point, the Employer and Contractor become restricted in the extent to which they can conduct their own affairs, without consent from relevant third parties.

After the trigger granting them rights, Purchasers and Tenants are still unable to veto any wish by the Parties to terminate or rescind their agreement, or to settle disputes. However, Clause 7A.3 obliges the Parties to seek consent before they may vary the vesting and 'consent' terms in Clause 7A, or the terms of Schedule 5,[30] in which Purchaser's and Tenant's rights are described.

Funders are given greater power, reflecting the influence they have on the project. Their interest lies, not simply in the quality of the work, but also in its timeous completion and

---

25 See Chapter 2, Section 2.1.2, dealing with Employer's implied duties.

26 See Schedule 5, Part 2, para. 14.2.

27 See Schedule 5, Part 1, para. 10.

28 See Chapter 1, Section 1.4.1.2 at 'Main differences between 'Simple' and 'Specialty' contracts', and at 'Attestation'.

29 The JCT have fixed the period during which action may be commenced, removing the normal extended exposure of a promisor under an indemnity, under which a cause of action may arise when the loss is incurred.

30 Schedule 5 Part 1 'P&T Rights'.

commercial success. It is in the interest of the Funder, who has invested in the project, to ensure that the project does not founder, and it is in this area that the approach to third party rights of the Funder, differs from those of Purchasers and Tenants. After rights are triggered by the Contractor's receipt of the Employer's notice, Clause 7B.2 provides that the Parties cannot agree to rescind the Contract, and without prior consent from the Funder, they are barred from varying the vesting and 'consent' provisions in Clause 7B, or the terms of Schedule 5,[31] in which the Funder's rights are described. Additionally, the Contractor cannot apply any entitlement to terminate his employment, or to treat the Contract as repudiated by the Employer, until he has given the Funder notice of his intention. The reason is to give the Funder an opportunity to resolve the matter and ensure continuity, if he can.

9.3.4 JCT 05 – Third Party Rights for Purchasers and Tenants (dealing with 'P&T Rights' under Schedule 5: Part 1)

Schedule 5, Part 1 sets out the rights (and their limitations) granted to Purchasers and Tenants, by the Contractor only. No rights are granted by the Employer.

Works to comply with the Contract and good practice: In paragraph 1.1, the Contractor promises, at practical completion (not before), that he has carried out the Works in accordance with the Contract. In this, the Contractor is simply repeating the contractual obligation he has to the Employer.

The Contractor also promises (para. 2) that, unless obliged to do so by the Contract or an Architect's instruction, he has not used any materials that do not comply with specified guidelines.[32]

Contractor's liability: The Contractor's liability in the event that the Works do not comply with the Contract, is described in paragraphs 1.1 to 1.4, and 9. The basic level of liability is for the reasonable costs of making good a defect. This includes the costs of repair or replacement, but only to the extent that the Purchaser or Tenant incurs or becomes liable for such costs, either directly or by way of contribution to others.

Rectification of defective work often involves losses for a building occupier, other than the direct cost of construction work. For example, a manufacturer who needs to move his equipment while repairs are carried out in the ceiling void, may lose production, and consequently, income (see Chapter 12, Section 12.6.4 dealing with Employer's additional losses, and also Chapter 3, Section 3.5.3). Such losses are expressly excluded unless included by an entry in the Contract Particulars (paras 1.1.2 and 1.2).

The Contractor's liability is limited to an equitable proportion of the total costs, related to his responsibility, and based on the premise that consultants and sub-contractors may also bear some responsibility (para. 1.3). The assumptions are:

- that Consultants named in the Contract Particulars are giving similar undertakings related to their services, either by a collateral warranty, or by third party rights, and that their liability is not limited under their consultancy agreement;

31 Schedule 5 Part 2 'Funder Rights'.
32 The current edition of 'Good Practice in Selection of Construction Materials', Ove Arup & Partners.

- that Sub-Contractors named in the Contract Particulars are giving similar undertakings related to their design in the Sub-Contract Works, and for which the Contractor bears no responsibility to the Employer;
- that the relevant Consultants and Sub-Contractors have paid, to the Purchaser or Tenant, their reasonable proportion of the losses.

The Contractor is not liable to Purchasers or Tenants for delay in completion of his work (para. 9).

The third parties have no greater entitlement than the Employer, and the Contractor is entitled to rely on any of the terms of the Contract in defence of a claim from a third party (para. 1.4).

Instructions: Purchasers and Tenants have no authority to instruct the Contractor in relation to the Contract (para. 3 – see Section 9.3.5 of this chapter to compare with rights of a Funder).

Use of Contractor's Design Document: Where a Purchaser's or a Tenant's part of the site is part of a Contractor's Designed Portion, the third party is given the same rights and licences regarding the Contractor's design as the Employer. This benefit is conditional on all monies that are due and payable under the Contract having been paid (Para. 4). The requirement for payment is not restricted to payment in respect of the CDP Works, but to any monies due under the Contract. The rights of Purchasers, and particularly Tenants, are likely to become effective during the Defects Rectification Period when retentions are still being held under the Contract, and the Final Account is still being calculated. That does not prevent the third party accessing his rights in principle, although, in practice, the amount 'due' at that stage may give rise to dispute.

Professional Indemnity Insurance: The Contractor promises to the third parties (para. 5) that, to the extent that he is required under the Contract to maintain Professional Indemnity insurance, he will do so for the period referred to in the Contract Particulars relating to Clause 6.11. If requested by a Purchaser or Tenant, the Contractor is obliged to produce evidence of his ongoing insurance, although a request for evidence must be reasonable. It is difficult to imagine in what circumstances such a request would be unreasonable, particularly as an annual premium could be withheld unilaterally by the Contractor at any time during the period of risk. Contractors specializing in 'design and build' are likely to maintain cover as a commercial norm, but traditional contractors, who only venture into design on an *ad hoc* basis to meet the needs of the occasional Contractor's Designed Portion, may have taken out special insurance, perhaps 11 years earlier if the Contract is under seal. In those circumstances the prudent Purchaser or Tenant will check regularly. The JCT remind contractors by a footnote,[33] to refer to the Guide if they do not normally carry Professional Indemnity insurance. However, the Guide provides little assistance, other than a general observation that the approach to insurance needs to be realistic.[34]

If the Contractor, having obtained cover, cannot maintain his insurance because of unrealistic commercial rates, he must immediately notify the Purchasers and Tenants so that a solution may be discussed to protect their respective interests.

---

33 JCT fn. [67].

34 See also Chapter 1, Section 1.5.2 dealing with Contract Particulars Part 1 – entry for clause 6.11, and also, Contract Particulars Part 2 – entry for clauses 3.7 and 3.9.

Assignment: (for comments on assignment generally, see also Chapter 8.2). P&T Rights may be assigned only twice in a chain by Purchasers or Tenants to others of the same class, without the consent of the Contractor (para. 6). That provision is clear, but the JCT have added, in order to avoid doubt, that no further assignment of either Purchaser's or Tenant's rights will be permitted. However, it seems that the JCT are not satisfied with that, for they go on to cloud the issue by adding 'and in particular [the third Purchaser, i.e. second assignee] shall not be entitled to assign these rights'. Unfortunately, having impliedly and expressly denied further assignment by both Purchasers and Tenants, the inclusion of yet greater emphasis on purchasers must leave the Parties wondering what loophole the JCT are closing.

The Contractor's obligations transfer when he is 'given' a notice. The Contract does not state who gives the notice, although in practice it is likely to be the assignor.

'Giving' of notices: Notices given by Purchasers or Tenants to the Contractor, or *vice versa*, are 'given' either when delivered by hand, or sent by special or recorded delivery post to the registered office of the other (para. 7). If the posting option is used, delivery is deemed to be 48 hours after posting, unless actual delivery is proved to be otherwise.

Enforcement of rights: (For comment on the forum for dispute resolution and limitation periods, see Section 9.3.3 in this chapter.)

9.3.5 JCT 05 – Third Party Rights for Funder (Schedule 5, Part 2: 'Funder Rights')

Schedule 5, Part 2 sets out the rights (and their limitations) granted to a Funder by the Contractor only. No rights are granted by the Employer, except acknowledgement where the Funder exercises 'step in rights' (see below).

Compliance with the Contract and good practice: In paragraph 1.1, the Contractor promises that he has complied with the Contract. The promise to the Funder is wider than that to Purchasers and Tenants, in that it is not limited to carrying out the Works, but applies also to any obligation including procedural matters, and it is not made as at practical completion of his work. Nevertheless, in making this promise, the Contractor is simply repeating the contractual obligation he has to the Employer.

The Contractor also promises (para. 2) that, unless obliged to do so by the Contract or an Architect's instruction, he has not used any materials that do not comply with specified guidelines.[35]

Contractor's liability: The Contractor's liability, in the event that he breaks his promise, is described in paragraphs 1 and 13. Unlike liability to Purchasers and Tenants, there is no 'basic' level, with loss of profit as an optional extra. This is to be expected for, whilst the nature of loss incurred by a Funder may possibly include costs associated with the

35 The current edition of 'Good Practice in Selection of Construction Materials', Ove Arup & Partners.

correction of building work, it is most likely to be financial;[36] that is the essence of a Funder's service.

The Contractor's liability is limited to an equitable proportion of the total claimable loss related to his responsibility, and based on the premise that consultants and sub-contractors may also bear some responsibility (para. 1.1). The assumptions are:

- that Consultants named in the Contract Particulars are giving similar undertakings related to their services, either by a collateral warranty, or by third party rights, and that their liability is not limited under their consultancy agreement;
- that Sub-Contractors named in the Contract Particulars are giving similar undertakings related to their design in the Sub-Contract Works, and for which the Contractor bears no responsibility to the Employer;
- that the relevant Consultants and Sub-Contractors have paid to the Funder their reasonable proportion of the losses.

The Contractor is not liable to Purchasers or Tenants for delay in completion of his work (para. 13), unless and until the Funder has exercised his right to step into the place of the Employer, under paragraph 5 or 6.4 (see below). However, if the Funder takes over from the Employer, the Contractor may become liable to the Funder for liquidated damages if he finishes late. In those circumstances, paragraph 13 expressly prevents payment of liquidated damages to the Funder, to the extent that the Employer has already recovered them from the Contractor.

A Funder has no greater entitlement than the Employer, and the Contractor is entitled to rely on any of the terms of the Contract in defence of a claim (para. 1.2).

Instructions: A Funder has no authority to instruct the Contractor in relation to the Contract, unless he steps into the place of the Employer under Paragraph 5 or 6.4 (see below).

'Step-in' rights: The most significant difference between P&T Rights and Funder Rights is the provision for a Funder to take over (or step into) the role of the Employer (paras 5, 6, 7). Whereas the grant of rights to Purchasers and Tenants may be viewed by Contractors as a gratuitous benefit (ignoring lump sum price principles), the grant of rights to a Funder is matched by reciprocal express benefits for the Contractor. A Funder has an interest in the timely completion of the project, which can be put at risk in the event of the Employer's insolvency, termination of the finance agreement by the Funder, or of a Contractor's intention to terminate his employment because of an Employer's behaviour. 'Step in rights' give the Funder an opportunity to avoid massive delay, and to maintain continuity by retaining the Contractor. To this end, having exercised his step-in rights, the Funder, or his appointee, may give the Contractor instructions. However, in doing so, the Funder is given greater powers than the Employer, in that the Employer under JCT 05 has no general authority to instruct at all.

---

36 See Chapter 12, dealing with the principles of damages claims generally.

For the Contractor, the benefit is security, particularly regarding payment previously outstanding from the Employer, for which the Funder now assumes liability (para. 7).

There are two provisions under which a Funder may step in, one activated solely by the Funder, and the other prompted by the Contractor.

Paragraph 5 allows the Funder to give notice to the Contractor requiring him to accept instructions regarding the Works. The notice is agreed to be conclusive evidence that the finance agreement is terminated, and the Employer acknowledges that the Contractor's acceptance of instructions from the Funder is not a breach of the contract with the Employer. The Funder has no authority unilaterally to change the terms of the Contract, and any instructions given are subject to the rules of the Contract.

Paragraph 6 prevents the contractor from terminating his employment under the Contract, or from treating the Employer's actions as repudiatory. In either case the Contractor must send to the Funder a copy of any warning notices sent to the Employer, and before action is taken following a warning notice, the Contractor must again notify the Funder, and wait for 7 days (or any other period entered in the Contract Particulars). This procedure gives the Funder an opportunity to save the situation, and to step in if he wishes. The stepping in procedure and effect (para. 6.4) is the same as that for stepping in on termination of the finance agreement. However, there is an important caveat, in that the Contractor is not indemnified against liability to the Employer for any breach of the Contract, or for wrongfully serving notice of termination or repudiation.

Use of Contractor's Design Document: Where part of the site is included in a Contractor's Designed Portion, the Funder is given the same rights and licences regarding the Contractor's design as the Employer. This benefit is conditional on all monies that are due and payable under the Contract, having been paid (para. 7). The requirement for payment is not restricted to payment in respect of the CDP Works, but to any monies due under the Contract. The rights of a Funder are likely to become effective during the Defects Rectification Period when retentions are still being held under the Contract, and the final account is still being calculated. That does not prevent the third party accessing his rights in principle, although, in practice, the amount 'due' at that stage may give rise to dispute.

Professional Indemnity Insurance: The Contractor promises to the Funder (para. 9) that, to the extent that he is required under the Contract to maintain Professional Indemnity insurance, he will do so for the period referred to in the Contract Particulars relating to Clause 6.11. If requested by the Funder, the Contractor is obliged to produce evidence of his ongoing insurance, although a request for evidence must be reasonable. It is difficult to imagine in what circumstances such a request would be unreasonable, particularly as an annual premium could be withheld unilaterally by the Contractor at any time during the period of risk. Contractors specializing in 'design and build' are likely to maintain cover as a commercial norm, but traditional contractors, who only venture into design on an *ad hoc* basis to meet the needs of the occasional Contractor's Designed Portion, may have taken out special insurance, perhaps 11 years earlier if the Contract is under seal. In those circumstances the prudent Funder will check regularly. The JCT remind Contractors by a footnote,[37] to refer to the Guide if they do not normally carry Professional Indemnity insurance. However, the Guide

37 JCT fn. [67].

provides little assistance, other than a general observation that the approach to insurance needs to be realistic.[38]

If the Contractor, having obtained cover, cannot maintain his insurance because of unrealistic commercial rates, he must immediately notify the Funder so that a solution may be discussed to protect their respective interests.[39]

Assignment: (for comments on assignment generally, see also Chapter 8, Section 8.2). Funder Rights may be assigned only twice in a chain by the Funder and another funder, without the consent of the Contractor (para. 10). No further assignment of Funder Rights is allowed. As with P&T Rights, the JCT add a further express bar: '*and in particular [the third Funder, i.e. second assignee] shall not be entitled to assign these rights*'.

The Contractor's obligations transfer when he is 'given' a notice. The contract does not state who gives the notice, although in practice it is likely to be the assignor.

'Giving' of notices: Notices given by Purchasers or Tenants to the Contractor, or vice versa, are 'given' either when delivered by hand, or when sent by special or recorded delivery post to the registered office of the other (para. 11). If the posting option is used, delivery is deemed to be 48 hours after posting, unless proved otherwise.

Enforcement of rights: (For comment on the forum for dispute resolution (para. 14) and limitation periods (Para. 12) – see Section 9.3.3 in this chapter.)

9.4 JCT 05 – Collateral Warranties

9.4.1 JCT 05 – Collateral Warranties generally

(See Section 9.1 for general introduction to the need for collateral warranties)

The Employer may wish the Contractor to grant rights to Purchasers, Tenants, and the Funder, but to do so by use of Collateral Warranties instead of by Third Party Rights. Clauses 7C and 7D provide the opportunity.

The Employer may also wish Purchasers, Tenants, the Funder, and the Employer to benefit from rights granted by sub-contractors. They may do so by collateral warranty under Clause 7E.

The collateral warranties specified for use are:

- Contractor Collateral Warranty for a Purchaser or Tenant (CWa/P&T);
- Contractor Collateral Warranty for a Funder (CWa/F);
- Sub-Contractor Collateral Warranty for a Purchaser or Tenant (SCWa/P&T);
- Sub-Contractor Collateral Warranty for a Funder (SCWa/F);
- Sub-Contractor Collateral Warranty for Employer (SCWa/E).

---

38 See also Chapter 1, Section 1.5.2 dealing with Contract Particulars Part 1 – entry for clause 6.11, and also Contract Particulars Part 2 – entry for clauses 3.7 and 3.9.

39 See Chapter 7, Section 7.3.6 dealing with Professional Indemnity insurance generally.

Collateral warranties are ancillary contracts, requiring the relevant parties to conclude a separate agreement.[40] Unlike the terms of the Third Party Rights options in Clause 7A and 7B, the terms of Collateral Warranties required under Clauses 7C to 7E do not form part of JCT 05 contract. The Collateral Warranty option is no more than an obligation to enter into a separate agreement upon specified terms.

Accordingly, the detailed terms of the various warranties are not dealt with here, save to observe that the warranties between Contractor and Purchaser or Tenant, and between Contractor and Funder, are intended by the JCT to have substantially the same effect as the terms granting third party rights in Schedule 5, which are dealt with in Section 9.3 of this Chapter. Where there is a significant difference between the effect of third party rights and the corresponding warranty, it is identified and dealt with below.

9.4.2 JCT 05 – Collateral Warranties from Contractor

Contractor's obligation to provide Collateral Warranty: Clauses 7C and 7D provide for the Contractor to grant rights to Purchasers, Tenants, and the Funder by collateral warranty, but substantially in the same terms as the rights described as 'P&T Rights' and 'Funder Rights' in Schedule 5. The use of Clauses 7C and 7D is triggered by an entry in the Contract Particulars, Part 2.

The promise given by the Contractor in the warranty, as with Third Party Rights, is given as at practical completion, so the request for the Collateral Warranty to be executed may be made at any time up to or after practical completion. On receipt of a notice from the Employer, the Contractor must, within 14 days, enter into a Collateral Warranty with the identified Purchaser or Tenant (JCT form CWa/P&T) or Funder (JCT form CWa/F).

Clearly, if the Contractor does not in fact enter into the required warranty, the third party cannot enforce the benefits. There are no express sanctions in the Contract for failure to comply with the Employer's notice. The obvious potential remedy would be in damages for breach of the main contract, although identifying who has suffered loss, and how much, will not be known until much later (if at all). Recovery of loss from the Contractor (say, in respect of a defective roof) would require the intended beneficiary, as the injured party, to establish a contractual entitlement against the Employer, who in turn would need to join the Contractor in the action. The outcome is fraught with uncertainty. Another possible, albeit unlikely, remedy is that of 'specific performance' (i.e. to ask the Court to force the Contractor to provide a signed warranty in accordance with his contractual obligation). The courts are reluctant to order performance that they cannot police, such as building work, and it has been held that an obligation to provide a bond was simply piecemeal performance of the contract.[41] However, it is within the courts' jurisdiction to enforce the obligation, and they may do so, provided the obligation is clear and damages would be an inadequate remedy.[42]

---

40 If a collateral warranty form is not completed by the relevant parties, the Collateral Warranty and its terms does not exist, and no reliance may be placed upon its terms by any party. See *Lang* v. *Cardinal Construction Pty Ltd* [2008] WASCA 244.

41 *South Wales Railway* v. *Wythes* (1854) 24 LJ Ch. 87, cited in *Hudson's Building and Engineering Contracts*, 11th edn, para 4.298.

42 See *Keating on Building Contracts*, 8th Edition, Para. 11-020 on this point.

In the absence of clear protection against failure by the Contractor to provide Collateral Warranties, the Employer may consider amendments to the standard form similar to those sometimes adopted by Contractors, discussed below.

Content of Contractor's Collateral Warranties: Whilst the terms of the Collateral Warranties are substantially the same as for the Third Party Rights option, there are two significant departures. First, the JCT have written third party rights out of all their collateral warranties. It must remembered here that, under the collateral warranty, the intended third party beneficiary is one of the parties to the warranty; the third parties excluded from the warranty are strangers who might otherwise try to construe a benefit.

The second departure concerns the forum for dispute resolution. Under the standard Collateral Warranties, disputes are referable to the courts, irrespective of whether the main contract provides for litigation or arbitration. The position changes if, under the Funder warranty, the Funder exercises his step-in rights. In that case a breach by the Contractor will leave the Funder as the damaged party under the main contract, and any dispute will then be referred to the courts or arbitration, or indeed adjudication, depending on the provisions of that contract.

9.4.3 JCT 05 – Collateral Warranties from Sub-Contractors

Contractor's obligation to obtain Sub-contractor's Collateral Warranty: Clause 7E is triggered by an entry in the Contract Particulars, Part 2. They provide for the Contractor to obtain promises from Sub-Contractors by standard form collateral warranty to Purchasers or Tenants (SCWa/P&T), the Funder (SCWa/F) and the Employer (SCWa/E).

Clause 7E provides for the standard form to be amended if a proposal by a Sub-Contractor to amend is approved by the Contractor and Employer, which approval must not be unreasonably withheld of delayed.

On receipt of a notice from the Employer, the Contractor must within 21 days 'comply with the requirements as set out in the Contract Documents as to obtaining such warranties in the form'. JCT 05 does not set out any particular requirements or procedures for obtaining warranties, other than the variables described in the Contract Particulars, so greater detail if required must be agreed by the Parties and inserted in the Bills or the Employer's Requirements.

The task of obtaining collateral warranties is not popular with Contractors, and for good reason. Once a sub-contractor has started work under his sub-contract, he may be less inclined to sign up for exposure to action from parties of whom he may never have heard. Unfortunately, this reaction is likely to be a result of previous bad experience with onerous collateral warranties created by an employer's legal team, and the fact that the JCT standard warranty is arguably fair to both parties may pass without consideration. Whatever the cause, contractors often find it necessary to impose special and oppressive terms in their sub-contracts in order to prise executed collateral warranties out of their sub-contractors. One term commonly used entitles the contractor to withhold payment; another sometimes adopted gives the contractor limited power of attorney. The power is to be exercised if the sub-contractor fails to provide a collateral warranty in accordance with the Contract, and enables the contractor to sign the warranty on the sub-contractor's behalf. Understandably, sub-contractors are reluctant to give the contractor such powers, but their reluctance is no greater than the contractor's in relying on the sub-contractor's promise to provide a warranty, without the protection of a guaranteed remedy.

The success achieved by the Employer in obtaining the required warranties from sub-contractors, relies entirely on the Contractor's own success. In the event that the Contractor fails, the Employer's position with regard to damages is much the same as that when he fails to obtain the Contractor's own Collateral Warranties. However the chance of obtaining an order for specific performance, already slim, is reduced further by the uncertainty of the terms of the subcontractors' collateral warranties, which the Contract, Clause 7C, provides may be amended at the request of the sub-contractor. For that reason the Employer may wish to adopt similar protective measures in the main contract to those taken by contractors in their sub-contracts.

Content of Sub-contractor's Collateral Warranties: The terms of the Purchaser or Tenant standard warranties are substantially the same as those set out in Schedule 5 relating to promises made by the Contractor to those parties; the terms of the Funder and Employer warranties are substantially the same as those made by the Contractor to the Funder (see Section 9.4.2 in this chapter). There are two main differences.

First, the JCT have written third party rights out of all their collateral warranties. It must be remembered here that, under the collateral warranty, the intended third party beneficiary is one of the parties to the warranty (i.e. Purchaser, Funder, Employer etc.); the third parties excluded from the warranty are strangers who might otherwise try to construe a benefit.

The second difference concerns the forum for dispute resolution. Under the Collateral Warranties, disputes are referable to the courts, irrespective of whether the main contract provides for litigation or arbitration. The position changes if, under the Funder or the Employer warranty, the Funder or Employer exercises his step-in rights. In that case a breach by the Sub-Contractor will leave the Funder or Employer as the damaged party under the Sub-Contract, and any dispute will then be referred to the courts or arbitration, depending on the provisions of that contract.

9.5 JCT 05 – Third Party Rights, or Collateral Warranties?

The method of conferring rights (i.e. by the use of Third Party Rights or Collateral Warranties) is a matter for agreement between the Employer and the Contractor, although in practice it is likely to be the Employer who makes the choice, based on personal preference.

There is no choice regarding rights from sub-contractors; the choice of a method is limited to rights conferred by the Contractor. The JCT have shied away from requiring third party rights under CRTPA to be provided by sub-contractors. The place for such rights to be conferred would have to be in the sub-contract, so the main contract (i.e. that between Employer and Contractor) could do no more than require the Contractor to enter into sub-contracts containing appropriate provisions. The JCT suggest in the Guide[43] that it would

43 Standard Building Contract Guide (SBC/G), Para. 96.

be unrealistic and overly prescriptive to require the Contractor to use the JCT standard sub-contract in every case. It might also be unrealistic to expect the Employer, or his Architect, to police the sub-letting to ensure compliance. As a result, JCT 05 describes the particular rights, and on whom, to be conferred by identifiable sub-contractors in specified forms of collateral warranty.[44]

Turning to rights to be conferred by the Contractor, and the use of either CRTPA, or conventional warranties, the Employer faces a difficult choice. In drafting Third party Rights in Schedule 5 of the Contract, the JCT have ensured that the promises described, together with the paragraph numbering, are substantially the same as those in their equivalent standard collateral warranties. Thus, in respect of the principal promises, there is no advantage in one form over the other. Even if a beneficiary wishes to amend the standard provisions, this is still so, for the amendments could be made in either form, and, in either case, would have to be made known to the Contractor at tender stage.

One criticism of CRTPA is that beneficiaries may not be aware of their rights. The ability to enforce relies on actual knowledge. JCT 05 does not require the Contractor to notify the third party, so it falls to the Employer to ensure that all beneficiaries are told that their rights exist, and in a form that the beneficiary will recognize. However, the same applies equally to collateral warranties in the beneficiary's favour, so there is no advantage in either method over the other.

There are four factors that may sway an Employer in his choice.

(1) Enforcement: There is a disparity regarding arbitration and access to adjudication, between those third parties with rights under the contract, and those receiving collateral warranties. Under the Contracts (Rights of Third Parties) Act, third parties have no more and no fewer rights than the contracting parties; if the contract provides, this would include the right to refer to an adjudicator,[45] or an arbitrator.[46]

Regarding arbitration, JCT 05 includes an option for parties to refer disputes to arbitration or the courts, and, if the arbitration provision is activated in the contract, the parties, and those with Third Party Rights, are obliged to take that route. However, the standard collateral warranties expressly give jurisdiction to the English courts, except where a Funder exercises step-in rights and can subsequently avail himself of the provisions in the building contract. Regarding adjudication, JCT 05 also includes an adjudication clause, which would be available to those beneficiaries with Third Party Rights under the Contract. However, the standard collateral warranties do not include express provision for adjudication, so unless the warranty could be construed as a 'construction contract' for the purposes of the Construction Act,[47] statutory adjudication would not be available to the beneficiary. Whether or not the collateral warranty falls under the Construction Act is arguable; at the time of writing, there is no case law on the issue. Clearly the promises made in the collateral warranty refer to matters that may, themselves, be construed as 'construction operations' under s. 105 of the Construction Act, unless they are excluded

---

44 See Section 9.4.3 of this chapter, and Chapter 1, Section 1.5.2 dealing with Contract Particulars Part 2 – Collateral Warranties from Sub-Contractors.

45 See Chapter 17, Section 17.4, generally.

46 See Section 9.2.3 of this chapter, referring to the case of *Nisshin Shipping Co. Ltd* v. *Cleaves & Co. Ltd* [2003] EWHC 2602 (Comm).

47 See Chapter 17, Section 17.4, generally.

under s. 105(2).[48] Likewise, the promises may relate to design and the execution of work for the purposes of the Construction Act,[49] but the warranty is not a contract for carrying out that work – it is a promise that the work under a separate contract has been done in compliance with that contract. The carrying out of the work has been done and paid for under the building contract. The warranty is more in the nature of an indemnity. However, whilst that point may be debatable, there is no doubt that a collateral warranty relating to work expressly excluded by the Construction Act, such as the design and installation of machinery in a food production facility, would not fall under CRTPA.

That leaves the Employer with the knowledge that all those with Third Party Rights under the building contract have a contractual right to refer to adjudication, but those with rights under collateral contracts have only a questionable entitlement under the Construction Act.

(2) Ease of obtaining: One great advantage of Third Party Rights over collateral Warranties, is that unlike warranties, they do not require collection. If an anticipated collateral warranty is not in fact executed, it does not exist,[50] and the beneficiary is left to prove his right to damages as best he can, in the event of a failure in thc work. Third Party Rights, however, come into existence on the issue of a notice from the Employer.

(3) Reduction of administration: It is tempting to view Third Party Rights in the contract as a means of avoiding the task of obtaining collateral warranties. Unfortunately, the saving is minimal, for the option only applies to rights conferred by the Contractor. When it comes to collecting collateral warranties together, the more numerous and more difficult to obtain are those from sub-contractors, and there is no option for the Third Party Rights alternative from sub-contractors. To put it into context, if rights were required to be conferred by ten sub-contractors on twenty potential purchasers or tenants, an employer and a funder, in a new shopping centre, there would be two hundred and forty separate documents to be signed in any event. That does not include warranties from the Architect and other designers. The only saving in time or effort if the Third Party Rights option is used, would be in avoiding collecting twenty-one warranties from the Contractor.

(4) Habit: Collateral warranties have been a common feature of the construction industry for many years, and the 2005 series of JCT warranties are drafted to be substantially in the same terms as previous editions.[51] Many of those employers who are frequent users of the industry will be familiar with the standard form warranties, and will have developed administrative machinery to deal with them. They may be unwilling to leave their tried and tested routines in favour of the unknown, particularly as the terms of the warranties are unchanged, and the scope of Third Party Rights is restricted to contractors.

---

48 See Chapter 1, Section 1.5.5.4.

49 Housing Grants Construction and Regeneration Act 1996, s. 104.

50 If a collateral warranty form is not completed by the relevant parties, the Collateral Warranty and its terms does not exist, and no reliance may be placed upon its terms by any party. See *Lang* v. *Cardinal Construction Pty Ltd* [2008] WASCA 244.

51 See 'Guidance Notes – General' published in JCT standard collateral warranties 'CWa/F' and 'CWa/P&T'.

10

Health and Safety obligations under the CDM Regulations

The Construction (Design and Management) Regulations 2007 (CDM 2007), which came into force on 6 April 2007, replaced the Construction (Design and Management) Regulations 1994 and the Construction (Health, Safety and Welfare) Regulations 1996. The Health and Safety Commission has produced an Approved Code of Practice (ACOP)[1] that provides guidance on compliance with the new Regulations. For all intents and purposes, it is part of the CDM Regulations themselves, as a court may treat failure to comply with the code as evidence of breach of the relevant health and safety legislation.

Clause 2.1 of JCT 05 imposes upon the Contractor a duty to comply with the Statutory Requirements, which are defined under Clause 1.1 to include any statute, statutory instrument or regulation which affects the Works or performance of any obligation under the Contract. The Contractor is therefore under a contractual obligation to comply with the CDM Regulations. It is also stated in Clause 3.25 that 'Each Party undertakes to the other that in relation to the Works and the site he will duly comply with the CDM Regulations…'. It is further provided in that Clause that the Employer has a duty to ensure that the CDM Co-ordinator and the Principal Contractor carry out their duties under the Regulations. Breach of these contractual duties by either party is a specified default for which the other would be entitled to terminate the Contract.[2]

The Employer and the Contractor are therefore required by the Contract to ensure performance of not only their own duties under the Regulations but also those of other dutyholders for whom they are responsible under the Contract or the Regulations. It is for this reason that this chapter outlines the duties of the various dutyholders under the Regulations. In most cases the Employer would be the CDM client in respect of the particular development.[3]

1 Health and Safety Commission, *Managing Health and Safety in Construction*, HSE Books, 2007 (ISBN 978 0 7176 6223 4).

2 See Clauses 8.4.1.5 and 8.9.1.4, which are discussed in Chapter 16, Sections 16.6.5 and 16.9.4, respectively.

3 As explained under Section 10.4 in this chapter, it is possible, although not advisable, for the CDM client to be a different entity.

As explained later in the chapter, he may also take on additional roles depending on his actions in relation to the project. The Contractor is primarily a contractor as defined under the Regulations. Depending on how the Contract Particulars are completed, he may also be the Principal Contractor or, in relation to any CDP component, a designer for the purposes of the Regulations.

The main State agency responsible for enforcing the Regulations is the Health and Safety Executive (HSE) although, in respect of construction work on rail projects, the Office of Rail Regulation (ORR) is the enforcing authority. Also, local authorities are responsible for health and safety breaches in premises such as offices, hotels, retail centres, and places of entertainment.

Any reference hereafter to the 'Regulations' is to the 2007 Regulation unless stated otherwise. Certain duties in relation to cooperation, coordination, competence and prevention on the project are imposed on every dutyholder. They are therefore described first.

10.1 Cooperation and coordination

Regulations 5 and 6 require cooperation and coordination among all CDM dutyholders on the Contract as well as those on neighbouring projects. Every dutyholder must seek cooperation from all other relevant dutyholders and also cooperate with them. They are also to coordinate their work with that of others in a manner which ensures, so far as reasonably practicable, the health and safety of all on the site as well as others affected by the construction work. The rationale behind this requirement is that such cooperation and coordination promotes timely identification and communication of risks to health and safety for appropriate action, including ensuring that everybody who needs to know of the risk has access to the relevant information. It is part of the duties of the CDM client to take reasonable steps to ensure that the management systems and procedures for the project adequately address the need for such cooperation and coordination.

10.2 Competence of CDM dutyholders

The Regulations treat competence as a key requirement of effective health and safety management systems of organizations and individuals. Regulation 4(1)(a) prohibits appointment or engagement of a CDM Co-ordinator, principal contractor, contractor or designer under the Regulations without taking reasonable steps to ensure that the appointee is competent to perform the duties imposed on him under the Regulations. Regulation 4(1)(b) prohibits acceptance of an appointment or engagement as a dutyholder unless the person concerned is competent. To be considered competent, an organization or individual must have: (i) sufficient knowledge of the specific tasks involved in the project and their associated risks; (ii) sufficient experience and ability to perform the relevant tasks.[4]

The ACOP recommends a two-stage procedure for assessing the competence of companies and individuals for appointment as dutyholders. In the case of a company the stages are: (i) assessment of the company's organization and arrangements to determine their

4 See Regulation 4(2) and para. 195 of the ACOP.

sufficiency for effective management of health and safety; (ii) assessment of the company's experience and track record of similar projects. In the case of individuals, the assessment entails seeking evidence that the individual has sufficient knowledge of the tasks he is required to perform and experience and track record in the performance of tasks of that type. Core assessment criteria are provided in Appendix 4 of the ACOP. It is important to note that competence requires availability of necessary resources to the company or the individual.

10.3 Prevention

Regulation 7 requires every person with duties in relation to the design of the project or the construction phase to ensure that the principles of prevention are applied in the performance of the relevant duties. The principles of prevention are defined in Regulation 2(1) by reference to Schedule 1 to the Management of Health and Safety at Work Regulations 1999 on which there is an approved code of practice.[5]

10.4 Duties of CDM client

A client, for the purposes of the Regulations, is widely defined in Regulation 2(1) as 'a person who in the course or furtherance of a business – (a) seeks or accepts the services of another which may be used in the carrying out of a project for him; or (b) carries out a project himself'. Such a person may be a corporation, a limited liability company or an individual. In some large developments, multiple parties may come within the definition and will be treated accordingly unless an election procedure allowed by Regulation 8 is undertaken. It entails one or more of the potential clients electing in writing to be treated as the client in respect of the Regulations. Thereafter, those of the potential clients who agreed in writing to the election cease carrying responsibilities of the client under the Regulations. However, the following duties remain with any person within the definition of a CDM client provided he has relevant information in his possession:

- to cooperate with those undertaking work on the project or any other project on adjoining sites (Regulation 5(1)(b));
- to provide designers and contractors with pre-construction information (Regulation 10(1));
- to provide the CDM Co-ordinator with pre-construction information (Regulation 15);
- to provide the CDM Co-ordinator with health and safety information (17(1)).

It is assumed in the rest of this work that the Employer is also the CDM client. Where this is not the case, there is need for the Employer to work closely with the client to ensure that the Employer can deliver his contractual undertakings in relation to the Regulations. Depending upon any wider role in relation to the project, the Employer's Agent (and,

5 A revised version was published in 2000.

Table 10.1 Duties of the CDM Client

Regulation	Description of duties
14	Appoint and maintain on the project a CDM Co-ordinator and a Principal Contractor until the end of the construction phase
4	Take reasonable steps to ensure that any CDM dutyholder he appoints is competent to perform his duties under the Regulations
10	Provide pre-construction information to designers and contractors
15	Provide the CDM Co-ordinator with pre-construction information including information on the minimum amount of time that will be allowed the Principal Contractor to carry out pre-construction planning and preparation
17	Ensure that the CDM Co-ordinator is provided with all health and safety information in his possession or reasonably obtainable that is likely to be needed for inclusion in the Health and Safety File
9(1)(a)	Take reasonable steps to ensure that: (i) the arrangements made by dutyholders for managing the project are suitable to ensure that construction work can be carried out, so far as is reasonably practicable, without risk to the health and safety of any person; (ii) there is allocation of sufficient time and resources in the arrangements
9(2)	Take reasonable steps to ensure that the arrangements are maintained and reviewed throughout the project
9(1)(b)	Take reasonable steps to verify that the arrangements are suitable to ensure that there are suitable welfare facilities on site during the entire construction phase
9(1)(c)	Take reasonable steps to ensure that there are suitable arrangements for compliance with the relevant provisions in the Workplace (Health, Safety and Welfare) Regulations 1992 in relation to the design of any structure to be used as a workplace
16	Ensure that the construction phase does not start without a suitable Construction Phase Plan and arrangements for suitable welfare facilities during the construction phase

indeed, the Architect or any other consultant on the project) may come within the definition of a CDM client. This uncertainty is easily cured by expressly addressing this issue in the relevant contract for professional services.

The bedrock principle underlying the entire CDM Regulations is that, of all the project participants, the client has the greatest influence on health and safety. The rationale behind this principle is that the client's procurement decisions determine the competence and teamworking skills of the organizations and individuals engaged to manage health and safety and the availability of the resources and time necessary to plan and deliver the project safely. To provide clients with sufficient incentive to exercise this influence properly, the Regulations impose certain duties on them. They are summarized in Table 10.1.

Many of the duties imposed on the client by the Regulations would normally be performed before the Contract is entered into. They also continue to apply during the construction phase, as Regulation 9(2) requires the client to take reasonable steps to ensure that the pre-construction arrangements made for the management of the project are maintained and reviewed throughout the entire duration of the project.

10.5 CDM Co-ordinator

Article 5[6] names the CDM Co-ordinator.[7] The Contract expects him to be the Architect or other named individual from a named organization. Having the Architect doubling up as the CDM Co-ordinator avoids creating an additional interface that would otherwise result in delays in managing the health and safety implications of variations and post-contract design and planning. Regulation 14(3) puts on the Employer a duty to ensure that there is, at all times until the end of the construction phase, a CDM Co-ordinator. The Employer must therefore appoint, as soon as reasonably practicable, a replacement if an incumbent CDM Co-ordinator ceases to act as such. As the Contract is silent on the Contractor's right to object, a contrast to the Contractor's express rights under Clause 3.5 to raise reasonable objections to the appointment of replacement Architects and Quantity Surveyors, the implication has to be that the Contractor has no such right in relation to the CDM Co-ordinator. However, this does not mean that the Contractor must not speak up if there are concerns about the suitability of the replacement.

The CDM Co-ordinator's general responsibility is to oversee and coordinate the health and safety aspects of the design and planning right from project inception to completion. There are three main tasks within this role.[8] First, he has the duty to advise and assist the client to comply with his duties under the Regulations. In particular, he must advise and support the client in verifying the competences of all CDM appointees and the adequacy of their management systems for the purposes of health and safety. Second, he must ensure effective communication and coordination among all the project participants. Finally, he must ensure that the Health and Safety File is prepared and delivered to the client. Specific duties under the Regulations are summarized in Table 10.2.

On the type of projects for which the JCT05 is suitable, most of the work of the CDM Co-ordinator is done at the pre-tender stage. For this reason, Regulation 14(1) requires the client to appoint to this role 'as soon as practicable after initial design work or other preparation for construction work has begun'. There is sanction against dilatoriness in making the appointment in that, for any period during which there is no CDM Co-ordinator appointed, the client is deemed to have appointed himself to that role. The duties of the CDM Co-ordinator would therefore apply in addition to the normal duties of the client. Although we are concerned mainly with his duties during construction, some understanding of his pre-construction duties is required as well, because many of them provide the necessary foundation for the duties during the construction stage.

10.5.1 Pre-construction duties

Specific pre-tender responsibilities include giving notice[9] of the project to the Health and Safety Executive, advising the client on competencies and resources of CDM dutyholders, and assisting the client to ensure that all of them fulfil their obligations under the

6 See Chapter 1, Section 1.4.1 (dealing with the Articles) and 1.5.2 (dealing with preparing the contract documents).
7 This role has replaced that of the Planning Supervisor under the 1994 CDM Regulations.
8 See para. 84 of ACOP.
9 The information to be provided in the notices is prescribed in Schedule 1 to the Regulations. There is a standard form, referred to as Form F10, for providing the information, although its use is not mandatory.

Table 10.2 Specific duties of the CDM Co-ordinator

Regulation	Description of duties
20(1)(a)	Give suitable and sufficient advice and assistance to the client in relation to the measures that he needs to take to comply with the duties of clients under the Regulations
20(1)(b)	Ensure that suitable arrangements are made and implemented for the coordination of health and safety measures during planning and preparation for the construction phase
20(2)(a)	Take reasonable steps to identify and collect pre-construction information and to advise the client to take action to fill any gaps in the available information
20(2)(b)	Provide promptly to every designer and contractor such of the pre-construction information in his possession as is relevant to each
21	Ensure that notice containing specified particulars is given to the HSE (or the ORR) as soon as practicable after his appointment
20(2)(c)	Take all reasonable steps to ensure that: (i) designers comply with their duties under the Regulations; (ii) designers provide sufficient information to assist the CDM Co-ordinator to comply with his duties under the Regulations
20(2)(d)	Take all reasonable steps to ensure co-operation between designers and the Principal Contractor during the construction phase in relation to any design or design change
20(2)(e)	Update or, where none exists, to prepare the Health and Safety File
20(1)(c)	Liaise with the Principal Contractor regarding the contents of the Health and Safety File, information needed to prepare the Construction Phase Plan and any design development which may affect the planning and management of construction work
20(2)(f)	Pass the Health and Safety File to the client at the end of the construction phase

CDM Regulations. He must compile a document referred to as 'pre-construction information' (PCI), which is a collection of information on health and safety issues in relation to the project (e.g. type of work, existing site conditions, access arrangements, traffic routes, security requirements and statements of risk assessments by designers). The information is acquired from many different sources such as the client and the designers. The CDM Co-ordinator's responsibility in relation to the production of the information is therefore largely one of coordination of the input of the necessary information from the various sources. Each tenderer is to be provided with the PCI as it has obvious pricing implications. Extracts from it are to be supplied to every dutyholder who needs them to perform its duties under the Regulations.

10.5.2 Duties during construction

A Construction Phase Plan must be developed before construction can start. This task is to be undertaken by the Principal Contractor although the responsibility to ensure that construction does not start until this has been done remains with the client. However, the CDM Co-ordinator may be under a duty imposed by his contract of engagement to oversee the development of the plan on the client's behalf and to advise whether a plan submitted by the Principal Contractor has reached the stage where construction can be lawfully started.

For the purposes of JCT 05, the CDM Co-ordinator has two main duties during the construction phase. First, he must continue to monitor and coordinate the health and safety aspects of any design and planning continuing during construction. Such design and planning may arise from variations, design work within the CDP or the appointment of sub-contractors with design responsibility. He must therefore be consulted on such matters and served with the information necessary to discharge this responsibility. Second, he must coordinate the production of a Health and Safety File for delivery to the client at the end of the construction phase. From Regulations 2(2) and 20(2)(e), the Health and Safety File is intended as a permanent record of those aspects of a construction project which might affect the health and safety of: (a) any person carrying out future construction, cleaning or demolition work upon the building; (b) any person occupying the building who may be affected by those carrying out work upon it. Typical contents of the file include 'as built' drawings, design criteria, inbuilt facilities for the maintenance of the building, operating and maintenance manuals for specialist plant and equipment and incoming services. Here again, the information required is scattered across the variety of participants on the project. Clause 3.25.3 requires the Contractor to comply with the CDM Co-ordinator's reasonable timetable for the provision of information for the Health and Safety File and to ensure similar compliance by all sub-contractors. The information is to be provided to the CDM Co-ordinator where the Contractor is also the Principal Contractor, but to the Principal Contractor if different from the Contractor.

It is important to note that, under Clause 3.25.1 of JCT 05, any failure of the CDM Co-ordinator properly to perform his duties under the CDM Regulations is in effect a breach of the Contract by the Employer for which the Contractor would be entitled to appropriate remedies, including termination of his own employment.[10] The Employer may in turn look to the CDM Co-ordinator for compensation against such liability to the Contractor.

10.6 Contractors under CDM

A 'contractor' under CDM 2007 is defined in Regulation 2 as 'any person (including the client, principal contractor or other person referred to in these Regulations) who, in the course or furtherance of a business, carries out or manages construction work'. The Contractor under JCT 05 is therefore clearly a contractor under the Regulations and, as such, is subject to the duties imposed on such contractors by the Regulations. They are summarized in Table 10.3. The Contractor would also be subject to the duties of a principal contractor under the Regulations if the Contractor is also appointed as the Principal Contractor under the contract.

10.7 Principal Contractor

The term 'principal contractor' under the Regulations is a creature of statute and he must not therefore be confused with the terms 'main contractor' or 'prime contractor'.

10 Under Clause 8.9.1.4 failure to ensure that the CDM Co-ordinator carries out his duties under the CDM Regulations is a default by the Employer for which the Contractor may terminate his employment under the Contract; see Chapter 16, Section 16.9.4 for more commentary.

Table 10.3 Duties of a contractor under CDM

Regulation	Description of duties
13(1)	Ensure that the client is aware of the duties applicable to him as client under the Regulations
13(6)	Not to start construction work on site unless reasonable steps have been taken to prevent unauthorized access to the site
19(1)	Not to carry out any work unless: (i) he has been provided with the names of the CDM Co-ordinator and the Principal Contractor; (ii) he has been given sufficient access to the part of the Construction Phase Plan relevant to his work; (iii) the project has been notified to the HSE/ORR
13(2)	Plan and manage construction work in a way which ensures that, so far as reasonably practicable, it is carried out without risks to health and safety
19(2)(a)	Provide the Principal Contractor with information relevant to the management of health and safety or for inclusion by the CDM Co-ordinator in the Health and Safety File
19(2)(b)	Notify the Principal Contractor of appointments of sub-contractors
19(2)(c)	Comply with the directions of the Principal Contractor on health and safety and any site rules
19(2)(d)	Provide promptly to the Principal Contractor information on any dangerous occurrence, injury, or death that must be reported under the Reporting of Injuries, Diseases and Dangerous Occurrences Regulations 1995
13(2)	Ensure that every sub-contractor that he engages on the project is informed of the minimum amount of time allowed the sub-contractor for planning and preparation before beginning construction work
19(3)(a)	Take reasonable steps to ensure compliance with the Construction Phase Plan in the carrying out of work
19(3)(b)	Take appropriate action to ensure health and safety in cases where it is impossible to comply with the Construction Phase Plan
19(3)(c)	Notify to the Principal Contractor any perceived need to review the Construction Phase Plan
13(3), 13(5)	Provide every construction worker under his control with the information and training he needs to carry out the work safely and without risk to health and safety
13(7)	Ensure that appropriate welfare facilities are available to every construction worker under his control for the duration of the construction phase

Regulation 14(2) requires the client to appoint a principal contractor as soon as practicable after appointing the CDM Co-ordinator and knowing enough about the project to be able to appoint a suitable person to that post. Article 6 of JCT 05 provides that the Contractor is also the Principal Contractor to take on the role of principal contractor under the Regulations unless the name and address of a different organization or person are completed against that Article.[11] Although the Contractor does not have to be appointed as the Principal Contractor, it is advised in the ACOP that having the Contractor as Principal

11 See Chapter 1, Section 1.4.1.

Table 10.4 Duties of the Principal Contractor

Regulation	Description of duties
13(1), 19(1)	Satisfy himself that: (i) the client is aware of his duties; (ii) the CDM Co-ordinator has been appointed; and (iii) the HSE is notified before work starts
23(1)	Prepare, implement and keep updated a Construction Phase Plan
22(1)(g)	Consult each contractor on the part of the Construction Phase Plan relevant to the work of that contractor before finalizing it
23(2)	Take reasonable steps to ensure that the Construction Phase Plan identifies not only risks to health and safety arising from the construction work but also suitable and sufficient measures for addressing them
22(1)(f)	Ensure that all contractors are inforned of the minimum amount of time allowed for them to carry out pre-construction planning
22(1)(a)	Ensure safe working, coordination or cooperation between contractors
22(1)(c)	Ensure availability of adequate welfare facilities on site
22(1)(l)	Take reasonable steps to prevent unauthorized access to the site
22(1)(d)	Prepare and enforce necessary site rules
22(1)(h) &(i)	Provide copies of (or access to) the Construction Phase Plan and other information to contractors in time for them to plan their work
22(1)(j)	Ensure that each contractor provides promptly information concerning that contractor's work that is needed by the CDM Co-ordinator for preparing and updating the Health and Safety File
22(1)(b)	Liaise with the CDM Co-ordinator and designers on design carried out during the construction phase
22(1)(e)	Give reasonable directions to contractors as necessary to comply with his own duties under the Regulations
22(2)	Ensure that all workers receive suitable health and safety induction and training
24(a)	Make and maintain arrangements which enable effective cooperation with workers engaged in the construction work that ensures their health, safety and welfare
24(b)	Ensure consultation with the workforce on matters of health and safety on the site
22(1)(k)	Ensure that notification of the project is displayed prominently on the site

Contractor represents good management practice that allows the management of health and safety to be incorporated into the wider management of project delivery.

The general responsibility of the Principal Contractor is to plan, manage, monitor and coordinate work during the construction phase in order to ensure that, as far as is reasonably practicable, the Works can be completed without risks to the health and safety of the people on the site as well as others likely to be affected by them. Specific duties are listed in Table 10.4. The term 'contractor' for the purposes of the Regulations is defined under Regulation 2(1) to include the Principal Contractor. The duties of contractors explained in Section 10.5 in this chapter therefore also apply to the Principal Contractor. The two Sections are therefore to be read together to develop understanding of the full set of duties applicable to this dutyholder.

More comment on the Construction Phase Plan is called for. It is defined in Regulation 2 as 'a document recording the health and safety arrangements, site rules and any special measures for construction work'. The term is also defined under Clause 1.1, which effectively adopts the statutory definition. It is therefore a different document from the master programme referred to in Clause 2.9. It must address issues such as the organization structure, site rules, and safety and emergency procedures. Throughout the construction phase, the plan is to be kept under continual review so that it adequately addresses evolving health and safety issues. Such issues may arise from design changes, instructions as to the expenditure of provisional sums, appointments of sub-contractors with new design input, work methods, and health and safety standards actually achieved. To make all these stringent requirements on planning meaningful, the Principal Contractor is also responsible for ensuring that the Contractor, sub-contractors and designers are competent and adequately resourced in relation to health and safety and well informed on the Construction Phase Plan.

Practice Note 27, which was published to accompany the 1994 Regulations, suggested that where the Contractor was also the Principal Contractor, the costs of reviewing the Health and Safety Plan[12] in response to variations or instruction as to the expenditure of provisional sums were to be added as part of the valuation of the relevant variation. Similarly, extensions of time considerations were to include the impact of any review work on progress with the Works. The same advice would also be applicable to reviews of the Construction Phase Plan. These entitlements would be subject to any separate agreement reached between the Contractor and the Employer regarding his role as the Principal Contractor.

When the 1994 Regulations applied, it was becoming common practice for the client to appoint one of his professional team to take on the role of the Planning Supervisor during the pre-contract phase of the project. Upon appointment of the Contractor, the Planning Supervisor changed over to the Principal Contractor. This arrangement offered the advantage of continuity in the development of the Health and Safety Plan. If this practice is to continue, it would entail the CDM Co-ordinator taking on the role of Principal Contractor as an additional responsibility because the 2007 Regulations require both dutyholders to be maintained throughout the construction phase.

10.8 Designers

The Regulations recognize that designers can play a major role by ensuring that they consider, in their design decisions, the health and safety of those who will construct, maintain, use and dismantle or demolish the structures that they design.[13] The question then is what amounts to design and who a designer is for the purposes of JCT 05 contract. Regulation 2(1) defines 'design' as including 'drawings, design details, specification and bill of quantities (including specification of articles or substances) relating to a structure, and calculations prepared for the purpose of a design'. A designer is defined under Regulation 2(1) as:

> any person (including a client, contractor or other person referred to in these Regulations) who in the course or furtherance of a business –
> (a) prepares or modifies a design; or
> (b) arranges for or instructs any person under his control to do so

12 This document is now called the 'Construction Phase Plan' under CDM 2007.
13 See ACOP para. 110.

relating to a structure or to a product or mechanical or electrical system intended for a particular structure, and a person is deemed to prepare a design where a design is prepared by a person under his control.

The professions known for providing design services, such as engineers, architects, landscape architects and building surveyors, are obviously within the definition. However, the scope of these definitions is so wide that any of the project participants may take on the role of designer without being aware. For example, the Contractor would be a designer in relation to temporary works, design of CDP work and any other contribution to the project that amounts to design as defined. Similarly, the Employer and the Employer's Agent would be designers if they specify work methods or materials or make decisions as to how alterations of existing buildings and other structures are to be made.

The philosophy underlying the Regulations that deal with design and designers is that a designer can influence health and safety positively in three ways: (i) identifying and eliminating risks by designing out the relevant hazards at source; (ii) reducing risks that cannot be eliminated by competent design; (iii) good communication with contractors, the CDM Co-ordinator and other designers so that they are sufficiently aware of the remaining risks to take appropriate action. To incentivize such a contribution, the Regulations impose specific duties on the designer. They are summarized in Table 10.5.

The importance of a designer's responsibilities under the Regulations cannot be overemphasized. Indeed, the first ever prosecution under the 1994 Regulations was of a

Table 10.5 Duties of designers

Regulation	Description of duties
11(1), 18(1)	Not to commence any work on the project unless the client is aware of his duties under the Regulations and the CDM Co-ordinator has been appointed
4(1)(b)	Ensure that he is competent and adequately resourced to deal with the health and safety aspects of his design
4(1)(a)	Not to appoint or engage any designer without taking reasonable steps to ensure that the person is competent
4(1)(c)	Not to instruct or arrange for any person to carry out any design without ensuring that the person is competent or suitably supervised
5	Cooperate with the CDM Co-ordinator, the Principal Contractor and any other designer in furtherance of project health and safety
11(3)	Avoid foreseeable risks to the health and safety of: (i) those who will carry out the construction of the works; (ii) people likely to affected by the carrying out of the works; (iii) users of the structure; (iv) people who will carry out maintenance or demolition activities on the structure
11(4)	Eliminate foreseeable hazards and minimize those that remain
11(5)	Take account of the provisions in Workplace (Health, Safety and Welfare) Regulations 1992 in his design of any structures to be used as workplaces
11(6), 18(2)	Provide with his design sufficient information about any aspect of the design, construction, maintenance, and demolition of the structure to assist the client, contractors, the CDM Co-ordinator and other designers in the performance of their duties under the Regulations

designer.[14] The firm of architects concerned were engaged to design and supervise the construction of an extension to a cash-and-carry store designed by them about 8 years earlier. During construction the contractor struck an 11,000 volt electricity cable, causing an explosion. As original designers of the store, the architects must have known of the existence and location of the cable. It was the early days of the 1994 Regulations when procedures were still being refined. Although the architects coordinated health and safety matters on site, they were never really appointed as the Planning Supervisor. If a Planning Supervisor had been appointed, it would have been his responsibility to have identified and flagged up the hazard in the Pre-tender Health and Safety Plan. The employers avoided prosecution for failing to appoint a Planning Supervisor because they never knew that they needed to make such an appointment. Although the architects could not be prosecuted as the Planning Supervisor, the firm were found guilty of contravening the Regulations by failing to advise their client to appoint a Planning Supervisor and were fined £500.

10.9 Enforcement of Regulations

Section 15 of the Health and Safety at Work Act 1974 (HASAWA) authorizes the Secretary of State for the appropriate Government Department to make health and safety regulations. The CDM Regulations were implemented by exercise of this power. They therefore carry effect as provided for under HASAWA. Contravention of any health and safety regulation is a criminal offence under s. 33(1)(c) and, as such, liable to prosecution under HASAWA. Also, where a breach results in any fatality the company's senior management or the individual responsible for it may be prosecuted for the common law offence of gross negligence manslaughter. Furthermore, as explained later, companies may also be prosecuted for the new statutory offence of corporate manslaughter.[15]

Enforcement actions available to relevant enforcement agencies include the issuance of improvement and prohibition notices and prosecution for breaches of the Regulations. They may also result in disqualification of directors in the company involved. Prosecution for manslaughter is the responsibility of the Crown Prosecution Service (CPS) although the Police, assisted by the HSE or the ORR, are responsible for investigating the relevant events to collect the information necessary for commencing prosecution.

10.9.1 Improvement/prohibition notices

A health and safety inspector may issue an improvement notice where he believes that a Regulation is being broken or is likely to be broken. Such a notice will usually specify steps that must be taken within a stated period. Where the breach creates a very hazardous situation, a prohibition notice may be issued. Such a notice is in effect an order to

---

14 See *Building* magazine, 19 January 1996, in which it was reported.

15 This legislation has abolished the common law offence of gross negligence manslaughter with respect to companies.

stop the activity giving rise to the breach. Where the prohibition covers all activities the site is effectively closed down.

10.9.2 Prosecution for criminal offence

As already explained, breach of the Regulations is a criminal offence under s. 37 of HASAWA. This would make the dutyholder concerned liable to prosecution by the HSE, ORR or relevant local authority. Depending on the nature of the offence, the Magistrates' Court may impose a fine of up to £20,000, a term of imprisonment not exceeding 12 months, or both.[16] The most serious offences are tried in the Crown Court, which has jurisdiction to impose a fine without limitation on the amount, or a prison term not exceeding 2 years, or both. Prosecutions will normally follow investigations into how the organizations and individuals concerned performed their duties under the Regulations. Such investigations may require the production of evidence of whether and how the duties were performed (e.g. how the competences and resources of appointees were assessed prior to their appointment, induction and training of workers, and health and safety management systems). Documents required to be produced (e.g. the Pre-Construction Information, information provided by designers, and the Construction Phase Plan) are also likely to be scrutinized for their adequacy in the particular circumstances.

A summary of prosecution data from April 1995 to March 2006 showed that there had been 276 successful prosecutions of breaches of the 1994 Regulations. Breaches by the client accounted for about 52% of the prosecutions whilst 7% applied to planning supervisors. One of the key changes made in the 2007 Regulations is the strengthening of the duty of the CDM Co-ordinator to advise and assist the client. It would therefore appear that, assuming a similar trend of prosecutions, there is likely to be a higher rate of successful prosecutions against CDM Co-ordinators for failure to provide adequate advice to clients.

Examples of successful prosecutions under CDM 94 include:[17]

- clients' failures to appoint planning supervisors;
- clients' failures to appoint principal contractors;
- clients allowing construction to start without a health and safety plan;
- clients' failure to provide information on risks, particularly information about the presence of asbestos;
- planning supervisors' failures to pass on information about the presence of asbestos;
- planning supervisors' failures to prepare pre-construction information;
- designers' failures to advise clients of their CDM duties;
- inadequate health and safety plans;
- contractors' failures to assess competence of subcontractors;
- principal contractors' failures to display the notification of the project.

The senior management of a company in breach of the Regulations may also be prosecuted personally under s. 37 of the HASAWA. To be liable to such prosecution it must be

16 See Health and Safety (Offences) Act 2008.

17 For an extensive database of prosecutions see http://hse.gov.uk/prosecutions/case, a website of the Health and Safety Executive.

shown that either the breach was committed with the consent or connivance of the senior manager or it was attributable to neglect on his part.

10.9.3 Disqualification of Directors

A director guilty under s. 37 may also be disqualified from serving as a company director for up to 2 years.[18]

10.9.4 Prosecution for common law gross negligence manslaughter

Manslaughter in this context refers to the killing of a human person by recklessness or gross negligence. An individual human person whose direct actions cause a fatality may be prosecuted for this offence. The maximum penalty for manslaughter is life imprisonment.

In principle, a company could also be prosecuted for manslaughter although it was notoriously difficult to convict because the requirement to establish a 'guilty mind' on the part of the company proved too high a hurdle to jump over in most cases. Public response to high-profile disasters such as the *Herald of Free Enterprise* accident in Zeebrugge, the sinking of *The Marchioness* in the Thames and the horrific rail accidents in Southall, Paddington and Hatfield was one of great outrage that big business and its 'fat cats' were getting away with 'murder'. In response to this concern, Parliament passed legislation the intended effect of which is discussed next.

10.9.5 Prosecution for corporate manslaughter

The Corporate Manslaughter and Corporate Homicide Act 2007, which came into force on 6 April 2007, creates the new offence of corporate manslaughter.[19] It carries the sanction of an unlimited fine. The offence is committed where an organization's activities caused death as the consequence of the quality of the management of the relevant activity by the organization's senior management falling far short of the reasonable standard to be expected. In the context of death from construction, breach of the Regulations would be an indication of such sub-standard management. The impact of this new legislation on poor health and safety management in the construction industry remains to be seen. To the disappointment of many, the Act does not provide for prosecution of the directors and senior managers of a company that is guilty of corporate manslaughter.[20] However, they may attract damaging publicity from the prosecution of their companies. Furthermore, the

---

18 See s. 2(1) of the Company Directors Disqualification Act 1986.

19 The Scottish Equivalent is corporate homicide.

20 Apparently, apart from the difficulty of coming up with an effective and fair procedure for holding an individual responsible for corporate acts, there were fears that legislating for the jailing of directors and senior managers would deter competent people from taking on such roles in companies.

court may issue a Publicity Order requiring a guilty company to advertise its mistake or write letters to inform its clients, suppliers and shareholders.

10.10 Civil liability for breaches of the Regulations

Breach of statutory duty may form the foundation for an individual bringing a civil action for compensation. Whether it is actionable depends on the provision in the statute that is broken. Regulation 45 provides that, with certain exceptions, breach of the CDM Regulations does not confer a right of action in civil proceedings by any person who is not an employee of the dutyholder in breach. This does not mean civil action cannot be brought on a different basis (e.g. breach of contract or breach of the common law duty of care).

10.10.1 Breach of contract

As already explained, JCT 05 contains mutual undertakings by the Employer and the Contractor to comply with the Regulations. They are also under obligations to ensure compliance by certain third parties. Breach of these undertakings and obligation therefore constitutes an actionable breach of contract, giving rise to damages. Serious breaches would also entitle the appropriate party to terminate the employment of the Contractor under the contract.

The contracts of employment between the Contractor or the Employer and their employees often contain express terms requiring the employer to take reasonable care of the safety of the employee. Such a term would be implied even if not so stated. Breach of the Regulations that causes personal injury or endangers the health and safety of the employee may also amount to breach of the contract of employment, thus allowing the employee to sustain a claim for breach of the employment contract.

10.10.2 Liability in negligence

Breach of the Regulations does not automatically amount to negligence. To succeed in an action for negligence, the claimant must prove that: (i) the defendant owed him a duty of care; (ii) the defendant acted in breach of that duty; (iii) the defendant's breach of duty caused injury to the claimant. Where the defendant acted in the course of a contract of employment, the defendant's employer would also be vicariously liable for the negligence of the defendant.

These requirements for liability in negligence have to be established on a 'balance of probabilities', the standard applicable to civil liability generally. Criminal liability has to be proven 'beyond reasonable doubt'. It is therefore possible to find a person liable for negligence even after failure of a criminal prosecution.

11

Delays, extension of time and liquidated damages

In most construction contracts, the contractor is expected to complete the contract works by a specified date. This date may be revised under the provisions of the contract in defined situations. A contractual provision which allows such revision is referred to as an 'extension of time' clause. Failure of the contractor to complete by the due date of completion would normally result in liability for damages for breach of contract. A common practice is to include a liquidated damages clause that states the amount payable in case of delayed completion. This chapter explains, first, the legal principles governing delays, extension of time and liquidated damages relating to construction contracts in general. The specific provisions of JCT 05 on these issues are then discussed.

In response to a high incidence of disputes concerning delays, extension of time and liquidated damages, a committee of the Society of Construction Law (SCL) developed and published the Society of Construction Law Delay and Disruption Protocol,[1] which is intended for use as guidelines on dealing with these issues on construction projects. This chapter also provides an overview of the relevant guidelines from the Protocol.[2]

11.1 Concept and application of liquidated damages

According to principles of the law of contract,[3] to succeed in a claim for damages for breach of contract, a claimant must prove to the satisfaction of the court that:

1. the defendant's breach of contract caused loss in the amount claimed;
2. the loss was not too remote at the time of formation of the contract;
3. the claimant took all reasonable steps to mitigate his loss.

1 Society of Construction Law, Society of Construction Law Delay and Disruption Protocol, October 2002.
2 See also Chapter 2, Section 2.3.7.1 on the Protocol's guidelines on the preparation of contractors' programmes.
3 See Chapter 12 for general principles on damages for breach of contract.

It is then the duty of the court to decide the amount in respect of which the claimant has furnished the required proof and to make an award in that amount. Damages so assessed by the courts are referred to as 'unliquidated' damages, which are also often referred to in the construction industry as 'common law' damages. This involves time-consuming and costly litigation. To avoid this difficulty, it has become common practice, particularly with construction contracts, to state expressly in the contract itself the amount that will be payable in the event of its breach. The amounts so stated are referred to as 'liquidated damages' whilst clauses in which they are stated are called 'liquidated damages clauses'.

11.1.1 Concept of liquidated damages

The basic essence of liquidated damages is that they represent a genuine pre-assessment of the likely loss that will flow from the breach of contract in question. This follows from the general principle that the aim of damages is to place the innocent party in the position he would have occupied had the contract been performed without a breach. Both parties then enter into the contract in full awareness of their monetary rights and liabilities in the event of a breach. When the breach occurs, the claimant can then recover his loss from the defendant without time-consuming litigation.

Construction workers often refer to liquidated damages as 'penalties'. This is a misnomer because penalty clauses are void at law and therefore unenforceable through the courts. The claimant will then have to go through the trouble of proving unliquidated damages. When a claim for liquidated damages is made, the defendant can challenge it on the grounds that the liquidated damages clause is, in fact, a penalty clause and must be ignored. In such an event, the courts apply well-established principles for distinguishing liquidated damages from penalties. These principles were stated by the House of Lords in the famous case of *Dunlop Pneumatic Tyres Co.* v. *New Garage and Motor Co.*[4] The facts were as follows. Dunlop sold tyres to New Garage under a contract that contained terms restricting New Garage as to the prices at which they could retail the tyres. There was an undertaking to pay Dunlop £5 per tyre sold in contravention of the terms. The House of Lords had to decide whether the undertaking was a genuine liquidated damages clause or a penalty. In deciding the case, Lord Dunedin put forward the following propositions which have now received universal acceptance as guiding principles for distinguishing between penalties and liquidated damages.

1. Whether the parties call a payment 'damages' or a 'penalty' is not conclusive. The court must determine whichever it is in truth.
2. The essence of a penalty is a payment of money stipulated *in terrorem* (Lat.: to frighten) of the offending party (i.e. its purpose is to strike terror on mere contemplation of the breach); the essence of liquidated damages is a genuine covenanted pre-estimate of likely damage. The Court of Appeal applied this principle in *Jeancharm Ltd* v. *Barnet Football Club*[5] in striking down a liquidated damages clause by a unanimous decision. The case arose from a contract for the supply of football kits to the football club which provided that late payment for the kits would attract 5% interest

4 [1915] AC 79.
5 (2003) 92 ConLR 26.

per week for the period of the delay. This provision amounted to an annual interest rate of 260%. Lord Justice Jacob described the interest payable as 'extraordinary'. In *Alfred McAlpine Capital Projects Ltd* v. *Tilebox Ltd*[6] Jackson J stated that the test of a 'genuine pre-estimate' is objective and does not turn on the genuineness or honesty of the party who made the pre-estimate. The discrepancy between the level of damages stipulated in the contract and the likely damages must be substantial and unreasonable if assessed objectively for the liquidated damages clause to be struck down for being penal. He added that, in such assessment, the court may still have some regard to the thought processes of the maker of the pre-estimate. In that case the liquidated damages were stated as £45,000 per week on a contract with a contract sum of £11,573,076. The judge examined the likely consequences of delayed completion as could reasonably have been foreseen at the time of entering into the contract. He concluded that, although £45,000 was at or slightly above the top of the range of possible weekly losses from delay, the gap was not wide enough to warrant striking down the liquidated damages clause.[7]

3. The question of whether a sum stipulated is a penalty or liquidated damages is one of construction to be decided upon the terms and inherent circumstances of each particular contract judged as at the time of making it and not as at the time of its breach. However, in *Philips Hong Kong* v. *The Attorney General of Hong Kong*[8] the Privy Council suggested what actually happened at the time of breach may provide evidence as to what could reasonably have been expected to be the likely loss at the time of contract formation.
4. It will be held to be a penalty if the sum stipulated is extravagant and unconscionable in comparison with the greatest loss that could conceivably be proved to have flowed from the breach. In the *Philips Hong Kong* case the Privy Council also provided some guidance on the application of this principle. In that case it was held that, to prove that a provision is penal, it will not normally be sufficient to identify a hypothetical situation where the application of the provision could result in a larger sum than the actual loss being claimed.
5. If the breach consists only of the non-payment of money and the amount stipulated is greater than the sum which ought to have been paid, it will be held to be a penalty.
6. Where a single sum is payable on the occurrence of one or more of several events, some of which may cause serious loss and others but trifling loss, there is a presumption that it is a penalty. However, a minimum figure for liquidated damages in contracts with variable liquidated damages is not necessarily penal.[9] It follows from this principle that the practice of stating liquidated damages as an amount 'per week or part thereof' of delay is open to challenge that the amount is a penalty. In the situations contemplated by such a provision, it is therefore better practice to convert the amount into damages per day of delay.[10]

---

6 [2005] EWHC 281 (TCC) (hereafter *McAlpine* v. *Tilebox*).

7 See *CFW Architects (A Firm)* v. *Cowlin Construction Ltd* [2006] EWHC 6 (TCC) in which HHJ Thornton QC applied this approach to uphold a liquidated damages clause.

8 (1993) 63 BLR 41 (hereafter *Philips Hong Kong*).

9 See *Philips Hong Kong* above.

10 For example, see *ICE Conditions of Contract, Measurement Version*, 7th Edition, Thomas Telford, 1999.

7. It is no obstacle to the sum stipulated being a genuine pre-estimate of loss that it is impossible to make a precise estimation of the loss consequent upon the breach. This principle is particularly helpful in situations where losses are not easily quantifiable (e.g. churches and public sector projects).

In modern times, liquidated damages clauses are increasingly perceived as having practical advantages, particularly in situations where the damages for the relevant breach are difficult to estimate with accuracy. Thus, as remarked in *Hudson's*,[11] they are looked upon with less disfavour than in the older cases. For example, in the *McAlpine* v. *Tilebox* case, Jackson J stated at para. 48:

> Because the rule about penalties is an anomaly within the law of contract, the courts are predisposed, where possible, to uphold contractual terms which fix the level of damages for breach. The predisposition is even stronger in the case of commercial contracts freely entered into between parties of comparable bargaining power.

In *Bath and North East Somerset DC* v. *Mowlem plc*[12] the Court of Appeal stated that, in a modern construction contract, liquidated damages are more likely to be a cap on the contractor's liability for damages for delay than a statement of a serious forecast of the maximum loss likely to be suffered by the employer.

The modern judicial approach to liquidated damages is also illustrated by the way the Scottish Court of Session, in *City Inn Ltd* v. *Shepherd Construction Ltd*,[13] decided a challenge to a condition precedent to obtaining extension of time. The contract from which it arose, an amended version of the JCT80 form, provided in Clause 13.8 a procedure for dealing with any Architect's instruction which, in the opinion of the Contractor, would require additional payment or extension of time. The Contractor was to serve notice of the time and cost consequences of the instruction within 10 working days. Also, the Contractor was not to comply with the instruction until the Architect had agreed the additional payment and extension of time with the Contractor. Clause 13.8.5 provided that, unless the Architect had waived the requirement for the notice, the Contractor was not entitled to extension of time if he failed to serve it.

The issue referred to the court was whether, having failed to serve the required notices, the Contractor was entitled to any extension of time. It was argued for the Contractor that the effect of Clause 13.8.5 was that liquidated damages of £30,000/week stated in the contract were a penalty because they were not a genuine pre-estimate of the loss likely to be suffered by the Employer as a consequence of the failure to serve the required notices. Both the first instance judge and the Inner House of the Scottish Court of Session held that the liquidated damages applied to the failure to complete by the completion date and not the failure to serve the notice. The parties were in agreement that the liquidated damages were a genuine pre-estimate of the loss likely from delayed completion. The Contractor's assertion that the liquidated damages clause was a penalty was therefore rejected.

---

11 Paragraphs 10.022–10.024. See also Ian Duncan Wallace 'Prevention and Liquidated Damages: A theory gone too far' (2002) 18 BCL 82; 'Liquidated Damages "Down Under": Prevention by Whom?' (2002) 7 Construction and Engineering Law 2.

12 [2004] EWCA Civ 115; [2004] BLR 153; 100 ConLR 111.

13 [2003] BLR 468.

11.1.2 Concept and sectional completion obligation

In construction contracts requiring completion of sections at different specified times, purported liquidated damages clauses have been construed as penalties by the application of the sixth principle cited above. Delay in completing only one section would not be as serious as delayed completion on all the sections. This means that if the liquidated damages are expressed as a single sum for the whole contract, it would be construed as a penalty on the grounds that the same sum is payable for both breaches even though they would have different consequences. The same problem arises with some contracts involving the construction of a number of standard units (e.g. housing), because delayed completion of only one unit would not be as damaging as delayed completion of every unit. The facts of *Bramall & Ogden Ltd* v. *Sheffield City Council*[14] illustrate the application of this principle to contracts involving sectional completion or multiple units:

> This case arose from a contract in the terms of the JCT 63 for the construction of 123 dwellings. Clause 22 provided for the deduction of liquidated damages for failure to complete the entire contract by the completion date. The Appendix [now referred to in JCT 05 as 'Contract Particulars'] stated the liquidated damages payable as £20 per week for each uncompleted house. The contract did not provide for sectional completion. This meant that if any number of dwellings remained uncompleted after the completion date, the Employer would have been entitled under the terms of the contract to deduct liquidated damages for the whole contract (i.e. in respect of all the houses (123 × £20/week)). Judge Hawsher QC decided that the liquidated damages clause was a penalty because the contract did not allow for the liquidated damages recoverable to be varied with the number of houses uncompleted. It was therefore unenforceable. Interestingly, the Employer's claim for liquidated damages in respect of only the uncompleted houses was also rejected on the grounds that the contract did not provide to that effect. The reasoning of the judge was that the liquidated damages clause was *prima facie* invalid and, having been struck down by the invalidity, there was no effective liquidated damages clause. No liquidated damages of any kind were therefore owed.

This case provides a good lesson on the importance of completing contract documents with due care.[15] It also highlights the fact that liquidated damages clauses are construed strictly *contra proferentem* in case of ambiguity.[16] There was a *Sectional Completion Supplement* for use with the JCT 98. This procedure for dealing with sectional completion has not been brought forward into JCT 05. Instead, sectional completion obligations are to be imposed by stating certain details on the sections in the appropriate places in the Contract Particulars. The required details are explained in the introduction to Chapter 3.[17]

Taylor Woodrow Holdings Ltd and Another v. *Barnes & Elliott Ltd*[18] highlights the importance of drawing up the descriptions of Sections and completing their details in Contract Particulars with care. In that case the contract provided for sectional completion.

---

14 (1983) 29 BLR 73; see also *Stanor Electric Ltd* v. *R. Mansell Ltd* (1988) CILL 399.

15 See Chapter 1, Section 1.5.2.4 dealing with the Contract Particulars entry for Clause 2.32.2.

16 See also *Peak Construction (Liverpool) Ltd* v. *McKinney Foundations Ltd* (1970) 1 BLR 1114; *Temloc Ltd* v. *Errill Properties Ltd* (1987) 39 BLR 30.

17 See also Chapter 1, Section 1.5.2.2 (dealing with the Sixth Recital) and 1.5.2.4 (dealing with entries for Clauses 1.1, 2.4, 2.32.2, 2.37, 2.38).

18 [2004] EWHC 3319 (TCC).

An arbitrator decided that the sections were not sufficiently detailed in the contract documents to allow the value of the work in a section to be quantified. He concluded, as a preliminary decision, that, as it was not possible to determine whether or not any particular section was completed by a particular date, it was not possible to determine the Contractor's liability for liquidated damages for delayed completion of any section. On an appeal to the Technology and Construction Court (TCC), HHJ Wilcox agreed with the arbitrator's decision that the liquidated damages clause was void for uncertainty.

11.1.3 Application of liquidated damages

Once the breach covered by a liquidated damages clause occurs, and provided the clause is neither invalidated by the Unfair Contract Terms Act 1977 nor considered a penalty, the only damages payable are the amounts stated in the clause. If the innocent party suffers no actual loss or even benefits from the breach, the damages would still be payable. This outcome follows from the principle that liquidated damages payable are assessed as at the time of the making of the contract rather than at the time of its breach.

> *Clydebank Engineering and Shipbuilding Co. Ltd* v. *Don Jose Ramos Yzquierdo y Castaneda*:[19] Clydebank contracted with the Spanish Government to supply them with four torpedo boats. Liquidated damages were set at £500 per week of delay. Delivery of the boats was delayed and the Spanish Government claimed liquidated damages for the period of the delay. Clydebank argued that, by delaying, they had done the Spanish Government a favour because soon after the due delivery date the American fleet had sunk the greater part of the Spanish fleet. The House of Lords rejected this argument and upheld the claim for damages.

Where liquidated damages are payable, they represent the total remedy available to the innocent party. Whether his actual losses exceed the amount of the liquidated damages is irrelevant.

> *Cellulose Acetate Silk Co. Ltd* v. *Widnes Foundry Ltd*:[20] the contract involved the construction and delivery of a chemical plant. Liquidated damages were set in the contract at £20 per week of delay. Delay of 30 weeks occurred. The purchasers stood to suffer actual losses of substantially more than the £600 to which they were entitled under the contract. The House of Lords rejected the purchasers' claim for their actual losses in the sum of £5850.

> *Temloc Ltd* v. *Errill Properties Ltd*:[21] this case is an extreme illustration of this principle and provides yet another lesson on the importance of completing contract documents with great care. '£NIL' was entered as liquidated damages in the Appendix [now referred to in JCT 05 as 'Contract Particulars'] of a JCT 80 contract for a shopping development. Completion was delayed and the Employer, a developer, was sued by his prospective tenants. The Employer's claim to recover his liability from the Contractor as general damages failed in the Court of Appeal.

The Scottish case of *John Maxwell & Sons (Builders)* v. *Simpson*[22] was decided along the lines of *Temloc*. More recently, it was applied by Ramsey J in the TCC in *Chattan*

---

19 [1905] AC 6; see also *BFI Group of Companies Ltd* v. *DCB Integration Systems Ltd* (1987) CILL 348.
20 [1933] AC 20.
21 (1987) 39 BLR 30 (hereafter *Temloc*).
22 (1990) SCLR 92.

Developments Ltd v. *Reigill Civil Engineering Contractors Ltd.*[23] In this case the Parties had agreed their Contract at a meeting. The result of the meeting was confirmed in a letter on behalf of the Employer which stated, among other things, 'liquidated and ascertained damages – n/a. All Relevant Events and List of Matters to remain unaltered'. An arbitrator determined that the Parties' Contract had incorporated certain terms of JCT 80 and that it was part of their agreement that no damages, liquidated or unliquidated, were to be recoverable. The arbitrator decided that, as a result of their agreement, there was no entitlement to unliquidated damages. Citing *Temloc*, Ramsey J held that the arbitrator had correctly applied the law to his findings of fact.[24]

However, the Temloc decision has been distinguished by some Australian courts. In *Baese Property Ltd* v. *R.A. Building Property Ltd*,[25] which concerned substantially the same facts, the Supreme Court of New South Wales distinguished it, principally on the fact that in *Temloc* the 'NIL' was entered against the heading 'damages for non-completion'. The judge treated this heading as evincing a common intention of the parties that the 'NIL' applied to not only liquidated damages but also general damages at common law. It was also distinguished by Jenkins J in the Supreme Court of Western Australia in *Silent Vector Pty Ltd (t/a Sizer Builders)* v. *Squarcini*.[26] The Contract in that case had an Annexure, which was a schedule of the Contract Particular in a format substantially the same as that of the Contract Particulars section in JCT 05. Guidance provided in the form of contract used required an amount to be completed against a reference in the Annexure to the liquidated damages clause. 'N/A' was entered instead. The Contractor contended that, as a consequence of the N/A entry, no damages of any kind were payable for delayed completion whilst the employer's position was that general damages were payable. Looking at the pattern of completion of N/As and NILs in the contract and contractual provisions highlighting the Contractor's duty to complete without delay, an arbitrator to whom the dispute about the effect of the N/A entry was referred agreed with the Employer. On an application for leave to appeal on a question of law arising from the arbitrator's determination, Jenkins J failed to see any evidence that, in his approach to the question, the arbitrator had made an error of law. The application was therefore dismissed.

In the light of this brief review of the effect of 'NIL' and 'N/A' entries for liquidated damages, it is submitted that the answer in any given case is a matter of the construction of the particular contract in all the surrounding circumstances such as related oral agreements, contractual provisions on timely completion, and the pattern if entries of data for the contract. Sponsors of contracts can help avoid the issue altogether by providing very prominent advice at appropriate locations in their contract against such entries.

An implication of the principle that liquidated damages represent an exhaustive remedy is that, where the cause of delay also constitutes a breach of contract over and beyond simple failure to complete on time and the breach causes unexpected consequential loss, there is no separate entitlement to recover such loss as unliquidated damages in addition to, or in place of, the liquidated damages stated in the contract.

---

23 [2007] EWHC 305 (TCC).

24 It is to be noted that, under the applicable arbitration law, there was no right of appeal against the arbitrator's finding of fact. The court therefore did not have to consider whether the arbitrator was right in concluding that the parties had agreed in their contract to exclude recovery of any damages for delay.

25 [(1989) 52 BLR 130.

26 [2008] WASC 246.

> *Biffa Waste Services Ltd* v. *Maschinenfabrik Ernst Hese GmbH*:[27] the contract from which this case arose was part of a private finance initiative project for the collection, recycling and disposal of domestic waste. The defendant was to design and construct the recycling plant within existing buildings and structures. The liquidated damages clause went on to state that any damages payable were to be 'the only monies due from the Contractor for such [delay]' and that the right to such damages was 'without prejudice to any other right or remedy of the Employer'. During the construction of the Works a fire broke out, causing substantial damage to the rest of the plant and delaying its operation by 7 months. Whilst the proceeds from insurance of the Works covered the cost of reinstatement, the consequential loss from the delay to the operation of the plant exceeded the liquidated damages stated in the contract. The claimant sought to circumvent the exhaustivity principle of liquidated damages by contending that the liquidated damages covered only 'simple' delay in completing the Works (i.e. without another breach of the contract), and that, where the event causing delay also caused consequential loss (business disruption in the case), there was a right to recover the consequential loss in addition to the liquidated damages. Ramsey J rejected the contention, emphasizing the exhaustive nature of liquidated damages. He treated the 'without prejudice to any other right or remedy of the Employer' proviso as referring to non-monetary rights.

In this case the claimant, in the alternative, claimed the consequential loss in tort. Also rejecting the claim, the judge explained that, although there was a concurrent duty in tort to take care to avoid the damage in question, the liquidated damages clause operated as a limit to the damages recoverable for breach of this duty.

11.1.4 Fixing liquidated damages in construction contracts

In most commercial situations, it is not very difficult to estimate with sufficient accuracy the loss that the Employer stands to suffer in the event of delayed completion. One approach would be to determine the profits that the Employer would have made in the period of delay. A second approach involves assessing the cost of alternative accommodation plus an amount for business disruption. However, in ecclesiastical and some public sector buildings, these approaches are not always suitable. For such projects, a common approach is founded upon the premise that, at about the time of the delay, a large proportion of the Contract Sum, typically 80–90 per cent, would have been paid to the Contractor. The Employer would therefore be out of pocket by this amount without the benefit of using his building. It is then argued that the interest that the Employer would have earned by putting that sum in a bank represents an acceptable measure of his loss. This approach is favoured by the Society of Chief Surveyors in Local Government. In *JF Finnegan Ltd* v. *Community Housing Association*[28] an Official Referee upheld a liquidated damages clause that required application of the same approach to 85 per cent of the estimated contract price to arrive at the liquidated damages payable. In the Australian case of *Multiplex Construction Pty Ltd* v. *Abgarus Pty Ltd*[29] the court accepted a similar approach to fixing liquidated damages. That approach involved determination of the damages by a formula that calculated the interest that would have been charged by the commercial banks on the actual payments made to the Contractor.

---

27 [2008] EWHC 6 (TCC); [2008] BLR 155 (hereafter *Biffa Waste*)
28 (1993) 65 BLR 103.
29 (1992) 33 NSWLR 504.

The facts of *Biffa Waste* illustrate a limitation to the formula approach to estimating liquidated damages. In that case the consequential loss flowing from the breach of contract that the delay was over and above the loss of the opportunity of putting the funds tight up in the incomplete project to alternative investment.

11.2 Effect of delay: general principles

The reality of construction projects is that a variety of events can prevent the contractor from completing the works by the agreed date. Some of these events may be due to the acts or omissions of the employer, or of people for whom the employer is responsible in law. Failure of the employer to give possession of the site of the works to the contractor is an example of an omission for which the employer would be directly responsible. An example of an omission of other parties for which the employer would be responsible is failure of his architect to give appropriate instructions in time. For the sake of simplicity both types of events are henceforth referred to as events for which the employer is responsible.

On the subject of delays, extension of time and liquidated damages, four general principles have to be borne in mind. Although these principles apply to all contracts that require performance within defined periods and entail liquidated damages, they are explained mainly in the context of construction.

The *first principle* provides an answer to the fundamental question of whether or not the contractor is under a strict obligation, irrespective of delaying influences, to achieve completion by the specified date. The position of the law is that the contractor is so obliged unless he is prevented by factors for which the employer is partly or solely responsible. Where the project is affected by such factors, the contractor is no longer obliged to complete by the specified date but within a reasonable time. In legal parlance, the date for completion is 'at large'.

> *Dodd* v. *Churton*:[30] this case concerned a contract which had variation and liquidated damages clauses. However, there was no provision for the revision of the agreed date for completion. Extra work was ordered and the builder was delayed as a result. Allowing a fortnight for the extras, the Employer purported to set-off for liquidated damages for that part of the delay beyond the fortnight. The Court of Appeal held that the liquidated damages clause no longer applied. The important issue involved here was that, as the contract did not contain a provision for extending the time for completion, no new date could be fixed under the contract. The date of completion therefore became at large.

However, there is an exception where the contractor expressly and unequivocally agreed to complete by the original completion date even if the employer commits acts of prevention. In such a case, the contractor could be held to his promise.[31] However, such one-sided provisions tend to be construed so *contra proferentem* the employer that if the actual cause of delay is not clearly within the acts of prevention contemplated in the contract, the contractor would be released from the obligation to complete by the due completion date.

---

30 [1897] 1 QB 562; see also *Holme* v. *Guppy* (1838) 2 M & W 387.

31 *Jones* v. *St John's College Oxford* (1870) LR QB 115; in the House of Lords' decision of *Percy Bilton* v. *Greater London Council* (1982) 20 BLR 1, at p. 3 Lord Fraser of Tullybelton accepted as correct a submission that the effect of the prevention principle can be modified by express terms in the contract.

The *second principle* is that where the delay is partly or wholly attributable to an event for which the employer is responsible, the liquidated damages clause may be kept alive only by the existence in the contract of an extension of time clause which covers that cause of delay.

> *Peak Construction (Liverpool) Ltd* v. *McKinney Foundations Ltd*:[32] Peak was the Main Contractor of Liverpool Corporation on a housing contract. The contract contained both liquidated damages and extension of time clauses. The extension of time clause empowered the Architect to grant extensions of time for, among other things, extras and additions, *force majeure* and 'other unavoidable circumstances'. After the defendant, nominated sub-contractors for piling, had finished their work and left the site, some of the piles were found to be defective. Everybody concerned agreed that work should be suspended pending investigation of the piles by a consulting engineer. The Corporation delayed not only in appointing the engineers but also in authorizing the report produced. The Court of Appeal held that, as between the employer and the main contractor, time for completion had become at large because part of the delay was caused by the employer. It is interesting to note that the court did not accept the argument that delays in making the appointment and authorizing the report constituted 'other unavoidable circumstances' for which time could be extended under the contract.

This principle is referred to as the 'prevention principle' or the 'Peak principle' after *Peak* v. *Mckinney*. As highlighted by Baker *et al.*,[33] it may be applicable in contexts where the delay is not caused by the employer's blameworthy conduct (e.g. delay by the employer's contract administrator to supply necessary drawings or information to the contractor).

A corollary of the prevention principle is that, where the contractor is delayed by an event for which the employer is responsible and that event is not covered by the extension of time clause, completion time becomes at large. It has often been thought that a catch-all provision which allows the architect to extend time in any circumstance is the most effective way of ensuring that a liquidated damages clause is always alive. This generalist approach does not always work because the courts are reluctant to allow the architect to extend time for employer-caused delays unless the extension of time clause is very specific on the cause of delay. For example:

> *Wells* v. *Army & Navy Co-operative Society Ltd*:[34] a building contract provided for extension of time for a list of specific causes of delay and 'other causes beyond the contractor's control'. The Court of Appeal ruled that the purported catch-all phrase could not include breaches of contract or other types of interference for which the Employer was responsible and that, consequently, liquidated damages were not recoverable.

Lord Justice Salmond in *Peak* v. *McKinney* stated that liquidated damages and extension of time clauses in printed forms of contract must be construed strictly *contra proferentem*.[35] It is often thought such an approach requires that any ambiguity as to whether a

---

32 (1970) 1 BLR 111 (hereafter *Peak* v. *McKinney*); see also *Rapid Building Group Ltd* v. *Ealing Family Housing Association Ltd* (1984) 29 BLR 5 and *Percy Bilton* v. *Greater London Council* (1982) 20 BLR 1.

33 Ellis Baker, James Bremen and Anthony Lavers, 'The Development of the Prevention Principle in English and Australian Jurisdictions' (2005) ICLR 197.

34 (1902) 86 LT 764; see also *Peak* v. *McKinney* in which the catch-all phrase 'or other unavoidable circumstances' failed.

35 *Peak Construction (Liverpool) Ltd* v. *McKinney Foundations Ltd* (1970) 1 BLR 111, at p. 121.

particular type of delay caused by the employer is captured by the wording of the extension of time clause is to be resolved in favour of not allowing extension of time, thus rendering completion time at large. The late Ian Duncan Wallace QC attributed this anti-employer construction of such clauses to early judicial dislike for liquidated damages, which were then perceived by the courts as penal;[36] pointing out that, in modern times, far from being penal and apart from the fact that they avoid disputes on quantum, liquidated damages are often a cap on the contractor's liability for delay. In his opinion, a legalistic interpretation of extension of time clauses to strike down genuine liquidated damages clauses would be a 'creative', or even wrong, application of the prevention principle.

In *Multiplex Constructions (UK) Ltd* v. *Honeywell Control Systems Ltd*,[37] Jackson J, after a review of the authorities on the prevention principle, agreed with Mr Wallace's commentary. Accordingly, he stated that the *contra proferentem* principle is more properly applied in favour of allowing contractors more time to complete. In that case, a dispute between the main contractor for the design of the Wembley National Football Stadium and a sub-contractor, it was contended on behalf of the sub-contractor that instructions issued by the main contractor were not within the extension of time clause in their contract and that, therefore, the consequence of delay resulting from compliance with the instructions was to put completion time at large. The sub-contract also required the sub-contractors, as a condition precedent to extension of time, to take certain steps in relation to the delay concerned (e.g. to serve all necessary notices and provide all supporting information). The sub-contractor argued further that the effect of its failure to take these steps was also that time for completion was at large. It drew support for the latter contention from the case of *Gaymark Investments Pty Ltd* v. *Walter Construction Group Ltd*,[38] a decision of the Supreme Court of the Northern Territory of Australia. The principle in *Gaymark* relied upon is that, where an extension of time clause lays down procedures to be complied with as a condition precedent to entitlement to extension of time, the Employer causes delay, and the Contractor fails to comply with the procedures, time for completion becomes at large. The judge rejected the sub-contractor's contentions. In arriving at his decision he distinguished *Gaymark* but not before expressing doubts as to whether it is in line with English authorities on the prevention principle.

In *Steria Ltd* v. *Sigma Wireless Communications Ltd*[39] HHJ Stephen Davies adopted the reasoning in *Multiplex* v. *Honeywell* in upholding the validity of a compliant notice of delay as a condition precedent to entitlement to extension of time under a contract in an amended MF/1 form for the development of a computer-aided despatch system.

The *third principle* is that, even with the existence of an extension of time clause covering the cause of a delay, time for completion could become at large if the extension of time to which the Contractor is entitled is not granted in accordance with the contract.

Miller v. *London County Council*:[40] an extension of time clause stated: 'it shall be lawful for the engineer, if he shall think fit, to grant from time to time, and at any time or times by writing under his hand, such extension of time for completion of the work and that either

36 Ian Duncan Wallace, 'Prevention and Liquidated Damages: A theory gone too far' (2002) 18 BCL 82; 'Liquidated Damages "Down Under": Prevention by Whom?' (2002) 7 Construction and Engineering Law, 2.
37 [2007] BLR 195 (hereafter *Multiplex* v. *Honeywell*).
38 (1999) NTSC 143 (hereafter *Gaymark*).
39 [2008] BLR 79 (TCC); *Waterfront Shipping Company Ltd* v. *Trafigura AG* [2007] EWHC 2482 (Comm) is another case in the growing line of cases supportive of enforcement of conditions precedent.
40 (1934) 151 LT 425.

prospectively or retrospectively, and to assign such other time or times for completion as to him may seem fit'. Another clause provided for liquidated damages for delays. Eight months after completion of the Works, the engineer issued a certificate granting extension of time and certifying the amount due the Employer as liquidated damages. Parcq J. held that the wording of the extension of time clause did not empower the engineer to extend time after the completion of the Works. The use of the word 'retrospectively' only allowed him to wait until the cause of the delay had ceased and then, within a reasonable time thereafter, to grant extension of time. The extension granted was therefore not in accordance with the contract, with the further consequences that time for completion was at large and liquidated damages irrecoverable.

Miller was distinguished in *Amalgamated Building Contractors Company Ltd* v. *Waltham Holy Cross UDC*[41] as limited to the special wording of the extension of time clause in the contract from which it arose. In the latter case the Court of Appeal upheld an Architect's extension of time decision made 3 months after the contractor had completed the works. In *Temloc Ltd* v. *Errill Properties Ltd*[42] Croom-Johnson LJ stated that the timetable in the JCT80 for the Architect to make a decision on the contractor's entitlement was only directory (i.e. a decision made beyond the timetable would be valid). However, *Miller* was applied by the Supreme Court of Western Australia in *MacMahon Construction Pty Ltd* v. *Crestwood Estates.*[43]

The Editor of the 11th edition of *Hudson's Building and Engineering Contracts* has doubted the correctness of *Miller* and *MacMahon* and suggests that they must have been influenced by the then existing antipathy of courts in common law jurisdictions to liquidated damages clauses.[44] He puts forward the alternative proposition that, except where the particular contract expressly and clearly provides that late extension of time decisions are to displace the liquidated damages clause, a delayed extension of time decision should be treated as a breach of contract for which the contractor may only claim damages. As already discussed, there is a growing judicial opinion, in not only England and Wales but also Scotland, that Mr Wallace's application of the prevention principle is more likely to be correct than the Australian cases to the contrary.

The *fourth principle* is that where time for completion becomes at large, any liquidated damages clause becomes ineffective and the employer is only entitled to unliquidated damages if the contractor fails to complete within a reasonable time.[45] Apart from the difficulty of proving his damages, there is the additional disadvantage that the employer may not be entitled to set-off such damages against interim certificates if the contractor objects.

It follows from the discussions so far that an extension of time clause is for the benefit of both parties. The advantage to the contractor is that his liability to pay liquidated damages is restricted to situations where the employer is not responsible for the delay. On the employer's side, it prevents the date of completion becoming at large when the contractor suffers delay for which the employer is responsible. The widespread belief in the construction industry that extension of time clauses are solely for the benefit of the contractor is therefore very much mistaken.

---

41 [1952] 2 All ER 452.

42 (1987) 39 BLR 30, at p. 39.

43 [1971] WAR 162 (hereafter *MacMahon*).

44 See para. 10.089.

45 *Rapid Building Group Ltd* v. *Ealing Family Housing Association Ltd* (1984) 29 BLR 5 (hereafter *Rapid Building* v. *Ealing*); *Peak Construction (Liverpool) Ltd* v. *McKinney Foundations Ltd* (1970) 1 BLR 111.

It is a source of common debate whether the amount stated in a failed liquidated damages clause still constitutes an upper limit to the unliquidated damages recoverable by the employer. In other words, is the employer 'hoist with his own petard'? In *Esley* v. *J.G. Collins Insurance Agencies Ltd*[46] the Supreme Court of Canada held that where liquidated damages are struck down for being penal, recoverable unliquidated damages cannot exceed the sum found to be penal. However, this point is still moot in English law so far as construction contracts are concerned. In *Cellulose Acetate Silk Co.* v. *Widnes Foundry Co. Ltd*[47] the House of Lords left the question open. The Court of Appeal did the same in *Rapid Building* v. *Ealing*.[48]

However, the circumstances in which a claimant can prove, to the satisfaction of the court, actual damages in excess of liquidated damages already determined by the court to be penal must be rare. Such a rare set of circumstances would be where the liquidated damages amount to a penalty if assessed as at the time of contract formation, as they should be, but would not be penal if assessed as at the time of the breach of contract. Also, liquidated damages clauses fail for reasons other than on grounds of being penal. For example, where they are rendered inapplicable because time for completion has become at large, the objection to actual damages being in excess of the stipulated liquidated damages would be less sustainable. In *Steria Ltd* v. *Sigma Wireless Communications Ltd*[49] the court rejected defences against a liquidated damages claim that relied upon the rule against penalties and the prevention principle. The court did not, therefore, have to decide the issue of whether the claimant was entitled to actual damages in excess of the liquidated damages in the contract. HHJ Stephen Davies, sitting as a judge of the High Court, stated that if he had had to decide the point he was inclined to the view that where a liquidated damages clause fails for being penal, the cap on recoverable damages disappears with it.

11.3 Extension of time and liquidated damages under JCT 05: an outline

The general scheme of the contract, as far as the timetable for the Contractor's performance is concerned, requires the timetable and the rate of liquidated damages for delay to be specified. The data required are normally assembled from decisions taken in the development of the project's procurement strategy and entered at the appropriate places in the Contract Particulars. As explained in the introduction to Chapter 3,[50] the required data vary depending on whether or not there are Sections to be completed to defined sectional completion timetables.

The Contractor is entitled to start carrying out the Works or each Section on the applicable Date of Possession and must complete it by the corresponding Completion Date (Clause 2.4). If the Contractor fails to achieve such completion he would be liable for liquidated damages at the relevant rate stated in the Contract. Where there is no requirement for sectional completion the Contract Particulars should specify one rate for recovery of liquidated damages in relation to the whole of the Works. Sectional completion requires completion of separate rates of recovery for each Section.

46 (1978) 83 DLR (3d) 1.
47 [1933] AC 20.
48 *Rapid Building* v. *Ealing*: see note 45.
49 [2008] BLR 79.
50 See also Chapter 1, Section 1.5.2.4 dealing with Contract Particulars entry for Clauses 1.1, 2.4, 2.37, 2.38.

The Architect is given the power to fix a new Completion Date from time to time when the carrying out of the Works or Section is delayed by a number of events referred to collectively as the 'Relevant Events'. For reasons of simplicity, delays, extension of time and liquidated damages are analysed and explained in terms of the Works as a whole. The same analysis and principles apply to each Section where the contract properly requires Sections to be completed by specified dates.

11.4 Relevant events

Clause 2.29 lists the Relevant Events as:

- (i) Variations and other matters to be treated as Variations (Clause 2.29.1);
- (ii) certain instructions of the Architect (Clause 2.29.2);
- (iii) deferment of possession of the site or a Section (Clause 2.29.3);
- (iv) execution of work covered by an Approximate Quantity in the Contract Bills (Clause 2.29.4);
- (v) suspension by the Contractor for non-payment by the Employer (Clause 2.29.5);
- (vi) impediment/prevention/default by the Employer, the Employer's Persons or any other agent of the Employer (Clause 2.29.6);
- (vii) the carrying out of work by a Statutory Undertaker (Clause 2.29.7);
- (viii) exceptionally adverse weather conditions (Clause 2.29.8);
- (ix) loss/damage due to the Specified Perils (Clause 2.29.9);
- (x) civil commotion or terrorism (Clause 2.29.10);
- (xi) industrial unrest (Clause 2.29.11);
- (xii) statutory intervention (Clause 2.29.12);
- (xiii) *force majeure* (Clause 2.29.13).

It is to be noted about the Relevant Events that they include not only events attributable to the Employer but also those outside his control (e.g. *force majeure*). Also, the mere fact that the Contractor is entitled to extension of time does not also carry a right to recover loss and/or expense unless the cause of the delay is also a Relevant Matter under Clause 4.24.

In principle, if the Contractor is delayed by any of these events, he would be entitled to extension of time. However, it must be pointed out that mere occurrence of any of these events does not automatically result in a right to an extension; the Contractor must actually be delayed by the event. In this regard, it is the Contractor's actual progress and not his planned progress which must be considered. It follows therefore that, if the Contractor is ahead of schedule at the time of the delaying event, he will not be entitled to extension of time although he may be entitled to recover his direct loss and/or expense for disruption. However, it is not uncommon for contractors to claim extension of time in those circumstances as protection against future delays for which extension of time may not be available.

It sometimes happens that an event, although a Relevant Event, would not have affected the Works had the Contractor not been subject to an earlier delay caused by his own fault. It would appear that the Contractor may still be entitled to extension of time in such event.

Walter Lawrence & Son v. *Commercial Union Properties Ltd*:[51] a Contractor under a JCT63 contract fell behind with his programme. The Architect refused to grant extension of time for

51 (1984) 4 ConLR 37.

adverse weather conditions on the grounds that had the Contractor followed the programme he would not have been affected by the weather conditions. Rejecting this argument, the court held that the Contractor was entitled to extension of time.

The right to extension of time is further qualified in three ways. First, under Clause 2.28.6.1 the Contractor must always use his best endeavours to prevent delay in the progress of the Works irrespective of the cause of potential delay. He is also to continue to use his best endeavours to reduce any delay that occurs despite the earlier endeavours. Unfortunately, there is little direct reasoned authority on the standard of endeavour in terms of what the Contractor must do. However, a number of cases on the use of the term in other contractual situations cast some light on the issue. In *Sheffield Railway Co.* v. *Great Central Railway Co.*[52] it was held that the term 'best endeavours' 'means what the words say; they do not mean second-best endeavours'.

More recent cases introduce the concept of reasonableness into the steps required. In *Terrel* v. *Mobie Todd and Co. Ltd*[53] Sellers J stated that a company's obligation to use 'best endeavours' to promote sales meant a duty on its directors 'to do what they could *reasonably* do in the circumstances' [emphasis added]. Although it suggests reasonableness, other cases[54] draw distinctions between 'reasonable endeavours' and 'best endeavours' whereby the latter is the more onerous obligation.

In *IBM UK Ltd* v. *Rockware Glass*[55] IBM contracted to buy a piece of land from the defendant under a contract which required IBM to make an application for planning permission and to 'use its best endeavours to obtain the same'. The Court of Appeal held that IBM were under an obligation to appeal to the Secretary of State against the local authority's refusal of the permission, a very expensive step. In their opinion, IBM was 'bound to take all those steps in their power which are capable of producing the desired result, namely the obtaining of planning permission, being steps which a prudent, determined and reasonable owner acting in his own interests and desiring to achieve that result would take'. In *Overseas Buyers Ltd* v. *Grandex*[56] Mustill J, as he then was, did not think that the Court of Appeal's formulation of the standard was different from 'doing all that can reasonably be expected'.

Clearly, therefore, the Contractor's obligation to use best endeavours to prevent delay may require expenditure not planned for originally but such additional cost does not have to be so substantial as to defeat the Contractor's business purpose. The Contractor does not have to use any means whatsoever within his power regardless of additional costs. Feasible schedule recovery strategies include redeployment of resources onto another activity where the activity they were originally planned for cannot proceed because of the delaying event. It is also arguable that the Contractor must consider any other schedule recovery measure for which there is entitlement to recovery of loss and/or expense under Clauses 4.23 to 4.25 provided the Architect would act in support (e.g. issue appropriate Variations).

The second qualification to the Contractor's entitlement to extension of time is that, by Clause 2.28.6.2, he must take all reasonable measures necessary to proceed with the Works. Such measures must be to the reasonable satisfaction of the Architect. Third, as will

52 [1911] 27 TLR 451.
53 [1952] 2 TLR 574.
54 For examples, see *Rhodia International Holdings Ltd* v. *Huntsman International LLC* [2007] EWHC 292 (Comm); *Jolley* v. *Carmel Ltd* [2000] 3 EGLR 68.
55 [1980] FSR 335.
56 [1980] Lloyd's Rep 608.

become apparent in the discussion of each Event, there are additional restrictions which apply to particular Relevant Events.

11.4.1 Variations (Clause 2.29.1)

Variations were treated in previous editions of the contract as part of one Relevant Event for delays caused by compliance with Architect's instructions generally. Treating Variations as a stand-alone Relevant Event is probably a consequence of acknowledgement that they are probably the most common cause of delays on construction contracts. This Relevant Event also covers delays caused by matters deemed to be Variations.[57]

11.4.2 Compliance with certain instructions of the Architect (Clause 2.29.2)

The instructions covered are those relating to:

- Clause 2.15: discrepancies within or between documents listed in the Clause;
- Clause 3.15: postponement of any part of the Works;
- Clause 3.16: expenditure of Provisional Sums in the Contract Bills or in the Employer's Requirements (instructions on the expenditure of Provisional Sums for defined work are excluded);[58]
- Clause 3.23: dealing with antiquities;
- Clause 5.3.2: Contractor's notice of disagreement to provide a Schedule 2 Quotation;
- Clauses 3.17 and 3.18.4: opening up work for inspection or testing which shows that the work is in accordance with the Contract.

11.4.3 Deferment of possession of site (Clause 2.29.3)

Where it is stated in the Contract Particulars that Clause 2.5 will apply, the Employer has a right under that Clause to defer giving the Contractor possession of site for any period not exceeding the stated maximum period of deferment (default maximum is 6 weeks). If the Employer exercises this right and the Contractor is delayed as a result, there is an entitlement to extension of time.

Deferment beyond the applicable maximum duration would constitute a breach of contract by the Employer. It could even be repudiatory[59] if the deferment is excessive. However, assuming continuation of the contract, the Architect would have the power to grant extension of time by treating the deferment as amounting to impediment/prevention/default by the Employer or his agents, the Relevant Event under Clause 2.29.6.

It is also arguable that the Architect may exercise his powers under the contract to achieve deferment of site without the Employer falling into breach of contract. For example, such deferment may amount to 'execution or completion of the work in any specific order', an

57 See Chapter 6, Section 6.3.2 for a list of such matters.

58 The rationale for excluding instructions on the expenditure of Provisional Sums for defined work is that the Contractor should have allowed for all the necessary work in his tender and thus his master programme. For commentary on the purpose and content of Provisional Sums, see Chapter 1, Section 1.4.5 and Chapter 6, Section 6.9.

59 See Chapter 16, Section 16.1 for explanation of repudiatory breach.

allowed Variation under Clause 5.1.2.4. Also, under Clause 3.15 he may postpone the carrying out of the Works although he has to be careful not to exceed the period of suspension stated in the Contract Particulars against Clause 8.9.2 for, as explained in Chapter 16, Section 16.10, suspension beyond the stated period constitutes a default by the Employer for which the Contractor may terminate his employment under the Contract.

11.4.4 Work with Approximate Quantity (Clause 2.29.4)

Though not expressly stated that the quantity in the Contract Bills must have been underestimated, the rationale behind this clause must be that the Contractor could not have allowed for the extra work in his programme.

11.4.5 Suspension for non-payment (Clause 2.29.5)

Section 112(1) of the Construction Act gives a party to any construction contract a right to suspend performance of his obligations under the contract where, without an effective Withholding Notice, the other party fails to make payment of a sum due under the contract by its final date for payment. This right is implemented into JCT 05 in the terms of Clause 4.14. The procedure for suspension of performance is described in detail in Chapter 15, Section 15.6.2.

Section 112(4) of the Act goes further to provide that 'any period during which performance is suspended in pursuance of the right conferred by [s. 112(1)] shall be disregarded in computing for the purposes of any contractual time limit the time taken by the party exercising the right or by a third party, to complete any work directly or indirectly affected by the exercise of the right'. Clause 2.29.5 states, as a Relevant Event, 'suspension by the Contractor under Clause 4.14 of the performance of his obligations under the contract'. It is arguable that, even where the Contractor was already behind programme at the beginning of non-payment, the effect of s. 112(4) is to exculpate the Contractor for some or even all of his earlier delay since he could have caught up on the programme during the period of suspension. Whether Clauses 2.28 and 2.29.5 together implement in full the right under s. 112(4) is therefore open to debate.

A Local Democracy, Economic Development and Construction Bill had just started its passage through Parliament at the time of writing. It will change the rights in relation to suspension for non-payment under the Construction Act in two ways. First, it is to be stated more clearly that the contractor may, in his discretion, suspend any or all of his obligations under the contract. Second, the period that must be ignored in the computation of the extended contract period is not only 'any period during which performance is suspended in pursuance of the right [to suspend]' but also 'any period during which performance is suspended in consequence of the right [to suspend]'. This means that any further period of suspension caused by the initial suspension must also be disregarded. For example, if the contractor responds to non-payment by refusing to place orders for resources with long lead times, any unavoidable period of delay flowing from the delayed order must also be disregarded even where the contractor had recommenced performance before the second period of delay.

There is also the question of what 'disregarding' a period means. Contractors under JCT 05 may argue that it means that the Architect should assume that all work was brought to a halt during the relevant period by the Employer even where the Contractor had suspended only a minor part of his obligations. The Contractor may therefore elect to claim extension of time for the Relevant Event under Clause 2.29.6 (impediment, prevention of default by the Employer).

11.4.6 Impediment/prevention/default by the Employer (Clause 2.29.6)

The 'impediment/prevention/default' provision under Clause 2.29.6 is intended as a 'catch-all' or 'sweep-up' Relevant Event designed to prevent the application of the prevention principle in any circumstances by giving the Architect the power to extend time to take account of delay attributable to the Employer or his agents however caused.

Some causes of Employer-generated delay that were listed as Relevant Events under JCT98 are not in the list under JCT 05. Those of the omitted events that may now be considered to be covered by this Relevant Event are:

- failure to comply with the Information Release Schedule (IRS);
- failure to provide further drawings and details;
- delay by nominated contractors and suppliers;[60]
- execution of work not forming part of the contract;
- Employer's supply or failure to supply agreed goods and materials;
- failure to give in due time ingress to or egress from the site;
- compliance or non-compliance by the Employer with his contractual obligations in respect of the CDM Regulations.

Failure to provide drawings and other information

The Fifth Recital contemplates, as an option, an Information Release Schedule (IRS) given by the Employer to the Contractor. It is a list of specified items of information that will be supplied to the Contractor by dates stated against the items. Under Clause 2.11, the Architect must ensure that the Contractor is supplied with two copies of any item of information referred to in the IRS by the relevant date. Clause 2.12.1 requires the Architect to provide the Contractor with any further drawings and details not mentioned in the IRS but reasonably necessary to explain or amplify the Contract Drawings. He must also issue instructions necessary for the performance of the Contractor's obligations. The Contractor's entitlement to these items of information and the timetable within which they are to be provided are discussed in detail in Chapter 2, Section 2.4.1.

The Architect and the Employer's other designers are under obligation to use reasonable skill, care and diligence to ensure that necessary drawings and other information are provided in time to enable the Contractor to plan and carry out and complete the Works in accordance with the contract.[61] Breach of these obligations entitles the Employer to damages, including loss of liquidated damages and payments to the Contractor in reasonable settlement of sustainable claims. There is a growing body of case law in which

---

60 This standard form no longer provides for nomination but this change in itself does not prevent nomination by users.

61 See *Royal Brompton Hospital National Health Trust* v. *Hammond (No. 4)* [2000] BLR 75, at p. 79.

employers have sought to recover from contract administrators, project managers and designers amounts paid to contractors in compromised settlement of such claims.[62] There is therefore an unavoidable conflict of interest associated with the Architect's duty to evaluate these claims. As a reflection of this conflict, there is an increasing tendency on the part of employers to find fault with contract administrators and their other design consultants concerning the timing of their supply of information to contractors. This increased risk of multi-party disputes makes the management of the supply of information to the Contractor of paramount importance. Issues often raised in such disputes, and which must therefore be properly understood and managed, concern:

- whether the Contractor was entitled to be supplied with the information in question or the Contractor was responsible for preparing or obtaining it;
- when the Contractor should have been supplied with the information;
- when the information was actually supplied;
- the delay to completion and/or extra cost caused by the delay in supplying the information;
- responsibility for coordinating the flow of information between designers and the Contractor.

Nominated contractors and suppliers

The omission of provisions on nominated sub-contractors and suppliers is one of the most remarkable changes made in this contract form in a very long time. Few would miss these provisions, as publicity about the pitfalls of nominated sub-contracting has resulted in considerable decline in the practice. Anybody contemplating having a go at making amendments to allow nomination must bear in mind that the pitfalls could be even more serious because of the absence of detailed provisions on nominations. In the context of extension of time, it is arguable that delay caused by a nominated sub-contractor is an act of prevention by the Employer unlikely to be caught by the catch-all Relevant Event. There is therefore the risk of time for completion being rendered at large by such delay.

Work not forming part of the contract

It is not uncommon for the Contractor to be responsible for only a part of a larger project, with the remainder to be carried out by the Employer himself, his employees or other contractors directly engaged by the Employer as he is entitled to do under Clause 2.7. Where this is the case and the Contractor is delayed by the manner of execution or non-execution of the remainder of the project, the delay would amount to impediment/prevention/default by the Employer.

Some statutory undertakers enter into commercial contracts for the execution of works which are distinct from those they are obliged to carry out by statute. In *Henry Boot* v. *Central Lancashire New Town Development Corporation Ltd*,[63] which arose from a JCT 63 contract, it was held that work carried out by statutory undertakers under commercial contracts constituted work not forming part of the contract. It may be concluded from this case that, under JCT 05, delays arising from the execution or non-execution of such work would be covered by Clause 2.29.6 rather than Clause 2.29.7.

---

62 See for example *Wessex Regional Health Authority* v. *HLM Design Ltd* (1995) 71 BLR 32.
63 (1980) 15 BLR 1.

Employer's breach of agreement to supply goods and materials

This was a specific Relevant Event under Clause 25.4.8.2 of JCT98. The reason for its omission from JCT 05 is probably that it is covered by the catch-all Relevant Event.

Employer's failure in respect of ingress to and egress from the site

JCT98 Clause 25.4.12 stated as a Relevant Event:

> ...failure of the Employer to give in due time ingress to or egress from the site Works or any part thereof through or over any land, buildings, way or passage adjoining or connected with the site and in the possession and control of the Employer, in accordance with the Contract Bills and/or the Contract Drawings, after receipt by the Architect/the Contract Administrator of such notice, if any, as the Contractor is required to give, or failure of the Employer to give such ingress or egress as otherwise agreed between the Architect/Contract Administrator and the Contractor.

The advantage of the JCT98 in relation to this possible cause of delay was the specificity. As it is not mentioned specifically as a Relevant Event under JCT 05, the only requirement is that it has such effects as to amount to an impediment/prevention/default by the Employer or by parties for whom the Employer is responsible.

Employer's undertakings in respect of CDM

Clause 3.25.1 imposes on the Employer an obligation to ensure that the CDM Co-ordinator and the Principal Contractor perform their duties under the CDM Regulations. Any act or omission by the CDM Co-ordinator or by the Principal Contractor (where this is different from the Contractor) in performance of their duties under the CDM Regulations that causes delay would amount to impediment/prevention/default by the Employer's Persons. Clause 6A.3 of the JCT98 provided that the Contractor was not entitled to extension of time for compliance with reasonable requirements of the Principal Contractor in the interests of the CDM Regulations. It is submitted that, in the absence of such express exclusion in JCT 05, the chances of the Contractor with this type of claim are greater.

11.4.7 Statutory undertakers (Clause 2.29.7)

This clause relates to the carrying out of, or failure to carry out, work pursuant to their statutory obligations. As already discussed, work arising from commercial contracts entered into with the Employer are not covered here but under Clause 2.29.6.

11.4.8 Exceptionally adverse weather conditions (Clause 2.29.8)

To succeed with a claim on this ground, the Contractor must produce evidence that the conditions complained of are exceptional for that time of year and location. Weather records covering a reasonable period as well as site diaries will normally be demanded by Architects.

11.4.9 Loss/damage from the Specified Perils (Clause 2.29.9)

The Specified Perils are defined in Clause 6.8 as consisting of:[64]

> ...fire, lightning, explosion, storm, flood, escape of water from any tank, apparatus or pipes, earthquake, aircraft and other aerial devices or articles dropped therefrom, riot and civil commotion, but excluding Excepted Risks...

The Excepted Risks are also defined in the same clause as consisting of:

> ...ionising radiations or contamination by radioactivity from any nuclear fuel or from any nuclear waste from the combustion of nuclear fuel, radioactive toxic explosive or other hazardous properties of any explosive nuclear assembly or nuclear component thereof, pressure waves caused by aircraft or other aerial devices travelling at sonic or supersonic speeds.

The Contractor is obliged to reinstate the Works if they are damaged by any of the Specified Perils. The reason for making such damage a Relevant Event is presumably to give the Contractor time to carry out the reinstatement. These are neutral events (i.e. neither party has any control over them). There is therefore some equity in the sharing of this risk, in that the Employer shoulders the consequences of delay whilst the Contractor bears any additional cost from disruption to regular progress.

A question often posed is whether or not the Contractor would still be entitled to extension of time if he is himself the cause of the Specified Peril (e.g. where an employee of the Contractor negligently starts a fire). Case law suggests that the answer is a matter of construction of the particular Specified Peril. In *Computer and Systems Engineering plc* v. *John Lelliott (Ilford) Ltd and Another*[65] a pipe in a sprinkler system was sheared off through the negligence of a sub-contractor. This caused discharge of 1600 gallons of water that caused damage to the Employer's property. The litigation was over whether the damage was caused by 'flood' under an equivalent definition of the Specified Perils. The Court of Appeal held that the word 'flood' in the definition suggests rapid accumulation of larger volumes of water from an external source and that it did not cover discharge from the sprinkler system. The Court also confined 'bursting of pipes' to damage from internal stresses.[66] This decision therefore suggests that damage to the Works from a Specified Peril caused by the Contractor's negligence is not a Relevant Event.

In addition, the combined tenor of Clause 2.28 and 2.29 is that the Contractor is not to benefit from his own default. Besides, the Architect is to grant extension of time if it is fair and reasonable to do so: Clause 2.28.1 and 2.28.4. The Architect may come to a view that the Contractor's negligence is a breach of his duty under Clause 2.28.6.1 to use his best endeavours to prevent delay and that, therefore, it is not fair and reasonable to grant extension of time. To avoid any doubt, the equivalent clauses in previous editions were often amended to limit expressly the right to extension of time to only those Specified Perils not caused by the fault of the Contractor or of his agents. Some employers delete it altogether.

---

64 Two changes have been made from the definition of Specified Perils in JCT 98: (i) tempest has been omitted from the list; (ii) 'bursting or overflowing of water tanks, apparatus or pipes' has been replaced with 'escape of water from any water tank, apparatus or pipes'.

65 (1990) 54 BLR 1.

66 See note 64 (ii). The case may well be decided differently under the new definition.

11.4.10 Civil commotion/terrorism (Clause 2.29.10)

To succeed with an extension of time claim in respect of civil commotion/terrorism, the Contractor would have to demonstrate: (i) the occurrence of civil commotion or threats or acts of terrorism; (ii) that any act of a relevant authority that caused the delay was to deal with them; (iii) the civil commotion or act or threat of terrorism or the response of the relevant authority caused delay. The following would qualify:

- disturbance to suppliers from such acts or threats;
- evacuation of the area covering the site;
- evacuation of areas in which work destined for the site was being carried out.

The meaning of the term 'civil commotion' has been considered in the case law but mainly in the context of its usage in insurance policies. In *London and Manchester Plate Glass Company* v. *Heath*,[67] after a review of the authorities, Vaughan Williams LJ described it as 'an insurrection of the people for general purposes, though it may not amount to a rebellion, where there is an usurped power'. Pointing out the connotations of the word 'commotion', he added that there must be turbulence, tumult, violence or intention to commit violence. The other members of the Court of Appeal identified the same ingredients to 'civil commotion'. In *Levy* v. *Assicurazioni Generali*[68] Luxmore LJ defined it in these terms:

> This phrase is used to indicate a stage between a riot and civil war. It has been described to mean an insurrection of the people for general purposes, though not amounting to rebellion, but it is probably not capable of any very precise definition. The element of turbulence or tumult is essential; an organised conspiracy to commit criminal acts, where there is no tumult or disturbance until after the acts, does not amount to civil commotion. It is not, however, necessary to show the existence of any outside organisation at whose instigation the acts were done.

In that case, a large number of women (the Suffragettes) simultaneously broke glass windows in different parts of London. Each woman went quietly to the police station when arrested. They created no disturbance in the street. No charge of riot or unlawful assembly was made against any of them. The Court of Appeal decided that, because of the absence of tumult and disturbance, the damage to the windows was not caused by 'civil commotion'. The claimant did not therefore succeed with a claim on an insurance policy against 'damage to plate glass caused directly by or arising from civil commotion or rioting'.

The meaning of the term 'terrorism' is explained in Chapter 7, Section 7.3.4.6.

11.4.11 Industrial actions (Clause 2.29.11)

The types of industrial action for which the Contractor is entitled to extension of time include:

> Strike, lock-out or local combination of workmen affecting any of the trades employed upon the Works or any of the trades engaged in the preparation, manufacture or transportation

67 [1913] 3 KB 411.
68 [1940] AC 791.

of any of the goods or materials required for the Works or any persons engaged in the preparation of the design for the Contractor's Designed Portion.

Strikes are the most commonly encountered of these problems although they have declined in recent years. The clause does not discriminate between official and unofficial strikes. However, it is believed that 'working to rule' does not belong to this Relevant Event. Neither does picketing or civil commotion by political agitators (e.g. peace demonstrators) because they are neither 'trades employed upon the Works or … engaged in the preparation, manufacture or transportation of any of the goods or materials for the Works' nor 'persons engaged in the preparation of the design'. Also, it would appear that the clause does not cover strikes by the employees of a statutory undertaker engaged directly by the Employer to carry out work not forming part of the contract. *Boskalis Westminster Construction Ltd* v. *Liverpool City Council*:[69]

Boskalis entered into a contract with Liverpool City Council for the construction of dwellings. The Contract incorporated the JCT 63 (1977 revision), Clause 23(d) of which provided for industrial action as a Relevant Event in almost identical wording. The question put to the court from an arbitrator's award was whether delays caused by the strike actions of the employees of a statutory undertaker employed by the Employer to carry out work not forming part of the contract was covered by the clause. The question was answered in the negative.

11.4.12 Statutory intervention (Clause 2.29.12)

The Relevant event is stated in Clause 2.29.12 as 'the exercise after the Base Date by the United Kingdom Government of any statutory power which directly affects the execution of the works'. The word 'directly' is operative. It follows therefore that if the effect of the governmental intervention is through a chain of events the Contractor would not be entitled to extension of time. Examples of this type of event include the imposition of a shorter working week and rationing of fuel. These events may also be covered by *force majeure*.

Change in Statutory Requirements, a Relevant Event under Clause 25.4.15 of the JCT 98, is not specifically mentioned in any of JCT 05 Relevant Events. However, delay from such a change is most probably part of this new Relevant Event.

11.4.13 Force majeure (Clause 2.29.13)

In general, this term refers to Acts of God or man-made events which are beyond the control of the parties. Examples include war, inundation, epidemics and strikes, or the closure by the authorities of roads surrounding the site, thus preventing access. However, as used in JCT 05, it must have a restricted meaning as several of the events normally classified under this term (e.g. strikes and lightning) are dealt with separately.

Shortages of labour, goods and materials that were unforeseeable at the time of tendering constituted a separate Relevant Event under Clause 25.4.10 of the JCT 98. Such shortage would not be a Relevant Event under JCT 05 unless it is caused by events amounting to *force majeure* or statutory intervention.

69 (1983) 24 BLR 83.

11.5 Pre-agreed Adjustment

There are two ways in which a new Completion Date can be fixed under the contract. The more common method occurs when the Architect performs his duty under Clause 2.28.1 to adjust the Completion Date for delay caused by a Relevant Event. The second method is where the Contractor submits a Schedule 2 Quotation that includes a request for a new Completion Date and the Architect issues a Confirmed Acceptance of the quotation by the Employer. The term 'Pre-agreed Adjustment' is defined under Clause 2.26.2 to refer to the adjustment of the Completion Date by the second method.[70]

Clause 2.28.6.4 provides that the length of any Pre-agreed Adjustment is not to be altered by the Architect in any extension of time decision unless there has been a Relevant Omission from the work covered by the relevant Schedule 2 Quotation.

11.6 Relevant Omissions

The term 'Relevant Omission' is defined under Clause 2.26.3 as 'the omission of any work or obligation through an instruction for Variation under Clause 3.14 or through an instruction under Clause 3.16 in regard to a Provisional Sum for defined work'. The Architect may by written notice reduce extension of time granted if it is fair and reasonable to do so having regard for any Relevant Omission instructed since the last occasion on which a new Completion Date was fixed.

There are limitations on the Architect's power to reduce extension of time. First, the Contractor should not be required to complete earlier than the Date for Completion stated in the Contract Particulars. Second, the 'length of any Pre-agreed Adjustment' is not to be altered 'unless the relevant Variation or other work referred to in clause 5.2.1 is itself the subject of a Relevant Omission'.

There is no express authority to require the Contractor to supply any information on the likely effects of a Relevant Omission. The Architect would therefore have considerable difficulty getting progress information to quantify the reduction in time that is fair and reasonable unless the Contractor cooperates. This task may be eased to an extent if there is subsequent delay, in which case the Architect can seize the opportunity to request the relevant information.

11.7 Administrative procedures

The discussion so far has been about the circumstances entitling the Contractor in principle to extension of time. In addition to defining these circumstances, JCT 05 lays down detailed procedures to be followed in the application for, and the granting of, extension of time.

11.7.1 Obligations of the Contractor

1. The Contractor is to provide the Architect with two copies of his master programme as soon as possible after the execution of the contract (Clause 2.9.1.2). However, the

70 See also Chapter 6, Section 6.13.7 dealing with problems associated with Schedule 2 Quotations.

Contractor is not contractually bound to carry out the work strictly in accordance with the submitted programme: *Glenlion Construction Ltd* v. *The Guinness Trust.*[71]

2. The Contractor is under an obligation to amend or revise the programme to take account of any extension of time granted by the Architect or arising from a Pre-agreed Adjustment and to supply the Architect with two copies of the amendments or revisions within 14 days of the award of the extension or date of the agreement (Clause 2.9.1.2).[72]
3. The Contractor is to give written notice not only when it becomes reasonably apparent that progress is being delayed but also when progress is likely to be delayed (Clause 2.27.1). The Contractor must comply with the notice requirements even if he does not intend to claim extension of time. The notice must:
   - give details of all the material circumstances surrounding the delay;
   - state the cause of the delay;
   - identify which of the causes the Contractor believes to be Relevant Events.
4. For each of the Relevant Events identified, the Contractor must in the notice, or in writing as soon as possible thereafter, provide:
   - particulars of the expected effects of the delay;
   - an estimate of any delay in completion arising from the delay (Clause 2.27.2).

   The estimate has to be made for each event notified as causing or likely to cause delay.
5. The Contractor must, once a notice of delay has been given, continue to monitor events and give such further notices as he thinks necessary, or as may be requested by the Architect (Clause 2.27.3).
6. He also has to notify the Architect of any material changes in the particulars and estimates previously given (Clause 2.27.3). Architects often complained that similar clauses in earlier editions of the standard form caused a lot of unnecessary paperwork. This observation by HHJ Seymour QC in *Royal Brompton Hospital NHS Trust* v. *Hammond (No.6)*[73] may not be too far from common practice:

> ...However, I have the distinct impression from the documentation to which I have referred that the approach adopted by [the Contractor] on the project was, every time anything altered or anything happened which could conceivably delay any individual activity, [the Contractor] gave a written notification of delay to the completion of the Works, regardless of whether it really thought that the alteration or the occurrence would cause delay to the completion of the Works. While, from a contractor's point of view, adopting such practice may have the advantage that he is covered, no matter how things turn out, it does make life needlessly difficult for those seeking to administer the contract. In addition, I have the impression that, for whatever reason, [the Contractor] made exceptionally pessimistic predictions of the extent of the likely delay caused by the matters which it notified....

7. Under Clause 2.28.6.1 the Contractor must constantly use his best endeavours to prevent delay and, where delay occurs despite such endeavours, further delay.[74]

---

71 (1987) 39 BLR 89.
72 No such duty existed under the JCT 98. This is a welcome change.
73 (2000) 76 ConLR 131, at para. 75.
74 See Section 11.4 of this chapter for commentary on the meaning of 'best endeavours'.

11.7.2 Duties of the Architect

Clause 2.28.2 defines a timetable within which the Architect must act on applications and notices for extension of time. If the Architect is of the opinion that any cause of delay notified is a Relevant Event and that it will cause delay in completion, he must then give a written decision adjusting the Completion Date. The new date is to be fair and reasonable. Where it is not fair and reasonable to grant extension of time, the Contractor is to be so informed in writing. Although it is the duty of the Contractor to serve notices and estimate the delay to completion of the Works, the Architect has an independent duty to assess the extension of time that is fair and reasonable. This means that the Architect is not entitled to refuse extension of time simply because the Contractor has failed to substantiate any extension claimed. He must grant the extension supported by the information provided by the Contractor as well as from his own sources. In the light of any failure by the Contractor to provide necessary information, which would be in breach of contract, the Contractor would have a hard task challenging the extension.

There are a variety of methodologies for assessing delay. The principal ones include: Global Impact Technique, As-Planned vs As-Built, Collapsed As-Planned, Collapsed As-Built, Window Analysis and Time Impact Analysis. Some of them are very complex, requiring the use of Critical Path Method software packages. The Architect does not have to apply any particular methodology but whatever approach he adopts must be methodical, logical and in compliance with the contract. In *John Barker Construction Ltd* v. *London Portman Hotel Ltd*,[75] which arose from a contract with similarly worded extension of time provisions, the court stated that the duty to estimate the extension of time that is fair and reasonable did not allow the Architect simply to make an impressionistic assessment without any logical analysis of the delays actually caused by the Relevant Events. Where the causal complexities of the delay are beyond the Architect's expertise, he should seriously consider advising the Employer of the fact so that appropriate expertise can be sought.

The timetable within which the Architect must reach his decision and act accordingly is the same regardless of the extension of time entitlement: not later than 12 weeks from receipt of the Contractor's notice or of reasonably sufficient particulars and estimates: Clause 2.28.2. Failure on the part of the Contractor to supply reasonably sufficient information for the Architect to make a fair judgement postpones the day from which the period of 12 weeks runs. The Contractor must therefore respond very quickly to the Architect's requests for further particulars and estimates. Where the time from the Contractor's notice to the Completion Date is less than 12 weeks the Architect must endeavour to make a decision and inform the Contractor not later than the Completion Date.

The Architect must state the time extension in respect of each Relevant Event in his written extension of time decision.[76]

As already explained in Section 11.2 of this Chapter, there is some uncertainty surrounding the effect of late extension of time decisions. Although it has been suggested that such delay is, at worst, only a breach of contract by the Employer, prudence demands

---

75 (1996) 83 BLR 31; 50 ConLR 43; (1996) 12 Const LJ 277.

76 Under JCT 98 the Architect was required to state the total time extension; he did not have to apportion it to each Relevant Event.

that the Architect complies with the timetable.[77] If the Employer suffers any loss because of negligent operation of the extension of time provisions, he could be entitled to recover damages from the Architect for negligence. The issue of the Architect's liability for professional negligence in relation to his decisions on extension of time is considered in Section 11.14 of this chapter.

11.7.3 Importance of the notices of delay

The nature of the decision-making required of the Architect highlights the importance of timely notices of delays. The aims of the stringent notice requirements are:

- to give the Architect the opportunity to take all reasonable steps available to him to minimize the effect of the delay (e.g. he may issue appropriate Variation orders);
- to alert the Architect to watch out for the reasonableness of the Contractor's endeavours to prevent or minimize delays in completing the Works;
- to alert the Architect to the effects of the delays as they occur;
- to allow the Architect to advise the Employer of likely delays so that the latter can rearrange his affairs accordingly.

The effect of a contractor's failure to serve notice of delay on contract administrators' power to grant extension of time has been a matter of great debate.[78] Fortunately, this problem does not arise in relation to JCT 05 because it states expressly in Clause 2.28.5 that, in his review of the Contractor's entitlement to extensions of time, the Architect must take into account all delays by Relevant Events including any delay in respect of which the Contractor failed to serve the required notice.

11.7.4 Review of extensions of time

The Contract authorizes three types of review of extension: (i) an optional review whenever the Architect is making an extension assessment; (ii) an optional review after the Completion Date if that date occurs before practical completion of the Works; (iii) a mandatory and final review within 12 weeks of practical completion.

Where any instruction for a Relevant Omission was issued after the last fixing of a new Completion Date, the Architect may fix a Completion Date earlier than that last fixed if it is fair and reasonable having regard to the omission (Clause 2.28.4).

In most cases, the exact effect of the delaying event cannot be known completely at the time the Architect is expected to grant extension of time. This may be because either the event is, at that time, a continuing one, or its effects lie in the future. For these reasons, the extensions of time granted before practical completion are generally only provisional. The aim of extension of time at this stage is to provide the Contractor with a rough but realistic Completion Date towards which to work. Towards the end of the contract, the Architect should be in a better position to appreciate the actual effect of all the delaying events for which extension of time is grantable. Under Clause 2.28.5, all extensions of time are therefore subject to review after practical completion of the Works.

---

77 It is explained in Chapter 14, Section 14.5 that failure of the Architect to operate the extension of time provisions has the consequence that the Contractor is entitled to price adjustment for cost fluctuations during the period of the Contractor's culpable delay.

78 See Section 11.2 of this chapter for discussion of this issue.

The final review must result in the Architect making any of three possible decisions open him:

1. He may fix a later Completion Date than that previously fixed if he had underestimated the delays and such later Date is fair and reasonable. The Architect may also take into account all Relevant Events (i.e. including those not notified by the Contractor (Clause 2.28.5.1)).
2. He may fix an earlier Completion Date than that previously fixed by the Architect's extension of time decision under Clause 2.28.1 or by a Pre-agreed Adjustment if it is fair and reasonable to do so having regard to any instructions for Relevant Omissions issued subsequently (Clause 2.28.5.2).
3. The Architect can confirm a Completion Date previously fixed by the Architect or a Pre-agreed Adjustment (Clause 2.28.5.3). It would appear that he must confirm the current Completion Date if it had been fixed by a Pre-agreed Adjustment and no Relevant Event has subsequently occurred. The rationale for this is probably that the Contractor should have considered all previous delays in the relevant Schedule 2 Quotation.

However, under no circumstances should a new Completion Date be fixed to occur at an earlier date than the original Date for Completion stated in the Contract Particulars (Clause 2.28.6.3). In addition, Clause 2.28.6.4 states that no review of extension of time 'shall alter the length of any Pre-agreed Adjustment unless the relevant Variation or other work referred to in Clause 5.2.1 is itself the subject of a Relevant Omission'.

11.7.5 Adjudicating on extension of time

The contract lays down a clear timetable within which the Architect must make a decision on adjustment of the Completion Date for delay caused by a Relevant Event. It would appear that, where a request for extension of time is made to the Architect, no dispute on the Contractor's entitlement can crystallize prior to the Architect's decision or failure to make one within the appropriate time window. In *R. Durtnell & Sons Ltd* v. *Kaduna Ltd*[79] Durtnell served a Notice of Adjudication contending that Kaduna was in breach of contract for a number of reasons and seeking extension of time under the contract. The adjudicator found that the Contractor had been delayed and ordered the contract period to be extended. However, in the subsequent enforcement proceedings, HHJ Seymour QC decided that the adjudicator did not have jurisdiction to make such a decision as the Architect had not yet made a determination and the time allowed in the contract for his determination in respect of Durtnell's application for an extension of time had not yet expired.

Durtnell argued that a decision of the Architect in relation to an application for an extension of time was not a condition precedent to the exercise of a right to adjudicate and submitted that that right was one exercisable at any time. Kaduna accepted that it was not a condition precedent to the jurisdiction of an adjudicator to determine whether, under the contract, Durtnell was entitled to an extension of time, that there should have been a decision of the Architect on those grounds. It was, however submitted that, where Durtnell had elected to seek a decision from the Architect, until there was such a decision or the Architect had failed to make it within the time allowed him, there was nothing for Durtnell to dispute; hence there was no 'dispute' capable of being referred to adjudication.

79 [2003] All ER (B) 281; [2003] EWHC 517 (TCC) (hereafter *Durtnell* v. *Kaduna*).

This is how at paragraph 42 the judge dealt with the rival arguments:

> ... in my judgment it cannot be said that there is a 'dispute' as to entitlement to extensions of time, or as to valuation of loss and expense consequent upon a grant of extensions of time, at a time at which the question of whether there should be any extension of time, or any further extension of time, has been referred to the architect for the purposes of the standard form, the time allowed by the standard form for him to make a determination has not expired, and no determination has been made. I readily accept that it is not, expressly, a condition precedent to any reference to adjudication of a dispute as to entitlement to an extension of time under a contract in the standard form that the dispute should first have been referred to the architect. However, it is not easy to see how a dispute as to entitlement to an extension of time could arise until that had happened and the architect had made his determination or the time permitted for doing so had expired. The reason is that under the standard form it is not for the employer to grant an extension of time or not. That function is entrusted to the architect who is under an obligation to act impartially in making his assessment. Until the architect has made his assessment, or failed to do so within the time permitted by the standard form, there is just nothing to argue about, no 'dispute'. ... It is nonsensical to suggest that a 'dispute' can exist between two parties as to a matter entrusted to a third party for independent decision in advance of the decision being known. For practical purposes, therefore, it seems to me that it is a condition precedent to the reference to adjudication of a 'dispute' as to entitlement to an extension of time and as to anything which is dependent upon such decision, such as a claim for payment of loss and expense in relation to an extension of time claimed but not granted, that the person to whom the making of a decision on the relevant issue is entrusted under the contract between the parties should have made his decision, or the time within which it should have been made has elapsed without a decision being made.

This extract suggests that, in so far as an Architect is at post and timeously performing his administrative duties in relation to decisions on extension of time and associated loss and/or expense claims, there is no right to refer this type of claim to adjudication until the Architect has taken the relevant decision.

11.7.6 Effect of an adjudicator's extension of time decisions

Such a decision is binding on the Parties until the relevant dispute is resolved finally by arbitration, litigation or agreement. Paragraph 23(2) requires the Parties to comply with the decision. In the case of an extension of time decision it would be implied that the Architect must adopt the adjudicator's decision as his decision under Clause 2.28 until final resolution of the dispute. The Employer, it is submitted, as an exception from his general duty to leave the Architect free to exercise his professional judgment, must instruct the Architect to do so, if he does not act voluntarily.

As already explained, an extension of time decisions are provisional until final review after practical completion. The question then arises whether the Architect may review the adjudicator's decision in a subsequent assessment of extension of time. On the authority of the Court of Appeal in *Quietfield Ltd* v. *Vascroft Contractors Ltd*,[80] where a subsequent extension

80 [2006] EWCA Civ. 1757; [2007] BLR 67; 114 ConLR 81; [2007] CILL 2425.

of time dispute covers grounds different from those in the dispute already decided by the adjudicator, it amounts to a different dispute which may be the subject of another adjudication. In short, the decision is only binding if the grounds are the same as in the earlier adjudication. It follows that the Architect has the power to consider extension of time claims based on causes of delay different from those in the dispute already decided by the adjudicator. The same principle applies to a claim based on the same causes of delay but with updated facts (e.g. evidence that the Contractor caught up on the programme or the delay turned out much shorter than had been forecast by the adjudicator).

11.8 Deduction of liquidated damages

Under Clause 2.32.1, the Employer is entitled to recover liquidated damages from the Contractor if he fails to complete by the Completion Date and two specified preconditions of such entitlement are satisfied. First, the Architect must have issued a Non-Completion Certificate. Second, the Employer must have properly notified the Contractor in writing that he has acquired the right to liquidated damages on account of the Certificate. This notice, which is hereafter referred to as the 'Notice of Liquidated Damages', must be served not later than 5 days before the final date for payment on the Final Certificate. The Employer does not have to leave servicc of the notice to the very last moment, which could be long after Practical Completion. Indeed, as next explained, he would be well advised to serve the notice as soon as possible after the issue of the Non-Completion Certificate if he intends to set off the liquidated damages against Interim Payment Certificates.

Liquidated damages supported by the Non-Completion Certificate and the Notice of Liquidated damages may be set-off against payment due under Interim Payment Certificates and the Final Certificate subject to service of a further notice in relation to the relevant payment certificate: Clauses 4.13.4 and 4.15.4. This notice, referred to as a 'Withholding Notice', is a requirement of the Housing Grants, Construction and Regeneration Act 1996.[81] It must be served not later than 5 days before the final date for payment pursuant to the relevant Certificate. This means that, in the case of an Interim Payment Certificate, the Employer has 9 days from the date of its issue within which to serve the required notice. With the Final Certificate, it is 23 days from its issue.

Clause 2.32.2 requires the Notice of Liquidated Damages to contain a statement to the effect that, for the period between the Completion Date and the date of practical completion, he requires the Contractor to pay or allow liquidated damages at the rate stated in the Contract Particulars and that he may recover the amount involved as debt or set-off it against Certificates. The total amount of liquidated damages recoverable from the Contractor at any point in time is determined by multiplying the rate of liquidated damages inserted in the Contract Particulars (e.g. £ *x*/week or day of delay) by the total period of delay. If the Employer decides to accept reduced liquidated damages (probably because the actual consequences of the delay are less financially damaging than was anticipated at the time of entering into the contract), he is to state the reduced rate in the Notice of

81 See Chapter 15, 15.2.4 for discussion of Withholding Notices.

Liquidated Damages. The purpose of this stipulation is probably to avoid challenges to attempted set-offs by the Employer on the argument that, because the amount involved is less than the liquidated damages due, it is for something else and therefore not authorized under the contract.

A Withholding Notice would meet the requirement for the Notice of Liquidated Damages if appropriately worded and served not later than 5 days before the final date for payment on the relevant Certificate.[82] However, it would be prudent to serve the latter as a separate notice upon receipt of the Non-completion Certificate. If there is no payment due the Contractor under the contract, the liquidated damages can be recovered as debt in the usual way. A separate Notice of Liquidated Damages protects this entitlement.

11.9 The Contractor's duty to update the master programme

Clause 2.12 requires the Contractor, within 14 days after any grant of extension of time or Pre-agreed Adjustment of the Completion Date, to update the master programme and to provide two copies of the new programme to the Architect.

11.10 Deduction of liquidated damages from money decided by an adjudicator

Where an adjudicator decides that the Employer must pay a stated amount to the Contractor, there is the question whether, by serving a Withholding Notice against the money due to the Contractor pursuant to the decision, the Employer can become entitled to deduct or withhold accrued liquidated damages from the money. The question was considered by HHJ Hicks QC in *VHE Construction plc* v. *RBSTB Trust Co. Ltd.*[83] The chronology of events leading up to the litigation was as follows. VHE submitted an application for payment of £1,037,898.05 to which RBSTB served neither a Payment Notice[84] nor a Withholding Notice. On 20 August 1999 VHE served notice to refer non-payment of the application to adjudication. On 5 October the adjudicator made a decision that VHE was to be paid the full amount in the application within 28 days after submission of a VAT invoice as required by the contract. RBSTB initiated a second adjudication to review the payment application. The second adjudicator made his decision revising the amount in the application down to £254,831.83.

Clause 24.2.1 of the contract[85] from which the case arose entitled the defendant, subject to appropriate prior notices, to deduct liquidated damages against 'any monies due or to become due to [VHE] under the contract'. The required notices were, first, one stating that VHE had failed to complete the Works by the due completion date and, second, a written requirement to VHE to pay liquidated damages for the delay. The first notice was

82 The final date for payment on an Interim Certificate is 14 days after its issue (Clause 4.13.1). For the Final Certificate it is 28 days after the date of its issue (Clause 4.15.4).
83 [2000] BLR 187 (hereafter *VHE* v. *RBSTB*).
84 See Chapter 15, Section 15.2.3 for explanation of the purpose of this notice.
85 This was the *JCT Standard Form of Building Contract with Contractor's Design*, 1981 Edition.

served before the submission of the payment application. After the second adjudicator's payment decision, but before the final date for payment on it, the defendant served the second notice in relation to the amount to be paid pursuant to the decision and refused to pay it. Instead, the defendant paid only £46,974.69, the difference between the amount in the decision and £207,857.14 claimed as liquidated damages.

In the proceedings to enforce the second decision without allowing liquidated damages, the position of RBSTB was that its obligation to comply with the decisions of adjudicators was qualified by 'without prejudice to [its] other rights under the contract'. It contended that another right under the contract was an entitlement under Clause 24.2.1 to set-off liquidated damages against any amount due or to become due under the contract. HHJ Hicks QC rejected the defendant's argument on two grounds. First, the 'without prejudice' qualification referred only to the entitlement of the parties to determine the dispute finally by litigation, arbitration or agreement. The second ground for the rejection was that money payable as a consequence of an adjudicator's decision was not money due or to become due 'under the contract' against which Clause 24.2.1 authorized set-off for liquidated damages.

In *M.J. Gleeson Group plc* v. *Devonshire Green Holding Ltd*[86] and *David McLean Contractors Ltd* v. *The Albany Building Ltd*[87] HHJ Gilliland QC applied the *VHE* v. *RBSTB* decision to similar facts. In the former, referring to the contractual scheme for adjudication in the contract from which it arose, he stated: '[a]n adjudicator's decision is meant to be enforced and complied with without, it seems to me, subtle arguments and detailed arguments as to other provisions of the contract'.[88]

In *The Construction Group Centre Ltd* v. *The Highland Council*[89] the Outer House of the Scottish Court of Session reached the same conclusion regarding entitlement to set-off liquidated damages against an adjudicator's payment decision. The Pursuer (claimant) submitted its payment application but the engineer under the contract declined to certify any amount as due. There was therefore no opportunity for the Defender to serve a notice to withhold liquidated damages. The dispute about the amount due was referred to adjudication. The Defender did not raise its entitlement to liquidated damages before the adjudicator. Apparently, the Defender did not raise the issue because of a presumption suggested in the case law that the adjudicator would not have had jurisdiction to consider any set-off not raised in a prior notice to withhold. The adjudicator made a payment decision in favour of the Pursuer without any reference to liquidated damages. The Defender served notice to withhold liquidated damages against the award.

Lord MacFadyen held that the requirement for a s.111 notice (Withholding Notice) applies to only normal payment certified or applied for under construction contracts and not payment due as a consequence of an adjudicator's decision. The judge accepted that an adjudicator's payment award constitutes a sum due under the contract but limited the application of s.111 to only sums due under the contract for which the contract provides a final date for payment. He stated that acceptance of the validity of a Withholding Notice against an adjudicator's decision would be 'destructive of the effectiveness of the institution of adjudication' as payers would do nothing about withholding in anticipation of putting forward the set-off for the first time at the enforcement stage of the adjudication.

---

86 TCC, 19 March 2004.
87 Salford District Registry, 10 November 2005.
88 At para. 20.
89 [2002] BLR 476.

Set-off of liquidated damages against an adjudicator's payment decision was allowed in *David McLean Housing Contractors Ltd* v. *Swansea Housing Association*.[90] In that case the claimant contractor served notice to refer non-payment of its application to adjudication. The defendant issued a certificate for the amount decided by the adjudicator in the Contractor's favour. The defendant then wrote to the claimant stating its intention to deduct liquidated damages from payment due under the certificate. The defendant wrote again to the claimant stating that the claimant was liable to pay liquidated damages of £130,359 and that this would be deducted from any money that became due under the contract. The defendant paid the amount in the decision less the amount of liquidated damages. The claimant applied for summary judgment for the full amount in the adjudicator's decision and the defendant counterclaimed for liquidated damages and sought summary judgment on the counterclaim.

The payment application in the dispute comprised several elements, including loss and/or expense. To determine the loss and/or expense element, the adjudicator had to consider the extension of time to which the Contractor was entitled under the contract. Thus, although the adjudicator had not been asked specifically to determine the extension of time issue, the adjudicator had had to do this because, in the words of HHJ Humphrey LLoyd QC, it was a 'necessary and indispensable precursor of direct loss or expense'. As is widely known within the construction industry, a decision about extension of time entitlement is also a decision on liability for liquidated damages if the time of actual completion is not contested. Thus, in effect, the decision of the adjudicator also included the defendant's entitlement to liquidated damages. The judge dismissed the claimant's application for summary judgment but allowed the defendant's application on its counterclaim.

Thus, rather than simply allowing set-off for liquidated damages against the decision of an adjudicator, the judgment gave effect to that very decision. *McLean* v. *Swansea* is therefore distinguishable from *VHE* v. *RBSTB* in that the latter concerned set-off for liquidated damages outside the adjudicator's decision. The distinction was highlighted by HHJ Seymour QC in *Solland International Ltd* v. *Daraydan Holdings Ltd*[91] which concerned set-off for liquidated damages and defects against an adjudicator's award. It was part of the adjudicator's decision that letters put forward as Withholding Notices in relation to liquidated damages and defects were invalid. The adjudicator decided that Daraydan should pay the amount in the claimant's invoice plus interest and the adjudicator's fees. Entitlement to liquidated damages and damages for defects was not part of the decision. Accordingly, the approach in *VHE* v. *RBSTB*, and not *McLean* v. *Swansea*, applied.

The mere fact that an adjudicator's decision included extension of time will not always support set-off of liquidated damages against the monetary part of the decision. The facts of *Balfour Beatty Construction* v. *Serco Ltd*[92] were similar to those in *McLean* v. *Swansea* but with two important differences. First, whilst the adjudication in *McLean* v. *Swansea* took place after practical completion the adjudication in *Balfour Beatty* v. *Serco* occurred prior to practical completion. Second, the adjudicator in *McLean* v. *Swansea* determined the final extension of time. The determination of extension of time in *Balfour Beatty* v. *Serco* was interim not only because the Works were not yet completed but also because

90 [2002] BLR 125 (hereafter *McLean* v. *Swansea*).

91 [2002] All ER (D) 203(Feb).

92 [2004] EWHC 336 (TCC) (hereafter *Balfour Beatty* v. *Serco*).

not all the delays had been referred to the adjudicator. Distinguishing *McLean* v. *Swansea*, Jackson J held that the contractor was entitled to enforce payment pursuant to the decision without set-off for liquidated damages.

It is concluded from the review of the case law that the Employer under JCT 05 would be entitled to deduct or withhold liquidated damages from the money decided by an adjudicator only where the Works have been completed and the Contractor's final entitlement to extension of time was properly referred to and decided by the adjudicator.

11.10.1 Refund of liquidated damages

Clause 2.32.2 provides that if a Non-Completion Certificate is invalidated by a subsequent extension of time or a Confirmed Acceptance of a Schedule 2 Quotation, the Employer is to repay the liquidated damages recovered on the strength of the superseded Certificate. Many contractors argue that any such refund must carry interest over the period during which the money was in the Employer's hands.

In *Department of the Environment for Northern Ireland* v. *Farrans (Construction) Ltd*[93] it was decided that interest was payable by the Employer. That litigation arose from a contract let on the JCT 63. This decision was founded on the assumption that deduction of liquidated damages which become refundable as a consequence of subsequent extension of time constituted a breach of contract for which interest was recoverable as damages under the principle in *Wadsworth* v. *Lydall*.[94] The correctness of this assumption has been doubted. As explained in Chapter 2, Section 2.4, the Employer does not warrant that in assessing quantum, be it payment or extension of time, the Architect will arrive at the figure the Contractor is properly entitled to under the contract. It is a breach of contract only where the Architect fails to operate the extension of time provisions at all.

In any case, it is arguable that *DOE* v. *Farrans* is not applicable to JCT 05 because Clause 2.32.2 clearly anticipates the possibility that extensions of time granted after deduction of liquidated damages may necessitate some refunds. This also follows from the general approach that extensions of time are only provisional until after the final review under Clause 2.28.5. Further, employers often argue that, as this contract provides for refund without any mention of interest as has been done in Clauses 4.13.6 and 4.15.6 in relation to delayed payment on certificates, the intention must be that the Contractor is not entitled to any interest on refunds.

However, the decision of the House of Lords in *Sempra Metals Ltd (formerly Metallgesellschaft Ltd)* v. *Commissioners of Inland Revenue and Anor*[95] casts new light on the issue. In that case the respondent company paid advance corporation tax prematurely. It was held by a majority of the House that the respondent was entitled to recover interest for the loss of the use of the money on a compound basis. In the light of this decision, contractors have a much better case than ever before for recovery of interest on refunds of liquidated damages on a compound basis. It is therefore for the drafters of this form to close off such entitlement with clear words if that is the consensus of the JCT.

---

93 (1981) 19 BLR 1 (hereafter *DOE* v. *Farrans*).
94 [1981] 2 All ER 401; see also Chapter 13, Sections 13.4.11 to 13.4.13.
95 [2007] UKHL 34; [2007] 3 WLR 354.

11.10.2 Non-Completion Certificate and Withholding Notices

The Employer acquires a right to withhold liquidated damages against payment due under an Interim Certificate if three conditions are satisfied: (i) the Architect has issued a Non-Completion Certificate; (ii) the Employer has served a Notice of Liquidated Damages; (iii) the Employer has served an effective Withholding Notice as required by Clause 4.13.4.[96] What is the effect of an extension of time granted by the Architect after service of the Withholding Notice but before the final date for paying on the relevant certificate? If the new Completion Date has already been passed, liquidated damages would be claimable but for the shorter period of delay. The contract is silent on the effect on the Withholding Notice although it states expressly that the Non-Completion Certificate is to be treated as cancelled.[97]

The proposition that the Withholding Notice falls with the Non-Completion Certificate carries the implication that the Employer must serve another Withholding Notice to be able to withhold the reduced liquidated damages. If the extension of time is granted later than 5 days before the final date of the certificate it would be too late to serve a valid Withholding Notice. The Employer must therefore pay the whole of the certified amount and wait for the next payment certificate, which could be a long while away. In any case, according to this proposition, if the Employer withholds the amount notified before the extension, the Employer would be committing the default of failing to pay, by the final date for payment in respect of a Certificate, an amount properly due to the Contractor. This conduct would therefore amount to the default under Clause 8.9.1.1 of JCT 05, which entitles the Contractor, subject to service of appropriate notices, to suspend performance or even terminate his employment.

This question was considered by the House of Lords in *Reinwood Ltd* v. *L. Brown & Sons Ltd*.[98] The JCT 98 contract, from which the case arose, stated that the Completion Date was 18 October 2004 or any date fixed under the Contract. Delays occurred and the Contractor applied for extension of time on 7 December 2005. On 14 December 2005 the Architect issued the Non-Completion Certificate. On 11 January 2006 the Architect issued an Interim Certificate for £187,988. The final date under the contract was therefore 25 January 2006. The Employer served the equivalents of the Notice of Liquidated Damages and the Withholding Notice on 17 January 2006, which was more than 5 days before the final date of the Certificate. On 20 January 2006 the Employer paid the amount certified less the liquidated damages notified (£61,629). On 23 January the Architect granted extension of time that had the effect of reducing the amount of recoverable liquidated damages notified to £12,326. The court of first instance held that the Withholding Notice fell with the cancelled Non-Completion Certificate and that, by failing to pay the certified amount without a valid Withholding Notice, the Employer had committed the default under the equivalent of Clause 8.9.1.1.

---

96 An effective Withholding Notice must: (i) state the amount to be withheld and the ground for the withholding; (ii) specify each ground where there is more than one ground; (iii) be served not later than 5 days before the final date for payment.

97 Note that under Clause 2.32.4 the Notice of Liquidated Damages remains valid (i.e. the Employer does not have to serve another Notice of Liquidated Damages in relation to the new Non-Completion Certificate).

98 [2008] UKHL 12.

The Court of Appeal,[99] by a unanimous decision, disagreed with the proposition that the Withholding Notice fell with the Non-Completion Notice. Dyson LJ explained that, by stating that the effect of extension of time is to cancel the Non-Completion Certificate and saying nothing about the effect on the Withholding Notice, the intention must be that the Withholding Notice may still be relied upon to deduct the notified liquidated damages. Nevertheless, the Employer must refund the excess liquidated damages withheld within a reasonable time of the extension of time decision. He added that upholding the Withholding Notice made better commercial sense than treating it as invalidated by the subsequent extension of time decision. He must have had in mind that, whilst upholding the notice would deprive the Contractor of the excess liquidated damages for only a matter of days, invalidation of the notice could deny the Employer recovery of liquidated damages for much longer.[100]

Adopting substantially the same analysis as Dyson LJ, the House of Lords dismissed an appeal by the Contractor. Lord Neuberger of Abbotsbury added some guidance on the timetable for repayment of any excess liquidated damages withheld by the Employer as a consequence of the new extension of time. As the contract does not specify the time for repayment where the Contractor applies for it, paragraphs 7 and 8 of the Scheme[101] apply. The consequence of such application is that the repayment becomes due after 7 days whilst the final date is 17 days thereafter.

In subsequent litigation, the Court of Appeal accepted the Contractor's argument that, because he had issued a notice of default for non-payment a year earlier, he was entitled to terminate for the subsequent non-payment without a new notice of default. Details of this aspect of the Reinwood/Brown litigation are provided in Chapter 16, Section 16.6.6.

11.11 Resisting liability for liquidated damages

Grounds for resisting claims for liquidated damages, some of which can be deduced from the discussion so far, include:

- that the liquidated damages are penal;
- that time for completion has become at large;
- programme specification defect in the liquidated damages clause;
- failure by the Architect to serve the Non-Completion Certificate;
- the Employer's failure to serve a valid Notice of Liquidated Damages;
- the defence of estoppel.

11.11.1 Penal liquidated damages provisions

The principles for distinguishing between genuine liquidated damages, which are recoverable, and a penalty, which is unenforceable, are explained in Section 11.1.1 of this chapter.

---

99 See *Reinwood Ltd* v. *L. Brown & Sons Ltd* [2007] EWCA Civ 601.

100 It may even be impossible for the Employer to refund the excess liquidated damages by the final date where the Completion Date is revised on the eve of the final date, with the consequence that the Employer commits the default under Clause 8.9.1.1 by acting in accordance with the contract.

101 The Scheme for Construction Contracts (England and Wales) Regulations 1998, SI 1998 No. 649; The Scheme for Construction Contracts (Scotland) Regulations 1998, SI 1998 No. 687.

11.11.2 Completion Date at large

It is explained in Section 11.2 of this chapter that if the Employer causes delay and there is no power under the contract to fix a new Completion Date to take account of that delay, the Contractor would be released from any obligation to complete by a definite date. The Employer's remedy for delayed completion is to claim the damages consequent upon the Contractor's failure to complete the Works within a reasonable time. In such a claim the burden is on the Employer to prove not only that the Contractor failed to complete within a reasonable time but also the amount of the damages flowing from the breach.

As explained in Section 11.4.6 of this chapter, the catch-all Relevant Event in Clause 2.29.6 (any 'impediment, prevent or default' by the Employer or his agents) is designed to avoid the Completion Date becoming at large.

11.11.3 Programme specification defect

The contract must properly require the Contractor to complete the Works or the relevant Section by the missed date relied upon in the liquidated damages claim. Unless this has been done the Contractor may successfully fend off liability.

In *M.J. Gleeson (Contractors) Ltd* v. *London Borough of Hillingdon*,[102] for example, the litigation arose from a contract for the construction of 300 dwellings. The contract was let in the terms of the then current version of the JCT 63. Under the terms of the printed conditions, the Contractor was only required to complete the whole of the Works by a stated completion date or other date properly fixed by the Architect. The standard form did not then contain any mechanism for specifying sectional completion. However, the Contract Bills purported to incorporate an annexed programme which entailed sectional completion. As already explained, the liquidated damage clause could be enforced only if the contract allowed for sectional completion. It was held that the programme was not effectively incorporated into the contract because Clause 12(1) (the equivalent of Clause 1.3 of JCT 05) provided that nothing in the Contract Bills could override the printed terms of the JCT 63.

It is submitted that JCT 05 mechanism for specifying sectional completion dates and liquidated damages is contractually sound and that, therefore, the problem highlighted in this case is unlikely to occur where the Contract Particulars are properly completed and the form is not amended.

11.11.4 No Non-Completion Certificate

The issue of this Certificate is clearly stated in Clause 2.32.1.1 as one of the preconditions for recovery of liquidated damages.

11.11.5 No Notice of Liquidated Damages

A contractor may challenge recovery of liquidated damages on grounds that the document actually received from the Employer is not the Notice of Liquidated Damages required

102 (1970) 215 EG 165.

under the Contract. Although *Jarvis Brent Ltd* v. *Rowlinson Construction Ltd*[103] would probably be decided differently under JCT 05, it nevertheless provides a good illustration of this type of challenge.

Jarvis carried out work as main contractors for Rowlinson on the terms of the JCT 80. After a Non-Completion Certificate had been issued, the Employer sent a letter enclosing a cheque to Jarvis. A letter sent to Rowlinson by the Quantity Surveyor was also enclosed. The cheque was in the amount due under an Interim Certificate reduced by an amount stated by the Quantity Surveyor in his letter to be recoverable as liquidated damages. Upon the issue of subsequent Interim Certificates, Rowlinson simply sent cheques which reflected deductions for liquidated damages. The claimant argued that a written requirement for liquidated damages was a condition precedent to deduction of liquidated damages and that the letter accompanied by the cheque did not constitute such a requirement. It was held that a written requirement was not a condition precedent. The judge emphasized that what was essential was that the Contractor should not be left in any doubt that the Employer was deducting liquidated damages from payment otherwise due under the Certificates. In this instance, the judge found that the Contractor was in no such doubt. It was further decided that, in any event, the letter and the reduced cheques constituted requirements in writing for the purposes of Clause 24.2.1 (the equivalent of Clause 2.32.1.2 of JCT 05).

On the question of validity of notices under JCT80 Clause 24.2.1, Judge Carr took a different approach in *JF Finnegan Ltd* v. *Community Housing Association Ltd.*[104] In that case, an Employer under a JCT 80 contract accompanied a cheque to the Contractor with written remittance advice that stated only the number of the certificate to which the cheque related, the amount deducted for liquidated damages and the difference between the value of the certificate and the amount of liquidated damages. This difference was also the amount in the cheque. Judge Carr held that the remittance advice did not meet the requirements of the notice required by Clause 24.2.1. He explained:

> In my judgment the 'requirement in writing' in Clause 24.2.1 should indicate at least the basic details which are being relied upon to justify the deduction including the period of overrun and the figure for deduction which is claimed. These details then become a matter of record and not left to the memories of men several years later as to why a specific figure was claimed for deduction. Such detail would not only concentrate the mind of the employer when making the deduction but also will allow the contractor to know precisely why that deduction is being made and give him an opportunity to challenge it, if he so desires.

The Court of Appeal rejected that the wording of the clause demands such detail even if it is commercially desirable.[45] According to Gibson LJ, a notice stating clearly the sum deducted or to be deducted and that the deduction is for the whole or part of liquidated and ascertained damages would be sufficient. However, prudence demands that the approach of Judge Carr is preferred. Indeed, JCT 05, Clause 2.32.2 now more clearly demands the same level of detail.

---

103 (1990) 6 Const LJ 292.
104 (1993) 65 BLR 103. Upheld by Court of Appeal (1995) 77 BLR 22.

11.11.6 Estoppel

According to the doctrine of equitable or promissory estoppel as stated by the House of Lords in *Hughes* v. *Metropolitan Railway*,[105] if a party to a contract leads the other party 'to suppose that the strict rights arising under the contract will not be enforced, or will be kept in suspense, or held in abeyance, the person who otherwise might have enforced those rights will not be allowed to enforce them where it would be inequitable having regard to the dealings which have thus taken place between the parties'.

This doctrine was invoked in defence of a liquidated damages claim in *London Borough of Lewisham* v. *Shephard Hill Civil Engineering Ltd*:[106]

> The Engineer under a contract that incorporated the 6th edition of the Institution of Civil Engineers Conditions of Contract carried out final determination of extension of time some two years after actual completion of the Works. An arbitrator to whom a dispute under the contract was referred found that the Employer had made representations to the Contractor that the liquidated damages clause would not be enforced and that the Contractor had relied upon it in several ways. In his award he held that the Employer had thereby lost the right to recover liquidated damages through the operation of the estoppel doctrine. In the response to an application for permission to appeal against the award, counsel for the Employer did not deny the applicability of estoppel in principle. Instead it was contended that the issue of estoppel had not been put to the arbitrator. HHJ Seymour QC refused the permission in a judgment that did not question the application of estoppel in principle.

It has to be noted that, as the Architect would not ordinarily have authority to waive any right of the Employer, in order for statements or conduct to give rise to estoppel, they must be the Employer's.

11.12 Concurrent delays

From a causation perspective delays may be categorized under the following types.

Delays caused by the Contractor: These include delays caused by parties for whom the Contractor is responsible in law. This type is often referred to as 'culpable delay'. Under most contracts the Contractor is neither entitled to extension of time nor recovery of loss and/or expense on account of culpable delay. Delays caused by the Contractor's Persons would fall into this category.

Delays caused by neutral events: Neutral events are those for which neither the Employer nor the Contractor is responsible (e.g. exceptionally adverse weather conditions). Most contracts allow the Contractor more time to complete but with no corresponding entitlement to recover any loss and/or expense caused.

105 [1877] 2 App Cas 439.
106 (2001) WL 825511.

Delays caused by the Employer: These include delays caused by parties for whom the Employer is responsible in law. In most standard forms, the Contractor is entitled to extension of time and recovery of loss and/or expense caused by this type of delay. Delays caused by the Employer's Persons in the course of discharging their roles in relation to the Contract belong to this category.

Excusable delay: Delays for which the Contractor is entitled to extension of time are referred to as 'excusable delay' in the sense that, by being entitled to extension of time, the Contractor is excused liability for liquidated damages which would otherwise be payable. The label 'Relevant Event' is used in the JCT family of contracts for excusable delay. Delays for which the Contractor is entitled to recovery of loss and expense are said to be 'compensable'.

Concurrent delay: the term 'concurrent delays' is used to describe the situation where a number of events, any one of which would cause delay if it occurred alone, overlap (e.g. non-receipt of information from the Architect and adverse weather conditions). This situation often gives rise to disputes concerning the extent to which each event was responsible for the overall project delay. The main approaches commonly adopted to deal with concurrent delays for the purpose of extension of time include the:

- first-in-line approach;
- 'but for' approach;
- dominant cause approach;
- apportionment approach;
- *Malmaison* approach.

Under JCT 05 Clause 2.28, on receiving notice of delay, the Architect is to grant extension of time that 'he then estimates to be fair and reasonable'.[107] It was suggested in *City Inn Ltd* v. *Shepherd Construction Ltd*[108] that this formulation of the Architect's duty would allow him to estimate the time extension by applying any of the dominant, apportionment or *Malmaison* approaches provided the result is fair and reasonable in the circumstances of the particular case.

11.12.1 The first-in-line approach

This approach assumes that the first event is the cause of the whole delay. On the one hand, this means that if the event is a ground for extension of time the Contractor gets the extension for the whole delay even if his subsequent actions compounded it. On the other hand, if his own delay was compounded by causes for which the Employer was responsible, the Contractor would not be entitled to extension of time.[109] There is some support

107 The laxity of this obligation is in contrast to the more onerous duty under Clause 4.23 to ascertain (or instruct the Quantity Surveyor to do so) loss and/or expense on account of regular progress being disturbed by Relevant Matters.

108 [2008] BLR 269; [2007] CSOH 190 (hereafter *City Inn No. 3*).

109 As the Employer is treated as not having caused the delay, the prevention principle is not applied.

for this approach in *Royal Brompton Hospital NHS Trust* v. *Hammond (No. 6)*[110] in which HHJ Seymour QC said at paragraph 31:

> [I]t is, I think, necessary to be clear what one means by events operating concurrently. It does not mean, in my judgment, a situation in which, work already being delayed, let it be supposed, because the contractor has had difficulty in obtaining sufficient labour, an event occurs which is a Relevant Event and which, had the contractor not been delayed, would have caused delay, but which in fact, by reason of the existing delay, made no difference. In such a situation although there is a Relevant Event, the completion of the works is not likely to be delayed thereby beyond the Completion Date.

In *City Inn No. 3* the Outer House of the Scottish Court of Session expressed some difficulty in accepting the distinction drawn by HHJ Seymour QC. In the opinion of Lord Drummond Young, it should not make any difference to the Contractor's entitlement whether any of the stated overlapping events predates the other.

11.12.2 The 'but for' approach

The 'but for' test is the common contractor response to the application of the first-in-line argument against its culpable delay. According to this test the delay is the responsibility of the Contractor if it would not have occurred but for the occurrence of the event for which the Contractor is responsible. This test exculpates the Contractor whenever events for which the Contractor is not responsible compound delay already caused by the Contractor. Application of the test to the other competing events would also lead to the result that none of the competing events caused the delay, a conclusion that defies common sense. Furthermore, it was stated in several court decisions that it had no application to the extension of time clauses in the predecessors of this standard form.[111] For these reasons, adoption of the 'but for' test is unlikely to be a winning strategy with contracts in this form.

11.12.3 The dominant cause approach

The dominant cause approach attributes the entire delay to the dominant event. There is support for this approach in *Keating*[112] on the general issue of causation in contract. In *City Inn No. 3* the court accepted that where it can be shown that either a Relevant Event or a contractor's risk event was the dominant cause of delay such event may be treated as the cause of the delay.

However, John Marrin QC[113] doubts its applicability in the determination of contractors' entitlement to extension of time. There are two problems associated with it. First, it breaks down if the events are of equal causative potency. Second, the implications for recovery of direct loss and/or expense could be unfair to one of the parties. For example, where

110 (2000) 76 ConLR 131.

111 See, for example, *City Inn Ltd* v. *Shepherd Construction Ltd* [2008] 269; [2007] CSOH 190 and other cases cited in it.

112 Furst, S. and Ramsey, V., 2006, *Keating on Construction Contracts*, 8th Ed., Sweet & Maxwell, at para. 8-019.

113 John Marrin, 'Concurrent Delay' (2002) 18 Const LJ 436.

the dominant cause of delay is also a ground for recovery of loss and expense, the Contractor would be entitled to recovery in respect of the whole delay though some of the contributory causes of the delay may not be grounds for recovery. For these reasons, the use of the approach was disapproved of in *H. Fairweather & Co. Ltd* v. *London Borough of Wandsworth.*[114]

11.12.4 The apportionment approach

The apportionment method attempts to distribute the total delay to the various contributing causes. It has been doubted whether, in the absence of an express power given by the contract, it is correct in law to adopt this approach. However, it would appear that under JCT 05 the Architect must follow this approach. Under Clause 2.27.2 the Contractor is expected to provide the Architect with estimates of the delay caused by each event that causes delay even where it is not a Relevant Event. Correspondingly, under Clause 2.28.3, the Architect must state in any extension of time decision the extension of time he has given for each Relevant Event and the reduction in time for each Relevant Omission covered by the decision.

11.12.5 The *Malmaison* approach

The so-called *Malmaison* approach stems from the judgment of Dyson J (as he then was) in *Henry Boot Construction (UK) Ltd* v. *Malmaison Hotel (Manchester) Ltd.*[115] In that case the judge recorded at paragraph 13, without questioning, agreement between the parties including:

> [I]t is agreed that if there are two concurrent causes of delay, one of which is a relevant event, and the other is not, then the contractor is entitled to an extension of time for the period of delay caused by the relevant event notwithstanding the concurrent effect of the other event. Thus, to take a simple example, if no work is possible on site for a week not only because of exceptionally inclement weather (a relevant event), but also because the contractor has a shortage of labour (not a relevant event), and if the failure to work during that week is likely to delay the works beyond the completion date by one week, then if he considers it fair and reasonable to do so, the architect is required to grant an extension of time of one week. He cannot refuse to do so on the grounds that the delay would have occurred in any event by reason of the shortage of labour.

The difference between this and the apportionment approach is that the Relevant Event is treated as the sole cause of the delay (i.e. the causative contribution of any cause of delay for which the contractor bears the risk is ignored). According to *Keating*,[116] the rationale for this proposition is that the parties, in agreeing to extension of time for certain events, must have contemplated that those events might occur simultaneously with other events and that time should still be extended regardless of the impact of the other events. In *Steria Ltd* v. *Sigma Wireless Communications Ltd*[117] HHJ Stephen Davies,

114 (1988) 39 BLR 106.
115 (1999) 70 ConLR 32.
116 See para. 8-021.
117 [2008] BLR 79; *Waterfront Shipping Company Ltd* v. *Trafigura AG* [2007] EWHC 2482 (Comm) is another case in the growing line of cases supportive of enforcement of conditions precedent.

sitting as High Court judge, treated the failure of Dyson J in *Malmaison* to express any adverse comments on the parties' agreement as strong indication that he considered the position to be correctly stated by the agreement.

The SCL Protocol recommends adoption of this approach for the purposes of assessment of extension of time.[118] However, on the question of compensation for prolongation from concurrent delays, its recommendation is that the contractor should recover the extra costs caused by the delay for which the employer is responsible but only if it is possible to segregate those costs from those caused by the contractor's own delays.[119] These recommendations are, of course, not applicable unless the parties' contract includes an agreement to adopt them whenever there are concurrent delays.

11.13 Delays within culpable delay

There has been some debate regarding the right of architects to extend time where a contractor already in culpable delay is delayed further by an event which is expressly a ground for extension of time under the contract. Assuming that there is a right to extend time, there is the further question regarding the method of assessing the amount of extension. To illustrate these problems, consider the situation where, after the award of all extensions of time due, a final completion date is fixed for the end of the 100th week. It is now the 120th week and the works are not yet complete. Exceptionally adverse weather conditions are then encountered and the contractor is delayed by 4 weeks as a result. If this cause of delay is a ground for extension of time, it may be argued that as the conditions would not have been encountered if the contractor had finished on time, he should not be entitled to an extension. Now suppose that the cause of delay is not weather conditions but a variation issued during the period of culpable delay. Can the architect extend time? If he has no jurisdiction then, on principles already discussed, time for completion *may* become at large. The word 'may' is emphasized because it is arguable whether completion time can become at large when it has already expired. Besides, the case law from which the principle of time being at large has been developed concerned delays occurring prior to the contractual completion date.

There is the further question whether, assuming time can be extended, the extension is to be 'net' or 'gross'. In the example a net extension would result in the new completion date being the end of the 104th week whilst the 124th week would be the case with a gross extension. Thus, a gross extension has the effect of exculpating the Contractor for his own earlier delays. *Dicta* in *Amalgamated Building Contractors Company Ltd* v. *Waltham Holy Cross UDC*[120] suggests that there is jurisdiction to extend and that net extension would be the more appropriate method. In that litigation Denning LJ (as he then was) said *obiter*:

> Take a simple case where the contractors, near the end of the work, have overrun the contract time by six months without legitimate excuse. They cannot get an extension of time for that period. Now suppose that the works are still uncompleted and a strike occurs and lasts a

---

118 See Section 1.4 of the Protocol.
119 See Section 1.10 of the Protocol.
120 [1952] 2 All ER 452.

> month. The Contractors can get an extension of time for that month. The architect can clearly issue a certificate which will operate retrospectively. He extends the time by one month from the original completion date, and the extended time will obviously be a date which is past.

The Commercial Court considered these chestnuts in *McAlpine Humberoak* v. *McDermott International Inc. (No. 1)*.[121] Responding to the argument that variations during culpable delay had the effect of rendering completion time at large, Lloyd LJ said:

> If a contractor is already a year late through his culpable fault, it would be absurd that the employer should lose his claim for unliquidated damages just because, at the last moment, he orders an extra coat of paint.

These questions were again put to the court in *Balfour Beatty Building Ltd* v. *Chestermount Properties Ltd*,[122] a case which arose from a contract substantially in the JCT 80 form of contract with approximate quantities. When the Contractor failed to complete the Works by a new completion date fixed by the Architect (9 May 1989), he issued a Certificate of Non-Completion under Clause 24.1. By agreement between the Contractor and the Employer, the Architect then issued variation orders between 12 February and 12 July 1990. Taking these variations into account, the Architect finally fixed the completion date as 24 November 1989 (note that this date is before the issue of the variations). Actual completion was achieved on 25 February 1991. Upon the issue of a Certificate of Non-Completion in respect of the new completion date, the Employer sought to recover liquidated damages.

In the litigation that ensued, two questions were put before the court. Did the issue of a variation order within the Contractor's culpable delay have the effect of rendering completion time at large? On this question, the Contractor argued that on the proper interpretation of Clause 25 a variation was a Relevant Event only if it was issued before the contractual completion date. It was further claimed that, as the variations were issued after the contractual completion date, time for completion was at large with the result that the Employer was not entitled to liquidated damages. Colman J. upheld the arbitrator's rejection of the argument. The judge explained that from the general scheme of Clause 25, in the absence of clear words to the contrary, it has to be inferred that a variation after the contractual completion date is a Relevant Event.

The implication of the case law for JCT 05 is that the Architect would be authorized to extend time on a 'net' basis if the Contractor is delayed during a period of his culpable delay by an event for which the Employer is responsible.

11.14 Extension of time and professional negligence

There is no doubt that the Architect owes the Employer a duty of care to perform his duties in relation to extension of time with the skill and care to be expected of a reasonably competent architect.[123] The principles on the standard for assessing the performance of the Architect are those in *Bolam* v. *Friern Hospital Management Committee*.[124]

---

121 (1992) 58 BLR 1.
122 (1993) 62 BLR 1.
123 See *Sutcliffe* v. *Thackrah* [1974] AC 727.
124 [1957] 1 WLR 582 (hereafter *Bolam*).

An implication of the *Bolam* principles is that, where there is a body of professional opinion among architects supportive of the Architect's alleged act, the Architect would not be in breach of duty even if such opinion is wrong. This proposition was applied by HHJ Seymour QC in *Royal Brompton Hospital NHS Trust* v. *Hammond (No. 6)*.[125] In that case the Architect was sued by the Employer for negligence in granting extensions of time on several grounds including the Employer's admitted failure to provide access to part of the site. The Employer contended that either the Architect should not have granted any extension at all or he should have granted shorter extensions. The judge accepted evidence of a common practice of architects acting as contract administrators to treat failure by an employer to provide access to the site in due time as giving rise to an automatic entitlement of the Contractor to an extension of time. The judge concluded that the Architect was not guilty of professional negligence in granting extension (he did what his peers would have done) although there was evidence that the Employer's delay in giving access did not cause delay to the completion of the project.

It is suggested in first-instance decisions in not only England and Wales but also Scotland[126] that, although there are many scientific methods of varying degrees of rigour for analysing delay for the purposes of extension of time, the Architect need do no more than decide the extension on his general impression of the delay caused.[127] Several epithets are used in industry (e.g. the 'seat of the pants' or 'I feel in my water' approach) to describe this approach. As remarked by HHJ Seymour in *Royal Brompton No. 6* in relation to the JCT 80, acceptance of this proposition would mean that an employer who contends that an architect has been negligent in granting extension of time for the completion of work under JCT 05 would have a very high evidential mountain to climb over.

However, the architect would be found negligent if, in his assessment of extension of time, he makes a factual error that no reasonably competent architect would make. An example of such an error in *Royal Brompton No. 6* was in relation to extension of time granted by the Architect for a variation instruction requiring the Contractor to use a special type of membrane in floors. It was clear from minutes of site meetings and progress reports that the Contractor had completed the flooring operation weeks earlier than shown on the programme to which the Contractor had been working. The judge held that no reasonably competent architect would have granted extension of time on that ground if he had applied his mind to the issue of causation and that, therefore, the Architect had been negligent.

11.15 The SCL Protocol

In addition to the subject of concurrent delay already referred to, matters on which the Protocol contains recommendations as to good practice include the preparation and

125 [2000] EWHC 39(TCC), (2000) 76 ConLR 131, 2000 WL 1841725 (hereafter *Royal Brompton No. 6*); in *Royal Brompton Hospital NHS Trust* v. *Hammond (No. 7)* [2001] EWCA Civ 206, (2001) 76 ConLR 148 the Court of Appeal refused both parties leave to appeal against the decisions of HHJ Seymour QC.

126 See, for example, *City Inn Ltd* v. *Shepherd Construction Ltd* [2007] CSOH 190.

127 cf. *John Barker Construction Ltd* v. *London Portman Hotel Ltd* (1996) 83 BLR 31; 50 ConLR 43; (1996) 12 Const LJ 277 in which HHJ Toulson QC held that an architect operating extension of time provisions in substantially the same terms as under JCT 05 was required to do more than make an impressionistic assessment of delay caused.

submission of contractors' programmes, record keeping, and extension of time procedures. For the principles advocated in the Protocol to apply to a project on which the adopted form of contract is JCT 05, they must have been incorporated into the contract through appropriate provisions in the Contract Bills.[128]

11.15.1 Preparation and submission of contractor's programme

A properly prepared and updated programme is a very powerful tool for managing a project, particularly where a method statement is not only provided with it but is also cross-referenced to it. Such a programme alerts the project participants to the timetable for performance of their various obligations; it also allows analysis and prediction of the impact of any event likely to cause delay for appropriate control action. JCT 05, like many standard forms of construction contract, does not specify the form the programme should take. It is recommended in the Protocol that, in all but the simplest projects, it should be prepared as a CPM network using commercially available project planning software.[129] Information to be provided on it includes:

- all relevant activities relating to design, manufacturing and procurement;
- on-site construction activities;
- the critical path;
- resource requirements;
- all major items of information the contractor requires from the contract administrator.

The contractor should be required to update the programme monthly or even more frequently where the project is very complex. Electronic copies of the programme, the method statement, their updates, and reports explaining the modifications made in the updates are to be downloaded and archived for future reference.

According to the Protocol, it is good practice for the contract administrator to be under a contractual duty to review and approve the contractor's programme, method statements and their updates. Although the *ICE* (Institution of Civil Engineers) *Conditions of Contract*, 7th edition, requires the contractor's programme to be accepted by the engineer before it is implemented, there are many in the construction industry of the view that how a contractor intends to carry out the Works is a matter entirely for the contractor. Likewise it is thought inappropriate for the contract administrator to have to police this obligation, thereby accepting some responsibility for the contractor's working methods and arrangements. JCT 05 does not give the Architect any vetting powers over the Contractor's programme submitted under Clause 2.9. It does not, however, mean that the Architect cannot comment on the programme if he has any concerns.

11.15.2 Record keeping

Generally, records kept should describe the activities, labour, plant and sub-contractors on the site, deliveries of materials, instructions received, and delays experienced. Actual

128 Care should be exercised to avoid the problems concerning Clause 1.3 explained in Chapter 1, Section 1.5.5.
129 Most of the software packages available allow bar charts to be produced from the network.

progress should be captured regularly by recording the actual start and finish dates of activities shown on the programme as well as percentages of ongoing activities completed. Ideally, the details and how often the records are collected (daily, weekly or monthly) should be specified in the contract so that the contract administrator can insist on compliance by the contractor.

11.15.3 Extension of time procedures

The contractor should comply with any contractual requirement to serve notices of delay and supply information reasonably required by the contract administrator. To determine the likely impact of the employer's delay, the programme must be updated to the point just before the occurrence of the delaying event. The programme is then modified to take account of reasonable re-planning to minimize the impact of the delay. A sub-network of the works affected by the event is prepared and entered into the programme after it is agreed with the contract administrator. Time analysis of the adjusted programme should then provide a measure of the delay to completion arising from the event.

One of the issues on which JCT 05 conflicts with the Protocol concerns when the extension of time decision should be made. According to the latter, the most appropriate strategy is to deal with the likely impact of any delay as soon as possible after the occurrence of the event. In any event the contractor's entitlement to extension of time should be assessed within 1 month after the contractor's application for it. The Protocol goes on to recommend that the contract administrator should grant the contractor the forecast delay as extension of time and not wait to observe how the impact of the event unfolds.[130] The contract administrator may increase any extension of time but not reduce it even if it transpires that the actual delay is much less than the forecast.

JCT 05 does not require the Contractor to make an application for extension as such, although the information required under Clause 2.27 would amount to most of the content of such an application. JCT 05 allows the Architect 12 weeks to carry out the assessment. Also, contrary to the position under the Protocol, the Architect may revisit the extension of time decision as explained in Section 11.7.4 of this chapter. JCT 05 provisions are, therefore, more advantageous to the Employer than the guidelines in the Protocol.

130 See para. 3.26.

12

Claims for damages and contractual claims: an overview

(See also Chapter 13: Contractual money claims under JCT 05)

12.1 Introduction

The term 'claim' has no precise meaning and conjures up a variety of emotions in employers, their professional teams, and contractors. In some sectors of the construction industry 'claims' are an unwelcome concept. They are often associated with aggressive predatory commercial practice by contractors. Experience of claims has even led some building clients to remove contractors from tendering lists solely on the grounds that they had received claims from those contractors on earlier projects. Even contractors sometimes distance themselves from claims, and market themselves as being not 'claims conscious'. However, some view claims as no more than a request for a contractual right, including payment for a variation;[1] for others a claim is a last resort, often born out of frustration at not being able to recover what they think is due to them – a last stage before formal proceedings. For the purpose of this book a money 'claim' is any request by the Contractor for payment, other than an application under Clause 4.12. Such claims are often categorized as:

1. loss and/or expense arising from matters provided for expressly in the contract (principally Clause 4.23 where JCT 05 applies);
2. damages arising out of a breach of contract, or a breach of duty in tort (e.g. negligence);
3. claims in Restitution;[2]
4. other matters.

---

1 See General Conditions of Contract for Water Industry Plant Contracts Form G/90.
2 See Chapter 1, Section 1.8.2.2: Quasi-contract and restitution.

Category One is commonly called a 'contractual claim' or 'loss and expense claim'. Category Two is often described as a 'common law claim' or 'damages claim', or sometimes 'unliquidated damages' to distinguish from liquidated damages. These two categories are dealt with below in Section 12.2.

Category Three is dealt with briefly in Chapter 1, Section 1.8.2.2, and is outside the normal authority of the Architect.

Category Four comprises mainly 'moral' or 'sympathy' claims, in which the contractor feels hard done by, but for which there is no breach or other actionable wrongdoing by the employer. Thus, generally, there is no entitlement to reimbursement either under the contract or as damages. An example might be a claim for extra costs incurred by the contractor entirely at his own risk and choice, but which in hindsight benefits the employer. In such cases the employer, out of gratitude, might contribute towards the contractor's costs as a goodwill gesture. The architect has no authority to deal with these matters unless expressly empowered by the employer. Such payments are often described as ex gratia payments.[3] However there are some circumstances where payment may be an obligation, where there is overlap between Categories Three and Four. Restitutionary claims based on unjust enrichment (dealt with generally in Chapter 1) rely on work not being intended as a gift, and on being carried out at the recipient's request. There may be circumstances where work may be carried out in the absence of an express instruction from the architect, and that work in hindsight may be of value to the employer. If the employer had become aware of the additional work and of the intention of the contractor to claim payment, and with that knowledge allowed the work to continue, he may be obliged under the principle of unjust enrichment to return the service provided, or more realistically, to pay a fair value.[4] If, on the other hand, the only reason for retaining the work is that it would be too disruptive to revert to the original contract specification, there may be no enrichment for the employer, and thus no obligation to return, or pay for, the service.[5]

12.2 'Contractual' and 'common law' claims: similarities

A claim for loss or expense, made under Clause 4.23 of the Contract, is broadly the contractual equivalent of damages for breach of contract. In *Wraight Ltd* v. *P. H. & T. Holdings*[6] it was said:

> There are no grounds for giving to the words 'direct loss and/or damage caused to the Contractor' any other meaning than that which they have, for example, in a case of breach of contract.

In earlier editions of the standard form many of the 'matters' specified giving rise to an entitlement under the contract, are matters which would also describe breaches by the Employer. For example, Clause 26.2.1 of JCT 98 expressly provides for the possibility of recovery by the Contractor in the event that the Architect issues information late; in the absence of such provision, late supply of information would anyway be a breach of the

3 Ex gratia: as a favour: *Osborn's Concise Law Dictionary* 10th edn (Sweet & Maxwell, 2005).

4 Not the same as payment due. In the absence of an agreed payment timetable, such payment may not become due to be made until a value can be established at the end of the relationship.

5 See *Bloor JS Ltd* v. *Pavillion Developments Ltd* [2008] EWHC 724 (TCC), dealt in Chapter 1, Section 1.8.2.3.

6 (1968) 13 BLR 26.

contract, giving rise to damages at common law. Similarly, under Clause 26.2.6 of JCT 98, failure by the Employer to allow access would also be a breach of contract. On the other hand, suspension of the Works by the Employer would be a breach of JCT 98 (pre Amendment No. 4)[7] giving rise to damages, but not grounds for the Architect to consider. JCT 05 'mops up' these breaches in a single sub-clause. Clause 4.24.5 provides that any impediment, prevention or default by the Employer (i.e. breach of contract) is a matter that may be considered in calculating loss and expense. Thus the number of potential breaches falling outside the ambit of loss and expense are reduced, and former breaches are brought within the Architect's authority to be dealt with under the Contract.

Some matters which create entitlement under Clause 4.23 would not be breaches of contract. The most common example is probably the effect of Variations. Whilst a late Variation might be hindrance by the Employer (i.e. a breach of contract), the issue of other Variations is not in itself a breach, because the Contract allows it; but the effect constitutes express grounds for extra payment under Clause 4.24.1.

12.3 'Contractual' and 'common law' claims: differences

12.3.1 Architect to ascertain

Whilst both common law claims and contractual claims are made against the Employer, the procedure for dealing with common law claims is not dealt with in the Contract. One of the principal characteristics of contractual claims is the ability of the Architect to ascertain the amount due. His authority and duty is set out in the Contract (see Chapter 13, Section 13.2.2); under JCT 05 it does not extend to dealing with claims from the Contractor for the Employer's breach of that contract (other than breaches dealt with in Clause 4.24), unless the Employer expressly gives him power as an agent. This can create confusion when a Contractor submits to the Architect a claim which describes itself as both a request for loss or expense and/or a claim for damages. In those circumstances the Architect should deal with the request to the extent that the claim is a *bona fide* request under Clause 4.23, and he should ignore the common law element other than advising his Client of the contents. On the other hand, the Contractor cannot expect the Architect to deal with such matters, and should submit his common law claim direct to the Employer.

12.3.2 Interim payment

A great advantage to the Contractor of framing his claim as a loss or expense claim under the Contract, as opposed to making a common law claim, is the entitlement to payment of loss and/or expense in interim certificates.

12.3.3 Notices

In return for the entitlement to recover the equivalent of damages as a right under the Contract, the Contractor is obliged to follow an administrative procedure in the form of

7 JCT 98 (P With), Amendment No. 4, incorporated October 2002.

applications (notices) to the Architect (see Chapter 13, Section 13.2). The procedures apply only to contractual claims; they do not apply to common law claims. Indeed it is this point which normally prompts the Contractor to submit his claim in the alternative (i.e. as a contractual claim and as a common law claim); this is particularly likely when the Contractor realizes that he has failed to comply properly with the requirement for notices of his loss or expense claim and anticipates rejection by the Architect.

12.4 'Contractual' and 'common law' claims: as alternatives

Some contracts do not have the equivalent of the loss or expense provisions in JCT 05.[8] Under those contracts, the Contractor would need to pursue a common law claim if the Employer were to commit a breach of the contract.[9] Most construction industry standard contracts, however, contain some form of provision to compensate for the Employer's breaches. Under those contracts the obvious and intended remedy is the use of the compensation provisions in the contract; indeed that might be the only remedy where the contract limits the rights of the parties to the contractual remedies.[10] That is not the case with JCT 05 which, at Clause 4.26, states that the provisions of Clause 4.23 are without prejudice to any other rights and remedies the Contractor might possess. That means the Contractor is not obliged to use Clause 4.23, so it does not necessarily prevent a common law claim instead of the contractual claim; nor does it preclude a common law claim being brought as an alternative, or even in addition to the contractual claim. The position was explained in *London Borough of Merton* v. *Stanley Hugh Leach Ltd*[11] in relation to Clause 24 of JCT 63 (the forerunner of the modern Clause 4.23):

> But the Contractor is not bound to make an application under Clause 24(1). He may prefer to wait until completion of the work and join a claim for damages for breach of obligation… under the contract. Alternatively, he can make a claim under Clause 24(1) in order to obtain prompt reimbursement and later claim damages for breach of contract, bringing the amount under Clause 24(1) into account.

It is open to the Contractor to choose which route he wishes to pursue,[12] or he may pursue both, but clearly he will not be entitled to recover the same damages twice.

12.5 Common principles

12.5.1 Cause and effect and global claims

It is tempting for a Contractor, when preparing a claim, to give notice that he is incurring both loss and expense, then simply to send a lengthy monetary calculation. The apparent basis is that he has incurred lots of additional costs (or damages), so it follows that

8 See for example JCT Minor Works Building Contract.
9 In *How Engineering Services Ltd* v. *Lindner Ceilings Floor Partitions plc*, QBD 24 June 1999; 17-CLD-10-21, it was held that the absence of a particular cause of loss from a loss and expense clause did not prevent a damages claim for Employer's breach.
10 See Clause 44.4 of MF/1 (Rev 3) Conditions of Contract, published by the Institution of Electrical Engineers.
11 (1985) 32 BLR 51.
12 See *Merton* v. *Leach* on this point.

the Employer must be liable for them. The claimant's effort is then put into calculating compensation, rather than establishing entitlement.

A common flaw in claims is the absence of a causal link between the matters giving rise to the claim and the compensation claimed. Similarly there is often no link identified between the matters described and a breach of contract or a matter described in the contract as giving rise to entitlement. The principle which such claims ignore is the maxim 'He who alleges must prove'. The paucity of causal links is understandable. Proving that a particular cost results solely from one particular cause can be difficult. It can be particularly difficult if causes are closely related, such as the progressive late issue of a number of drawings. Each drawing may be a cause in its own right, and the task of showing which portion of each head of claim results from each delayed drawing is daunting to say the least. It may not even be possible, but the Contractor is required to try. Indeed the main task of the Contractor in a contractual claim is to provide sufficient information to enable the Architect to ascertain the amount due.

Likewise the Contractor may have difficulty when two or more causes of loss occur concurrently. The problem is often associated with delay, particularly when the contract expressly provides for reimbursement in respect of one concurrent cause but not in respect of another. One example would be delay caused by late information at the same time that the contractor delays himself by (say) relaying foundations that were installed negligently (i.e. there are two breaches, one by each party). Another example would be delay caused by exceptionally adverse weather running concurrently with delay caused by variations. It may, in those circumstances, be that the start of one cause preceded the other, so if a variation pushed work back into bad weather, loss may be recoverable under principles used in 'winter working' claims.[13] There are a number of alternative ways in which such delays can overlap,[14] but the overriding principle is that damages are losses which would not have been incurred in any event.

Whilst Contractors are often criticized for failing to demonstrate causation, Architects are often equally criticized for showing dogged resistance to any claim, and for demanding more and more information in the hope that it will not be forthcoming. This may seem a cynical view, and obviously it does not apply to all contractors and architects; but it is a common aspect of dealings on the battleground known as the 'global claim'.

Global claims gained notoriety amongst architects, and popularity amongst contractors, following the reporting of the decision in *Crosby* v. *Portland UDC*,[15] in which a 'rolled up claim' was said to be acceptable in some limited circumstances. Contractors seemed to take that to be a general rejection of the requirement to establish cause. After a number of cases looking into the extent of the need to identify each causal link the position was neatly summed up by the court in the case of *Mid Glamorgan County Council* v. *J. Devonald Williams and Ptnrs*:[16]

1. A proper cause of action has to be pleaded.
2. Where specific events are relied upon as giving rise to a claim for moneys under the contract then any preconditions which are made applicable to such claims by the

---

13 In *Ellis-Don Ltd* v. *The Parking Authority of Toronto* (1985) 28 BLR 98, an employer's delay preventing summer work being carried until the winter entitled the contractor to be paid extra for the winter working.

14 See Chapter 11, Section 11.12; See also Egglestone, B., *Liquidated Damages and Extension of Time in Construction Contracts*, 2nd edn, paras 14.2, 14.5.

15 (1967) 5 BLR 121.

16 [1992] 29 ConLR 129; (1992) CILL 722; 17 September 1991.

terms of the relevant contract will have to be satisfied, and satisfied in respect of each of the causative events relied upon.

3. When it comes to quantum, whether time based or not, and whether claimed under the contract or by way of damages, then a proper nexus should be pleaded which relates each event relied upon to the money claimed.
4. Where, however, a claim is made for extra costs incurred through delay as a result of various events whose consequences have a complex interaction that renders specific relation between event and time/money consequence impossible or impracticable, it is permissible to maintain a composite claim.

The problem was reviewed again, albeit in the Scottish Court,[17] in *John Doyle Construction* v. *Laing Management (Scotland) Ltd*[18] in which the Court set out the basic position. It was stressed that all events contributing to the loss must be culpable events (i.e. it would be unjustifiable to claim for loss caused by events that were not the defender's fault), but that causation should be treated as a matter of common sense; apportionment could then be made if the evidence contained sufficient detail. The claim would not be struck out simply because it was made on a global basis. This decision was confirmed on appeal by the Inner House of the Court of Session.[19] The Inner House added pragmatically that if one of the events for which the Employer was responsible could be identified as the dominant cause of a particular item of loss, it would be enough to establish liability, and even if a dominant cause could not be identified it may still be possible to make some apportionment. However, it was also said that if the loss is related to delay, and the contractor is himself partly responsible for that delay, he should not recover any loss for that period.[20] Whilst the Court did not put it in such simple terms, this latter point is simply a reminder that loss is something that would not have been incurred in any event.

Common sense, and the burden of proof in civil cases (i.e. proof on the balance of probabilities), was the theme in the Court of Appeal case of *Drake* v. *Harbour*.[21] A fire occurred in an unoccupied bungalow in the early hours of the morning, during re-wiring. The fire was found to have started in the roof space where temporary lighting had been installed, but there was no evidence as to which of several causes was to blame. Lord Justice Toulson said:[22]

> ... where a claimant proves both that a defendant was negligent and that loss ensued which was of a kind likely to have resulted from such negligence, this will ordinarily be enough to enable a court to infer that it was probably so caused, even if the claimant is unable to prove positively the precise mechanism.

The pragmatic approach to deciding liability was also used by the House of Lords in *Allbright & Wilson UK* v. *Biachem*.[23] The correct chemicals ordered from two suppliers

---

17 Decisions in the Scottish courts are not binding in England, Wales, and Northern Ireland. However, like decisions from other common law jurisdictions (e.g. Australia), they are of great persuasive authority.

18 [2002] BLR 393 (CS), [2002] 85 ConLR 98.

19 11 June 2004.

20 In *London Underground Ltd* v. *City Link Telecommunications Ltd* [2007] EWHC 1749 (TCC), at para 41 it was confirmed that the proper approach to global claims was that set out in the *Doyle* case.

21 [2008] EWCA Civ 25, 31 January 2008.

22 At para. 28.

23 [2002] UKHL 37.

were provided with the wrong delivery notes by a haulier company, and were subsequently delivered to the wrong location at the purchaser's premises, where one was mixed by the purchaser with its existing stock, causing an explosion. The purchaser commenced action against both suppliers, although both suppliers had supplied what had been ordered from them. Lord Hoffman viewed the problem as a decision between breach of a contract for chemicals with the appropriate papers, and breach of a contract for the correct papers with the appropriate chemicals. It was held that the supplier of the chemical that caused the explosion was to blame, even though both suppliers had to some extent breached their contract. It was a matter of common sense that only one supplier's breach had resulted in the explosion.

Whilst the apparent swing to pragmatism may suggest a growing laxity, the courts are still prepared to require proof of a connection between losses claimed and the alleged cause. In *Petromec Inc.* v. *Petroleo Brasileiro SA*[24] the contractor, upgrading a vessel, claimed the difference between its actual and anticipated costs in dealing with instructions to vary the work. The Court of Appeal decided that the claim was of a global nature, and that it was not sufficient for the claimant to simply allege that its costs were reasonable, and then leave it to the defendant to prove otherwise. There had to be a connection proved between the cost and the instruction that caused it.[25]

Any principles emerging from these cases are established in the context of litigation, and apply to any claim made by the Contractor, whether as a contractual claim or as a common law claim and would certainly apply to a claim in arbitration or adjudication. However, it must be remembered that contractual claims are intended to be paid in interim certificates. The Architect cannot expect to receive on a month-to-month basis, the same amount of information, records, and analysis as would be presented for formal proceedings. In short, the Architect is required to do what he can with the information available to him at the time, and the Contractor is required, at the Architect's request, to provide only sufficient to enable the Architect to form an opinion. The practicality of the situation during the course of a project was recognized by the court in *London Borough of Merton* v. *Stanley Hugh Leach Ltd*:[26]

> If application is made for reimbursement of direct loss and/or expense attributable to more than one head of claim and at the time when the loss or expense comes to be ascertained it is impracticable to disentangle … the part directly attributable to each head of claim, then … the architect must ascertain the global loss directly attributable to the two causes … To this extent the law supplements the contractual machinery which no longer works in the way it was intended to work, so as to ensure that the contractor is not unfairly deprived of the benefit which the parties clearly intend he should have.

If the information provided by the Contractor is insufficient to form any opinion, the Architect may request more, stating what he needs. This extra information might need to be only brief, since the Architect 'is not a stranger to the work'.[27] Indeed, in *Merton* v. *Leach* the judge continued by observing that in some cases, 'the briefest and most uninformative

---

24 [2007] EWCA Civ 1371; [2007] EWHC 1589 (Comm).

25 But see also commentary regarding *Amec Process and Energy Ltd* v. *Stork Engineer & Contractors BV (No. 3)* in Chapter 13, Section 13.4.5.

26 (1985) 32 BLR 51.

27 Per Mr Justice Vinelott in *Merton* v. *Leach* (1985) 32 BLR 51.

notification of a claim would suffice … for instance where the architect was well aware of the contractor's plans'.

It seems the Architect cannot just sit back and wait for the Contractor to meet the highest standards of notification and analysis; he must act pragmatically. Nevertheless, the prudent Contractor will ask what the Architect requires, and keep proper and adequate records; and still give as much information as he can in the time available, to meet the timetable of interim payment certificates.

12.5.2 Measure of recovery

The basic object of damages, whether recovered either as loss or expense under the Contract, or as damages for breach of the Contract, is to put the claimant back 'so far as money can do it … in the same situation … as if the contract had been performed'.[28] This is subject to any limitation agreed between the parties, such as express agreement that liability may be 'capped', or limited to 'costs' – thus excluding 'losses'.[29]

Neither loss or expense, nor common law damages, are intended to enhance profits; nor are they to be treated like added turnover, unless the contract treats them as such. Some engineering contracts provide for profit to be added to costs in the manner of Variations,[30] but JCT 05 does not.

However, not all losses or costs are recoverable; the law allows only those amounts that would not be incurred in any event, that are not too remote or extravagant, and that do not unnecessarily improve the claimant's position.

1. Losses or costs that would be incurred in any event

If a cost would have been incurred in any event, there is no loss suffered. The impact of circumstances, such as intervening factors between a breach and the loss, that would have occurred, even though not anticipated, must be taken into account. For example, in the shipping case of *Golden Strait Corporation* v. *Nippon Yusen Kubishka Kaisha*,[31] a long term charter contract was terminated early by the charterer, giving rise to loss of income over the remaining four years. The House of Lords held that war starting in Iraq between the time of the breach of contract and the calculation of damages, had to be taken into account. The loss that would have been incurred in any event, because of the war, could not be recovered.

2. Remoteness

The damages that may be recovered must not be too remote from the cause; they must be a foreseeable result. The rules are stated in the case of *Hadley* v. *Baxendale*:[32]

> The damages which the other party should receive in respect of such breach should be such as may fairly and reasonably be considered as either arising naturally, i.e. according

28 *Robinson* v. *Harman* (1848) 1 Ex. 850.
29 E.g. ICE Conditions, 7th edn, which limits liability to 'costs' defined in Clause 1(5).
30 See Conditions of Contract MF/1 (rev 3) published by the Institution of Electrical Engineers, Clause 41.2.
31 [2007] UKHL 12.
32 Per Baron Alderson (1854) 9 Ex 341.

> to the usual course of things, from such breach of contract itself, or such as may reasonably be supposed to have been in the contemplation of both parties at the time they made the contract as the probable result of the breach ... If special circumstances ... were communicated ... to the defendants ... the damages ... would be the amount ... which would ordinarily follow from a breach of contract under the special circumstances so known and communicated.

The two concepts emerging from this statement are known as 'general damages' (the first rule), and 'special damages' (the second rule). An example of the first rule in the context of a construction project would be site establishment costs, which everyone would reasonably expect to occur naturally as a result of site delay. An example of the second rule would be a claim for loss of profit suffered as a result of losing a particularly lucrative contract elsewhere.[33] A common example of putting special circumstances into contemplation at the time the contract is formed can be seen in the practice of advising a sub-contractor of the liquidated damages applying on the main contract. The forewarning enables a Contractor to claim from a sub-contractor (in principle) the damages paid to the Employer; it prevents the sub-contractor from alleging that the damages are 'special' and not in contemplation at the time of entering the sub-contract.

The rules in *Hadley* v. *Baxendale* have been refined on a number of occasions to deal with difficult cases, where the contemplation of likelihood of the resulting damages is questioned.[34] Each instance of claim has to be considered on its own merits, but the general test for whether a particular loss was in the parties' contemplation, and could therefore be claimed as damages, was considered by the House of Lords in *Transfield Shipping Inc. of Panama* v. *Mercator Shipping Inc.*,[35] in which Lord Hoffmann said:

> 12. It seems to me logical to found liability for damages upon the intention of the parties (objectively ascertained) because all contractual liability is voluntarily undertaken. It must be in principle wrong to hold someone liable for risks for which the people entering into such a contract in their particular market, would not reasonably be considered to have undertaken.
>
> 15. In other words, one must first decide whether the loss for which compensation is sought is of a 'kind' or 'type' for which the contract-breaker ought fairly to be taken to have accepted responsibility.
>
> 18. That seems to me in accordance with the careful way in which Robert Goff J stated the principle in *Satef-Huttenes Albertus SpA* v. *Paloma Tercera Shipping Co SA* (*The Pegase*) [1981] Lloyd's Rep 175, 183, where the emphasis is upon what a reasonable person would have considered to be the extent of his responsibility:
>
>> The test appears to be: have the facts in question come to the defendant's knowledge in such circumstances that a reasonable person in the shoes of the defendant would, if he had considered the matter at the time of making the contract, have contemplated that, in the event of a breach by him, such facts were to be taken into account when considering his responsibility for loss suffered by the plaintiff as a result of such breach.

---

33 See *Victoria Laundry (Windsor) Ltd* v. *Newman Industries Ltd* [1949] 2 KB 528.
34 See *Victoria Laundry* case at n. 19; and *Czarnikow* v. *Koufos* [1969] 1 AC 350.
35 *The Achileas* [2008] UKHL 48, 9 July 2008.

3. Extravagance (the 'duty' to mitigate)

Strictly there is no duty to mitigate losses, unless the contract expressly requires, for a contractual claim. The injured party can incur whatever he wishes; the so-called 'duty' simply means the wrong-doer cannot be expected to compensate the injured party for his extravagance. In mitigating his loss, the claimant must take reasonable steps to minimize his loss, and not take unreasonable steps that increase it.

Under a construction contract, the cost of alternative methods of repair or reinstatement, the timing, and the views of experts,[36] are all factors that may be taken into account. However, it is for the injured party to decide by what method he needs to repair the injury, and in *Mirant Asia-Pacific Construction (Hong Kong) Ltd* v. *Ove Arup & Partners International Ltd*[37] it was held that the courts should be slow to accept objection by the wrong-doer to the method adopted.

Delay in having remedial work done does not necessarily indicate a failure to mitigate, even though the delay may lead to greater loss. In *Saunders* v. *Terry Williams, Peter Guidotti and Kim Guidotti*[38] it was held that because it was the builder's responsibility to put right his damage to a party wall, the claimant was justified in waiting for eight years, and loss of opportunity to charge rent during that period was recoverable as damages.

Whilst mitigation suggests spending less, it is permissible to spend money in order to save money, and recover the expenditure as mitigation. This can be so, even if the resulting losses actually turn out to be greater than the losses that the expenditure was intended to reduce, provided the decision was a reasonable decision to make at the time it was made.[39] In building terms, this approach can sometimes be seen in the use of acceleration measures to reduce time overrun and resultant prolongation costs (see Chapter 13, Section 13.4.8).

4. Betterment

The general purpose of damages for a breach is to put the injured party back in the position he would have been in, but he should not be enriched by an award of damages (see above). There are times when the correction of a breach either requires something different from the original intention, or provides an opportunity to change or improve the claimant's position.

An example of something different being required would include replacement of a defective boiler with a new model when the original was no longer available.[40] The change is necessary, and there is no reasonable alternative but to install the new model. The general principle that the claimant should not be enriched may not apply where the 'enrichment' cannot reasonably be avoided. In *Harbutt's 'Plasticine' Ltd* v. *Wayne Tank & Pump Co Ltd*,[41] the full cost of rebuilding a factory was awarded without any allowance for the factory being new and more valuable; the factory owner had no alternative but to rebuild. In *C J Elvin Building Services Ltd* v. *Noble*[42] the replacement of a door of higher

36 *The Board of the Hospitals for Sick Children and Anr* v. *McLaughlin & Harvey plc and Ors.* (1987) 19 ConLR 25.

37 [2007] EWHC 918 (TCC).

38 25 April 2002, CA [2002] 18 BLISS 15; also *Building Magazine* 22 November 2002, p. 55.

39 See *Melachrino* v. *Nicholl & Knight* [1920] 1 KB 693.

40 *Halsbury's Laws of England*, 4th edn Reissue, Butterworths 1998, Vol 12(1), para 983, discusses 'betterment' in terms of new for old, and only arising when there is an increase in benefit value.

41 [1970] 1 QB 447.

42 [2003] EWHC 837 (TCC), 3 April 2003.

quality was said to be capable of being betterment, but it was unavoidable. Similarly, in *Pegler Ltd* v. *Wang (UK) Ltd*[43] in which it was clear to the purchaser that the supplier was not going to fulfil the contract provisions, it was held that the mere fact that an alternative computer system contained additional features, and would last longer, did not require an allowance for betterment. However, in extreme cases betterment may be considered; in *Bacon* v. *Cooper (Metals) Ltd*[44] it was said that if absurdity resulted from ignoring betterment, then justice may require it to be taken into account.

An example of opportunity to improve would be replacement of a defective gas-fired boiler with solar panels. Here the claimant is taking advantage of necessary work to improve his overall position, by changing the original specification to solar heating. He is not replacing new for old. This, arguably, is better categorized as a failure to mitigate.[45]

5. Actual loss – whether costs actually incurred

There are situations in which the injured party may have incurred no actual cost or loss, but nevertheless claims damages in the expectation that he may. For example:

Where remedial work, or other cure, has not yet been carried out: The normal method of assessing damages when, for example, construction work is defective, is by calculating the cost of the cure (i.e. the cost of making it good). There are many reasons why the remedial work may not have been carried out; the injured party may be short of funds, or it may be that a developing defect has not yet fully materialized and it is too early.[46] Failure to have corrected the defects does not prevent recovery, provided it can be shown that there is good reason for the failure, and it is intended to be done. This was put succinctly by Lord Clyde in *Alfred McAlpine Construction Ltd* v. *Panatown Ltd*:[47]

> … this approach can be seen as identifying a loss upon the innocent party who requires to instruct the remedial work. That loss is, or may be measured by, the cost of repair. The essential for this formulation appears to be that the repair work is to be, or at least is likely to be, carried out.

In *Forsyth* v. *Ruxley Electronics & Construction Ltd*,[48] the House of Lords considered the damages due to the building owner when a swimming pool was built nine inches shallower than the contract specification. It was held that in principle it was irrelevant what the plaintiff intended to do with his damages, although in that case it was also held that it would be unreasonable to rebuild the pool, and the owner was awarded only a nominal amount for loss of amenity (see Section 12.6.2: Failure to complete the Works).

---

43 [2000] BLR 218.

44 [1982] 1 All ER 397.

45 This point was considered in *Skandia Property (UK) Ltd and Vala Properties BV* v. *Thames Water Utilities*, ORB; 30 July 1997, in which the obligation to give credit under 'betterment' principles was differentiated from the principles in 'mitigation' to act reasonably.

46 The injured party may be obliged to start proceedings before completing remedial work if a limitation period is about to expire.

47 [2000] BLR 331 at 344.

48 (1995) 73 BLR 1, HL.

Where remedial work carried out, or other liability incurred, but not yet paid for: A common argument for resisting payment of losses, or very often a contractor's loss and expense claim, is that there may be no proof of actual payment. It may be that the cost has not yet been notified to the injured party, or that an invoice has not been paid due to lack of funds; but that is irrelevant, at least in principle. This is not so much a question of liability, but a matter of proof, and a tribunal may be reluctant to award damages where there is a dearth of evidence.[49] Nevertheless, when there is a liability to pay a third party which has not yet been discharged, the late payment which would, when paid, represent a cost incurred does not remove entitlement to recover that cost as damages. It was said by the Court of Appeal in *Chaplin* v. *Hicks*:[50]

> The fact that damages cannot be assessed with certainty does not relieve the wrongdoer of the necessity of paying damages.

It is for the injured party to demonstrate that, on the balance of probabilities a cost has been or will be incurred.

6. Diminution

Sometimes, although only rarely, the proper measure of value of damages is the effect on the value of property, rather than the cost of rectifying the breach. Whilst the normal measure in the case of construction defects is the cost of repair, or cure, diminution may apply when there is no intention of carrying out remedial work, and particularly where the works are commissioned with the intention of selling on completion.[51]

12.6 Employer's claims

Claims made by the Employer against the Contractor are governed by the same principles as those applying to Contractor's claims (i.e. under the rules of the contract where a claim is made under the contract, and under general damages principles where the claim is made in respect of a breach).

There are several situations in which the Employer has entitlement to claim against the Contractor under the contract; but each may also give rise to alternative or additional rights in common law. The most common are:

- delay in completion of the Works;
- failure to complete the Works;
- failure to comply with an Architect's instruction;
- the correction of defects.

---

49 In *Tate & Lyle Food & Distribution Ltd* v. *GLC* [1983] 2 AC 509, the court confirmed that Tate & Lyle were entitled to recover the cost of management time spent in problem-solving, but they were awarded nothing because they could not prove the cost to them. This was a strict approach which has since been tempered in later cases – see Chapter 13, Section 13.4.3.

50 [1911] 2 KB 786 at 792.

51 See also Section 12.6.4 of this chapter – dealing with 'Limitations'.

12.6.1 Delay in completion of the Works

The Employer's remedy under the Contract lies in the deduction of liquidated damages (see Chapter 11). The remedy is exhaustive.[52] When entering his liquidated damages sum in the Contract Particulars, the Employer is expected to have included all the costs and losses he might expect to suffer as a result of delay, so he cannot then deduct liquidated damages and add a claim for other delay related matters. However, he can still maintain a claim for damages for breach if the liquidated damages clause fails.[53] This can occur when the clause is challenged due to a legal defect in construction[54] or if the liquidated damages amount is successfully challenged as being a penalty (see Chapter 11 for principles of liquidated damages). Whether the Employer is entitled to recover more in general damages than he would have received under the failed liquidated damages clause is uncertain. It has been held that the figure agreed as liquidated damages does not always prevent a party from recovering a larger sum;[55] but any such claim would be subject to the rules on remoteness, so the level of liquidated damages could be taken into account when considering contemplation under the second rule in *Hadley* v. *Baxendale*.[56]

12.6.2 Failure to complete the Works

The Employer's contractual remedy under JCT 05 in the event of failure to complete is the right to termination under Clause 8.4, to have the work completed by others, and to adjust the contract sum accordingly (see Chapter 16). Nevertheless, Clause 8.3.1 provides that such rights are without prejudice to any other rights or remedies that the Employer may possess. Such rights would encompass similar rights in common law and entitlement to damages, including the additional cost of completing the work plus any other damage suffered. In some circumstances the proper measure of damages might be depreciation of property value,[57] or an amount for loss of amenity.[58] Such damages would be subject to the general rules on remoteness, mitigation, and betterment.

12.6.3 Failure to comply with Architect's Instructions

Under Clause 3.11 the Employer may employ others to carry out an Architect's instruction which the Contractor has refused to do after receiving a notice to comply. All costs incurred in such employment are deducted from the Contract Sum.

---

52 See *Temloc Ltd* v. *Errill Properties Ltd* (1987) 39 BLR 30, 12 ConLR 109; and *Biffa Waste Services Ltd* v. *Maschinenfabrik Ernst Hese GmbH* [2008] EWHC 6 (TCC).

53 Per Lord Justice Croom-Johnson in *Temloc* v. *Errill*.

54 E.g. *Peak Construction Ltd* v. *McKinney Foundations Ltd* (1970) 1 BLR 111, in which it was held the Employer lost his rights to liquidated damages when an Employer's breach put time at large.

55 Per Scrutton LJ in *Cellulose Acetate Silk Co. Ltd* v. *Widnes Foundry (1925) Ltd* [1933] AC 20.

56 See Section 12.5.2 of this chapter. See also *Temloc* v. *Errill* in which liquidated damages were inserted as 'NIL' and the Employer was unable to recover any general damages.

57 See also Section 12.6.4 of this chapter – dealing with 'Limitations'.

58 In *Forsyth* v. *Ruxley* (1995), 73 BLR 1, HL, it was held there was no diminution in value of the property caused by a swimming pool being nine inches shallower than the contract requirement, but it was an option for consideration. £2500 was awarded for loss of pleasurable amenity by the trial judge, which was criticized by Lord Mustill in the House of Lords as being a large amount, although the quantum had not been challenged.

In the absence of such a clause it is a matter of debate whether the Employer would have a similar right in common law. The difficulty lies in the principle that the Contractor has until the Completion Date in which to complete his obligations. Failure to do any particular work by a particular time, if not dealt with in the Contract, arguably is not a breach until the Contractor has failed at the Completion Date, since the Contractor will inevitably claim the right to correct his own errors up until Practical Completion.[59]

12.6.4 Correction of defects: limitations and Employer's additional losses

The Contract, at Clause 2.38, provides that any defects resulting from work or materials not in accordance with the contract (i.e. a breach) appearing during the Rectification Period shall be corrected by the Contractor at no cost to the Employer. Many contractors look upon this arrangement as no more than an obligation, but its important function is to entitle the Contractor to put right his breach at his own cost, instead of the cost which he would have to bear if the Employer engaged others to do the work.[60]

Limitations: If the Employer fails to allow the Contractor to correct his work, and instead has the work done by others, the Employer will be in breach; he cannot then recover all his losses. In *Pearce & High Ltd* v. *John P. Baxter and Mrs A. Baxter*[61] the Court of Appeal held that a notice required by a contract[62] to be given within a stipulated time, advising the Contractor of the need to correct defects, was a condition precedent to the Contractor's obligation. However, it was also stated[63] that the Employer does not lose his right to recover entirely. This confirmed the Court's decision in *William Tompkinson & Sons Ltd* v. *The Parochial Church Council of St Michael-in-the-Hamlet*,[64] that if the sum claimed from the Contractor is greater than the amount which the Contractor would have incurred, had he been allowed to correct his own work, the claim will be limited to that lesser amount.

One difficulty for the Employer in these situations is the extent to which he can put right work to provide what he expected from the contract. If the Contractor provides something less than the contract specification, it would seem only natural that the Employer should be able to make good the Contractor's breach of contract, and recover the reasonable cost incurred as damages. The case of *Forsyth* v. *Ruxley Electronics & Construction Ltd*[65] (see also Section 12.5.2 of this chapter dealing with costs actually incurred) provides a good example. A swimming pool was built nine inches shallower than the contract

59 See paper by Ellis Baker and Anthony Lavers, 'Temporary Disconformity', Society of Construction Law, June 2005.

60 In *Maersk Oil UK Ltd* v. *Dresser-Rand (UK) Ltd* [2007] EWHC 752 (TCC), it was held that the employer was not in breach for having rectification work done by others where there was no express obligation to allow the contractor to return.

61 15 February 1999, 1999 CILL 1488; referring to JCT Agreement for Minor Building Works but equally applicable to JCT 98.

62 JCT Minor Works Building Contract.

63 Per Evans LJ (obiter), at 104.

64 (1990) Const LJ 319.

65 (1995) 73 BLR 1, HL.

specification. Since the pool could not be deepened, the building owner claimed for the cost of replacing the whole pool to the depth required. The House of Lords held that where reinstatement of defective work is necessary, such reinstatement must be a reasonable thing to do; in short it would be wrong to spend large sums of money to obtain a new pool which gave no additional benefit over that which is replaced, simply to achieve the strict specification (it appears that if the pool had been required for diving competitions, it may then have been reasonable to reinstate). The *Ruxley* case was considered in the Scottish case of *McLaren Murdoch & Hamilton Ltd* v. *The Abercromby Motor Group*.[66] A complete new heating system to a car showroom was replaced, although the designer maintained that the inadequate underfloor heating could have been augmented. After considering all relevant factors including running costs, it was held that replacement was not manifestly out of proportion to the benefit, and the cost was allowed as damages. Clearly, building owners need to give careful consideration to their actions; whilst bringing defective work up to the standard promised is an entitlement in principle, extravagance is not condoned, and may not be recoverable.

In some circumstances the Employer may claim diminution of property value, although he may also receive no more than nominal damages for loss of amenity. In *Earl Freeman* v. *Mohammed Niroomand*,[67] the Court of Appeal upheld the decision of the trial judge who had said:

> In some circumstances it might be the diminution in value of property. In other circumstances it might be the cost of remedial work. In this case neither is ... [applicable] ... since there is no evidence of any diminution ... and ... [no evidence of intention to correct the work]. Therefore the only remaining touchstone is the evidence of [the builder] that it would have cost him an additional £130, if at the time of the original construction, he had complied with the drawing.

Whether diminution or remedial costs apply will depend on individual circumstances, although the trend seems to be towards remedial costs, when remedial work is a reasonable course. If property is intended for sale or leasing, then any diminution in value may be relevant. If remedial work would be unreasonable and provide no benefit, and if diminution is not present, then damages may be limited to a nominal amount for loss of enjoyment or amenity.

Employer's additional losses: The Contractor's obligation to correct defects is not an exhaustive remedy. It may be that, in allowing the Contractor necessary access to do the work, the Employer suffers other losses. For example, correcting a leaking pipe at high level over a production area may cause disruption to the Employer's production schedule; major work in offices, or inhabited premises, may even require temporary accommodation. Additional costs such as these may be recovered by the Employer from the Contractor, in a damages claim. In *H.W. Nevill (Sunblest) Ltd* v. *William Press & Son Ltd*[68] the employer claimed both the cost of delays caused to a second contractor on site, when the first contractor returned to repair defective drains, and also the cost of late opening. It was argued by the contractor that the obligation to correct his own work was his only

---

66 Outer House, Court of Session 22 November 2002; (2003) SCLR 323, 100 ConLR 63.

67 CA: 8 May 1996: CCRTF 95/0660/C: BLISS 4 November 1996 p. IB 43/2.

68 (1981) 20 BLR 78. See also *Raflatac* v. *Eade & Ors* [1999] BLR 261 (Comm) where it was held that employees' costs in clearing up where recoverable.

obligation. The court held that the remedies under the contract were not exclusive, that the defective work was a breach of contract, and that the employer was entitled to recover his losses from the contractor in addition to having the work corrected. The Employer's right to recover such addition losses will be a claim for damages and subject to the rules on remoteness, mitigation, and betterment.

12.7 Excessive claims: criminal liability

Whilst it is the duty of the Architect to ascertain loss and expense under the Contract, Contractors often produce applications in formal claim documents. Such claims are usually presented after the Contractor has become dissatisfied with the Architect's ascertainment. Consequently claims may be viewed by Contractors both as a last chance to 'get everything in' and also to create a negotiating position. Occasionally the latter can lead to the claim being inflated to a level higher than the Contractor really thinks it is worth. It was said by one distinguished judge in 1704:[69] 'Shall we indict one man for making a fool of another?' He had just decided it was not stealing when the defendant had pretended to be authorized to collect money on behalf of someone else. But contractors should be very wary; there are many Employers who do not take an eighteenth-century view of things, and who are not aware of what is seen by some as the unspoken ground rules adopted by building trades. To them, inflating the claim could be a criminal act, and the modern courts agree. Under the Theft Acts 1968 and 1978 it is an offence to obtain money or avoid payment by deception or false accounting. If a grossly excessive price is quoted and charged, where there is a relationship of trust between customer and tradesman, an offence may be committed under the Theft Act 1968 s. 15(1).[70] In *R* v. *Williams (Roy)*,[71] a builder dishonestly over-billed for work done. He was convicted of obtaining by deception and sentenced to over five years' imprisonment.

Likewise, it is an offence to fabricate or change records relied on to support a claim for loss or a Variation; this would extend to daywork sheets.

However, the principle works both ways. Employers and their consultants, including the Architect and Quantity Surveyor, should also avoid overstated counterclaims,[72] made with the sole intention of cancelling or reducing payment to a Contractor.

---

69 Holt CJ in *Jones* (1704), *R* v. *Jones* (1795) 91 ER 330; Salk 379.

70 E.g. *R* v. *Silverman*; 31 March 1987, The Times 3 April 1987, 33; (CA).

71 [2001] Cr.App.R 23.

72 See paper by Neill Stansbury and Catherine Stansbury, 'Unethical Behaviour and Criminal Acts', March 2005, Society of Construction Law, for extensive list of typical examples of possible criminal acts by employers, certifiers and contractors in the execution and administration of construction contracts. See also *Building*, 2 February 1997 p.33; Tony Bingham discusses a case in which the director of a building company received a prison sentence after his company avoided paying a sub-contractor's account by fabricating an excessive counterclaim for alleged defects.

13

Contractual money claims under JCT 05

(For general principles, see also Chapter 12: Claims for damages and contractual claims: an overview)

13.1 Introduction

Losses incurred by the Contractor resulting from deferred possession of the site, or disruption to the progress of the Works caused by specified matters, may be recovered by the Contractor from the Employer under Clause 4.23. It is for the Architect to ascertain, or to arrange for ascertainment of the amount due, and to certify payment to the Contractor in interim certificates. The successful operation of the Clause 4.23 provisions relies on a procedure triggered by the Contractor, in which he first makes application to the Architect, together with details if he wishes, followed by information which the Architect may request. Recovery under JCT 05 is described as 'direct loss and/or expense',[1] that has been held to be the equivalent of damages for breach of contract[2] (i.e. to put the injured party back – so far as money can do it – into the position he would have been in had the contract not been broken). The general principles governing the level of damages recoverable apply equally to 'loss and/or expense' (see Section 13.2.2 (3) of this chapter).

An important provision, often overlooked by Contractors and Architects, is the express reservation of the Contractor's rights under Clause 4.26. The provisions for reimbursement of loss and expense in the Contract do not affect other rights and remedies which the Contractor may possess. In practice this usually means that the Contractor may choose to claim reimbursement under the Contract, or to claim damages, or both. However, he will not get his reimbursement twice (see Chapter 12, Section 12.4 dealing with contractual and common law claims as alternatives).

---

1 Some construction contracts limit recovery (e.g. ICE *Conditions*), 7th edition, which defines 'Cost'.

2 See Chapter 12 for comparison with claims for damages, and general principles applying to both loss or expense, and damages.

13.2 Procedures

13.2.1 Application by Contractor (notices)

Clause 4.23 states that if the Contractor makes an application to the Architect that he has incurred, or is likely to incur, direct loss and/or expense for which he would not be reimbursed under other provisions, either as a result of deferred possession or as a result of specified matters, the Architect must then take action (see Section 13.2.2 below). The Contractor 'may make written application', but the Contract does not oblige the Contractor to make application; it simply states the effect if he does; it is his choice. However, if the Contractor does not make an application, the Architect has no obligation to take any action (see Condition Precedent below, and Section 13.2.2 of this chapter), with one exception. Under the provisions for dealing with antiquities, Clause 3.24 requires the Contractor to do no more than inform the Architect, or the Clerk of Works, that antiquities have been found on site, and for the Architect to ascertain loss and expense if he thinks loss has been incurred; it is for the Architect to detect the loss, then reimburse it.

An application by the Contractor (often referred to as a 'notice') must conform to several conditions:

1. The application must be in writing and made to the Architect by the Contractor.[3] An oral application would not be sufficient, even if made at a site progress meeting and recorded in the minutes.[4] Neither would a Contractor's valuation application for an interim certificate under Clause 4.12 constitute good notice, even if it identified the relevant matters, since such a valuation is submitted to the Quantity Surveyor.
2. Under Clause 4.23.1, the application should be made as soon as it becomes (or should reasonably become) apparent to the Contractor that regular progress of the work has been or is likely to be affected. It should be noted here that it is the recognition of disrupted progress that triggers an application, not the recognition of the monetary effect. Identifying the likely effect requires a certain amount of crystal ball gazing, and perhaps the Contractor should be excused if he takes a conservative approach by giving notice at the slightest prospect. One of the Contractor's difficulties lies in spotting the point at which a possible effect on progress becomes a likely effect. Another difficulty lies in who it is in the Contractor's organization that should do the spotting. The uncertainty of this provision almost encourages contractors to send off standard notices every time any instruction is issued, in the hope that it comes good retrospectively. Unfortunately, that approach is likely to overwhelm the Architect with frivolous notices, and the result can be a desperate plea from the Architect to stop sending all notices.[5] The temptation to make such a plea should be resisted by the Architect, if only because it could open a 'waiver' defence[6] for

---

3 *Steria Ltd* v. *Sigma Wireless Communications Ltd* [2008] BLR 79 (TCC).

4 In the Scottish case of *John Haley Ltd* v. *Dumfries and Galloway Regional Council* (1988) 39 GWD 1599, it was held that minutes were not good notice of delay. See also the *Steria* v. *Sigma* case.

5 This effect was seen following the Court of Appeal decision in *Minter* v. *WHTSO* (1980) 13 BLR 1, when contractors started sending what became known colloquially as 'Minter Notices' (i.e. a notice every month on all of their JCT jobs to protect their position regarding financing charges).

6 A position where one party alleges that by its actions the other party had clearly represented it did not wish to rely on certain contractual rights, and the alleging party had relied on that representation. See *Hughes* v. *Metropolitan Railway Co.* (1877) 2 App Cas 439 at 448; see also *Tameside MBC* v. *Barlow Securities Group Services Ltd* [2001] BLR 113 in which the Court of Appeal rejected a waiver defence on the ground that the claimed representation was not suficiently clear and unequivocal.

the Contractor in the resulting absence of valid notices, and put the Architect in breach of his contract with his Client.

3. The Contractor is not obliged to provide any information with his application, but must do so on the request of the Architect (Clause 4.23.2). The level of information is that which would reasonably enable the Architect to form an opinion. This seems to require objective consideration by the Contractor, and dispute can arise over what is reasonable. The prudent Contractor will ask the Architect in his application what information is required. Unfortunately, it is common for Architects to forget that the Contractor provides information on request. The presentation of claim documents often results from stagnation of a Contractor's application, with the Contractor and the Architect each awaiting action from the other.
4. The Contractor may provide his details or quantification of loss and expense if he wishes; but there is no obligation until requested by the Architect under Clause 4.23.3 to provide such details as are necessary for the Architect, or the Quantity Surveyor, to make his calculations. It is not clear whether it is the duty of the Architect or the Quantity Surveyor to make the request for details of loss or expense; it is simply the obligation of the Contractor to provide the information 'on request'. The prudent Contractor will ask the Architect in his application what information is required. As with information about disruption or delay causing loss, it is common for Architects to forget that the Contractor provides information about his loss on request, again leading to stagnation as the parties wait for each other.
5. The Contractor's application under Clause 4.23 must be in respect of 'material' effect on progress. This would exclude disruption of a trivial nature, but a series of individual disruptions, each trivial in nature, could amount to a significant disruption and be termed 'material'. The extent of disruption required to become 'material' is not specified, but Contractors are only likely to make application when a disruption or delay is great enough to come to their attention.
6. The effect on progress identified in the application must be caused either by deferred possession or by one or more of the matters listed in Clause 4.24. The words 'affected by any of the Relevant Matters' seems to allow the Contractor to roll up causative matters, at least in his application, in conflict with the general rules on causation (see Chapter 12, Section 12.5.1). Any matters outside those parameters are not things which the Architect is obliged, or indeed entitled, to consider under this clause, although the introduction by the JCT of Amendment 4 in JCT 98, repeated in JCT 05 as Clause 4.24.5 provides a very wide choice, and covers any matter of a preventative nature that affects progress, for which the Employer would be culpable.

Condition Precedent

The term 'condition precedent' simply means that certain rights are conditional upon a specified event; for example the giving of a notice in specified terms. In *Bremer Handelsgesellschaft mbH* v. *Vanden Avenne-Izagem PVBA*,[7] the House of Lords said that a provision for a notice was unlikely to be a condition precedent unless it was expressly stated to be so, or the effect of failing to comply was clearly stated.

The importance of the notice is often overlooked by contractors. It is sometimes difficult to supply information in the form required, or even to recognize the cause of a

7 [1978] 2 LLoyd's Rep 109.

loss until long after the event, particularly when there are interacting problems, but the Contractor should at least give what warning he can. The danger of failing to give notice can be fatal to entitlement.

In *City Inn Ltd* v. *Shepherd Construction Ltd*[8] the Scottish court held that where the contract expressly provided a machinery for warning, giving the Architect an opportunity to avoid or limit delay or additional cost, the absence of a notice deprived the Contractor of his entitlement. The Court decided that the Contractor was not in breach of the contract by failing to give notice, but that it was the Contractor's choice whether or not he wished to avail himself of the protection offered by the clause. In the *City Inn* case, the clause specifically barred the contractor from proceeding with the relevant work until he had given notice, so it is not surprising that the provision was construed as a condition precedent. However, the earlier case of *Minter* v. *WHTSO* (see Sections 13.2.2(3) and 13.4.11 below) also provided a salutary lesson when the absence of required notices regarding ongoing financing costs was held to prevent the Contractor from recovering those costs. In JCT 05 the requirements concerning notice for loss and expense do not include reasons why notice must be given, although it would be natural to conclude that notice enables the Architect to consider ways of reducing or avoiding problems by positive management, including reversing an instruction if that is the cause. It is likely that notice of loss and expense will be viewed as a condition precedent to the Architect ascertaining entitlement under the contract, and the Contractor ignores the provision at his risk. In the absence of notice, and the resulting inability of the Architect to act, contractors can then only fall back on their common law rights, and claim damages in lieu of loss and expense. The significant differences between loss and expense and damages, are considered in Chapter 12.3.

13.2.2 Architect to ascertain

The role of the Architect is set out in Clauses 4.23:

> If the Contractor makes such application … if, and as soon as, the Architect is of the opinion that the regular progress has been or is likely to be materially affected as stated in the application … or that direct loss and/or expense … (is incurred) … due to such deferment … the Architect shall from time to time thereafter ascertain, or instruct the Quantity Surveyor to ascertain, the amount of … (loss incurred) …

and in Clause 4.25:

> Any amounts from time to time ascertained … shall be added to the Contract Sum.

1. 'If the Contractor makes such application': There is no common law obligation on the Architect to consider loss or expense in the absence of an application from the Contractor. Clause 4.23 states the procedure to be followed if the Contractor starts the process by making an application. It has been held that the Architect would be negligent in his duty to his Employer if he certified loss and expense in the absence of a written application.[9]

---

8 (2002) SLT 885, 2003 WL 212363394 (2 Div), Inner House of Court of Session.

9 *Turner Page Music Ltd* v. *Torres Design Associates Ltd,* Judgment 12 March 1997, ORB 0237; 1997 CILL 1263.

2. 'If, and as soon as the Architect is of the opinion': There is no absolute obligation on the Architect to ascertain the amount due; the obligation is to consider the matter fairly and to form an opinion, if the Contractor starts the process by making an application. This issue was considered in a series of engineering cases[10] concerning the duty of an Engineer to certify interim amounts. In *Kingston-upon-Thames* v. *Amec Civil Engineering Ltd*[11] it was said:

> The use of the word 'opinion'... implies that there may be a degree of latitude... It may not be practicable to produce an exact valuation in 28 days; and the ... valuation has to be made on the basis of the contractor's statement ... On matters of opinion ... there may well be room for difference of opinion ... it is manifestly implicit that in arriving at his opinion he must correctly apply the provisions of the contract.

The Architect may actually form an opinion that nothing is due; if he does then he may have discharged his duty under Clause 4.23, provided he has correctly applied the provisions of the Contract. However, if he fails to apply his mind properly or at all, he will put the Employer in breach, and the Contractor will be entitled to damages.[12] Such a claim would be in common law and not within the Architect's normal authority.

3. 'Direct loss and/or expense': Clause 4.23 removes any doubt about the nature of the compensation which may be claimed. The use of the word 'direct' prevents reimbursement of consequential loss. Some construction contracts provide only for reimbursement of 'costs' which may be defined in the contract;[13] JCT 05 allows for both costs and losses, and to avoid pedantic argument, expressly allows either or both to be ascertained. Neither of the terms 'loss' or 'expense' are defined in the contract, but the Court of Appeal in *F.G. Minter* v. *W.H.T.S.O.*[14] held that 'direct loss and/or expense' is the same as damages arising naturally in the ordinary course of things as described in the first rule in *Hadley* v. *Baxendale*[15] (see also Chapter 12, Sections 12.2 and 12.5.2). Similarly, in *Saintline* v. *Richardson Westgarth & Co.*[16] it was said: 'Direct damage is that which flows directly from the breach without intervening cause and independently of special circumstances, while indirect damage does not so flow'. In *British Sugar plc* v. *NEI Power Projects Ltd*,[17] the Court of Appeal decided 'consequential loss' was loss coming under the second rule in *Hadley* v. *Baxendale* (i.e. special damage) (see Chapter 12, Section 12.5.2 dealing with remoteness). In *The Simkins Partnership* v. *Reeves Lund and Co. Ltd*[18] actual knowledge of a security risk led to the conclusion that loss from a breach of security was the sort of loss that came within the first limb of *Hadley* v. *Baxendale*.

---

10 Relating to ICE *Conditions of Contract*, 5th edition.
11 (1993) 35 ConLR 39.
12 See *Croudace Ltd* v. *London Borough of Lambeth* (1985) 6 ConLR 70, in which it was said 'it necessarily follows that Croudace must have suffered some damage as a result of there being no one to ascertain the amount of their claim'.
13 See ICE *Conditions*, 7th Edition, Clause 14(8) dealing with delay caused by late information; see also definition of 'cost' in Clause 1(5).
14 (1980) 13 BLR 1.
15 (1854) 9 Ex 341.
16 13 [1940] 2 KB 99.
17 (1998) 87 BLR 42.
18 (2003) EWHC 1946, QBD.

The task for the Architect is to satisfy himself that the heads of claim under consideration are caused by the matters alleged (i.e. that they arise naturally without intervening cause). That would exclude the type of loss that would not have been in the parties' contemplation when they entered into the contract.

4. 'Regular progress or… materially affected as stated in the application or that… loss and/or expense… (is due) to such deferment': The Architect does not have any discretion to consider matters which are not referred to in the application. Indeed he would probably be in breach of his contract with his client if he were to do so, unless the client had given express permission.

When forming an opinion, the Architect is required to establish that the loss or expense is due to the matters set out in the application. In order to do this he needs to apply the normal rules of causation (see Chapter 12, Section 12.5.1).[19]

5. 'For which (the Contractor) would not be reimbursed by a payment under any other provision': The Contractor is only entitled to apply for additional amounts. This provision makes Clause 4.23 a final mopping-up clause. The main effect should be to remind the Architect that the Contractor may be receiving payment towards his site establishment costs and overheads through Variations (whether they be based on rates in the Contract Bills or dayworks), insurance claims, or as amounts in lieu of loss or expense in an accepted Schedule 2 Quotation.

6. 'The Architect shall from time to time… ascertain, or instruct the Quantity Surveyor to ascertain, the amount of (the loss)': It is for the Architect to calculate the amount due; he is required to ascertain as far as he can with the information provided, to find out for certain.[20] Alternatively the Architect is empowered to instruct the Quantity Surveyor to perform the duty. However, the Quantity Surveyor's authority in carrying out this duty is limited. The Architect does not have authority to pass on the job of forming an opinion whether or not disruption occurred; that duty is retained by the Arcitect. The authority of the Quantity Surveyor is simply to calculate the loss.[21]

The duty to calculate the amount of reimbursement due is a continuing obligation. Having decided that loss or expense has been or is likely to be incurred the Architect must 'from time to time' update his calculations. There is no need for the Contractor to repeat his notices provided the cause of continuing loss remains the same, although he will be expected to provide updated information about his loss, but only at the request of the Architect (Clause 4.23.3).

13.2.3 Relevance of extensions of time

The Contractor's entitlement to recover his loss or expense under Clause 4.23 is not directly linked to extensions of time under Clause 2.28. Both clauses independently set

19 See summary in *Mid Glamorgan County Council* v. *J Devonald Williams & Ptnrs* (1992) CILL 722 and the *Doyle* case, cited in Chapter 12, Section 12.5.1.
20 *Alfred McAlpine Homes Northern Ltd* v. *Property and Land Contractors* (1996) 76 BLR 59.
21 See also JCT 05 Clauses 4.23.2 and 4.23.3 dealing with provision of information.

out events and matters giving rise to entitlement. An extension of time does not automatically bring with it the right to extra payment. The fundamental difference is that extensions are related to a delayed completion date, whereas loss or expense is related to disrupted, or prolonged, progress. In addition some grounds for extensions of time (e.g. 'neutral events' such as exceptionally adverse weather) do not appear as relevant matters for the purpose of Clause 4.23.

Under Clause 2.28, the Architect is obliged to notify the Contractor of any extensions of time granted in respect of each of the Relevant Events applicable. Knowledge of which events the Architect has used to adjust the Completion Date may point the Contractor towards losses for which the Employer may be liable, but the Employer's liability will still depend on a causal link between the matters cited in the application and the money damage (if any) suffered by the Contractor. Conversely, knowledge of events causing delay to the Completion Date may distract from the fact that a delay, discarded as having no effect on the Completion Date, may still have caused loss.

Delay to a single activity of work may cause disruption or prolongation of that activity and those associated with it, albeit the delay may be absorbed in the overall programme. In those circumstances there would be an entitlement to recover the loss and expense resulting from the discrete prolongation and disruption. A simple example would be where the Architect delayed some remote hard landscaping, requiring the Contractor to keep his setting out engineer on site longer than otherwise necessary, and for his sub-contractors to carry out work in several visits rather than the anticipated single visit.

13.3 Matters giving rise to entitlement

13.3.1 Deferment of giving possession of the site

Clause 4.23 provides that the Contractor may apply (see Section 13.2.1 of this chapter) to the Architect if possession of the site is deferred under Clause 2.5. That clause allows the Employer to delay the start of work for up to 6 weeks or such lesser period as is entered in the Contract Particulars. However, it is important for the Architect to remember that Clause 2.5 is optional, triggered by deleting the relevant option in the Contract Particulars. If the deferred possession option is not operative, or if possession is deferred longer than the operative period (i.e. the deferment is not within the parameters of Clause 2.5), the Contractor is still entitled to loss or expense under Clauses 4.23 and 4.24.5, provided it is caused by an act or omission by the Employer or those for whom he is responsible. Clause 4.24.5 has the effect of making preventative breaches of contract by the Employer grounds for reimbursement under the contract, and thus to be dealt with by the Architect.

However, on a very narrow interpretation, Clause 4.24.5 does not appear to cover causes which do not result from acts or omissions by the Employer or his team. For example, if the site became occupied by squatters just before possession by the Contractor was due, it may not have resulted from a culpable act or omission by the Employer, and would not therefore seem to fall within the Clause 4.24.5 grounds. However, that is not to say the Contractor has no entitlement at all, but his claim may be for common law damages arising out of the Employer's failure to give possession as promised.[22] In those circumstances the Architect has no authority to deal with the matter unless expressly authorized by the Employer.[23]

13.3.2 List of Relevant Matters affecting regular progress

Clause 4.24 identifies the matters which may affect the regular progress of the Works. The following is a brief summary with notes where necessary:

Clause 4.24.1: Variations arising from Architect's instructions, including 'deemed variations'[24] (other than an accepted Schedule 2 Quotation).

The Architect must constantly keep in mind that loss or expense may be reimbursed elsewhere, and in particular through the Schedule 2 Quotation procedure. In many cases where a Schedule 2 Quotation is accepted it will be difficult for the Architect later to separate the individual heads of claim relating to individual causes; in those circumstances he may need to resort to a simple abatement of amounts agreed under Schedule 2 against Clause 4.23 entitlement. Unfortunately, the complexity of this exercise may encourage the Architect to refrain from agreeing amounts in lieu of loss and/or expense in Schedule 2 Quotations.

Variations in this context includes any matters which the Contract requires to be treated as Variations. A lacuna exists regarding the finding of antiquities, dealt with in Clauses 3.22 to 3.24. Instructions regarding antiquities are not expressly Variation instructions, so any loss and expense resulting from finding antiquities does not fall under Clause 4.23, but is covered by the curtailed terms of Clause 3.24, under which there are no 'notice' provisions concerning loss.

Clause 4.24.2.1: Instructions regarding postponement under Clause 3.15, and instructions for expenditure of Provisional Sums.

With regard to adjustment of Provisional Sums the Contractor is entitled to reimbursement only in respect of loss relating to undefined sums. In the case of Provisional Sums for defined work, the Contractor is deemed to have made provision in his price.

Clause 4.24.2.2: Instructions for opening up work which is found to be in accordance with the Contract, and unless the cost is provided for elsewhere in the contract.

Clause 4.24.2.3: Instructions relating to discrepancies in or divergence between documents.

Clause 4.24.3: Suspension by the Contractor resulting from the Employer's failure to pay the full amount due by the Final Date for Payment, provided the suspension was not frivolous or vexatious.

---

22 In *Rapid Building Group Ltd* v. *Ealing Family Housing Association Ltd* (1985) 1 ConLR 1, the Court of Appeal held that despite the Employer obtaining an Eviction Order, the Employer's failure 'for whatever reasons' to remove squatters by the date for possession amounted to a breach by the Employer of the term that possession would be given to the Contractor.

23 See Chapter 12, Section 12.3.

24 See Chapter 6, Section 6.3.2.

The right to suspend was introduced into JCT contracts to comply with the Construction Act.[25] The entitlement to loss and/or expense as a result of such suspension is a matter also incorporated by the JCT but is not a requirement under the Act. Suspension may be valid under the Act even though the sums involved may be minor, but in order to maintain entitlement to compensation the Contractor must ensure his suspension is a deserving response to the payment default. What may be considered frivolous will depend on the circumstances. Whilst a Contractor would be entitled under the Act (and therefore under the Contract) to suspend for underpayment of ten pounds, it would be considered frivolous to incur thousands of pounds in extra costs in suspending a multimillion-pound project for such an amount. In those circumstances he would be entitled under Contract to an extension of time, but not to compensation. The Contractor's claim would then be in common law, although he might have difficulty showing the expenditure was reasonable.[26] What might constitute vexatious suspension is unclear, particularly if the sum outstanding is considerable, and the Architect must not lose sight of the fact that the Contractor has a statutory right to suspend if money is not paid when it should be.

Clause 4.24.4: Approximate Quantities in the Bills which are not a reasonably accurate forecast of the work required.

The nature of Approximate Quantities is that the work cannot be accurately determined, so the Contractor might expect some adjustment; but that is not to say it should be at his risk. However, there is no guidance as to what reasonable accuracy might mean, and it will be for the Contractor to show how the change in quantity affected his work.

Clause 4.24.5: Acts or omissions by the Employer and his team hindering the Contractor, except to the extent that the Contractor or any of the Contractor's Persons contribute. The Employer and his team include the Architect, Quantity Surveyor and others engaged by the Employer.

This sub-clause, at least in spirit, was introduced by Amendment 4 to JCT 98 in January 2002. The effect is to entitle the Contractor to recover loss and expense for what otherwise may only have been recoverable as damages. The result is to bring damages claims into the scope of the Architect's ascertainment. This does not prevent damages from being claimed as an alternative, but it avoids the need to claim damages in cases where there may have been a breach without express grounds to claim loss and expense. Typical examples would be loss resulting from the Employer's failure to maintain the engagement of necessary CDM duty holders or failure to provide necessary information timeously.

It should be noted here that one frequent cause of delay on site (i.e. delay by Statutory Undertakers) is treated as a 'neutral' event, the relative costs of delay being borne by the two Parties. Clause 4.24.5 would catch delays by Statutory Undertakers only when caused by failures on the part of the Employer or Contractor, such as failure to place an order in time, or failure to allow the undertaker access. Statutory Undertakers in this context refers only to those undertakings when they are carrying out statutory obligations. If the Statutory Undertaker carries out work other than a statutory obligation, liability for any

25 Housing Grants Construction and Regeneration Act 1996, Part II, s. 112.

26 See Chapter 12, Section 12.5.2 dealing with mitigation.

relevant delay or disruption lies with either the Employer or the Contractor, depending on who commissioned that work.[27]

The wide wording of Clause 4.24.5 encompasses many of the matters which, in previous editions of the standard form, were spelt out separately. They include:

1. Failure by the Architect to comply with an Information Release Schedule, or failure to provide information under Clause 2.12.2 at a time necessary to enable the Contractor to complete by the Completion Date, having regard to progress if the Works are behind programme. Unlike early editions of this form (prior to JCT 98), JCT 05 no longer makes requests in writing for information a condition of entitlement to loss or expense. However, a prudent Contractor will probably still provide a list of information required, so that he can at a later date establish the effect on his progress in support of his Clause 4.23 application.
2. Employer's own work or materials supply.
3. Failure to give ingress or egress. There can be situations where the site is accessed over property in the Employer's possession and control. This may commonly arise where the work is an extension or refurbishment to an existing facility, on a site where the Employer's operations are continuing. It makes no difference whether access provisions are described in the Contract or not; if the Employer impedes the Contractor, then Clause 4.24.5 will apply. However, if the Contract Documents provide for the Contractor to give notice requiring access, then he must give such notice to the Architect to maintain his entitlement.
4. Matters arising out of the Employer's obligation to ensure that duty-holders appointed by him (other than the Contractor) under the CDM Regulations perform their duties. CDM duty holders are not specifically included in the definition of Employer's Persons, but fall into the class described as 'all persons employed, engaged or authorized by the Employer'. The Employer expressly undertakes in Clause 3.25.1 to ensure that the CDM Co-ordinator carries out all his duties.

13.4 Heads of claim

13.4.1 Introduction

Whilst it is for the Architect to ascertain loss and/or expense, the Contractor will often feel moved to prepare a calculation to guide him. Thus the notion of the Architect (or the Quantity Surveyor) ascertaining the amount due tends to be a fiction, and it is far more likely in practice that the Architect will check the Contractor's figures.

Claims from Contractors are often split into delay (or prolongation) heads, and disruption (or uneconomical working) heads. Typical heads include:

- Site establishment costs, and other project-related overheads;
- Head office overheads;
- Visiting head office staff;

27 *In Henry Boot Construction Ltd* v. *Central Lancashire New Town Development Corporation* (1980) 15 BLR 1, it was held that the Employer was liable for the costs of delays caused to the Contractor by the Statutory Undertaker where the undertaker was carrying out work for the Employer other than its statutory obligation.

- Uneconomical working;
- Uneconomical procurement;
- Loss of profit;
- Acceleration;
- Third party settlements;
- Inflation;
- Financing other heads of claim;
- Financing retentions;
- Interest;
- Cost of producing claim;
- VAT on damages.

13.4.2 Site establishment costs, and other-project related overheads

Sometimes misleadingly called preliminaries (from association with site establishment allowances in the Contract Bills), this head includes such items as hutting, electricity, standing plant, small tools, site supervision, and non-productive labour including cleaning operatives and those in attendance on domestic sub-contractors.

Recovery of preliminaries, where related directly to a Variation, are expressly required under Clause 5.6.3.3 to be valued with the Variation under the applicable rules (see Chapter 6).

Loss and expense claims will sometimes be made (and for convenience, paid) on the basis of Bill allowances, but, with the exception of preliminaries items valued under Clause 5.6.3.3, that is not what the Contractor is entitled to recover. For a damages claim (or loss and/or expense) the entitlement is the extra cost incurred.

Hutting and other plant: When the Contractor does not own the hutting and plant, his entitlement will normally be based on presentation of hire invoices, including charges from a sister company.

A difficulty arises when the Contractor owns his own huts and plant. It has been the habit of many Architects and Quantity Surveyors, when faced with a claim for contractor-owned items, to pay a reasonable hire rate (i.e. the commercial rate which the Contractor would have paid if he had hired the equipment). This course is practical, particularly for inclusion in interim certificates, but it is not what the Architect is authorized to certify. The position was clarified in *Alfred McAlpine Homes North Ltd* v. *Property and Land Contractors Ltd*,[28] in which the Court considered an arbitrator's award based on reasonable hire charges:

> Ascertainment on the basis of hire charges might not have been questioned if there had been a finding that (the contractor) would have hired out this plant but there is no such finding … Only if there had been such a finding could the … award have been justified as representing … the valuation of lost opportunity… The question of law implicit in this part of the appeal [is] … that in ascertaining direct loss or expense under Clause 26 of the JCT conditions[29]

28 (1996) 76 BLR 59.
29 Clause 26 of JCT 80 is the forerunner and equivalent of Clause 4.23 of JCT 05.

> in respect of plant owned by the contractor the actual loss or expense incurred by the contractor must be ascertained and not any hypothetical loss or expense that might have been incurred whether by way of assumed or typical hire charges or otherwise.

The costs which the Contractor may claim are depreciation, maintenance, and additional fuel (if any). Different companies write off plant in different ways, but there seems no reason to calculate the depreciation in any way other than that normally used by the Contractor.[30]

The Contractor is entitled only to those costs actually incurred. Thus the cost associated with a time-related claim are not normally those incurred during the period of extended time; they are those incurred from time to time, at the points where the delay occurred. The point is emphasized by the Society of Construction Law in its Delay and Disruption Protocol.[31]

Once it is established that compensation is due, the evaluation of the sum due is made by reference to the period when the effect of the Employer's risk event is felt, not by reference to the extended period at the end of the project.

A convenient method for calculating the relevant costs is to identify 'time-slices' of delay caused by the Employer. The costs incurred during each 'time-slice' then form the basis of the Contractor's total recovery. This system is sufficiently flexible for different breaches to be identified against individual time-slices, providing valuable assistance in overcoming the difficulties of establishing causation. The Society of Construction Law suggests[32] that the 'time-slice' (i.e. time impact analysis)[33] method is their preferred technique to resolve complex disputes related to delay and compensation. In *Balfour Beatty Construction Ltd* v. *The Mayor and Burgesses of the London Borough of Lambeth*[34] the Court referred to the order of preference in the protocol as a reasonable hierarchy, although in that case it was not possible to make an assessment on the basis of time impact analysis.

Supervisory staff and non-productive labour: The principles are the same as for hutting and other plant. Hired staff could be claimed against invoices, but employed staff must be claimed on the basis of cost. Cost in this instance, taken from the Contractor's accounts and wages records, will include wages, benefits such as health insurance, and any statutory payments by the company in respect of employment.

Other project-related overheads: In some instances, parts of general overheads can be identified directly with a particular project. Typical examples would include special premiums for a professional indemnity insurance where taken out for one contract, or for a bond, or the hire of off-site storage facilities. The general principles described for other heads in this section apply equally here. Any increase in value affecting premiums would normally be deemed to be covered by the overhead and profit element in the valuation of Variations, but a prolonged construction period may increase the premiums paid. In those circumstances the Contractor would be entitled to reimbursement as a prolongation cost, based on actual additional cost incurred.

---

30 In *McAlpine* v. *Property and Land* case (1996) 76 BLR 59 referred to in Section 13.4.2 of this chapter above, the Court allowed the Contractor to rely on its normal manner of trading with only one company.

31 October 2002, *Delay and Disruption Protocol*, The Society of Construction Law, para. 1.11.1.

32 October 2002, *Delay and Disruption Protocol*, The Society of Construction Law, para. 3.2.11.

33 'Appendix A, Definitions and Glossary' – '*time impact analysis: Method of delay analysis where the impacts of particular delays are mapped out at the point in time at which they occur, allowing the discrete effect of individual events to be determined*'.

34 [2002] BLR 288.

13.4.3 Head office overheads

Head office overheads relate either to the cost associated with running the Contractor's business, or the contribution required by the Contractor from each of its contracts towards such cost. The former should be considered as expense, and the latter as loss. On JCT contracts the difference is immaterial,[35] since the entitlement is to loss and/or expense. Items falling within this head include offices, support overheads such as buying and accounts departments, rent, rates, heating and telephone bills, indeed anything going towards the cost of maintaining the business operation as a whole, as opposed to an individual project.

Claims for such costs are notoriously difficult to establish, the main problem for the Contractor being to show entitlement in principle. The basis of claim is that because of delay on the project, the company workforce was deprived of the chance of earning contribution or recovering its overhead costs from elsewhere. Small companies with limited staff resources can sometimes demonstrate by reference to correspondence that they have had to turn away new work; but it is difficult for a large national company to show that a delay on a single project had a significant effect on the whole company's ability to accept new work.[36]

Once a claim is established in principle, a contractor is then obliged to provide information to enable the loss or expense to be calculated. In practice the calculation is normally prepared by the Contractor for checking by the Architect. A popular means of calculation among contractors is the use of a formula. Formulae in common use are the 'Hudson formula', the 'Emden formula', and the 'Eichleay formula'.

The 'Hudson formula'[37] calculates loss as an average overhead and profit allowance per week, based on the contract period and percentage mark-up included in the Bills, then multiplied by the length of delay. The formula can be criticized on the grounds that it relates to tender allowances (i.e. value), rather than actual costs.

The 'Emden formula'[38] differs from the Hudson formula only in that the overheads are taken as an average percentage from the Contractor's accounts. As a means of calculation once entitlement in principle is established, the Emden formula has received some apparent approval in the courts.[39]

The 'Eichleay formula'[40] approaches the problem from a different direction, arguably based on a shortfall in contribution.[41] The average weekly contribution necessary to run the company is calculated from the company accounts, and is multiplied by the period of delay. The product, which represents the total contribution required from all income in order to run the business during the delay period, is reduced *pro rata* to reflect the share

35 Under contracts such as those published by the ICE which refer only to 'cost', the difference may be significant.

36 In *Whittal* v. *Chester-le-Street D C* (unreported) 3 July 1984 (Mr Recorder Percival QC), cited in *JF Finnegan* v. *Sheffield City Council* (1988) 43 BLR 124, it was found as a fact that work was available which could not be taken.

37 Set out in *Hudson's,* 11th edn, at para. 8.182.

38 Set out in *Emden's Construction Law,* 8th edn, rev. A. J. Anderson, S. Bickford-Smith, N. E. Palmer and R. Redmond-Cooper (Butterworth, 1990) 57, Aug. 1999, vol. 2, at p. N/46.

39 In *JF Finnegan Ltd* v. *Sheffield City Council* (1988) 43 BLR 124, the judge preferred the 'Hudson formula' to one of the Contractor's own making, although he then went on to describe a form of the 'Emden formula'. See also *Norwest Holst Construction Ltd* v. *Co-operative Wholesale Society Ltd and Another,* (unreported) 17 February 1998, case No. 1997 ORB 466–468; *Harvey Shopfitters Ltd* v. *ADI Ltd,* EWCA Civ 1757, 13 November 2003.

40 Based on the first case in the USA to use the formula; see Duncan Wallace, I. N., *Construction Contracts: Principles and Policies in Tort and Contract* (Sweet & Maxwell, 1986), paras 8-30 to 8-33 for detailed comment.

41 This point was argued in *Alfred McAlpine Homes Northern Ltd* v. *Property and Land Contractors* (1996) 76 BLR 59.

of the total contribution required from the project in delay to pay its way. The *pro rata* is made by comparing the turnover of the project in delay during the period with total turnover (value) on all projects during that period. This formula was used in *Alfred McAlpine Homes North Ltd* v. *Property and Land Contractors Ltd,*[42] but again the Court emphasized the need to establish entitlement before any form of calculation should be employed:

> There may be some loss as a result of the event complained of, so that in the case of delay to … completion … there will be some 'under recovery' towards the cost of fixed overheads as a result of the reduced volume of work … but this state of affairs must of course be established as a matter of fact. If the contractor's overall business is not diminished during the period of delay … (due to increased contribution from variations etc.), or if as a result of other work, there is no reduction in the overall turnover so that the cost of fixed overheads continues to be met from other sources, there will be no loss attributable to the delay.

Similarly, in *Amec Building Ltd* v. *Cadmus Investment Co. Ltd*[43] it was said:

> It is for Amec to demonstrate, in respect of the individuals whose time is claimed that they spent extra time allocated to the particular contract. This proof must include the keeping of some form of record that the time was excessive and their attention was diverted in such a way that the loss was incurred. It is important … that the plaintiff places some evidence before the court that there was other work available which but for the delay he would have secured … thus he is able to demonstrate that he would have recouped his overheads.

If a formula claim for overheads is allowed, care must still be taken to remove any duplication with other heads, such as overlap with site supervision costs which may be included as overheads in the company accounts or other sources of income such as Variations.

One alternative to calculation by formula is by identifying, where possible, the cost of managerial time spent in problem-solving. An example can be seen in *Tate & Lyle Food and Distribution Ltd* v. *GLC*[44] in which a claim for damages, albeit in tort and not contract, was enhanced by the addition of 2.5 per cent to cover managerial and supervisory resources. It was held that the time spent was a proper head of damage, but in this case nothing was proved. It was not sufficient to add a percentage; in an organization such as Tate & Lyle there should be records available to demonstrate the loss:

> While I am satisfied that this head of damage can properly be claimed, I am not prepared to advance into an area of pure speculation when it comes to quantum.
>
> I have no doubt that the expenditure of managerial time in remedying an actionable wrong done to a trading concern can properly form a subject matter of a head of special damage … I would … accept that it must be extremely difficult to quantify. But modern office arrangements provide for the recording of the time spent by managerial staff on particular projects.

It is clear that the 'Tate & Lyle method' is likely to produce a more accurate result than any of the formula methods, but it relies on accurate record-keeping.

However, the Contractor can face difficulty in identifying the cost of general overhead resources, such as the time spent by the managing director. Indeed it may be questioned whether there is a loss or cost at all, because the managing director is not paid any more,

---

42 (1996) 76 BLR 59.
43 (1997) 51 ConLR 105; (1997) 13 Const LJ No. 1 p.50.
44 [1982] 1 WLR 149.

and thus the company does not increase its outlay. This issue was dealt with in *Euro Pools* v. *Clydeside Steel Fabrications Ltd*[45] In coming to the conclusion that the managing director's time in problem-solving was a loss, the Court considered the effort that should properly be put into developing new products and market initiatives, or administering the company's affairs, rather than dealing with (in this instance) defective work from a supplier. It was said:

> That in my opinion clearly represents a loss to the company. It may not leave the company out of pocket, in the sense of having to pay more… nevertheless, the company will inevitably be deprived of part of the services that it would normally expect.

When it came to establishing the loss, the managing director was unable to produce time records for the 20 hours being claimed. The Court took the pragmatic view that a managing director will not normally keep timesheets, but documents such as a diary, correspondence etc., may go a long way to support oral evidence. In the absence of written evidence, oral evidence may still be accepted, but damages may be reduced to reflect the uncertainty.[46]

As with claims using the formula method described above, a claim based on time records is also reliant on the Contractor's demonstrating that the diversion of resources caused disruption to the business. In *Standard Chartered Bank* v. *Pakistan National Shipping Corporation*,[47] an employee was sent abroad, diverting him from his normal duties. A claim was made for travel expenses plus a portion of his salary. The Court of Appeal allowed the expenses, but not the portion of salary, on the grounds that there was no disruption to the bank's business or loss of profit.

In *Aerospace Publishing and another* v. *Thames Water Utilities Ltd*,[48] it was argued that the claimant should demonstrate that the diverted activities of staff in problem-solving resulted in loss of revenue for the business. The Court of Appeal set out three principles:[49]

- The fact and extent of diversion of staff time has to be properly established with evidence (records) that it would be reasonable to expect to be kept;
- It has to be established that the staff diversion caused significant disruption to the business;
- It is reasonable for the tribunal to assume that the normal activities of the staff would be expected to generate at least as much revenue as the cost of their employment.

The need to provide records that it would be reasonable to expect, places a special burden on the Contractor. There are times when it is not known that a legal action will follow, until after the loss is incurred. In those circumstances, evidence is likely to be the type of record which should normally be kept, depending on the type, size and sophistication of the organization. However, losses on construction contracts often continue for many months, particularly those related to prolongation. When the Contractor gives notice to the Architect under Clause 4.23, he immediately puts himself on notice of the need for reasonable records, sufficient to prove his case.

45 17 January 2003, Scottish Court of Session [2003] 4 BLISS 20.
46 In *Bridge UK.Com Ltd* v. *Abbey Pynford plc* [2007] EWHC 728 (TCC) when relying solely on oral evidence, the sum awarded was discounted by 20% to reflect the uncertainty.
47 [2001] EWCA Civ 55.
48 [2007] EWCA Civ 3, 11 January 2007.
49 Per Lord Justice Wilson, at para. 86.

13.4.4 Visiting head office staff

One head of claim that straddles site project costs and head office costs is the additional time spent by head office personnel visiting site, either in problem-solving or in prolonged involvement. A typical example would be the Contractor's surveyor who may be required to visit the site regularly to deal with domestic sub-contractors and internal cost accounts. A disrupted site may result in greater involvement each week, whereas a delayed site may prolong the visits over a longer period. Claims of this nature fall squarely into the type considered in the *Tate & Lyle*, *Euro Pools* and *Aerospace Publishing* cases dealt with above.[50] The difficulty for the Contractor is in establishing what resources would have been spent in any event, which may require evidence of the normal level of resource required for that type of project. Actual costs for staff who keep time records should be comparatively easy to identify, and include salaries, travelling expenses, car hire or depreciation, fuel, accommodation etc., under the same principles as those referred to above dealing with Site Establishment.

13.4.5 Uneconomical working

Uneconomical working claims are often made under the heading of 'disruption'. Delay or disruption on site may manifest itself as loss of motivation leading to loss of productivity. The difficulty with disruption is that everyone can see the effect of its presence on site, but the type of information required to demonstrate quantification and causation are rarely available. It has been said[51] that 'the quantification exercise is neither precise nor undertaken using clearcut methods of measurement since human activities, unlike physical phenomena, are not susceptible to precise methods of measurement, and the units and process of measurement can be somewhat subjective'. However, if the Contractor is able to demonstrate that events caused uneconomical working and wasted costs, the amount of such costs may be calculated either by comparing actual hours expended with those acticipated, or by comparison of work output during disrupted and undisrupted periods.

Comparison of actual hours expended with those anticipated: Calculating wasted time costs by simply deducting anticipated hours from actual hours is often looked upon by practitioners in the industry with deep suspicion, the principal objections being that the tender price may have been underestimated, and the actual costs may be inflated by the Contractor's own inefficiency. Nevertheless, if the Contractor can show that the tendering method and level is similar to that used for other jobs, and in those jobs he was able to work economically without loss, he may have a basis for recovery. In *Amec Process and Energy Ltd* v. *Stork Engineers and Contractors BV (No. 3)*[52] the Court was satisfied that the tender was built up from norms which were the product of much experience of the output that could be achieved, and that it was a reasonable prediction of the number of hours that would have been incurred. It was then necessary to calculate the hours expended to isolate the wasted or lost time, and to put a cost to it. The Court saw the

---

50 See Section 13.4.3 of this chapter.

51 *Amec Process and Energy Ltd* v. *Stork Engineers and Contractors BV (No. 3)*, 15 March 2002, 1997 ORB 659, (unreported); per HH Judge Anthony Thornton QC, at para. 707.

52 15 March 2002, 1997 ORB 659, (unreported), at para. 804–805.

exercise as the answer to four simple questions:[53] (1) How many hours did the relevant operatives actually work? (2) How many of those hours are paid for? (3) In respect of how many of the residual hours should the Contractor bear the costs itself? (4) At what rates should the hours be remunerated?

In the *Amec* case, Stork argued that the costs included the Contractor's inefficiencies, but in the absence of evidence of inefficiency, the Court accepted the costs as reasonable.

Comparison of work output during disrupted and undisrupted periods: The comparison of work output method was considered in *Whittal Builders Co. Ltd* v. *Chester-le-Street D.C.*[54] The value of work output per man week while the Employer's breaches continued was compared with the value of work output after the breaches had ceased. An indication of willingness to find an answer can be seen in the words of the judgment:

> It seemed to me that the most practical way of estimating the loss of productivity, and the one most in accordance with common sense and having the best chance of producing a real answer was to take the total cost of labour and reduce it in the proportion which those actual production figures bear to one another – i.e. by taking one third of the total as the value lost by the contractor.

The calculation of one third wasted costs in this case came from comparing disrupted production value of £108 per man week with undisrupted value of £161 per man week.

Contractors should have no difficulty in obtaining figures; costs or man weeks may be taken from accounts records, and value may be taken from interim valuation calculations. The pitfall comes in finding two periods in which the work is sufficiently similar to be compared like with like, avoiding comparison between repetitive work and work requiring a learning curve.

It is tempting for architects to resist ascertainment on the grounds that evaluation cannot be made scientifically and in minute detail; but it is worth remembering the pragmatic approach in the *Whittal* case[55] and the judgment in *Chaplin* v. *Hicks*,[56] where it is held that if the calculation is difficult the Court must do its best.

13.4.6 Uneconomical procurement

If a project is delayed or disrupted, it is possible that procurement of materials may be affected. Such claims are rare but may be sustained if the purchase of materials in large quantities is prevented, or if materials on long-term delivery have to be accepted out of sequence. A claim could include the loss of special bulk discounts, storage costs, and funding capital expenditure until the materials are included in a certificate.

---

53 15 March 2002, 1997 ORB 659, para. 712.

54 (Unreported) 3 July 1984 (Mr Recorder Percival QC), cited in *JF Finnegan* v. *Sheffield City Council* (1988) 43 BLR 124.

55 See also *Penvidic Contracting Co. Ltd* v. *International Nickel Co. of Canada Ltd* (1975) 53 DLR (3d) 748, in which the contractor was awarded the difference between what he tendered and what he would have tendered if he had foreseen the conditions caused by the Employer's breach.

56 [1911] 2 KB 786.

13.4.7 Loss of profit

Loss of profit claims fall under the same general principles as head office overheads. As with overheads it is for the Contractor to establish that he was deprived of the opportunity to earn profit elsewhere. It is not sufficient to add a percentage to the net value of other heads of claim.[57] If the Architect omits work from the contract in order to give it to someone else, the Employer will be in breach of contract. The damages will be the loss of profit which the Contractor included in his tender.[58]

13.4.8 Acceleration

Employers will sometimes require their finished project by the original Completion Date, or by some other fixed date, irrespective of the Contractor's entitlement to extensions of time. Contractors may then propose accelerating their work if they are reimbursed their additional costs, or if the Employer offers inducement in a bonus. Occasionally an Employer may try to influence the Architect (or the Architect may act on his own) to deny extensions of time in order to coerce the Contractor into finishing earlier than his entitlement; or the Contractor might simply decide to accelerate of his own accord, in both his own and the Employer's interests.

Only the last of these three categories fall within the authority of the Architect without additional express powers being given by the Employer.

1. Agreement to accelerate/bonus: Where a building owner expressly proposes to the Contractor that he would like the project finished at a particular time before the contractor's strict entitlement, he may suggest some form of acceleration, which is then a matter for the parties to agree.

The Society of Construction Law[59] suggests that where acceleration is agreed, the basis of payment for acceleration measures should also be agreed before acceleration is commenced. The parties should, however, ensure that such agreements are made between persons having the requisite authority.

If an agreement on acceleration costs or a bonus is made between the Contractor and the Employer (or with the Architect if the Employer gives him the authority), the terms of the agreement need to be considered carefully. The Employer is at risk if the Architect issues further Variation instructions under the Contract. If the Contractor fails to meet the deadline because of such Variations, or if the Architect in any other way prevents the Contractor from meeting his target, the Employer will be in breach of an implied 'non-hindrance' term. In *John Barker Construction Ltd* v. *London Portman Hotels Ltd*,[60] the Employer and Contractor agreed on a bonus to be paid if the Contractor achieved completion by the agreed date. The agreement was construed as a collateral contract. The Contractor missed the agreed date due to additional work ordered by the Architect in Variation instructions. Damages awarded were one half of the promised bonus.[61] Whilst there was no certainty that

---

57 See Section 13.4.3 of this chapter, dealing with the *Tate & Lyle* case.

58 See Chapter 6.4 referring to *Carr* v. *J. A. Berriman Pty Ltd* (1953) 879 ConLR 327; *Amec Building Ltd* v. *Cadmus Investments Co. Ltd* [1997] 51 ConLR 105.

59 October 2002, *Delay and Disruption Protocol*, The Society of Construction Law, p. 31.

60 [1996] 50 ConLR 43; (1996) 12 Const LJ 277.

61 In *Bournemouth & Boscombe Athletic FC* v. *Manchester Utd FC*, 1974 B. No. 1531, CA, Judgment Wednesday 21 May 1980, *The Times* 22 May 1980, a bonus as part of a transfer deal was awarded as damages (less only an allowance for the possibility of time lost through injury) when a footballer was prevented by his new club from scoring his target.

the Contractor in that case would have met his target date, it was held that he was deprived of the chance by being hindered (i.e. there was breach of an implied non-hindrance term in the collateral agreement); the hindrance was the issue of *bona fide* Variations given under the building contract. A claim by the Contractor under this head is not a claim under the Contract, but is a common law claim. If the Employer wishes such matters to be dealt with by the Architect, the acceleration/bonus agreement needs to clearly set out the administrative arrangements, together with the effect of further Variations.

Sometimes an acceleration agreement is struck, and in hindsight the Architect decides that the contractor was not entitled to an extension of time. The Contractor would then be doing no more than his existing contractual duty to finish on time, and the agreement may be unenforceable for lack of any consideration by the contractor.[62] However, the agreement may still be binding if the employer 'benefits' from the deal. In *Williams* v. *Roffey Bros*,[63] a sub-contractor in financial difficulty agreed with the main contractor to finish on time in return for bonus payments. Although the sub-contractor had an existing contractual obligaton to finish on time, it was held that the agreement was binding on the grounds that the contractor avoided the dis-benefit of paying liquidated damages and having to arrange for another sub-contractor to finish the work.

2. Constructive acceleration: The Contractor is faced with a dilemma if he believes he is entitled to an extension of time, but he is deprived of his entitlement either by the Employer's breach of contract in deliberately preventing the award of an extension, or by the Architect's failure to form an opinion. The Contractor has to decide whether to hope his entitlement will eventually be recognized (and to continue at a natural pace), or whether to avoid exposing himself to liability for liquidated damages (and accelerate to meet the current contractual date). Acceleration in these circumstances is said to be 'constructive'.[64] The Contractor's claim is for common law damages arising from an alleged breach, and is persuasive in principle, although the UK courts have been slow to accept the proposition.[65] In *Motherwell Bridge Construction Ltd (t/a Motherwell Bridge Storage Tanks)* v. *Micafil Vakuumtechnik and Another*,[66] the contractor was given design changes involving additional work. He then incurred additional costs by working night shifts in an attempt to achieve the scheduled completion date, when the building owner refused to grant extensions of time. It was held that the additional costs were recoverable, although the judgement in that case did not set out reasons sufficiently for a principle to be established. There are two major obstacles for the Contractor to negotiate. The first is in establishing a failure to grant extensions when there is an obligation to do so, and the second is satisfying the causation and remoteness requirements of the general principles applying to damages.[67]

Contractors tend to view any failure of the Architect to award extensions as a breach of duty, but it has been held that the Architect is under no duty to provide a completion date

62 In England, Wales and Northern Ireland, an agreement requires consideration from each party to be enforceable.

63 *Williams* v. *Roffey Bros & Nicholls (Contractors) Ltd,* (1989) 48 BLR 69.

64 Where the law implies a right without reference to the intention of the parties. See *Osborn's Concise Law Dictionary.*

65 See also The Society of Construction Law (2002), *Delay and Disruption Protocol*, p. 31, which suggests that a contractor who accelerates of his own accord should not be compensated.

66 31 January 2002 [2002] 81 ConLR 44, (2002) CILL 1913.

67 See Chapter 12, Section 12.5.1, Cause and Effect, and Section 12.5.2, Measure of Recovery.

at which the Contractor can aim.[68] Clearly, a breach can be established if the Contractor can demonstrate that the Employer has interfered by instructing the Architect. Deliberate intervention by the Employer is probably rare, although where the Employer is a Local Authority and the Architect is one of the Authority's officers, there is often conflict between the officer's duties under the Contract, and his attempts to bide by rules set by the Authority for its employees. However, it is more likely that the Architect has either failed to address his mind to the matter, or he has acted after forming an opinion, albeit that opinion might later change, or be found by a tribunal to be wrong. Unfortunately for Contractors, simply forming a wrong opinion is not in itself a breach.[69] The essential act is in forming an opinion, as required by the Contract (see Section 13.2.2 (2) of this chapter). Thus failure to consider the matter at all, provided the Contractor has fulfilled his obligations, would be a breach giving rise to entitlement to compensation. In *Perini Corporation* v. *Commonwealth of Australia*,[70] it was held that the certifier had to give his decision in a reasonable time in the absence of express periods. On this premise an administrator may be in breach for not forming his opinion at the proper time. That is unlikely to be the position under JCT 05.

Under JCT 05 the Contractor cannot expect an accurate correction of the completion date before Practical Completion. Although it has been held by the Court of Appeal[71] that the 12-week review period under a JCT 80 contract was no more than 'directory',[72] and the Architect did not put the Employer in breach if he took longer, it is still the case that the Contractor knows that the Architect has an opportunity to fix the completion date three months after Practical Completion. The whole premise of constructive acceleration that the Contractor would be put in breach which he must avoid, is undermined.

Provided an entitlement could be established in principle, the Contractor's task then is to identify the effect. If the acceleration is simply working longer shifts, the cost is comparatively straightforward to calculate. However the effect may be more complicated if acceleration is achieved by substantial increase in the number of operatives, and the Contractor may be entitled to yet further recovery if, as is likely, the increase in labour resources causes disruption in the whole site effort, resulting in wasted labour costs.[73]

3. Acceleration in mitigation: Contractors entitled to an extension of time for reasons which also give entitlement to loss or expense, are required under general principles to mitigate their costs or losses if they intend to seek reimbursement (see Chapter 12, Section 12.5.2 dealing with extravagance and 'duty' to mitigate). Under those same principles the Contractor is entitled to spend money in order to save money. If the Contractor's estimation of the costs involved in a prolonged contract period could be reduced by working overtime or by introducing extra labour to complete earlier, then he is entitled to take such action. Provided the decision to 'accelerate' was a reasonable decision at the time it was made, the Contractor would be entitled to recover his acceleration costs; this would be so even if the eventual cost exceeded the estimated cost it was intended to reduce. There is no strict obligation on the

---

68 See *Amalgamated Building Contractors* v. *Waltham Holy Cross UDC* [1952] 2 All ER 452.
69 See *S. Pembrokeshire D C* v. *Lubenham Fidelities and Wigley Fox Partnership* (1986) 33 BLR 39.
70 (1969) 12 BLR 82.
71 See *Temloc* v. *Errill Properties Ltd* (1987) 39 BLR 30.
72 At p. 39; the context in which Lord Justice Croom-Johnson uses 'directory' makes clear that the term is used in the sense of guidance, giving advice or direction.
73 For general discussion on effect of accelerated working see Horner, R. M. W. and Talhoune, B. T. (1995) *Effects of Accelerated Working Delays and Disruption on Labour Productivity*, Chartered Institute of Building.

Contractor to seek approval for such a decision, but the prudent Contractor would notify the Architect for evidential reasons, simply to show at a later date that the action came from a positive decision. Since acceleration in this form is claimed as mitigation of loss and/or expense, it is a matter which the Architect can (and must) consider.

The only real risk for the Contractor in this course seems to be the possibility of a poor decision to accelerate when he is later found to be not entitled to an extension; clearly the decision is still in his own interest if it is a good decision, whether or not he later receives an extension, since he will have minimized his own losses.

13.4.9 Third party settlements

Whenever a project is delayed or disrupted, it is likely that the Contractor's sub-contractors will be affected. It is rare for Contractors to carry out all the work themselves, indeed it is more likely that the majority of the work is carried out by sub-contractors. It follows that many of the costs or liabilities borne by the Contractor, in reality, are liabilities to sub-contractors. A difficulty arises for Contractors when a sub-contractor presses its claim long before the Contractor's entitlement is finally ascertained. Rather than face long legal battles, the Contractor may reach a settlement with its sub-contractor, then turn to the Architect for reimbursement under Clause 4.23.

The Contractor will probably be faced with argument that the settlement is irrelevant to the Contractor's entitlement, or may even be put to the task of proving the sub-contractor's claim. In this latter task he is likely to fail since he will not have the necessary records. However, the courts have considered the problem on a number of occasions,[74] and in *Oxford University Press* v. *John Stedman Design Group*,[75] the following principles were identified (paraphrased):

- The law encourages reasonable settlement;
- The cost of pursuing litigation is relevant in deciding whether a settlement was reasonable;
- A party who relies on the settlement as the basis for compensation must prove that it is reasonable; but he does not have to prove strictly the claim against him in all its particulars;
- In establishing the reasonableness of a settlement, it is relevant that it was made under legal advice, although the advice itself may not be relevant or admissible.

It is still for the claiming party to establish that the third party settlement was reasonable[76] (that does not mean simply showing it was reasonable to settle). The case of *J Sainsbury* v. *Broadway Malyan and Ernest Green Partnership*[77] provides a warning to those making settlements with the intention of passing on the agreed amount to another as damages. A settlement between an architect and the building owner over losses from a design error

74 See *Biggin & Co. Ltd* v. *Permanite Ltd* [1951] 1 KB 422; [1950] 2 All ER 859; *Fletcher & Stewart Ltd* v. *Peter Jay & Ptnrs* (1976) 17 BLR 42 (CA). In *Royal Brompton NHS Trust* v. *Hammond (No. 1)* [1999] BLR 162; 1999 CILL 1464, it was held that the principles in the *Biggin* case applied equally to cases where there were several defendants from whom contributions are sought.

75 (1993) 34 ConLR 1; (1990) CILL 590.

76 See *P&O Developments Ltd* v. *The Guy's and St Thomas' NHS Trust & others* [1999] BLR 3.

77 [1998] 61 ConLR 31.

could not be passed on in full by the architect to others. The Court decided that the settlement had been generous, and that the architect had not been liable for a high proportion of the agreed costs. By contrast, in *John F Hunt Demolition Ltd* v. *ASME Engineering Ltd*,[78] a claim was made by the employer against the contractor for which there was later found to be no liability, but nevertheless was settled, then passed on as a third party settlement in a claim against a subcontractor. It was held in this case, on the particular facts, that the settlement was not reasonable,[79] but that the absence of liability did not necessarily make the settlement unreasonable. If at the time the contracts were formed it was reasonably foreseeable that such a claim might be settled, the third party settlement could be reasonable.[80] In *Bovis Lend Lease Ltd* v. *RD Fire Protection Ltd*,[81] the contractor settled a counter-claim for remedial costs from the building owner, and then pursued its sub-contractor for the full claimed remedial costs. The claim failed. It was held that the claim against the sub-contractor had been made as though the settlement with the building owner had never happened, and that the contractor could only recover the actual loss suffered. Unfortunately for the contractor the loss caused by the sub-contractor could not be identified from the settlement with the owner, which included other things. This may lead a contractor to protect his position by wording a third party settlement in appropriate terms, but he must do so honestly. In *Durabella Ltd* v. *J. Jarvis & Sons Ltd*,[82] the contractor settled a claim with the building owner; the settlement agreement contained a statement that nothing was included in respect of the sub-contractor's work. The contractor then relied on that statement to avoid paying the sub-contractor. The court took the view that the statement did not protect the contractor as it had been included to mislead.

Whilst a reasonable settlement with a third party may be construed by a contractor as a reasonable amount to claim and recover, it is only the quantum of the claim. The matter of the Employer's liability still needs to be addressed, and the fact that the contractor has reached a reasonable settlement with (say) a sub-contractor will not, in itself, establish any liability in principle on the part of the Employer.[83]

The difficulty for the Architect, when dealing with claims that include amounts in respect of liability for which a settlement has been made with a third party, is that he must ascertain the loss. He must take into account the pragmatic principle of encouraging settlements to avoid expensive dispute, whilst at the same time ensuring that he does not commit his client to payment of sums that are not due; but he cannot avoid ascertaining the head of claim purely on the grounds of a demand for the Contractor to prove the case.

13.4.10 Inflation

If a non-fluctuating price project is delayed or disrupted due to an Employer's delay, the Contractor is entitled, in principle, to claim the costs resulting from work being carried out at a later time. The calculation should not be based on the Completion Date, but on the

78 [2007] EWHC 1507 (TCC).
79 See also *Axa Insurance UK plc* v. *Cunningham Lindsey United Kingdom* [2007] EWHC 3023 (TCC) in which a claim was settled at a high figure to avoid publicity, but which was rejected by the Court as unreasonable.
80 See also *Comyn Ching & Co. Ltd* v. *Oriental Tube Ltd*, [1979] 17 BLR 47.
81 [2003] EWHC 939 (TCC), [2003] 89 ConLR 169.
82 1998 ORB 33, [2002] 83 ConLR 145.
83 For more detailed analysis of this topic see Judge Peter Coulson QC, *Catching Water in a Net: the Elusive Concept of 'A Reasonable Settlement'*, paper 149, Society of Construction Law, September 2008.

comparable dates of the various activities. Thus, the extra cost of carrying out brickwork in (say) week 50, instead of (say) week 30 is the cost claimed. The evaluation may be based on invoiced costs compared with anticipated costs from price lists, and a similar calculation for labour cost. Alternatively, a notional calculation may be made by calculating the costs which would have been incurred over the original contract period by using published indices, and comparing it with the extra costs incurred over the actual period, again using indices. In order to carry out this type of calculation it is first necessary to establish the anticipated 'cashflow' of the original project for application to the indices. The 'actual' indices can then be applied to the Quantity Surveyor's interim valuations. Whilst this latter method cannot be said to be proof of the actual loss incurred, it does attempt, by application of the facts related to time and by some science related to evaluation, to get to a fair assessment.

13.4.11 Financing other heads of claim

In times of high borrowing rates, financing costs can be a significant part of the Contractor's total loss and expense claim. Financing is normally looked upon as an 'expense', incurred by way of interest charges paid by the Contractor on an increased overdraft, although it may also be a 'loss', equivalent to lost income on reduced capital invested. When a Contractor incurs additional expense for which he is entitled to reimbursement under Clause 4.23, he has to fund that expense until he is paid under an Architect's certificate. It is part of that extra funding which is claimed as financing cost, and it should not be confused with interest charged to the Employer for late payment of a debt (see Section 13.4.13 of this chapter).

The principles governing financing to be reimbursed as loss or expense are to be found in two cases. The first, *F. G. Minter Ltd* v. *WHTSO*,[84] established that financing was an entitlement as loss or expense, and the Architect was obliged to consider it if proper notice had been given. The second case is *Rees & Kirby Ltd* v. *Swansea City Council,*[85] which dealt with the method of calculation and timing. It was held that:

- The Contractor's notice of the primary loss (the head of claim being funded) must be given within a reasonable time of the loss being incurred;
- Financing costs cannot be considered parasitic to the primary loss, and therefore some mention of the further expenditure in financing must be made in the notice;
- Financing is due from the date of loss and expense being incurred;
- Financing is due up to the date of the last application made before the issue of the certificate in respect of the primary loss (i.e. the payment of the head of claim on which financing is incurred);
- Financing charges incurred during a period when an independent cause operated should not be recoverable (this would include delay in ascertainment caused by the Contractor's failure to make application or failure to provide information after being requested);[86]
- The date of practical completion is irrelevant to the calculation;
- The rate of interest to be applied is the actual borrowing rate paid by the Contractor, provided it is reasonable;
- The calculation should be made on the same basis as that used by the banks (i.e. simple interest compounded at quarterly intervals). In *Amec Process & Energy Ltd* v. *Stork*

84 (1980) 13 BLR 1.
85 (1985) 5 ConLR 34.
86 In the *Rees & Kirby* case the contractor's delayed notice was not held to prevent recovery, since the contractor had relied on ongoing negotiations in not submitting notices in strict compliance with the contract.

Engineers & Contractors BV (No. 4),[87] it was held that the cost of borrowing from a parent company was recoverable on a compound basis, but recovery could not exceed the costs that would have been incurred when borrowing from a bank;
- Where the Contractor is 'cash rich', the rate is that earned by the Contractor on money placed on deposit;
- Account should be taken of payments made progressively.

The calculation of financing is not difficult once the relevant periods have been established. The most elusive of these periods are the dates when the primary losses are incurred, and the dates on which the Contractor's application ought reasonably to be made. Since the losses are incurred progressively an interest calculation becomes a rolling calculation, and the cost of preparing the calculation can easily exceed the amount of the claim. A practical solution is to calculate financing separately either as a part of the costs relating to a discrete time period (time-slice) (see Section 13.4.2 of this chapter above), or on each head of primary loss, starting at the date on which the 'centre of gravity' of the primary loss falls; the 'centre of gravity' is found by adding together all the costs under the head of claim, and identifying the date on which the mid-point of cost occurs. As to identifying the reasonable date of notice, the Architect will need to take into account the requirements of the Contract (i.e. Clause 4.23.1: the Contractor's application shall be made as soon as it has become, or should reasonably have become, apparent to him that the regular progress of the Works... has been or is likely to be affected... (to cause loss or expense)). The Contractor may not be aware of disruption or possible loss or expense immediately, so notice given responsibly may be some considerable time after the first signs. Insistence by the Architect on notice on the off-chance that disruption or additional costs may occur, is likely to produce a standard *pro forma* response to all directions and instructions; such an approach is not to be recommended since it clouds the important issues, and prevents the Architect taking preventive action to minimize the Employer's risk.

13.4.12 Financing retentions

When the Contractor tenders, he is deemed to have included for his risks under the Contract. He anticipates that he will be paid at certain times and he can make due allowance for the cost of funding the project. One factor to take into account is that he will be obliged to finance the retention monies until practical completion, when he will receive an injection of funds as monies are released. The Contractor may accordingly claim the financing costs of later retention release, delayed to the extent that the Employer is culpable under Clause 2.28. The calculation follows the same principles as those described for financing other heads of claim above.

13.4.13 Interest

Where money is not paid on time the Contractor may claim interest in four ways: (i) by claiming an entitlement to interest on a debt, by use of a term in the Contract, or (ii) by claiming interest under by a statutory provision, or (iii) by claiming the cost of borrowing (i.e. interest paid to others) as damages, or (iv) by claiming for the restitution of the value gained by the Employer in having use of the Contractor's money.

---

87 15 March 2002, (2002) CILL 1883.

***Contractual right to interest*:** Under Clause 4.13.6 the Contractor is entitled to simple interest on late or non-payment of an Architect's certificate at the rate of 5% over the Bank of England Base Rate (see Chapter 15, Section 15.6.1). He is not entitled to interest under that clause for late issue of a certificate. Interest under Clause 4.13.6 is not a matter for the Architect.

***Statutory right to interest*:** Under the Late Payment of Commercial Debts (Interest) Act 1998, and in the absence of a substantial contractual remedy, a company supplier of goods and/or services has a right to claim interest on a debt. The Act does not apply where the contract contains substantial provision for interest like that in Clause 4.13.6 of JCT 05 (see Chapter 15, Section 15.6.1), although the term 'substantial' must be construed reasonably. In *Jeancharm (t/a Beaver International) Ltd* v. *Barnet Football Club Ltd*[88] an interest rate of 5% per week was held to be unenforceable as a penalty, by applying the rules in the *Dunlop* v. *New Garage* case.[89]

***Interest as damages*:** The Contractor may replicate part of his contractual loss or expense claim, including 'financing', as an alternative common law claim for damages. In the case of *Wadsworth* v. *Lydall*,[90] the Court of Appeal held that if it can be proved that special damage[91] is incurred by way of interest payments on an overdraft, which in turn is caused by late payment of a debt, then it may be recovered. The principle did not extend to interest as a primary loss prior to the Employer's breach of contract. In *Amec Process & Energy Ltd* v. *Stork Engineers & Contractors BV (No 4)*,[92] it was held that financing costs, claimed in the alternative as damages, could have been foreseen through knowledge gained during the tendering procedure, and so satisfied the test for special damages. In this case it was also held that the cost of borrowing from a parent company was recoverable on a compound basis, but since recovery should not exceed what would have been the cost if borrowing from a bank, compounding was calculated at quarterly intervals.

Entitlement to damages was widened further in *Sempra Metals* v. *IR*,[93] in which the House of Lords held that in the modern commercial world, interest was a loss arising 'in the ordinary course of things',[94] thus bringing it into the category of general damages.[95] The benefit for the Contractor is the assumed entitlement to damages, rather than the need to prove entitlement in principle first. However, it is still necessary to prove the amount of damages. Lord Nichols explained:[96]

> The claimant would have to show, if his claim is for ancillary interest, that his actual losses were more than he would recover by way of interest under statute.

It seems that the Contractor is entitled to claim statutory or contractual interest, and then present a claim to have it topped up if his losses are not compensated.

---

88 16 January 2000; [2002] EWCA Civ 58.
89 See Chapter 11, Section 11.1.1.
90 [1981] 2 All ER 401.
91 See Chapter 12, Section 12.5.2, dealing with the rules in *Hadley* v. *Baxendale*.
92 15 March 2002; 1997 ORB 659; (2002) CILL 1883.
93 *Sempra Metals Ltd (formally Metallgesellschaft Ltd)* v. *Her Majesty's Commissioners of Inland Revenue & Anr*, [2007] UKHL 34.
94 Per Lord Hope of Craighead, at para. 16.
95 See Chapter 12, Section 12.5.2, dealing with the rules in *Hadley* v. *Baxendale*.
96 At para. 17.

Lord Scott added:[97]

> ...interest losses caused by a breach of contract... should be held to be in principle recoverable, but subject to proof of loss, remoteness of damage rules, obligations to mitigate damage and any other relevant rules relating to the recovery of alleged losses.

In the *Sempra* case, it was considered fair to award compound interest, but it is clear that a Contractor wishing to claim damages for the loss of use of money must still prove his loss in the same way that he would prove his financing costs (see Section 13.4.11 of this chapter). It is not simply a case of adding a notional rate of compound interest.

Interest as restitution: An alternative claim may be made in the law of restitution.[98] Restitutionary claims are not reliant on a breach of contract, but are based on unjust enrichment, and an obligation to pay back a benefit. An example would be the use of money held properly as liquidated damages, but returned when the Architect awards a further extension of time to the Contractor. In that situation, there is no breach of contract, but nevertheless the Employer would have had the use of the Contractor's money; it is the time-value of that use, or benefit, that the Contractor would seek to recover. Whereas financing and damages relate to costs incurred by the Contractor when funding another head of loss until compensation is received, restitution relates to the benefit of the use of the money by the payer, until it is paid. Claims in restitution for the return of the actual benefit are likely to be rare, if for no other reason than that it may be difficult for a Contractor to discover the value of benefit, and the Employer is likely to resist disclosing his benefit unless it is small. In *Sempra Metals* v. *IR*,[99] Lord Mance said:[100]

> Using their discretion, courts will be able to keep equitable claims seeking to investigate and recover any actual benefit … within sensible bounds … Courts should be able to discourage or refuse expensive demands for discovery … hoping to investigate precisely what interest benefit a defendant may have made.

In the *Sempra* case the claimant was paid compound interest as being just, because the special relationship between the Government and the Bank of England made calculation of actual benefit difficult. In a restitutionary claim from the Employer, it would be for the Contractor to demonstrate that an award of compound interest would be a just alternative to the return of the actual benefit gained by the Employer. However, if the Employer can show that the use of the money was of no benefit to him, the Contractor's claim could fail.

There are many situations where money is paid back, or paid 'late', innocently. The case of liquidated damages properly deducted and paid back when an extension of time is awarded, is one. Another is the situation where an architect certifies an interim payment, for a Variation, which later is valued much higher. In these examples the architect had formed a *bona fide* opinion, which was found to be wrong, so was corrected later.

97 Lord Scott of Foscote, at para. 132.

98 See Chapter 1, Section 1.8.2.2 for further comment on restitution.

99 *Sempra Metals Ltd (formally Metallgesellschaft Ltd)* v. *Her Majesty's Commissioners of Inland Revenue & Anr*, [2007] UKHL 34.

100 At para. 240.

The employer had held the contractor's money innocently. Nevertheless, the contractor was inconvenienced by the shortfall in payment, and the employer benefited by the use of the money. The contractor cannot claim his loss as damages, because there is no breach of contract, but he may consider pursuing the return of the benefit in a restitution claim. The *Sempra* case seems to provide fresh opportunity to test the ingenuity of the parties to find new claims. Only time, and the courts, will tell how such claims are allowed to develop.

13.4.14 Cost of producing claim

One head which invariably appears in a Contractor's claim is the cost of preparing the claim. The cause normally given is that the Contractor was put to the trouble and expense of preparing a claim by the failure of the Architect to ascertain the loss and expense properly. In many cases the claim is no more than the Contractor is obliged to provide as information given under the procedures in the Contract. One case often mistakenly cited is *James Longley & Co.* v. *S W Regional Health Authority*,[101] in which a claims consultant's fees were reimbursed in part. However, the portion paid related only to the work done in preparing schedules to be annexed to the Points of Claim in arbitration, and which were of assistance to the arbitrator. Similarly in *Amec Process & Energy Ltd* v. *Stork Engineers & Contractors (No. 4)*[102] the collation and analysis of evidence carried out by the Claimant's own staff was held to be recoverable, on the grounds that the cost would be recoverable if carried out at greater expense by solicitors; but that too was in respect of recovering cost in prosecuting formal proceedings. However, the likely success of such claims is uncertain, and the cases do not give clear guidance. In *Aerospace Publishing & Anr* v. *Thames Water Utilities Ltd*,[103] the cost of engaging former employees on a freelance basis just before legal proceedings were started, and included time writing a witness statement, were disallowed by the Court of Appeal. Contractors sometimes annotate their claim with a note to the effect that it has been prepared in contemplation of arbitration. This is an attempt to bring the costs of preparing the claim into the ambit of the *James Longley* case, but it is not a matter within the Architect's power, and is not an entitlement. It is no more than evidence for judicial consideration at a later date, when applying discretion in awarding costs.

It has been argued that if the Contractor can show that the Architect has failed to carry out his duty to ascertain,[104] and as a result the Contractor was obliged to prepare a detailed and fully documented claim as a result of a breach, he may be able to recover.[105] In particular, a small Contractor company with limited staff expertise and without specialist knowledge might be able to demonstrate that he bought in management expertise, and prove his expenditure. The *Tate & Lyle* case[106] is sometimes cited as authority for entitlement to recovery. However, it is unlikely that the cost of preparing a claim would fall into the category of managerial time in problem-solving, and the principle of keeping records

101 (1983) 25 BLR 56.
102 15 March 2002; 1997 ORB 659; (2002) CILL 1883.
103 [2007] EWCA Civ 3, 11 January 2007.
104 See Section 13.2.2 of this chapter dealing with the Architect's duty.
105 See Powell-Smith, V., 'Architect Must Tot Up Contractor's Losses', *Contract Journal,* 30 July 1992, p. 7.
106 See Section 13.4.3 of this chapter.

is not unique to *Tate & Lyle*; it is an evidential principle of general application. In *Milburn Services Ltd* v. *United Trading Group (UK) Ltd*,[107] the judge said:

> I do not regard the Tate & Lyle case as enshrining any principle of law… beyond the principle that the burden is on the plaintiffs to prove their case, but it is an interesting approach to a factual matter by an experienced and respected judge.

Nevertheless, the faint possibility of success produced by cases like *James Longley* and *Amec* v. *Stork*, encourages contractors and their consultants to continue to pursue the cost of producing claims. However a side effect of adjudication as the first resort for dispute resolution, is to reduce, in practice, the occasions when preparation costs could be recovered in any event. Many claims are produced in order to establish the existence of a dispute, and to show that the issues have been crystallized before prosecuting an adjudication under Clause 9.2. Thus the costs are expended in pusuit of adjudication. Unlike litigation and arbitration, adjudication rules do not normally provide for the successful party to be awarded its costs.

Claims made under this head are usually for breach of contract, and are outside the normal authority of the Architect unless he is given express powers by the Employer.

13.4.15 VAT on damages

Payment in respect of damages (i.e. arising from a breach of contract) is not considered to be payment for supply of goods or services, and therefore falls outside the requirement for the addition of VAT. However, loss and expense, where paid as an amount due under a contract provision, does attract VAT. Entitlement may depend on how a claim is framed, or more particularly how a claim is met, when it is made in the alternative as a claim for loss and expense, or for damages. It does not matter what the claim is called – it is categorized by what it is – a contractual claim, or a claim against a breach.[108]

---

107 9 November 1995; 1993 ORB 534; (1995) CILL 1109.

108 See *Pring & ST Hill Ltd* v. *CJ Hafner T/a Southern Erectors* [2002] EWHC 1775 (TCC), per HHJ Humphrey LLoyd QC, at para 42.

14

Fluctuations

In construction contracts with long contract periods, tenderers are not normally expected to make allowances for inflation in their tender prices. They are usually required to tender on only costs current at the time, with provisions being made for adjustment of the contract price for inflation during the performance of the contract. Such contracts are referred to as 'fluctuating price' contracts whilst the term 'fluctuations' refers to the amount of adjustment to the contract price. This chapter examines JCT 05 provisions on fluctuations. Although the term covers both increases and reductions in payment caused by variation in costs, for simplicity and clarity, the discussion is mainly in the context of increases in costs.

There are three alternative sets of provisions on fluctuations contained in Schedule 7: Fluctuations Options A, B and C. Clause 4.21 provides that fluctuations under the Contract are to be dealt with in accordance with whichever one of these is stated in the Contract Particulars as applicable. Option A applies if no choice is indicated. The general approach in Options A and B is determination of fluctuations as the net amount arising from changes in certain types of costs and is sometimes referred to as the 'traditional approach' whilst that in Option C is the 'formula approach'. The formula approach attempts to determine what the Contract Sum would have been if the Contractor, assuming the same pricing level, had taken into account, in his tender, the inflation actually experienced.

14.1 Fluctuations under Option A

Generally recovery of fluctuations under Option A is limited to changes in the Contractor's tax liability and other liability of a statutory nature incurred in respect of the Works. The Contractor has to bear any additional costs for which he is not statutorily liable. For these reasons, contracts incorporating this clause are sometimes referred to as 'firm' or 'fixed' price contracts. The specific recoverable or allowable costs are in respect of:

1. changes in rates and types of contributions, levies and taxes paid or payable by the Contractor in his capacity as an employer of workers (paragraphs A.1.1 to A.1.9);
2. changes in duties and taxes payable by the Contractor as a consequence of the procurement, use or disposal of 'materials, goods, electricity, fuels, materials taken from the site as waste or any other solid, liquid, or gas' necessary for the execution of the Works (A.2);

3. the above items in respect of sub-contract works (A.3.1 to A.3.2);
4. an optional percentage addition to each of the above fluctuations (A.12).

14.1.1 Statutory contributions, levies and taxes in respect of employees

The Contractor is deemed to have allowed in the Contract Sum for any statutory contributions, levies and taxes which he, as an employer, was obliged at the Base Date to pay in respect of: (i) his 'workpeople' engaged upon or in connection with the Works either on or adjacent to the site; (ii) his workpeople directly employed by the Contractor and engaged upon the off-site production of materials and goods for the Works. 'Workpeople' is defined under paragraph A.11.3 as 'persons whose rates of wages and other emoluments (including holiday credits) are governed by the rules or decisions or agreements of the Construction Industry Joint Council (CIJC) or some other wage-fixing body for trades associated with the building industry'.[1] In *Murphy & Sons Ltd* v. *London Borough of Southwark*[2] the Court of Appeal held that labour-only sub-contractors were not included in a similar definition of 'workpeople' in the JCT 63. It is submitted that the decision is equally applicable to JCT 05.

It is stated in paragraph A.11.2 that 'materials' and 'goods', as used in Option A, include timber used for formwork but excluding other consumable stores, plant and equipment. The full list of items in relation to which fluctuations in duties and taxes are recoverable is 'materials, goods, electricity, fuels, materials taken from the site as waste or any other solid, liquid or gas'.[3] For simplicity, all these items are hereafter referred to collectively as 'relevant materials'.

The net increase or decrease in the Contractor's financial liability arising from any changes in the rates or types of these statutory liabilities from the position at the Base Date is recoverable or allowable by the Contractor. Similar entitlements apply in respect of other employees engaged upon or in connection with the Works but who do not fall under the definition of 'workpeople' – e.g. secretarial, administrative and managerial staff (para. A.1.3). Employees of this type are hereafter referred to as 'ancillary staff'. However, for the purpose of recovering the fluctuations, ancillary staff are to be treated as craftsmen with the following provisos in paragraph A.1.4: (i) the employee must have worked on or in connection with the Contract for at least 2 whole working days in the week to which the fluctuation relates; (ii) periods of less than a whole working day must not be aggregated into a whole working day; (iii) the highest rate for the Contractor's (or sub-Contractor's) craftsmen is to be used.

Under paragraph A.1.5 the Contractor is deemed to have included in the prices in the Contract Bills any refunds of statutory contributions, levies and taxes and other statutory payments in respect of the relevant workpeople payable by the Contractor as an employer at the Base Date. If there are any changes in the type or rate of refund or payment after the Base

---

1 This definition is repeated in para. B.12.3 of Fluctuations Option B.
2 (1983) 22 BLR 41.
3 Under JCT 1998 fuels were to be included in the fluctuations only if it was so stated in the Contract. This was in effect an option to the Employer to omit recovery of fluctuations on fuels. This option has been dropped from JCT 05. There is also now a general provision for recovery of fluctuations in respect of taxes on disposal of waste. The specific provision for Landfill Tax has therefore been dropped.

Date, the net amount representing the changes is to be paid to or allowed by the Contractor (para. A.1.6). For the purpose of calculating fluctuations, employees who have contracted-out of the state pensions scheme are to be treated as if they have not done so (para. A.1.8).

14.1.2 Duties and taxes in respect of materials

The Contractor is deemed to have priced in the Contract Bills for any duties and taxes payable at the Base Date on account of the procurement, use or disposal of the relevant material. The Contract Sum is to be adjusted for any changes in the duties and taxes so payable by the Employer (para. A.2.2). VAT is not included because it is dealt with elsewhere in the Contract.

14.1.3 Percentage for additional payment

The Contractor is allowed to indicate in his tender a percentage to be applied to the net amount payable under the heads already discussed to arrive at an additional payment (para. A.12). The applicable percentage is to be stated in the Contract Particulars. The general intention behind paragraph A.12 is to compensate the Contractor against cost increases not expressly recognized as subject to fluctuation.

14.1.4 Notices, evidence and calculations

By paragraph A.4.1 the Contractor must, as a condition precedent to recovery of fluctuation entitlements, give written notice of any changes in respect of which adjustment for fluctuation is applicable. He is also required to supply the Architect, or Quantity Surveyor, with all evidence and calculations reasonably necessary to determine the amount of fluctuations. Types of evidence commonly required to be submitted include:

- invoices from suppliers;
- invoices/receipts from bodies to which contributions, levies and taxes are payable;
- take-off of materials and labour;
- miscellaneous supporting calculations.

Furthermore, he must, on a weekly basis, provide a certificate of the validity of any evidence and calculations supplied in respect of employees who do not fall under the definition of workpeople.[4] These requirements must also be complied with in respect of fluctuations on the work of sub-contractors.

Paragraph A.5 provides that the Quantity Surveyor and the Contractor may agree the amount of adjustment to make in respect of any events giving rise to fluctuation entitlements. In *John Laing* v. *County and District*[5] it was held that this type of provision in a previous edition of this form of contract did not authorize the Quantity Surveyor to ignore the provisions of the Contract such as giving notices of the events as a condition precedent to recovery.

4 Intentionally issuing a false certificate could amount to theft; see Chapter 12, Section 12.7.
5 (1982) 23 BLR 1.

14.1.5 Exclusions from fluctuations under Fluctuations Option A

Paragraph A.10 excludes recovery of fluctuations on work done on a dayworks basis and changes in VAT. The reason for the first exclusion is that dayworks are paid for on the basis of current prices. Changes in VAT are covered elsewhere in the Contract.

14.2 Fluctuations under Option B

Where Fluctuations Option B applies, fluctuations are recoverable under the following heads:

1. labour costs (paragraphs B.1.1 to B.1.6);
2. statutory contributions, levies, and taxes and refunds/receivable premiums relating to 'workpeople' and ancillary staff (B.2.1 to B.2.8);
3. costs of materials and associated duties and taxes (B.3.1 to B.3.3);
4. the above costs in respect of sub-contract works (B.4.1 and B.4.2);
5. an optional percentage addition to the total of the above items of fluctuations (B.13).

14.2.1 Labour costs

The Contractor is deemed under paragraph B.1.1 to have included in the Contract Sum wages, other emoluments and other expenses payable to workpeople[6] and the cost of related employer's liability and third party insurance. This deeming provision expressly covers: (i) workpeople directly employed upon or in connection with the Works either on or adjacent to the site; (ii) workpeople employed elsewhere by the Contractor in the production of materials for the works (e.g. employees in an off-site workshop). The wages, emoluments and other expenses referred to above are only those determined at rates and prices governed by the rules, decisions, bonus schemes or other agreements of the CIJC or other appropriate wage-fixing body promulgated as at the Base Date.[7]

Fluctuations after the Base Date in wages, emoluments and other expenses of 'workpeople' arising from alterations in the rules, decisions and agreements already referred to are recoverable or allowable by the Contractor. Similar fluctuations apply in respect of ancillary staff (para. B.1.3). However, for the purpose of recovering fluctuations, such employees are to be treated as craftsmen with the following provisos under paragraph B.1.4: (i) the employee must have worked on or in connection with the Contract for at least 2 whole working days in the week to which the fluctuation relates; (ii) periods of less than a whole working day must not be added up into a whole working day; (iii) the highest rate for the Contractor's craftsmen is to be used. Consequential changes in the cost of employer's liability and third party insurance in relation to such ancillary staff also qualify as relevant fluctuations.

---

6 For discussion of the definition of 'workpeople' see Section 14.1.1 in this chapter.

7 The Contractor is deemed to have allowed for decisions and the like promulgated as at the Base Date even if they were then not yet in force.

The Contractor is also to recover or allow fluctuations in respect of costs of transporting workpeople for the purposes of carrying out the Works (paragraphs B.1.5 and B.1.6).[8] The baseline from which changes apply is a list of basic transport charges attached to the Contract Bills or figures determined from the rules of a recognized wage-fixing body. Where there is a change in the transport charges, or the wage-fixing body fixes new rates of reimbursement of fares to workpeople, they are recoverable or allowable by the Contractor.

In *Sindall (William)* v. *N.W. Thames Regional Health Authority*[9] the House of Lords held that increases in costs attributable to a voluntary bonus scheme operated by the Contractor were not recoverable under the equivalent provisions of the JCT 63. It is submitted that recovery of fluctuations under JCT 05 would be similarly restricted.

Good practice requires the Contractor and sub-Contractors to draw up and maintain up-to-date schedules of relevant workpeople and ancillary staff, their wages, other emoluments and related expenses.

14.2.2 Statutory contributions, levies or taxes

Paragraph B.2 broadly mirrors paragraph A.1. The Contractor is deemed to have allowed in the Contract Sum for any statutory contributions, levies and taxes which he, as an employer, is obliged to pay in respect of his workpeople (para. B.2.1). The net amount from any changes in the rates of these statutory liabilities from their figures at the Base Date is recoverable or allowable by the Contractor (para. B.2.2). The same principle applies to ancillary staff (para. B.2.3). There are similar provisions in paragraphs B.2.4 and B.2.5 regarding changes in tax refunds and receivable premiums in respect of employees engaged in connection with the Works.

For the purpose of calculating fluctuations, employees who have contracted out of the state pensions scheme are to be treated as if they have not done so. However, there is an exception where the private pension scheme is one established and operated by the CIJC or other recognized wage-fixing body. In such a case, the Contractor's contributions are to be treated as an element of wages, emoluments and other expenses under paragraph B.1 and, as such, recoverable or allowable as fluctuations.

14.2.3 Materials

The Contractor's tender is deemed to have been prepared on the basis of the market prices at the Base Date of 'materials, goods, electricity, fuels or any other solid, liquid or gas necessary for the execution of the Works'. The Contractor is expected to attach a list of these market prices to his tender. This list is normally referred to in the industry as the 'List of Basic Prices'. If during construction the Contractor incurs additional costs or makes savings on these resources because of changes in their market prices, the net difference is recoverable or allowable by the Contractor.[10] To qualify as fluctuations, any change in costs must satisfy two conditions. First, it must be in respect of items on the

8 There is no provision for recovery of similar fluctuations in respect of ancillary staff.

9 [1977] ICR 294.

10 Strictly speaking, the difference is to be determined by reference to prices payable when the Contractor bought the relevant materials or fuel.

List of Basic Prices. Second, the change must be due solely to changes in market forces, including the effects of changes of statutory taxes and duties on these items.

14.2.4 Sub-contractors and suppliers

The Contractor is to include in sub-contracts provisions that mirror the main contract fluctuation provisions: paragraph B.4. The net amount arising from fluctuations under the sub-contract is recoverable or allowable by the Contractor under the main contract.

14.2.5 Percentage addition

Where a percentage is stated in the Contract Particulars against paragraph B.13 the net amount of fluctuations determined under the heads discussed above is to be increased by that percentage. This is designed to allow additional payment to compensate the Contractor against cost increases not expressly recognized as subject to fluctuation (e.g. construction plant, consumable stores, head office overheads and profit).

14.2.6 Exclusions from fluctuations under Option B

Paragraph B.11 is the same as paragraph A.10, which is explained in Section 14.1.5 in this chapter.

14.2.7 Notices, evidence and calculations

These matters are covered in paragraph B.5 and are broadly similar to paragraph A.4.1 already discussed in Section 14.1.4 in this chapter. To avoid repetition, the reader is therefore referred to that section.

14.3 Fluctuations under Option C

Fluctuations Option C provides for the calculation of fluctuations by a formula requiring the use of indices[11] that reflect movements in national wages and prices. Neither the costs upon which the Contractor built up his tender nor the actual costs incurred by the Contractor are used in the calculation. There is therefore no need to specify the resources subject to fluctuation or their basic prices. Unlike the traditional approach, the formula method allows recovery of additional overheads and profit.

The formula method is incorporated into the 'With Quantities' and 'With the Approximate Quantities' variants of JCT 05 by selecting Fluctuations Option C in the Contract Particulars (i.e. by deleting Options A and B). The formula method is based on

11 These indices have traditionally been referred to as the 'NEDO Series 2 Indices'. For reasons given later, they are now referred to as the 'BERR Series 2 Indices'.

a classification of general building work into Work Categories. An index for each Work Category is calculated from indices tracking variations in costs of labour, plant and materials and weightings of the resources required to carry out the work in that Work Category. There is a Working Group on Indices, with wide representation from relevant sectors of the construction industry, which is responsible for overseeing the compilation of the indices and reviewing them from time to time. The Construction Directorate of the Department for Business, Enterprise and Regulatory Reform (BERR) is responsible for compiling and maintaining the indices. This responsibility is discharged by providing the Working Group with a Technical Secretariat for performing the task. Two types of indices are maintained: Series 2 Indices with 1976 as base date and Series 3 indices with 1990 as base date. These indices are still published in parallel and it is for contractual parties to make their choice from them in their contract.

Option C incorporates into JCT 05 the *Formula Rules*,[12] which sets out the formulae and defines their use.[13] The effect of the definition of 'Work Category' in Rule 3 is that the Series 2 indices must be used. The Work Categories are listed in Table 14.1.[14] The primary source of the indices is a monthly bulletin available as part of the online information services of the Building Cost Information Service of the Royal Institution of Chartered Surveyors. Indices for any month are provisional when first published. This is because the collection of the information to compile the indices takes considerable time and effort if they are to be accurate. The firm indices, which are published when sufficient information becomes available, are then substituted for the provisional figures in the valuation following their publication as firm.

Application of Fluctuations Option C requires a Schedule of Fluctuations, which indicates to which of the following categories each item in the Contract Bills is allocated:[15]

- one of the Work Categories;
- 'fix-only' work;
- contractor's specialist work;
- Balance of Adjustable Work;
- work excluded from formula adjustment;
- work covered by Provisional Sums subject to formula adjustment.

The Schedule is normally completed by a quantity surveyor, who is often, but does not have to be, the Quantity Surveyor under the Contract, and sent out as part of the tender documents to each tenderer. Sometimes, rather than include a schedule in the tender documents, each item in the tender Bills of Quantities is annotated with whichever of the above classifications it belongs to and the Schedule prepared from the annotation after contract formation. Once the Contract is entered into, the Schedule is binding upon both parties.

Specialist work includes electrical installations, heating, ventilating and air-conditioning installations, lift installations, structural steelwork, and catering equipment installations. For these types of work, there are no Work Categories except structural steelwork

12 Joint Contracts Tribunal Ltd, *Formula Rules*, Sweet & Maxwell, London, 2006 (hereafter *Formula Rules*).
13 They are part of a group of formulae referred to collectively as the 'BERR Price Adjustment Formulae', which include formulae for price adjustment on construction projects other than building.
14 The 49th Work Category in the original Series 2 Indices has been omitted from the *Formula Rules*.
15 See Rule 11a of the *Formula Rules*.

Table 14.1 Work Categories

No.	Description of work
1	Demolition and alteration
2	Site preparation, excavating and disposal
3	Hardcore and imported filling
4	General piling
5	Steel sheet piling
6	Concrete
7	Reinforcement
8	Structural precast and prestressed concrete units
9	Non-structural precast concrete components
10	Formwork
11	Brickwork and blockwork
12	Natural stone
13	Asphalt work
14	Slate and tile roofing
15	Asbestos cement sheet roofing and cladding
16	Plastic-coated steel sheet roofing and cladding
17	Aluminium sheet roofing and cladding
18	Built-up felt roofing
19	Built-up felt roofing on metal decking
20	Carpentry, manufactured boards and softwood flooring
21	Hardwood flooring
22	Tile and sheet flooring (vinyl, thermoplastic, linoleum, and other synthetic materials)
23	Jointless flooring (epoxy resin type)
24	Softwood joinery
25	Hardwood joinery
26	Ironmongery
27	Steelwork
28	Steel windows and doors
29	Aluminium windows and doors
30	Miscellaneous metalwork
31	Cast iron pipes and fittings
32	Plastic pipes and fittings
33	Copper tubes, fittings and cylinders
34	Mild steel pipes, fittings and tanks
35	Boilers, pumps and radiators
36	Sanitary fittings
37	Insulation
38	Plastering (all types) to walls and ceilings
39	Beds and screed (all types) to floors, roofs and pavings
40	Dry partitions and linings
41	Tiling and terrazzo work
42	Suspended ceilings (dry construction)
43	Glass, mirrors and patent glazing
44	Decorations
45	Drainage pipework (other than cast iron)
46	Fencing, gates and screens
47	Bituminous surfacing to roads and paths
48	Soft landscaping

which may be treated as part of Work Category 27 (steelwork). However, there are specialist engineering formulae for dealing with them. These formulae are not described in this book. Readers are referred to the *Formula Rules* for their details.

'Balance of Adjustable Work' covers that part of the Contractor's work which ranks for adjustment but which is not allocated to Work Categories or specialist work. Examples include preliminaries, water for the works, and insurance. Where there is a significant value of work in respect of items of a 'fix-only' nature, it may be appropriate to create a weighted index covering such work. The weighted index is called a 'Fix Only' index.

Fix Only work items may also be dealt with by allocating them to appropriate Work Categories, or including them in the Balance of Adjustable Work. The *Formula Rules* require that where the method of dealing with Fix Only items has not been specified in the Contract, the method to be adopted should be agreed with the Contractor.

14.3.1 Calculation of fluctuations

There are different formulae for each type in the Schedule of Fluctuations. The appropriate formula is applied to each type as next described. The net total of the fluctuations for all elements constitutes the adjustment for that Valuation Period. The net total of the fluctuations for all Valuation Periods then forms a component of the 'total value of work properly executed' under Clause 4.16.1.1.

14.3.2 Work Categories/Work groups

The formula for work allocated to a Work Category is stated in the *Formula Rules* as:

$$C = \frac{V(I_v - I_o)}{I_o}$$

where, C = the amount of fluctuation for the Work Category to be paid to or allowed by the Contractor; V = the value of work executed in the Work Category during the Valuation Period; I_v = the index number for the Work Category for the month during which the mid-point of the Valuation Period occurred; I_o = the Work Category index number for the Base Month.

An alternative form of this formula may be expressed as follows:

$$C_g = \frac{V \times I_v}{I_o}$$

where C_g is the gross amount due on the valuation (i.e. the sum of the value of work in the Work Category executed in the valuation period and the amount of fluctuation in respect of that Work Category); the other symbols have the same meaning as before.

Work Categories may be combined into a Work Group for which a weighted index number can be calculated and used in the manner already described for a Work Category. For example, where a priced Activity Schedule was provided, index numbers for each activity on the Schedule may be calculated. In the same way, if trade bills are used, suitable trade groupings may be established. The *Formula Rules* require that where this procedure is to be adopted the following should be complied with:[16]

- it should be stated in the Contract Bills that Part II of the *Formula Rules* shall govern the formula adjustment;
- the Contract should define what Work Categories are to be included in each Work Group;
- the Base Month should be stated in the Contract Bills;
- the Schedule of Fluctuations should show the total value of each Work Category within each Work Group.

For the Balance of the Adjustable Work the formula is:

$$C = \frac{V_o \times C_c}{V_c}$$

where, C = the amount of fluctuation for the Balance of Adjustable Work to be paid to or recovered from the Contractor; V_o = the value of work in the Balance of Adjustable Work in the Contract Bills; C_c = the total amount of fluctuation for all other Work Categories to be paid to or allowed by the Contractor for the Valuation Period; V_c = the total value of work in the other Work Categories for the Valuation Period.

For the purpose of calculating fluctuations, a Valuation Period is defined as commencing on the day after that on which the previous valuation was done and finishing on the date of the succeeding valuation. The index numbers for the Valuation Period are those for the month in which the mid-point of the Valuation Period occurs. If the Valuation Period has an odd number of days, then it will be the middle day of the Period. However, if it contains an even number, the mid-point will be the middle day of the period remaining after deducting the last day. It should be clear from the description of the way the formula method works that it cannot be used without valuation of work completed within each Valuation Period. Without such valuation, it would be impossible to determine the work in each Work Category completed within each Period. For this reason, Clause 4.11 and paragraph C.2 in Fluctuations Option C make valuation before each Interim Certificate mandatory.

14.3.3 Fluctuations after Practical Completion

The formula for adjusting the value of work which is included in Interim Certificates issued after the practical completion is given in Rule 28 as:

$$C = \frac{V \times C_t}{V_t}$$

where C = the amount of the fluctuations to be paid to or recovered from the Contractor; V = the value of work executed; C_t = the net total of the formula adjustment included in all previous certificates excluding fluctuations on specialist engineering work; V_t = the

16 See Rules 30 and 34.

total value of work but excluding the Contractor's specialist work included in previous Certificates.

This formula therefore has the effect of applying the average fluctuation rate over the contract period to the additional work involved.

14.3.4 Imported articles

Under Rule 4(ii) of the *Formula Rules* formula adjustment does not apply to articles imported for direct incorporation into the Works without prior processing. Paragraph C.3[17] of Fluctuations Option C requires the Contractor to attach to the Contract Documents a list of such articles indicating the market prices of their delivery to the site at the Base Date. Any change in these prices at the time of actual delivery is recoverable or allowable by the Contractor.

14.3.5 Delay, suspension and cessation of publication of the indices

Paragraph C.5 deals with the eventuality of delay in, or suspension or cessation of publication of the indices. The Contractor and the Employer are to continue with the normal procedures for determination of the fluctuations (e.g. interim valuations of the work in each Work Category for each Valuation Period). Whenever calculation of the fluctuations is due but the indices are not available for whatever reason, the Contract Sum is to be adjusted on a fair and reasonable basis (e.g. using the last published indices increased by the monthly rate of general inflation in the economy). Whenever indices are published after such reasonable adjustment but before the Final Certificate, the estimated adjustment must be replaced with an adjustment by the formula in the next payment certificate.

14.4 Errors, certification and retention

From the description of the processes and calculations involved in determining the amount to be recovered on account of fluctuations, the possibility of errors either in the allocation of work to the Schedule or in the calculations is only too real. To enable the Quantity Surveyor to agree fluctuation accounts speedily, well-organized contractors have well-designed proformas and computer systems for the calculation of fluctuations. In examining the Contractor's statements of fluctuations under Option A or B for inclusion in valuations, the Quantity Surveyor will normally check that:

- any cost item included is allowed under the appropriate clause;
- all conditions precedent have been complied with (e.g. notices and proper evidence from the Contractor);
- the calculations are arithmetically accurate;
- quantities of items in respect of which fluctuations are claimed are not in excess of the requirements of the Contract.

17 Rule 4(ii) refers to para. C.2. This must be a mistake.

The *Formula Rules* allow the Quantity Surveyor to correct the following types of error: (i) arithmetical errors in the calculation of the adjustment; (ii) incorrect allocation of the value of Work Categories to Work Groups; (ii) incorrect allocation of work as contractor's specialist work; (iv) use of incorrect index numbers.[18]

Fluctuations recovered or allowed under Fluctuations Option A or B are not subject to retention (Clause 4.16.2.3) whilst those under Fluctuations Option C are subject to retention (Clause 4.16.1.1).

14.5 Extension of time, delayed completion and fluctuation

By paragraph A.9.1 of Fluctuations Option A, if the Completion Date (as stated in the Contract Particulars or an Accepted Schedule 2 Quotation, or fixed by the Architect under Clause 2.28) is overrun, the amounts of fluctuations recoverable are frozen at levels which were operative at the Completion Date. However, under paragraph A.9.2, if any of the provisions in Clauses 2.26 to 2.29 (provisions on adjustment of the Completion Date) is amended, or the Architect fails to grant properly extension of time to which he considers the Contractor entitled, paragraph A.9.1 ceases to have effect. Paragraphs B.10.1 and B.10.2 of Fluctuations Option B contain corresponding provisions to the same intended effect where that option applies. Under paragraph C.6.1 of Fluctuations Option C, the formula is applied to the value of work executed after the Completion Date (or extended date of completion) but using the indices applicable to the month in which the date falls. This has the same effect as if all the work outstanding after the Completion Date was carried out during the month in which the Completion Date occurred. Similarly, this paragraph does not apply where any of the provisions on adjustment of the Completion Date is amended or the Architect fails to grant properly extension of time to which the Contractor is entitled.

Two main implications may be drawn from these provisions. First, Clauses 2.26 to 2.29 must not be amended without corresponding deletion of paragraphs A.9.2, B.10.2 and C.6.2 in Schedule 7, whichever is applicable. Second, they are yet another illustration of the general principle explained in Chapter 11, Sections 11.2 and 11.7.2, that an Architect does the Employer no favours by refusing to grant extensions of time in accordance with the Contract.

18 See Rule 5.

15

Payment

In Article 1 of the Articles of Agreement the Contractor undertakes to carry out and complete the Works in return for the Employer's promise in Article 2 to pay him a named sum: the Contract Sum. This sum, adjustable in defined circumstances, is to be paid to the Contractor at times and in a manner specified in the Conditions. The Employer may also make an advance payment to assist the Contractor's mobilization for the project and his cashflow. This chapter covers the following relevant issues most of which are covered in Section 4 of the Contract:

- the advance payment;
- Interim Certificates;
- responsibility for valuation and certification;
- valuation to determine the amount due;
- unfixed materials;
- failure to pay on a certificate;
- the treatment of retention funds;
- final adjustment of the Contract Sum;
- the Final Certificate;
- the Architect's liability for negligent certification;
- bonds that may be stated to be required in the Contract Particulars.

15.1 Advance Payment

The Employer must pay any amount entered against Clause 4.8 in the Contract Particulars as advance payment on the stated date. It is reimbursed to the Employer by deducting cumulatively the instalments specified in the Contract Particulars in the calculation of the amount to be stated in Interim Certificates as payable to the Contractor.[1]

Failure to make the advance payment by its due date amounts to a breach of contract by the Employer for which the Contractor would be entitled to interest either under the Late Payment of Commercial Debts (Interest) Act 1998 or as damages at common law. Details on the basis of these entitlements are provided in Section 15.6.1 in this chapter.

---

1 See Section 15.4 for details on the relevant calculations.

The Contract Particulars may be completed requiring the Contractor to provide an Advance Payment Bond in favour of the Employer as a condition precedent to the Employer's obligation to make the Advance Payment. The nature, purpose and other details of the Advance Payment Bond are explained in Section 15.10.1 in this chapter.

The Contract does not state the consequences of delay by the Contractor in providing the required bond. A possibility is that the Contractor's right to the advance payment is lost forever unless the Employer elects to accept the late bond, in which case there is a variation to the Contract that would determine the date when the advance payment becomes due again. Implication of a continuing duty to pay the amount upon provision of the bond even if late is also arguable.

15.2 Interim Certificates

Cashflow is the very lifeblood of the construction industry; its stoppage on any project will often bring it to a complete standstill.[2] Most construction contracts recognize this fact by allowing for the making of payment on account to the contractor before the works are complete. Under s. 109(1) of the Housing Grants, Construction and Regeneration Act 1996 (referred to hereafter as the Construction Act), a qualifying construction contract must provide for periodic payment to the contractor. Where there is a failure to comply with this requirement, relevant provisions in a Scheme for Construction Contracts, which provide for periodic payment, apply. Other sections of the Act make more detailed provisions on the parties' rights and obligations in respect of periodic payment. A major impetus towards the production of JCT 98 was a need to redraft JCT 80 to comply with the relevant provisions of the Act. This compliance has been carried forward to JCT 05.

By Clauses 4.9.1 and 4.9.2, from time to time, the Architect is to issue Interim Certificates stating the amounts to be paid to the Contractor. Generally, the amount in an Interim Certificate is an instalment of the Contract Sum reflecting the accomplishment of the Contractor's obligations since the previous Interim Certificate. As with every Certificate of the Architect, Interim Certificates are to be issued to the Employer, with a copy to the Contractor (Clause 1.9). The amount in an Interim Certificate is to be determined by applying valuation rules in Clauses 4.10, 4.16 and 4.20 to work done up to and including a date not more than 7 days before the date of the Certificate. This means that the amount does not include monies earned under the Contract in the 7 days immediately preceding the date of the Certificate.

15.2.1 Timing of issue of Interim Certificates

Clauses 4.9.1, 4.9.2 and 4.15.1 of JCT 05 together specify dates when the Architect is to issue to the Employer Certificates stating the amounts then due to the Contractor.

The date of issue of the first Interim Certificate determines the dates of subsequent Interim Certificates. It is to be noted that, in deciding this date, it is often desirable that account is taken of any organizational and financial constraints of the Employer regarding

2 For examples of judicial notice of this feature of the construction industry see *Dawnays Ltd* v. *F. G. Minter* [1971] 2 All ER 1389, as per Lord Denning at p. 1393; *Gilbert-Ash (Northern) Ltd* v. *Modern Engineering (Bristol) Ltd* [1973] 3 All ER 195 (hereafter *Gilbert-Ash*), per Lord Diplock at pp. 215–216.

payment. For example, if the Employer's organization issues cheques only within certain periods of the month, then the date must be set in such a way that payments on Certificates are not held up by any such restrictions. Failure to consider such factors may lead to delayed payment and, therefore, breach of the Contract by the Employer.

15.2.1.1 Timetable up to Practical Completion

From the Date of Possession up to the date of issue of the Practical Completion Certificate, the Architect is to issue the Interim Certificates on dates determined from a formula provided in the Contract Particulars (Clause 4.9.2). The first Interim Certificate is to be issued on the 'first date' stated in the Contract Particulars. It is advised in a footnote against the entry for this date that it should not be more than one month after the Date of Possession.[3] Thereafter, an Interim Certificate must be issued on the same date of each month adjusted to the nearest working day in that month. For example, if the date stated in the Contract Particulars is 31st of the month and that date for a particular month falls on a Sunday, the certification date would be Friday the 29th of that month. In contrast, if the first date were the 1st of the month the certification date would be Monday the 2nd of the month. This process must continue up to the date of practical completion or to within one month thereafter (Clause 4.9.2).

A footnote in the Contract Particulars suggests the last day of the month as an alternative to a specific date being stated for the first certificate. For example, if the first month on site is September, the alternative entry would be: '*the last day of September* and thereafter the *last day* in each month or the nearest Business Day in that month'. 'Business Day' is defined in Clause 1.1 as 'any day which is not a Saturday, a Sunday or a Public Holiday'. 'Public Holiday' is also defined in the same Clause as 'Christmas Day, Good Friday or a day which under the Banking and Financial Dealings Act 1971 is a bank holiday'.

It is also stated in the Contract Particulars that if the date of the first Interim Certificate is not stated in any of these ways, Interim Certificates are to be issued at intervals not exceeding one month up to the date of Practical Completion or to within one month after that date. In that event, the first Interim Certificate is to be issued within 1 month of the Date of Possession.

15.2.1.2 Timetable after Practical Completion

After the issue of the Practical Completion Certificate, Interim Certificates are to be issued as and when further amounts are ascertained as payable to the Contractor by the Employer (Clause 4.9.2). Examples of such amounts include: settled claims for loss and/or expense; additional fluctuations payable as a consequence of publication of firm fluctuations indices that are different from their provisional figures applied in previous certificates; and final releases of retention in respect of parts taken over before the issue of the Practical Completion Certificate in respect of the whole of the Works.[4]

An Interim Certificate must be issued upon the expiry of the Rectification Period or upon issue of the Certificate of Making Good in respect of the whole of the Works, whichever occurs later. As an implication of the rules under Clause 4.20 on the ascertainment of the Retention, the second half of the Retention is to be released in this Certificate. The flexibility in the timing of the issue of Interim Certificates after Practical Completion

---

3 The contract does not offer corresponding advice for the situation where the Works are to be completed in Sections.

4 See Section 15.4.2 for explanation of the timetable for release of retention.

is subject to the proviso that the Architect is not obliged to issue an Interim Certificate within 1 calendar month of having issued a previous Interim Certificate.

15.2.2 Payment on Interim Certificates

At the time of writing, a Local Democracy, Economic Development and Construction Bill to amend the Construction Act had just started its passage through Parliament. It is anticipated that the final changes will be implemented into the Construction Act towards the autumn of 2009. One of the main areas of changes concerns payment. In the rest of Section 15.2, we explain the procedures for interim payment under the existing Act and their implementation into the current JCT 05. The proposed interim payment procedures, which the JCT will have to implement by appropriate revisions to JCT 05 after enactment of the changes, are explained in Section 15.11. The Construction Act in the form at the time of writing is hereafter referred to as the original Construction Act.

Section 110 (1)(b) of the original Construction Act states that 'every construction contract shall provide for a final date for payment in relation to any sum which becomes due'. It is important to note that the Act only requires the parties to agree a final date and that it does not specify when this should be. The parties are therefore at liberty to fix whatever final dates they wish to apply to their contract.[5] Clause 4.13.1 implements the requirement for a final date in relation to payment on an Interim Certificate by providing that it is 14 days from the date of its issue.

The payment procedures under the Contract entail two types of notices referred to in this book as 'Payment Notice' and 'Withholding Notice'. Both notices are required under the Contract to be in writing. The House of Lords have pointed out on two occasions[6] that the purpose of these notices in construction contracts is to reduce the incidence of set-off abuse by formalizing the payment process so that a contractor knows immediately and with sufficient clarity the payment that should be made and why any of it is being withheld. This knowledge gives the Contractor the opportunity to seek early resolution of any disputed aspects by adjudication.

The timetable for the procedures is summarized in Figure 15.1.

15.2.3 Payment Notice

Section 110(2) of the original Construction Act states:

> Every construction contract shall provide for the giving of notice by a party not later than five days after the date on which payment becomes due to him under the contract, or would have become due if:
>
> (a) the other party had carried out his obligations under the contract, and
> (b) no set-off or abatement was permitted by reference to any sum claimed to be due under one or more other contracts,

---

5 This freedom has been the subject of some criticism of the Construction Act: it is feared that parties in strong bargaining positions (e.g., employers and main contractors) may impose unreasonably distant final dates for payment just to comply with the letter of the Act but to defeat its spirit, which includes speedy payment for work done.

6 See *Melville Dundas Ltd (In Receivership)* v. *George Wimpey UK Ltd* [2007] BLR. 257; 112 ConLR 1; [2007] CILL 2469; [2007] 1WLR 1136; [2007] 3 All ER 889 and *Reinwood Ltd* v. *L Brown & Sons Ltd* [2008] UKHL 12.

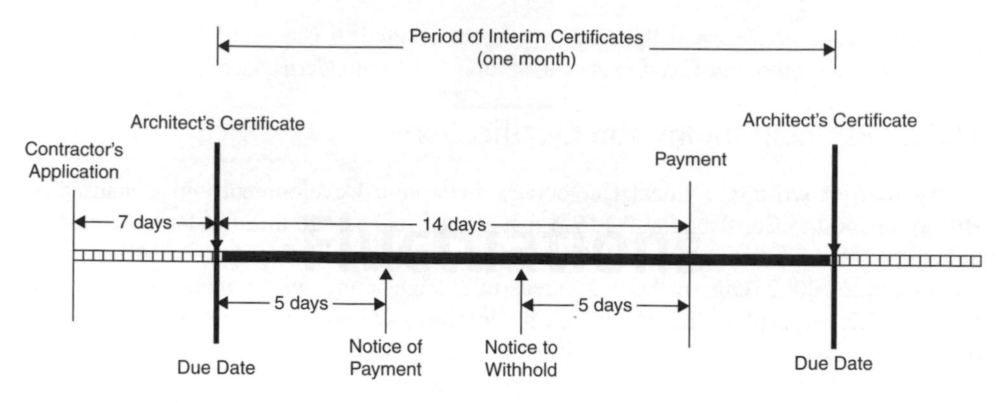

Fig. 15.1 Certification and payment timetable

specifying the amount (if any) of the payment made or proposed to be made, and the basis on which that amount was calculated.

This section is implemented in Clauses 4.13.3 and 4.15.3, which require the Employer to serve the Contractor with this notice within 5 days after the issue of an Interim Certificate and the Final Certificate, respectively. This is the type of notice referred to in this book as a 'Payment Notice'. In relation to the amount stated in an Interim Certificate or the Final Certificate JCT 05 requires the Architect to specify the basis on which it was calculated. No details as to the precise content of the accompanying statement are provided. With an Interim Certificate it would be good practice to state that it was calculated in accordance with Clause 4.10 and include the following information:

- project title;
- project reference number;
- names of the parties;
- names of the Architect and the Quantity Surveyor;
- the total value of work properly executed by the Contractor since commencement;
- total loss and/or expense granted by the Architect;
- value of materials and goods on the site;
- value of relevant materials and goods off site;
- total adjustment for price fluctuation;
- total retention deducted;
- cumulative value of instalments of any advance payment to be reimbursed;
- amount payable under the previous Interim Certificate;
- amount payable under the present Interim Certificate.

In the case of the Final Certificate the appropriate information would include the final ascertainment of loss and/or expense and final adjustment of the Contract Sum in accordance with Clauses 4.3 and 4.5.

As the Construction Act does not prescribe the detailed contents of a Payment Notice, one to the effect that the payment is for work executed under the Contract and that the basis of the calculation of the amount to be paid is as set out in the Architect's statement accompanying the relevant Certificate would be sufficient. Indeed, the Guidance Notes to Amendment 18, which introduced the requirement for Payment Notices into the JCT 80, suggested that a Payment Notice is not necessary where the Employer intends to pay on time the full amount stated in a Certificate.[7]

Where the Employer intends to pay less than the amount certified the Payment Notice must set out in sufficient detail the basis of the Employer's calculation of the amount due.

15.2.4 Withholding Notice

Section 111(1) of the original Construction Act states:

> A party to a construction contract may not withhold payment after the final date for payment of a sum due under the contract unless he has given an effective notice of intention to withhold payment.

This is the notice referred to in this book as the 'Withholding Notice'. Section 111(2) of the Act specifies the contents of an effective Withholding Notice as: (i) the total amount to be withheld, (ii) the grounds for the withholding, and (iii) itemization of the total amount to be withheld by individual grounds if there is more than one ground. Section 111(1) also states that a Payment Notice with the required content may also serve as Withholding Notice. This means that if a Payment Notice also contained the information required in a Withholding Notice, the latter notice may be dispensed with. The use of the term 'withhold' in s. 111(1) is unfortunate because it has no defined legal meaning. To decipher the effect of the section, we must therefore apply its ordinary everyday meaning. The *Concise Oxford Dictionary* (7th edition) defines it as 'hold back'.

The most common grounds upon which employers and main contractors refuse to pay on a certificate in full include abatement and set-off[8] which, as a matter of strict law, are different from holding back. It has been argued that technically, therefore, a would-be payer may lawfully refuse to pay in full without serving a Withholding Notice if the failure to pay in full does not amount to 'holding back'. Under Clauses 4.13.5 and 4.15.4, the Employer is to serve a Withholding Notice not later than 5 days before the final date for the relevant payment. The use in these Clauses of the wording 'the Employer may give written notice to the Contractor which shall specify any amount proposed to be *withheld and/or deducted*' was probably intended to avoid this type of argument on semantics.

15.2.5 Contractual grounds for withholding

The clauses that expressly give the Employer the power to deduct certain sums from payment due or to become due to the Contractor are:

- **Clause 2.32.1** – liquidated damages;
- **Clause 6.4.3** – the amount paid or payable by the Employer in respect of premiums for insurance taken out by the Employer to remedy the Contractor's defaults on his obligation to take out or maintain insurances required under Clause 6.4.1;

---

7 The drafting of the requirement for payment notices was severely criticized by the House of Lords in *Melville Dundas Ltd (In Receivership)* v *George Wimpey UK Ltd* [2007] BLR 257; 112 ConLR 1; [2007] CILL 2469; [2007] 1 WLR 1136; [2007] 3 All ER 889.

8 See Section 15.2.6.

- **Paragraph A.2 of Insurance Option A of Schedule 3** – the amount paid or payable by the Employer in respect of premiums for insurance taken out by the Employer to remedy the Contractor's defaults on his obligation to take out or maintain All Risk Insurance required under paragraph A.1.

The Employer can therefore deduct such sums against payment required under a Certificate provided valid Withholding Notices are served. It is to be noted that the Architect and the Quantity Surveyor are not to take account of these sums in valuations or Certificates. They are matters entirely for the Employer to deal with although the Architect or the Quantity Surveyor may be under a contractual obligation to advise him accordingly.

It is to be noted that, compared to JCT 98, the matters for which the Employer is expressly allowed to make deductions against payment certificates have been reduced. Matters allowed under JCT 98 but not under JCT 05 are:

- **Clause 2.10** – deduction of an appropriate amount where the Architect instructs that the Contractor's setting out errors are not to be amended;
- **Clause 2.38** – an appropriate deduction where, after the issue of the Practical Completion Certificate, the Architect instructs that any defects and the like for which the Contractor is responsible are not to be made good;
- **Clause 3.11** – the Employer's cost of implementing an Architect's instruction with which the Contractor has failed to comply;
- **Clause 3.18.2** – an appropriate deduction where, before the issue of the Practical Completion Certificate, the Architect instructs that any defects and the like for which the Contractor is responsible are not to be made good.

All these deductions are now to be made by the Architect against the Contract Sum. Many Employers are likely to welcome the convenience of having all these matters wrapped up in certificates rather than complying with the bureaucracy surrounding withholding of payment certified.

15.2.6 Other grounds for withholding

In *Gilbert-Ash*,[9] the House of Lords held that a payer of a certificate under a construction contract is entitled to raise any set-off available to him under the general law unless the contract itself expressly removes that right. JCT 05 does not remove this right of the Employer in relation to his obligation to pay on Interim Certificates. It follows therefore that the Employer would be entitled to deduct from Certificates not only sums expressly authorized under the Contract but also set-offs available under the general law. An example of a right of set-off under the general law is set-off through the defence of abatement at common law.[10] This set-off involves the contention that the Contractor's performance is worth much less than he is contractually obliged to achieve (e.g. because of defects). This right is appropriate where the Employer wants to rely on the defects to make a deduction against the certified sum. Liquidated mutual debts may also be set against each other even if they arise from different contracts.[11] For example, at common law, the Employer

9 *Gilbert-Ash (Northern) Ltd* v. *Modern Engineering (Bristol) Ltd* [1973] 3 All ER 195.

10 *Mondel* v. *Steel* (1841) 8 M. & W. 858; (1976) 1 BLR 106; see also s. 53 of the Sale of Goods Act 1979; *Barrett Steel Buildings Ltd* v. *Amec Construction Ltd* (1997) 15 CLD-10-07; *Mellowes Archital Ltd* v. *Bell Projects Ltd* (1997) 87 BLR 26 (damages for delays cannot be the basis of abatement).

11 *Hargreaves (B) Ltd* v. *Action 2000 Ltd* (1992) 62 BLR 72.

would be entitled to deduct liquidated damages incurred on another contract. In equity, the Employer may raise a cross-claim (counterclaim) by way of set-off (e.g. damages for delay). Finally, under s. 323 of the Insolvency Act 1986, where the Contractor is insolvent, the Employer may be entitled to set-off the contingent liability of the Contractor as a result of termination of the Contractor's employment (e.g. estimated damages of completing the project with another contractor).[12]

The general right of set-off recognised in *Gilbert-Ash* was applied in *Acsim (Southern) Ltd* v. *Danish Contracting Ltd*,[13] which concerned withholding of payment by a main contractor under a sub-contract. The sub-contract expressly stated that the parties' rights to set-off 'are fully set out in these conditions and no other rights whatsoever shall be implied as terms of this sub-contract relating to set-off'. The Court of Appeal decided that the exclusion was not wide enough to exclude the defence of abatement at common law.[14]

15.2.7 Effect of failure to serve the notices

Clause 4.13.5 of JCT 05 states:

> Subject to any notice given under Clause 4.13.4 [Withholding Notice], the Employer shall no later than the final date for payment pay the Contractor the amount specified in the notice given under Clause 4.13.3 [Payment Notice] or, in the absence of a notice under Clause 4.13.3 [Payment Notice], the amount stated as due in the Interim Certificate.

Clause 4.15.5 makes corresponding provision in relation to any balance stated in the Final Certificate as due to the Contractor. Before these provisions are examined in the light of litigation on the construction of similarly worded contractual provisions, case law on the general question of the effect of failure to serve Payment and Withholding Notices is analysed to highlight the underlying issues in such disputes.

The Construction Act does not state what the consequences of failing to serve a Payment Notice should be. It was probably intended as a statement of good practice on early identification of differences between the parties regarding the quantification of the amount due. It is stated in s. 111(1) that, without an appropriate Withholding Notice, the paying party 'may not withhold payment after the final date for payment of a sum due under the contract'.[15] The effect of this provision has been one of the most controversial issues concerning adjudication. However, it is fairly uncontroversial that an adjudicator should not enter into inquiry about certain types of set-off if they are not covered by a valid Withholding Notice. Examples of this type of uncontroversial set-off include liquidated damages for delayed completion and other sums which, under the terms of the contract, the Employer is entitled to deduct from payment due under the Contract or to recover from the Contractor as debt.

---

12 *Willment Brothers Ltd* v. *North West Thames Regional Health Authority* (1984) 26 BLR 51.

13 (1989) 47 BLR 55; see also *GPT Realisations Ltd (in Administrative Receivership & in Liquidation) Ltd* v. *Panatown Ltd* (1992) 61 BLR 88.

14 For a recent review of the authorities on set-off by way of abatement in construction contracts see *Multiplex Construction (UK) Ltd* v. *Cleveland Bridge UK Ltd and Another* [2006] EWHC 1341 (TCC); 107 ConLR. 1.

15 As explained in Section 15.6.2, s. 112(1) also provides that if the paying party withholds the whole of the amount or any part of it beyond the final date without an effective Withholding Notice, the payee is entitled to suspend performance of its obligations. The Contractor may choose to terminate his employment under Clauses 8.9.1.1 and 8.9.3.

The question at the centre of the controversy has been whether, when a contractor makes a claim for payment and the Employer fails to serve a Withholding Notice, the Employer is entitled, in any ensuing adjudication, to argue that the amount claimed is not due under the Contract. The bases of these arguments have been given such labels as 'abatement', 'counterclaim', 'cross-claims' or even 'set-off'.

The decisions and judicial comments on this question can be categorized into two groups: the pro-contractor approach[16] and the pro-employer approach.[17] According to the pro-contractor approach, if the paying party failed to serve any notice within the relevant periods the Contractor becomes entitled to payment, in full, of the amount in the application. And, if the dispute is referred to adjudication, the adjudicator must not enter into any inquiry as to whether the amount claimed in the application is the amount due under the contract. The underlying argument in the pro-employer approach is that, on a literal construction of s. 111(1), the Employer cannot be said to be 'withholding' if the reason for not honouring a payment application is that the amount claimed is not that due under the Contract. Accordingly, the Employer would be entitled to put forward to the adjudicator the amount that the Employer believes to be the amount due and the adjudicator must then adjudicate between the rival contentions about the amount due.

Most of the debate concerning the propriety of an adjudicator entering into an inquiry as to whether the amount claimed is the amount due arose from contracts that did not provide for certification of the amount due by an independent third party such as the Architect under JCT 05. The effect of such a role was considered in the Scottish case of *Clark Contracts Ltd* v. *The Burrell Co. (Construction Management) Ltd.*[18] It concerned the effect of failure by an employer to serve a Withholding Notice on his obligation to pay on a payment certificate issued by an architect under a contract which provided for certification and payment in essentially the same terms as JCT 05. Clause 30.1.1.1 stated:

> The Architect shall from time to time as provided in Clause 30 issue Interim Certificates stating the amount due to the Contractor from the Employer and the Contractor shall be entitled to payment therefor within 14 days from the issue of each Interim Certificate.

Relying on Lord Hoffman's *dictum* in *Beaufort Developments (NI) Ltd* v. *Gilbert Ash NI Ltd*[19] that an Architect's payment certificate was not conclusive evidence that the Works had been carried out in accordance with the Contract, the Employer contended that it was entitled to defend payment on the certificate by arguing that the amount certified was not 'the amount due under the contract'. The Employer also relied on Lord MacFadyen's suggestion in *S L Timber*[20] that where the paying party's case for not paying is that the amount claimed is not the amount due because the work was either not done or not measured and valued properly, there would be no need for a Withholding Notice. The

---

16 Cases supportive of the pro-contractor construction include *VHE Construction plc.* v. *RBSTB Trust Co. Ltd* [2000] BLR 187; [2000] 2 TCLR 278; *Northern Developments (Cumbria) Ltd* v. *J & J Nichol* [2000] BLR 158; *KNS Industrial Services (Birmingham) Ltd* v. *Sindall Ltd* (2000) CILL 1652; (2001) 17 Const LJ 170.

17 *Woods Hardwick Ltd* v. *Chiltern Air Conditioning Ltd* [2001] BLR 23; *S L Timber Systems Ltd* v. *Carillion Construction Ltd* [2001] BLR 516 (hereafter *S L Timber*); *Millers Specialist Joinery Company Ltd* v. *Nobles Construction Ltd*, TCC Case No. TCC 64/00; (2001) CILL 1770.

18 *Clark Contracts Ltd* v. *The Burrell Co (Construction Management) Ltd*, (No.1) (2002) SLT (sh. Ct) 103.

19 *Beaufort Developments (NI) Ltd* v. *Gilbert Ash NI Ltd* [1999] 1 AC 266; (1998) 88 BLR 1; 39 ConLR 66.

20 *S L Timber Systems Ltd* v. *Carillion Construction Ltd* [2001] BLR 516.

Contractor distinguished *S L Timber* as applicable only where there was no third party certification of the amount due. This is how Sheriff Taylor dealt with the arguments:

> There was no dispute that the architect had issued an interim certificate. It therefore seems to me that [the pursuers] became entitled to payment of the sum brought out in the interim certificate within 14 days of it being issued. In my opinion that is an entitlement to payment of a sum due under the contract. In order to reach the figure in the interim certificate one has made use of the contractual mechanism. To use the words deployed by Lord MacFadyen in para 20, the issue of an interim certificate was the occurrence of 'some other event on which a contractual liability to make payment depended.' This situation falls to be contrasted with the position in *SL Timber Systems* where, before the adjudicator, there had been no calculation of the sum sued for by reference to a contractual mechanism and which gave rise to an obligation under the contract to make payment. There had been no more than a claim by the pursuers which claim had not been scrutinized by any third party. Thus, in my opinion, if The Burrell Co (Construction Management) Ltd wished to avoid a liability to make such payment because the works did not conform to the contractual standard they would be withholding payment of a sum due under the contract. In order to withhold payment they would require to give notice in terms of section 111(1) of the Act. No such notice was given.
>
> The interim certificate is not conclusive evidence that the works in respect of which the pursuers seek payment were in accordance with the contract…That however does not preclude the sum brought out in an Interim Certificate being a sum due under the contract. The structure and intent of the Act, as I understand it, and accepted by [counsel for the pursuers], is to pay now and litigate later. Accordingly the defenders would not be precluded from suing the pursuers should the works carried out by the pursuers and for which the pursuers have been paid turn out faulty. Indeed, as I understand matters that is what the defenders seek in the counterclaim. However, in my opinion the pursuers have pled all that is required of them given the omission on the part of the defenders to serve a notice under section 111.

Sheriff Taylor's analysis of s. 111 in the context of a contract with an independent third party certifier was fully endorsed by the English Court of Appeal in *Rupert Morgan Building Services (LLC)* v. *David Jervis and Harriet Jervis*.[21] The contract employers were the appellant and the builder was the respondent. The Contract between the parties was for certain works to be done on their cottage. Payment was required to be made on certificates issued by the employers' architect. A dispute arose in connection with payment certificate No. 7, which was for £44,000 odd plus VAT. The employers contended that only £17,000 of this amount was payable but failed to serve notice to withhold payment of the balance within the prescribed period before the final date for payment. The builder argued that, by virtue of section 111(1), the effect of this failure was that the employers were to pay the entire certified sum. The builder, therefore, applied for summary judgment. In reply, the employers argued that it was open to them to defend the summary judgment application on the grounds that the unpaid balance was for work that was either not done or was already paid for. On when payment becomes due, Jacob LJ said at paragraph 11:

> In this ASI contract, the sum is determined by the *certificate* [His Lordship's emphasis]. Clause 6.1 provides that "payment shall be made to the Contractor only in accordance with

21 [2003] EWCA Civ.1563; [2004] BLR 18; [2004] TCLR 3; 91 ConLR 81.

the Architects certificate'. Clause 6.32 defines the sum – essentially the approved gross value of work done less retention and amounts previously paid. Clause 6.33 says when it is to be paid: "the employer shall pay to the Contractor the amount certified within 14 days of the date of the certificate, subject to any deductions and set-offs due under the contract'. So it is not the actual work done which either defines the sum or when it is due. The sum is the amount in the certificate. The due date is 14 days from certificate date. The certificate may be wrong – the architect may (though this is unlikely because he will be working from the builder's bill) have missed out work done (which would operate against the contractor) or he may have included items not in fact done or items already paid for (which would operate against the client). In the absence of a withholding notice, s. 111(1) operates to prevent the client withholding the sum due. The contractor is entitled to the money right away. The fundamental thing to understand is that s. 111(1) is a provision about cash-flow. It is not a provision which seeks to make any certificate, interim or final, conclusive. Analysed this way one sees that there is something inconsistent about the clients' argument here. Their duty to pay now and the sum they have to pay arise only because of the certificate.

In summary, the case law thus suggests that the effects of failure to serve a Payment Notice (required under Clauses 4.13.3 and 4.15.3) and/or a Withholding Notice (required under Clauses 4.13.4 and 4.15.4) are as follows.

1. The Employer must pay the full amount certified within 14 days if none of the notices is served. However, *Collins (Contractors) Ltd* v. *Baltic Quay Management (1994) Ltd*[22] highlights practical difficulties the Contractor may encounter in enforcing such payment where the Contract incorporates a valid arbitration agreement. The Contract from which that litigation arose provided for certification and service of notices in identical terms to those in JCT 05. The Contract Administrator under the Contract issued an Interim Certificate which the Employer failed to pay by the final date but without service of any of the required notices. When the Contractor brought court proceedings for summary judgment to enforce payment the Employer applied to stay them to arbitration under section 9 of the Arbitration Act 1996. The Court of Appeal, by a unanimous decision, held that, on the authority of *Halki Shipping Corp.* v. *Sopex Oils (The Halki)*,[23] the Employer was entitled to such a stay.
2. If only a Payment Notice is served the amount stated in the notice is the Employer's assessment of the amount due. A dispute therefore crystallizes if the Contractor disagrees with the Employer's assessment. Either party, but the Contractor more usually, may refer the dispute to adjudication. The adjudicator in such a reference would be acting correctly to enter into inquiry as to whether the amount certified is the amount due. Meanwhile the Employer must pay the amount specified in the Payment Notice by the final date of the certificate. Depending on the adjudicator's decision, the Employer may have to make further payment to the Contractor in respect of the Certificate. Such additional payment should carry interest from the final date of the Certificate.
3. If only a Withholding Notice is served the Employer's obligation to pay by the final date applies to the difference between the certified amount and the sum to be withheld

---

22 [2004] EWCA Civ 1757; [2005] BLR 63; [2005] TCLR 3; 99 ConLR. 1 (hereafter *Collins* v. *Baltic Quay*).
23 [1998] 1 WLR 726.

stated in the notice. The Contractor may challenge the validity of the Withholding Notice in adjudication and the Employer would have to make further payment to the extent that the challenge is successful.

4. Where both notices are served, the Employer must pay the amount stated in the Payment Notice less the sum to be withheld pursuant to the Withholding Notice. Similarly, any of the notices can be challenged in adjudication as already explained.

In most of these disputed payment situations the Referring Party would be the Contractor. However, before the Contractor has done so, the Employer may himself refer to adjudication upon being informed that the Contractor disputes the relevant notice.

15.3 Responsibility for valuation and certification

The responsibility for issuing Interim Certificates is the Architect's but he may delegate the task of producing the supporting valuation to the Quantity Surveyor whenever he considers it necessary (Clauses 4.9.1 and 4.11). Where fluctuations are to be recovered under Schedule 7 Fluctuations Option C (adjustment by formula), valuations are mandatory (Clauses 4.11 and paragraph C.2 of Fluctuations Option C).

Clause 4.12 allows the Contractor to make an application setting out the Gross Valuation according to rules stated in Clause 4.16. If the Contractor makes such an application, the Quantity Surveyor must make an interim Gross Valuation. If the valuation of the Quantity Surveyor differs from the Contractor's, he must submit a copy to the Contractor. There is a further requirement that, in case of such difference, the valuation of the Quantity Surveyor must have the same level of detail as that of the Contractor and identify the areas of differences. This is really a formalization of the common practice whereby valuations are prepared jointly by the Quantity Surveyor and the Contractor's quantity surveyor.

The Architect is responsible for the correctness of the sum stated in Certificates although he is entitled to rely on the measurements and valuations of the Quantity Surveyor unless they are obviously wrong.[24] The Architect therefore has the power to adjust any valuation prepared by the Quantity Surveyor where he considers it necessary. As quantity surveyors are more expert in valuation, most Architects would be reluctant to embark upon any major adjustments of the valuations without good cause.

The Conditions do not expressly provide for the possibility of the Architect failing to issue an Interim Certificate when he should have done so. However, a number of cases involving other building contracts which were also silent on the point have been before the courts.

> *Compania Panamena Europea Navigacion Ltd* v. *Frederick Leyland & Co. Ltd*:[25] a contract for repair of a ship provided that the repairers were to be paid their expenditure certified by the owner's surveyor. The surveyor contended that his jurisdiction under the contract went beyond the quality of the work of repair to include questions of economy of

24 *R. B. Burden* v. *Swansea Corporation* [1957] 1 WLR 1167; the liability of the Architect in respect of certificates is discussed in Section 15.9.

25 [1947] AC 428 (hereafter *Panamena*); this was applied by the Supreme Court of New South Wales in *Perini Corporation* v. *Commonwealth of Australian* (1969) 12 BLR 82.

the repairs. He then refused to issue a certificate unless the repairers produced full information on which he could consider the issue of economy. The owners were in concurrence with this stance. The House of Lords held that the surveyor's interpretation of the contract was incorrect and that the repairers' action to recover the value of the repairs without a certificate should succeed. The basis for the decision was that it was an implied term of the repair contract that certificates would be issued in accordance with the contract. It was also suggested that where a certifier failed to perform his role, the employer has a positive duty to take action to ensure his due performance and even to dismiss and replace him if his defaults continue.

Croudace Ltd v. *London Borough of Lambeth*:[26] the Architect under a contract in the terms of the JCT 63 Form was the Chief Architect of the Council. When he retired, the Council failed to appoint a replacement, with the consequence that the Contractor's claims for loss and/or expense could not be considered. The Court of Appeal held that the failure of the Council to appoint or to instruct anyone to assess the Contractor's claim was a breach of an implied duty to take reasonable steps to ensure that the Contractor's claim was ascertained.

It is submitted therefore that failure to issue an Interim Certificate under JCT 05 puts the Employer in breach of contract with the Contractor.[27] The Contractor's damages for the breach would include the amount that should have been payable if the Architect had certified. The Contractor can therefore enforce payment even in the absence of a Certificate from the Architect. In addition, depending upon the nature and extent of the failure, the Contractor may argue that this type of breach goes to the root of the Contract and therefore that it entitles him to treat the Contract as terminated and to sue for damages.

Failure to certify is different from wrong certification. In *Lubenham Fidelities and Investments Co. Ltd* v. *South Pembrokeshire District Council and Another*[28] the Court of Appeal decided that an Employer under the JCT 63 who became aware of shortcomings in the Architect's performance of his function as certifier was not under a duty to stop the shortcomings or tell the Architect how to do the job properly. In that case, in calculating the amount due under a certificate, the Architect made deductions not authorized under the Contract. It was also stated that, as applies to most standard construction contracts, an Employer under the JCT 63 did not warrant that the contract administrator would certify the correct amount due under the Contract. Any errors were to be pointed out to the Architect for correction in the next certificate or resolved through the contractual dispute resolution processes. The Contractor's contention that he was entitled under the principle in *Panamena* to recover the amount that should have been certified was rejected. That case was distinguished on the grounds that, unlike the JCT 63, the Contract in *Panamena* did not contain an arbitration clause. It is submitted that there is no material difference between JCT 63 and JCT 05 on the issue of the Employer's obligations in respect of the Architect's performance of his role as certifier, particularly where Article 8 applies.[29]

---

26 (1986) 33 BLR 20.
27 cf *Penwith District Council* v. *VP Developments Ltd* [1999] EWHC 231 (TCC).
28 (1986) 33 BLR 39.
29 Article 8, which is optional, contains an arbitration agreement.

15.4 Valuation for certification

The amount to be stated as due under an Interim Certificate is to be quantified in accordance with Clauses 4.10, 4.16 and 4.20. Some of the calculations involved may be unnecessary where a schedule of stage payments is prepared and incorporated into the Contract. In such a situation, the Contract would have to be amended accordingly. In view of the pitfalls associated with amending JCT 05, due care has to be exercised.[30]

The amount to be certified is given by the expression:[31]

$$C = G - R - A - P$$

where C is the amount of the Certificate; G is the Gross Valuation; R is the Retention (as explained in Section 15.10.3 of this chapter, no deduction is to be made in respect of retention where the Contract Particulars state that Clause 4.19 applies and the Contractor provides to the Employer a Retention Bond in the form set out in Part 3 of Schedule 6[32]); A is the cumulative value of reimbursements due on any advance payment;[33] and P is the total of the amounts stated as due in previous Interim Certificates.

15.4.1 The Gross Valuation

The Gross Valuation can be determined from the formula:[34]

$$G = g_r + g_o - g_c$$

where g_r = total value of work subject to retention (see Section 15.4.2); g_o = total value of work not subject to retention (see Section 15.4.3); g_c = the total deduction in pricc to be allowed by the Contractor (see Section 15.4.4).

15.4.2 Sums subject to retention

The following are subject to retention:

1. the value of work properly executed by the Contractor, including Variations (Clause 4.16.1.1);[35]
2. fluctuations where Fluctuations Option C (the formula method) applies (Clause 4.16.1.1);[36]

---

30 See Chapter 1, Section 1.5.5.
31 See Clause 4.10.
32 See Section 15.10.3 of this Chapter for details on the Retention Bond.
33 See Sections 15.1 and 15.10.1 of this Chapter for details on advance payment and its reimbursement by instalments.
34 See Clause 4.16
35 This figure may be determined by: (i) physically measuring work items completed, multiplying the quantity of each item completed by its rate in the Contract Bills and summing up for all items; (ii) assessing the stage the Works have reached and taking the applicable payment from a stage payment schedule if it applies; (iii) where a priced Activity Schedule has been provided, assessing the percentages of the activities completed, taking the same percentage of the relevant price in the Schedule. The value of work in Variations is determined in accordance with the relevant provisions in Section 5.
36 Details on this method of recovering fluctuations are in Chapter 14.

3. the total value of materials and goods on the site and intended for incorporation into the Works (Clause 4.16.1.2);
4. the total value of Listed Items (Clause 4.16.1.3).

By Clause 4.20, the Retention Percentage is 3% or such other rate stated in the Contract Particulars. The Retention to be deducted is determined by applying the Retention Percentage to the listed elements as follows. The full Retention Percentage is applied to the value of work which forms part of the Works or a Section yet to reach practical completion (Clause 4.20.2.1). Where the Works have reached practical completion but a Certificate of Making Good is yet to be issued for the Works, the percentage to be applied is one half of the Retention Percentage (4.20.3). Similarly, half of the Retention Percentage is to be applied to the work in any Section that has reached practical completion but for which a Section Certificate of Making Good has not been issued. Upon the issue of a Certificate of Making Good for a Section the value of work in that Section ceases to be subject to retention. Where the Employer takes over part of the Works or Section under Clause 2.33 the Relevant Part is deemed to have reached practical completion by the Relevant Date. Only half of the Retention Percentage is to be applied to the value of work in that Part.

The result of the operation of Clause 4.20 is therefore that half of the Retention (or the part of it for completed Sections or Relevant Parts taken into possession by the Employer before Practical Completion of the whole of the Works) is released and paid to the Contractor in the Interim Certificate that follows practical completion of the Works or Section. The second half of the Retention in respect of each completed Section is released in the Interim Certificate after the issue of the Certificate of Making Good for that Section. Clause 4.9.2 requires an Interim Certificate to be issued upon whichever of the following occurs last:

- expiry of the Rectification Period or of the last of the corresponding periods for Sections;
- the issue of the Certificate of Making Good for the Works or of the last of the corresponding Certificates for the Sections.

The remainder of the Retention is to be released in this Interim Certificate (i.e. no deduction is made for retention).

15.4.3 Sums not subject to retention

They are listed under Clause 4.16.2 as:

1. costs incurred by the Contractor in respect of:
   (i) additional premiums paid on account of early use or occupation of the Works or the site by the Employer (Clause 2.6.2);
   (ii) fees and charges legally demandable as a consequence of the Statutory Requirements (Clause 2.21);[37]

37 Such fees and charges are deemed already included in the Contract Sum where they are priced in the Contract Bills or they relate solely to the Contractor's Designed Portion. There is therefore no entitlement to recover such costs as an extra.

(iii) the Contractor's liability for infringement of royalties and patent rights (Clause 2.23);
(iv) compliance with AI requiring the opening up of work for inspection or the carrying out of any tests where the results of the inspection/tests showed that the work or materials were in accordance with the Contract (Clause 3.17);[38]
(v) the taking out and maintenance by the Contractor of insurance against damage to property from the Employer's risks defined under Clause 6.5.1 (Clause 6.5.3);
(vi) increase in premiums for Terrorism Cover under paragraph A.5.1 of Insurance Option A in Schedule 3;
(vii) premiums paid for remedial insurance by the Contractor in response to any default by Employer in respect of his obligation to take out and maintain All Risks Insurance of the Works as required under paragraph B.2.1.2 or C.3.1 of Schedule 3;

2. loss and/or expense under Clauses 3.24 or 4.23 (Clause 4.16.2.2);[39]
3. the value of work of restoration, replacement or repair of loss or damage to the Works from the insured risks under Insurance Options B or C and disposal of debris (Clause 4.16.2.2);[40]
4. increase in costs under Fluctuations Option A or B (Clause 4.16.2.3).[41]

15.4.4 Deductions to be allowed by the Contractor

Clause 4.16.3 lists the following deductions to be made as part of the pre-certification valuation process:

1. Deductions in respect of the Contractor's setting out errors which are not to be corrected (Clause 2.10).
2. Deductions in respect of defects, shrinkages and other faults that appear within the relevant Rectification Period for which the Contractor is responsible but which, by agreement, are to remain without rectification (Clause 2.38).
3. The Employer's cost of implementing an Architect's instruction with which the Contractor failed to comply (Clause 3.11).
4. Deductions in respect of work, materials and goods not in accordance with the Contract but which are by agreement to remain without rectification (Clause 3.18.2).
5. Cost reduction under Fluctuations Option A or B as applicable.

15.4.5 Negative Interim Certificates

Over-valuation can result in over-payment of the Contractor by the Employer through Interim Certificates. The situation may also arise where, regardless of whether or not there

38 Where such work is provided for in the Contract Bills it is to be paid for as part of the value of the Works and is subject to retention.
39 The subject of loss and/or expense is discussed in detail in Chapter 13.
40 Paragraphs B.3.5 and C.4.5.2 of Schedule 3 provide that such work is to be treated as a Variation.
41 These methods of recovery of fluctuations are explained in detail in Chapter 14.

has been any wrongdoing by the Architect or the Quantity Surveyor, latent defects are detected in work already paid for. The Architect can, and indeed must, correct such valuation errors in subsequent Interim Certificates if additional work has been done. Where the over-payment cannot be fully recovered by practical completion, there is the question whether the Architect may issue a negative Interim Certificate which can have the effect of requiring the Contractor to pay the amount stated to the Employer. The wording of Clause 4.9 on Interim Certificates refers to 'the amount due to the Contractor from the Employer'. Only the provisions on the Final Certificate contemplate expressly an amount due to the Employer from the Contractor. An implied authority to issue a negative certificate imposing an obligation on the Contractor to pay the stated amount over to the Employer may however be arguable on grounds of business efficacy.

Interpreting the express terms strictly and absent an implied term, it would therefore appear that, where additional work done after the over-payment is insufficient to allow its recovery in full before practical completion, the Employer must wait until the Final Certificate to recover any shortfall within the normal administrative machinery of the contract. A possible way forward for the Employer would be to challenge the certificate through adjudication although prior payment notices may allow estoppel defences depending on their wording. The Employer may also make an unjust enrichment claim against the Contractor[42] but it is arguable that such a claim, having its foundations in equity rather than contract, cannot give rise to a dispute 'under the contract' capable of being referred to adjudication.

The Contractor's case for holding on to the over-payment until the Final Certificate is not without merit. He could contend that the express provision for a negative Final Certificate goes against implying a term for enforceable negative certificates in the sense that providing expressly for a negative Final Certificate whilst omitting to do the same for Interim Certificates must have been a deliberate expression of allocation of the cash-flow risk from over-certification to the Employer. Considering that the Employer not only appoints the Architect but also has the opportunity to correct any mistakes by service of appropriate notices, few would find such allocation unfair.

The answer to this issue is by no means clear, and depending on the cause of the over-certification, there is likelihood that the Employer will suffer some injustice and loss. His natural reaction may be to turn to the Architect, or even the Quantity Surveyor, who may be liable for the loss.[43]

15.5 Unfixed materials and interim Certificates

Materials intended for construction contract works but not yet incorporated into them fall into two categories. The first category covers those delivered to the site, and these are usually referred to as 'materials on site'. The second category consists of all those materials intended for but not yet delivered to the site (e.g. goods of mechanical and electrical

42 On the authority of the House of Lords decision in *Sempra Metals Ltd (formerly Metallgesellschaft Ltd)* v. *Her Majesty's Commissioners of Inland Revenue and Anor* [2007] UKHL 34; [2007] 3 WLR 354 the Employer may also claim compound interest on the overpayment.

43 See *Sutcliffe* v. *Thackrah* [1974] AC 727.

installation and materials for temporary works stored in the contractor's workshops or on the premises of suppliers and sub-contractors). This second category is usually referred to as 'off-site materials'. The two categories are often referred to collectively as 'unfixed materials'.

Most construction contracts contain provisions allowing the contractor to be paid for unfixed materials in interim certificates. Such payment exposes the employer to the risk that the materials may never actually be used for the Works. For example, the employer may be paying for materials ownership of which never did vest in the contractor in the first place. The contractor, suppliers and sub-contractors and their holding companies may still have lien in the materials after they have been paid for by an employer. In addition, the materials and goods may be stolen, lost, damaged or even diverted by the contractor to another project. JCT 05 contains elaborate provisions designed to protect the Employer against these risks. They are contained mainly in Clauses 2.24, 2.25, 3.9.2, 4.16 and 4.17.

15.5.1 On-site materials and goods

In respect of materials delivered to the site and intended for incorporation into the Works, the important provisions are as follows:

1. They are not to be removed from the site for any purpose other than the carrying out of the Works without the written consent of the Architect. This consent must not be unreasonably withheld (Clause 2.24). A corresponding provision must be included in every sub-contract (Clause 3.9.2.1).
2. Their total value must be included in the Interim Certificate provided they are reasonably, properly and not prematurely delivered to the site, and are adequately protected against loss or damage (Clause 4.16.2).
3. After their inclusion in Interim Certificates and payment on the Certificate by the Employer, the materials become the property of the Employer (Clause 2.24). A term is to be included in every sub-contract to the effect that the sub-contractor concerned shall not deny the transfer of title to the Employer where the sub-contractor's materials are affected (Clause 3.9.2.1.1).
4. The Contractor is to include a term in every sub-contract to the effect that where the Contractor pays for the materials before he is paid by the Employer ownership in the materials is to pass to the Contractor (Clause 3.9.2.1.2).
5. They must be insured against loss or damage from the risks covered by the Clause 6.8 definition of 'All Risk Insurance' (paragraph A.1/B.1/C.2 of Schedule 3 as applicable).
6. The Contractor is responsible for any loss or damage to the materials even after ownership in them has passed to the Employer (2.24).

The primary intention behind the conditions that must be satisfied before unfixed materials are included in an Interim Certificate is to ensure that they do become the property of the Employer upon payment on the Certificate. Unfortunately, complete compliance with the conditions does not necessarily guarantee this end. This is because it is not uncommon for contracts for the supply of materials to contain stipulations to the effect that the materials remain the property of the supplier until paid for in full by the buyer or even until all debts owed by the buyer to the supplier are settled in full. Such clauses are referred to as 'Retention of Title' (ROT) clauses or *Romalpa* clauses after a famous case in which the court upheld the effectiveness of such clauses against purported transfer of title to third parties.[44] In any such situation, no matter the terms of JCT 05, the Contractor cannot effectively transfer title in the materials and goods to the Employer as required by some of its provisions. The Latin maxim *nemo dat quod non habet* (you cannot give away what you have not got!) best expresses this position of the law.

This problem came to prominence in the construction industry through the litigation in *Dawber Williamson Roofing Ltd* v. *Humberside County Council*.[45]

> Main contractors on a JCT 63 contract for the construction of the Council's school sub-contracted the roofing to the claimant sub-contractor who then delivered 16 tons of roofing slates to the site. After receiving payment for the slates through a certificate issued under the main contract, but before paying the claimant for them, the main contractor became insolvent. The sub-contract provided that materials brought onto the site by the sub-contractor remained the property of the sub-contractor until they were paid for by the contractor. Relying upon a clause of the JCT63 that unfixed materials and goods paid for under an Interim Certificate become the property of the Employer, the Council claimed ownership in the slates. It was held that ownership of the slates had not passed to the Contractor. The sub-contractors were therefore entitled to damages for wrongful detention and use of the materials by the Council.

It is important to note that *Dawber Williamson* applies to contests between an employer and the contractor's sub-contractors for ownership of unfixed materials and that such a contest between an employer and the contractor's supplier may well result in a different outcome. The difference arises from the fact that the Sale of Goods Act 1979 applies to supply of materials but not to sub-contracts, which are contracts for work and materials rather than sale of goods contracts. Section 25(1) of Sale of Goods Act states:

> Where a person having bought or agreed to buy goods obtains, with the consent of the seller, possession of the goods or the documents of the title to the goods, the delivery or transfer by that person, or by a mercantile agent acting for him, of the goods or documents of title, under any sale, pledge or other disposition thereof, to any person receiving the same in good faith and without notice of any lien or other right of the original seller in respect of the goods, has the same effect as if the person making the delivery were a mercantile agent in possession of the goods or documents of title with the consent of the owner.

The effect of this section on the competing interests of the contractor's suppliers and the Employer, where clauses in main contracts purported to transfer ownership of unfixed materials, was considered in *Archivent Sales & Development Ltd* v. *Strathclyde Regional Council*.[46]

> This case arose between an Employer under a JCT 63 contract and a supplier of ventilators to the Contractor. The supply contract provided that 'until payment of the price in full is received by the company the property and the goods supplied shall not pass to the customer'. Clause 14 of the JCT 63, along the same lines as JCT 05 Clause 2.24, provided that unfixed materials must not be removed without the consent of the Architect and that

44 *Aluminium Industrie Vaasen BV.* v. *Romalpa Aluminium* [1976] 2 All ER 552.
45 (1979) 14 BLR 70 (hereafter *Dawber Williamson*).
46 (1984) 27 BLR 98.

ownership of materials passed to the Employer upon their inclusion in an Interim Certificate and payment on it by the Employer. After the Contractor had been paid for the ventilators by the Employer but before the Contractor had in turn paid for them, the Contractor went into receivership. The sellers took proceedings against the Employer claiming ownership of the ventilators. The Scottish Court of Session (Outer House) held that the effect of s. 25(1) of the Sale of Goods Act was to confer good title on the Employer.

An essential requirement for application of s. 25(1) is that the second buyer was not aware of any ROT clauses in the original supply contract. The principle in this case is therefore likely to apply to suppliers only where the Employer and the Architect are not aware of the terms of the relevant supply contracts.

15.5.2 Off-site materials and goods

Clause 4.17 distinguishes between uniquely identified items (e.g. a boiler from a specified supplier), and those not uniquely identified (e.g. aggregates). Regardless of whether the item is uniquely identified or not, the Architect must include its value in Interim Certificates provided the following conditions are met:

1. The item must be included in the Listed Items, defined under Clause 1.1 as 'materials, goods, and/or items prefabricated for inclusion in the Works which are listed as such items by the Employer in a list supplied to the Contractor and annexed to the Contract Bills'.
2. The item must be in accordance with the Contract.
3. Where materials belonging to that type of item are in the premises of their manufacture, assembly or storage, they must either be set apart or suitably marked to identify the Employer as the party who ordered them and that they are destined for delivery to the Works.
4. The Contractor has insured them for their full value against loss or damage from the Specified Perils under a policy which covers both the Employer and the Contractor and has provided the Architect with proof of the cover. The period of cover is to be from when ownership in the materials passed to the Contractor until they are delivered to the site. As explained in Chapter 7, Section 7.3.4, upon delivery they become part of Site Materials and covered by the Contractor's All Risk Insurance of the Works and Site Materials.
5. The Contractor has provided the Architect with reasonable proof that ownership of the relevant materials is vested in the Contractor so that, as provided for in Clause 2.25, they become the property of the Employer after payment on the relevant Certificate. As a minimum, the Architect must check relevant supply contracts to ensure that under their terms the Contractor has acquired ownership of the relevant materials.
6. The Contractor has procured from a surety approved by the Employer any bond stated as required in the Contract Particulars. This type of bond is referred to in this work as an 'Off-site Materials Bond' and is explained in detail in Section 15.10.2.

15.6 Remedies for non-payment by the Employer

The remedies against failure of the Employer to meet his payment obligation include claiming interest on the amount not paid, suspension of performance, or termination of the

Contractor's employment. Some of these remedies are available concurrently. For example, the Contractor may suspend work and claim interest at the same time. Regardless of whether the Contractor has decided to suspend performance or terminate his own employment, he would still be interested in ways of compelling the Employer to pay on the Certificate, including interest, if the Employer is solvent. This section therefore also considers the appropriateness of arbitration, litigation and adjudication. Whichever of these is most appropriate depends upon the dispute resolution provisions in the Contract and the reasons for the failure to pay.

15.6.1 Interest

Failure by the Employer to pay on a Certificate by its final date for payment is a breach of the Contract for which the Contractor may claim payment of interest under one of three heads: (i) a claim under the Contract; (ii) a claim under the Late Payment of Commercial Debts (Interest) Act 1998 (LPCDIA); (iii) a claim for damages at common law.

Clause 4.13.6 provides that the Contractor is entitled to recover simple interest on the amount the Employer failed to pay for the period of non-payment beyond the final date. The applicable rate of interest is referred to as the Interest Rate, which is defined under Clause 1.1 as 'a rate 5% per annum above the official dealing rate of the Bank of England current at the date that a payment due under the Contract becomes overdue'.

The LPCDIA applies if the Employer and the Contractor both entered into the Contract in the course of a business.[47] This Act implies into any qualifying contract a term that any debt under it carries simple interest at a rate fixed by the Secretary of State. The current rate is 8 per cent above the Base Rates of the Bank of England. There may be a problem with the 5 per cent stipulated in the Contract because s. 8(4) of LPCDIA provides that any contract terms are void to the extent that they purport to 'vary the right to statutory interest so as to provide for a right to statutory interest that is not a substantial remedy for the late payment'. Section 9(1) of LPCDIA provides that a remedy shall be regarded as a substantial remedy unless:

1. the remedy is insufficient for the purpose of compensating the supplier for the late payment; or
2. it would not be fair and reasonable to allow the remedy to be relied on to oust or vary the right to statutory interest that will otherwise apply in relation to the debt.

Under s. 9(3), the reasonableness test involves having regard to the benefits of commercial certainty, the relative bargaining positions of the parties, whether the term was imposed by one party to the detriment of the other, whether the supplier received an inducement to accept the term. It may be argued that only percentages over the current figure set by the Secretary of State would be considered substantial. However, it is submitted that, as JCT 05 is an industry-wide standard form, the 5 per cent is likely to be considered a substantial remedy.

47 The LPCDIA applies to all qualifying contracts made after August 7, 2002. Before then its provisions could be enforced by only small businesses.

For a long time the law has been that interest cannot be awarded as damages[48] although it can be awarded as special damages if they are pleaded and proved.[49] This position may well have changed after *Sempra Metals Ltd (formerly Metallgesellschaft Ltd)* v. *Her Majesty's Commissioners of Inland Revenue & Anor*.[50] In that case the House of Lords held by a majority of three to two that the court had jurisdiction to award interest, simple and compound, as damages on claims for non-payment of a debt.

15.6.2 Suspension of performance/termination

At common law, there is no implied right to suspend work on account of non-payment by an employer.[51] Ss 112(1) to (3) of the Construction Act state:

(1) where a sum due under a construction contract is not paid in full by the final date for payment and no effective notice to withhold payment has been given, the person to whom the sum is due has the right (without prejudice to any other right or remedy) to suspend performance of his obligations under the contract to the party by whom payment ought to have been made ('the party in default').
(2) The right may not be exercised without first giving the party in default at least seven days' notice of intention to suspend performance, stating the ground or grounds on which it is intended to suspend performance.
(3) The right to suspend performance ceases when the party in default makes payment in full of the amount due.

Clause 4.14 implements the statutory right to suspend performance of contractual obligations. The Employer is therefore not entitled to terminate the Contractor's employment on grounds of suspension where the reason for it is the Employer's failure to pay on a Certificate by the final date for payment without a valid Withholding Notice.

A Local Democracy, Economic Development and Construction Bill had just started its passage through Parliament at the time of writing. The Bill will amend the Construction Act to state more clearly that the Contractor may, in his discretion, suspend any or all of his obligations under the contract for non-payment. Even if the JCT does not amend this form accordingly after enactment of this change, the Contractor may still rely on the amended Act in support of partial suspension.

The Contractor's right to suspend performance ceases when the Employer pays in full. The Contractor must then therefore recommence the carrying out of the works unless, in the

48 *London Chatham and Dover Railway* v. *South Eastern Railway* [1893] AC 429; *President of India* v. *La Pintada Compania* [1985] AC 104.

49 *Wadsworth* v. *Lydall* [1981] WLR 598 – a decision of the Court of Appeal which was approved by the House of Lords in: *President of India* v. *La Pintadaa Compania* [1985] AC 104.

50 [2007] UKHL 34; [2007] 3 WLR 354.

51 However, see *C.J. Elvin Building Services Ltd* v. *Noble* [2003] EWHC 837 (TCC) in which it was stated that an employer's refusal to honour payment obligations or threats not to pay further sums due in accordance with the contract is capable of amounting to repudiation. *Lubenham Fidelities and Investment Co. Ltd* v. *South Pembrokeshire District Council and Another* (1986) 33 BLR 39 (CA); see also *Perini Corporation* v. *Commonwealth of Australia* (1969) 12 BLR 82 (a decision of the Supreme Court of New South Wales) and *Canterbury Pipelines* v. *Christchurch Drainage* (1979) 16 BLR 76 (a decision of the Court of Appeal of New Zealand).

meantime, his employment has been validly terminated. Unfortunately, the Contract is silent on how soon after the payment in full the Contractor must recommence the performance of his contractual obligations. It is submitted that an obligation to recommence within a reasonable time of the payment would be implied. What is a reasonable time would depend on the surrounding circumstances of the relevant obligation (e.g. the period of non-payment), whether the Contractor has removed his resources from the site, and other work taken on as a result of the suspension. For example, where the Contractor stopped insuring the Works, it may be quicker to reinstate the insurance cover than to re-start physical work on site.

It is essential that the Contractor follows the specified procedures for suspending performance. Failure to do this would leave the Contractor open to charges of repudiation. The procedure is as follows. When the Contractor reaches a decision to suspend after the Employer has failed to pay in accordance with the Contract, he is to give the Employer written notice of his intention to do so on account of the non-payment. A copy of the notice to suspend must be served on the Architect. The right to suspend crystallizes after 7 days of continued failure to pay after the Contractor gave the notice. It is provided in Clause 1.7 that notice is deemed duly given or served if: (i) it is addressed and given by actual delivery or (ii) it is addressed and sent by pre-paid post to the Party at the appropriate address. The appropriate address is that stated in the Contract Particulars or such other address as may be agreed by the parties from time to time. If there is no such stated or agreed address the appropriate address is the registered or principal office in the case of corporate parties. In other cases it is the last known principal business address.

The Contractor may also terminate his employment under Clauses 8.9.1.1 and 8.9.3 as explained in detail in Chapter 16, Section 16.9.1.

15.6.3 Adjudication

Invoking adjudication would be the Contractor's best course of action where the Employer puts up any defence to his liability to pay; for example, that there are defects in work included for payment, or that the Quantity Surveyor failed to apply the valuations rules correctly, or that he is entitled to set-off. Dispute resolution by adjudication is outlined in Chapter 17.

15.6.4 Litigation

The *Beaufort Developments (NI) Ltd* v. *Gilbert Ash (NI) Ltd and Others*[52] decision reasserted the jurisdiction of the courts to open up and review and revise certificates and decisions of contract administrators. It has been suggested that, as a consequence of that decision, litigation may be a more attractive option to some project participants than arbitration. There are two powerful weapons available in litigation that the Contractor can apply for on commencement of proceedings: Summary Judgment and/or Interim Payment under the Civil Procedures Rules 24 and 25, respectively.

The essence of an application for Summary Judgment is that there is no realistic defence to the claim for payment. It enables the applicant to obtain final judgment without a full trial, thereby saving time and costs. Usually both the evidence to support the

52 (1998) 88 BLR 1.

application and that required to rebut it are provided in the form of sworn statements of truth. Such statements may give rise to prosecution for the crime of contempt of court if they contain falsehoods.

Interim Payment is ordered where the court determines that a full trial is required but is nonetheless satisfied that, should the trial be held, the applicant would be awarded a substantial amount. In such circumstances, the court can order immediate payment of the part of the claim clearly due pending the final trial. However, if the outcome of the trial is that the amount paid was not owed in part or at all, the court can order an appropriate refund.

The appropriateness of litigation depends upon whether or not Article 8, JCT 05 arbitration agreement, applies. If it does, the Employer is likely to seek stay of any legal proceedings to arbitration. Section 9(4) of the Arbitration Act 1996 states that, on such an application, the court must stay the proceedings 'unless the arbitration agreement is null and void, inoperable or incapable of being performed'. This was construed by the Court of Appeal in *Halki Shipping Corporation* v. *Sopex Oils Ltd (The 'Halki')*[53] to imply that the court cannot exercise jurisdiction to order summary judgment unless the Employer admits that the sum is due. This means that if the Employer puts forward any argument against liability, no matter how flimsy, the court has no choice but to stay the proceedings. This decision was applied by the Court of Appeal in *Collins (Contractors)* v. *Baltic Quay Management (1994) Ltd*,[54] a case which highlights a major shortcoming of arbitration agreements in construction contracts.

15.6.5 Arbitration

Arbitration is an option if the Contract Particulars are completed to indicate that Article 8 applies or the parties, after the dispute has arisen, agree in writing to resolve their dispute by arbitration. It is often argued that current arbitration law gives arbitrators sufficient powers to provide summary remedies similar to summary judgment and interim payment. This would be helpful to the Contractor only if the appointed arbitrator is adequately skilled and is prepared to use his powers in that way.[55]

15.7 The treatment of retention funds

Each Interim Certificate is to be accompanied by a statement specifying the retentions of the Contractor (Clause 4.18.2). The statement should also specify any set-off against the retention.

The primary purpose of retention in construction contracts is to provide an employer with security against latent defects and other failures of the Contractor to perform his obligations. However, many Employers use retention funds as working capital in the running of their businesses. Apart from the risk of misappropriation of the Contractor's retention by the Employer, there is also the risk that in the event of the Employer's insolvent

53 [1998] 1 Lloyd's Rep. 465.
54 [2004] EWCA Civ 1757; [2005] BLR 63; [2005] TCLR 3; 99 ConLR 1.
55 See John Uff '100-day arbitration: is the construction industry ready for it?' (2005) Const LJ 3.

liquidation or bankruptcy, the retention funds will have to go into the pot for distribution among his creditors.[56] To be able to follow our examination of how these risks are addressed in the Contract, there is a need to understand the concept of a 'trust'.

15.7.1 The concept of a trust

The essence of a trust is that the legal owner (the trustee) of property is only holding it for the benefit of another (the beneficiary). There is therefore a separation between legal ownership and beneficial ownership. The trustee is said to stand in a fiduciary relationship (i.e. one of the utmost confidence) to the beneficiary. This means that the trustee must not derive any personal benefit from holding the property unless the terms of the trust provide for that benefit. In addition, in the trustee's performance of his responsibility, he must not put unacceptable risks on the beneficiary's interests in the property.

A trust can be created expressly by statute, the operation of the law or a declaration of the legal and beneficial owner (settlor) during his lifetime. We are concerned here with creation of trusts by declaration. Although no specific words are required, the words actually used must show that: (i) the settlor intended to create a trust (certainty of intention); (ii) the intended beneficiaries are identifiable (certainty of objects); (iii) the property subject to the trust is specific or ascertainable (certainty of subject matter).[57]

15.7.2 Trusts of retention funds

Clause 4.18.1 states that 'The Employer's interest in the Retention is fiduciary as a trustee for the Contractor (but with no obligation to invest)'. This clause raises the question whether it creates a trust of the retention funds in favour of the Contractor. The effect of a similar clause in the JCT 63 was considered in *Rayack Construction Ltd* v. *Lampeter Meat Co. Ltd.*[58] Clause 30(4) of that form stated:

> The amounts retained by virtue of sub-clause (3) of this condition shall be subject to the following rules: (a) the Employer's interest in any amounts so retained shall be fiduciary as trustee for the Contractor (but with no obligation to invest)...

Mr Justice Vinelott decided that the effect of that provision was to impose on the Employer an implied obligation to set up the Retention as a separate trust fund in favour of the Contractor. Until the Employer does this by putting the Retention into a separate bank account, there is no trust because the mixing of the Retention with other monies of the Employer, which would otherwise occur, contravenes the principle of certainty of subject matter.

Setting up a trust fund deprives the Employer of the opportunity of putting the Retention to other alternative uses more to his advantage. This means that, in the event of the Employer's insolvent liquidation, the Retention will be held by the liquidator upon trust for the benefit of the Contractor. The creditors cannot therefore touch it. It is important to note that for the Retention to be unavailable to the Employer's creditors in insolvency,

56 See *Re Jartray Development Ltd* (1983) 22 BLR 134.

57 *Knight* v. *Knight* (1840) 3 Beav. 148.

58 (1979) 12 BLR 30 (hereafter *Rayack* v. *Lampeter*); this was followed by the High Court of Hong Kong in *Concorde Construction Co. Ltd* v. *Colgan Co. Ltd* (1984) 29 BLR 125.

the trust fund must have been effectively set up prior to the commencement of proceedings to put the Employer into liquidation or bankruptcy.

Imposition of an implied obligation to set up a trust means that, if the Employer fails to set it up, the Contractor can apply for a mandatory injunction compelling him to do so. In the *Rayack* v. *Lampeter* case, the Contractor successfully applied for such an injunction. For reasons relating to future business opportunities with the Employer, many contractors are reluctant to use this weapon unless there is clear evidence of imminent insolvency. However, it has to be cautioned that an application for a mandatory injunction after the commencement of liquidation will fail. In *Re Jartray Developments Ltd*,[59] which arose from a JCT 80 Contract, the application was made only after the appointment of a liquidator. The injunction was refused because to have held otherwise would have given unfair preference to the Contractor as against other unsecured creditors, a contravention of the *pari passu* principle of insolvency law.

Case law suggests that the application may be rejected even if the Employer is only in informal insolvency (i.e. events other than formal winding up such as administration and administrative receivership). In *Mac-Jordan Construction Ltd* v. *Brookmount Erostin Ltd*[60] the Contract incorporated the JCT 81 Form, Clause 30.4.2.1 of which was equivalent to Clause 4.18.1 of JCT 05. The Employer gave a floating charge over his assets including his interests in the building contract to a bank. After the Bank had appointed administrative receivers of the Employer's assets, the Contractor applied for an injunction to compel the Employer to put the Retention into a separate account. The Court of Appeal held that the Contractor was not entitled to the injunction because, as at the time of crystallization of the Bank's charge no trust existed over the Retention, the Bank had already obtained good title to the Retention by the time of the application.

JCT 05 Clause 4.18.3 gives the Contractor the power to direct the Employer to pay the Retentions into a separate bank account but only where the Employer is not a Local Authority. If such a request is made, the account must be set up within a reasonable time after the request. In *Bodill & Sons (Contractors) Ltd* v. *Harmail Singh Mattu*[61] Akenhead J stated that, considering the commercial realities of the banking world, a reasonable period would be 2 to 3 weeks. At the date of payment under each Interim Certificate, the Employer must pay the retention component into the trust account and certify such compliance to the Architect with a copy to the Contractor. Interest earned on the trust account belongs to the Employer.

The effect of the requirement in Clause 4.18.3 that the account should be 'so designated as to identify the amount as the Retention held by the Employer on trust as provided in Clause 4.18.1' is probably that the title of the account must make it clear that it is a trust account for the benefit of the Contractor.[62] This should alert third parties such as receivers, liquidators and administrators that they have no access to those funds. If this is not done, there is always the risk that these third parties may, without any other notice of the trust position, lawfully make use of the funds. Although the Contract does not specify

59 (1983) 22 BLR 134; see also *GPT Realisations Ltd (in Administrative Receivership & in Liquidation) Ltd* v. *Panatown Ltd* (1992) 61 BLR 88.
60 (1992) 56 BLR 1.
61 (2007) EWHC 2950 (TCC); [2008] CILL 2553.
62 See *Bodill* v. *Mattu* above.

that the request to set up the trust account should be in writing, it would be good management practice to do so in writing. In *JF Finnegan Ltd* v. *Ford Sellar Morris Developments Ltd*,[63] which arose from the equivalent in the JCT 81 Form of JCT 05 Clause 4.18.3, the court rejected the Employer's contention that the Contractor must make the request each time retention is deducted (e.g. after the issue of each Interim Certificate). It was stated that only one request is necessary and it may be made at any time.

As an injunction is a discretionary remedy, the court may refuse to grant it if a grant would have unfair consequences. For example, in *Henry Boot Building* v. *The Croydon Hotel*[64] the Contractor's application for a mandatory injunction to the Employer to set up retention monies as a trust fund failed because liquidated damages for non-completion were in excess of the retention monies.

Wates Construction v. *Franthom Property*[65] suggests that the Contractor would be entitled to demand a separate account for the Retention even in the absence of Clause 4.18.3. It arose from a contract in the terms of the Private version of the JCT 80 but from which the clause equivalent to JCT 05 Clause 4.18.3 had been deleted. The Court of Appeal rejected the Employer's contention that the deletion demonstrated a common intention that the Employer would not be under any obligation to open a separate trust account for the retention monies. The Contractor was therefore granted an injunction to enforce the setting up of the trust. The only practical effect of Clause 4.18.3 therefore appears to be that a Local Authority Employer cannot be compelled to set up a trust account for the Retention.

15.7.3 Set-off against retention

Clause 4.13.2 provides that the obligation of the Employer to hold the Retention in trust does not negate his right to exercise set-off rights allowed under the Contract against it. Whenever the Employer makes deductions against the retention funds, he is required, in addition to supplying a Statement of Retention, to specify to the Contractor how much of it has been deducted (Clause 4.18.4).[66]

15.8 The final adjustment of the contract sum

Although JCT 05 is a lump sum contract, the Contract Sum is adjustable for a variety of events. The effect, or even occurrence, of these events cannot always be determined accurately at the pre-contract stage. Clauses 4.3 and 4.5 describe in detail when and how the statement of the final adjustment of the Contract Sum is to be produced. This statement is referred to in the construction industry as the 'Final Account' and is normally drawn up as a formal document in a sectionalized format. The first section, usually on one side of A4,

---

63 (1991) 53 BLR 38 (hereafter *Finnegan* v. *Ford Sellar*).
64 (1985) 36 BLR 41.
65 (1991) 53 BLR 23.
66 This information can be provided as part of the relevant Withholding Notice required under clause 4.13.4.

contains a summary of the account and states the Contract Sum and various additions and deductions. This section is followed by separate sections setting out the details of:

- Variations;
- loss and/or expense;
- fluctuations;
- provisional sums;
- miscellaneous items.

Valuations of Variations agreed by the Contractor and the Employer under Clause 5.2.1, amounts stated in accepted Schedule 2 Quotations, and variations in premium for Terrorism Cover, may be additions to or deductions from the Contract Sum. However, adjustments for some matters are always additions to, whilst others are always deductions from the Contract Sum. Deductions are summarized in Table 15.1 whilst Table 15.2 covers additions. It is important to bear in mind that, contrary to practice with the final accounts of civil engineering contracts, the work in a JCT 05 contract is not to be re-measured for final accounts (i.e. for each work item the Contractor is to be paid only the amount for it in the Contract Bills unless it was affected by a Variation, in which case, there could be additions or deductions as explained above).

Table 15.1 Deductions from the Contract Sum

Clause	Deduction
Clause 4.3.2.1	(i) all Provisional Sums in the Contract Bills and the value of all work for which an Approximate Quantity is provided in the Contract Bills or the Employer's Requirements
Clause 4.3.2.2	(ii) valuation under 5.6.2 of work omitted from Contract Bills by a Variation (iii) valuation under 5.8.3 of work omitted from CDP Analysis by a Variation (iv) the amount included in the Contract Bills or the CDP Analysis for work which has been so affected by Variations or the execution of work covered by an Approximate Quantity that it has itself to be re-valued as a Variation under the Contract
Clause 4.3.2.3	(v) amounts in respect of the Contractor's setting out errors which the Architect has, pursuant to Clause 2.10, instructed to remain without correction (vi) amounts in respect of defects, shrinkages, and other faults which appear within the relevant Rectification Period for which the Contractor is responsible but which the Architect has, pursuant to Clause 2.38, instructed to remain without correction (vii) Employer's cost of giving effect to an Architect's instruction with which the Contractor has failed to comply (Clause 3.11) (viii) amounts in respect of materials or work not being in accordance with the Contract but which the Architect has, pursuant to Clause 3.18.2, instructed to remain without correction (ix) where the Contractor, without reasonable cause, failed to proceed regularly and diligently with Remedial Measures specified by insurers of the Works for breach of the Joint Fire Code, the Employer's cost of carrying them out (Clause 6.15.2) (x) price decreases from the fluctuation provisions
Clause 4.3.2.4	(xi) any other amount which is required by the Conditions to be deducted from the Contract Sum (anticipates amendments stipulating other deductions)

Table 15.2 Additions to the Contract Sum

Clause	Addition
Clause 4.3.3.1	(i) amounts payable to the Contractor by the Employer in respect of statutory fees and charges under Clause 2.21 (ii) amounts payable to the Contractor by the Employer in respect of infringement of patents rights under Clause 2.23 (iii) amounts payable to the Contractor by the Employer in respect of testing and inspection under Clause 3.17 (iv) Contractor's cost of taking out and maintaining insurance required under Clause 6.5
Clause 4.3.3.2	(v) the Valuation of additional or substituted work under Clauses 5.2 and 5.3 (vi) the Valuation of work which is not varied itself but, under Clause 5.9, has to be treated as a Variation
Clause 4.3.3.3	(vii) amount of Valuation of work executed by the Contractor in accordance with an Architect's instruction as to the expenditure of a Provisional Sum included in the Contract Bills or in the Employer's Requirements (viii) the amount of any disbursement by the Contractor in respect of an Architect's instruction as to the expenditure of a Provisional Sum included in the Contract Bills or in the Employer's Requirements (ix) amount of the Valuation of executed work covered by an Approximate Quantity included in the Contract Bills or in the Employer's Requirements
Clause 4.3.3.4	(x) amounts ascertained as loss and/or expense under Clauses 3.24 and 4.23
Clause 4.3.3.5	(xi) insurance premiums paid or payable by the Contractor under Clauses B.2.2 and C.3.2 of Insurance Options B and C, respectively on account of the Employer's defaults in taking out or maintaining the required insurance cover (xii) additional contract works insurance premiums paid or payable by the Contractor on account of early use or occupation of the Works or the site
Clause 4.3.3.6	(xiii) price increases from the fluctuation provisions
Clause 4.3.3.7	(xiv) any other amount which is required to be added to the Contract Sum (anticipates amendments stipulating other additions)

15.8.1 Timetable for final accounts and the Final Certificate

The timetable governing the preparation of the final accounts and the issue of the Final Certificate are explained in Chapter 3, Sections 3.7 and 3.8. As explained in Section 15.3, preparation of the final accounts is the responsibility of the Quantity Surveyor whilst the issue of the Final Certificate is that of the Architect.

15.8.2 The effect of the Final Certificate

This subject is covered in Chapter 3, Section 3.9.

15.9 Architect's liability for negligent certification

The law on the liability of contract administrators to employers for negligence in the performance of their administrative responsibility has been settled since *Sutcliffe* v.

Thackrah[67] in which the House of Lords decided that an Architect was liable for the Employer's loss arising from negligent over-certification.

Whether the Contractor has a cause of action for negligent under-certification or other decision against the Architect is less clear. As there is usually no contract between the Contractor and the Architect, there is no contractual basis for such a claim. That leaves the Contractor with only a possible cause of action in tort. A number of cases suggest an affirmative answer. For example, in *Arenson* v. *Casson Beckman Rutley & Co.*[68] Lord Salmon said *obiter* that, considering the importance of cashflow to a builder, an architect who negligently certified less money than was payable could be successfully sued by the builder for the damage caused by being wrongfully starved of money in that way. In *Lubenham Fidelities and Investments Co. Ltd* v. *South Pembrokeshire District Council and Another*[69] the Court of Appeal accepted the possibility of the special tort of procuring breach of a contract to which the defendant is a stranger arising from deliberate under-certification by the Architect. The court concluded that the proper remedy available to the Contractor is for him 'to request the Architect to make the appropriate adjustment in another certificate or if he declines to do so, to take the dispute to arbitration'.

However, in *Pacific Associates Inc.* v. *Baxter*[70] the Court of Appeal held that supervising engineers were not liable to contractors for under-certification. In that case the claimant contractor was engaged by the Ruler of Dubai under a FIDIC (International Federation of Consulting Engineers) contract to carry out dredging work. Condition 86 of their contract stated, *inter alia*, that the Engineer was not to be in any way liable to the Contractor for the performance of the Engineer's duties under the Contract.[71] There was also an arbitration clause typical of construction contracts. Claims for unforeseen conditions were rejected by Halcrow, the consulting engineers appointed to act as the Engineer. Part way through subsequent arbitration proceedings, the claim was settled at a fraction of its value by agreement between the claimant and the Ruler. However, Pacific Associates sought to recover the balance of their claim from Halcrow in tort. The issue that the Court of Appeal was required to decide was whether Halcrow owed Pacific Associates a duty of care in tort to avoid causing that loss. The existence of the duty was argued on the voluntary assumption of responsibility principle and the proximity/foreseeability/policy test. It was held not to exist for two main reasons: (i) the effect of the condition exonerating the Engineer meant that there was no voluntary assumption of responsibility; (ii) because of Condition 86 and the arbitration clause, it was not fair, just or reasonable to impose a duty of care. According to Purchas LJ at p. 53:

> ...where the parties have come together against a contractual structure which provides for compensation in the event of failure of one of the parties involved, the court will be slow to superimpose an added duty of care beyond that which was in the contemplation of the parties at the time that they came together.

Pacific Associates has been applied or considered without disapproval in some other common law jurisdictions to deny engineers' liability in tort to contractors in relation to their

67 [1974] AC 727; see also: *Townsend* v. *Stone Toms & Partners* (1984) 27 BLR 26; *West Faulkner Associates* v. *London Borough of Newham* (1994) 71 BLR 1.
68 [1977] AC 405, at 437G.
69 (1986) 33 BLR 39.
70 (1989) 44 BLR 33.
71 There is no equivalent provision in the current edition of the FIDIC Red Book.

roles within the traditional model for the procurement of construction. In *Leon Engineering & Construction Co. Ltd (in Liquidation)* v. *Ka Duk Investments Co. Ltd*[72] the High Court of Hong Kong held that, in the administration of a contract with respect to claims and certificates, an architect under a form of contract very similar to JCT 05 did not owe the contractor a duty of care in tort. In *Spandeck Engineering (S) Pte Ltd* v. *Defence Science & Technology Agency*[73] the Court of Appeal of Singapore rejected a contractor's claim in tort against a contract administrator for systematic under-certification of payment.

Despite the reliance placed upon *Pacific Associates* to deny liability in these jurisdictions, the general principle for which it is authority has been a matter of some debate among commentators.[74] A decision of the Supreme Court of Canada suggests that it may not be applicable where there is no express disclaimer of the engineer's liability. In *Edgeworth Construction* v. *Lea & Associates*[75] the Appeal Court of British Columbia decided that design engineers were not liable in negligence to road contractors for their loss arising from design defects and errors in drawings. This was reversed by the Supreme Court of Canada, which suggested that had the engineers expressly excluded their liability to the contractors in the contract documents, the decision might have been different.

Does the Architect under JCT 05 owe the Contractor a duty of care in his role as certifier? The Contract does not provide expressly that the Architect is not liable to the Contractor in respect of negligent performance of his duties under the Contract. However, it is arguable that the availability of adjudication for challenging certificates makes the contract structure argument against imposition of a duty still applicable even where the Contract does not incorporate the arbitration agreement in Article 8. If *Pacific Associates* is distinguished as limited to its facts, the issue has to be determined by the proximity/foreseeability/policy test for the existence of a duty of care. Since it is difficult to deny that the ingredients of proximity and foreseeability exist in this context, the issue would then be essentially one of policy. Unfortunately, as highlighted in the published case commentaries already referred to, opinions are divided here. For example, while the Supreme Court of Canada in the *Edgeworth Construction* case stated that there is no policy reason for denying liability, the editor of *Hudson's Building and Engineering Contracts*, 11th edition, argues against putting contract administrators in a position where they will be "shot at by both sides".[76] It has also been suggested that professional indemnity insurance premiums are likely to be significantly increased if such a duty is recognised to exist.[77]

15.10 Bonds

On many projects the contractor is required to procure, in favour of the employer, financial instruments of a type referred to as a 'bond'. A bond, in the context of a construction

72 (1989) BLR 139; (1989) 5 Const LJ 288.

73 (2007)114 ConLR 166.

74 For example, see: Nicholas Lane, *Pacific Associates*: *Charter for Professional Negligence?* (2003) 19 Const LJ 311; I.N.D. Wallace, *Pacific Associates Revisited: a Rejoinder* (2003) 19 Const LJ 304; Nicholas Lane, *Constructive Acceleration* (2000) 16 Const LJ 231; Timothy Trotman, *Pacific Associates* v. *Baxter*: Time for Re-consideration (1999) 15 Const LJ 449; Duncan Miller, The Certifier's Duty of care to the Contractor – *Pacific Associates* v. *Baxter* Reconsidered (1993) ICLR 173.

75 *Edgeworth Construction Ltd* v. *Lea & Associates* (1991) 54 BLR 11 (Court of Appeal of British Columbia); (1993) 66 BLR 56 (hereafter *Edgeworth Construction*).

76 Paragraph 1.302.

77 Ian Duncan Wallace, *Charter for the Construction Professional?* (1990) 6 Const LJ 207.

project, is an undertaking by a bank or other financial institution, hereafter referred to collectively as the 'surety', to make payment to the employer up to a stated aggregate amount (the bond amount) in defined circumstances. To provide the bond, the surety charges the contractor a bond fee, which the contractor would normally take into account in the pricing of the project. Where the circumstances in which the employer may demand payment on a bond amount to breaches of contract by the contractor, it is said to be a 'conditional' bond (i.e. the employer's right to be paid by the surety is conditional on a breach of contract by the contractor). Other types of bonds require only a simple demand by the employer. The latter type is known as an 'on-demand' bond and, in effect, isn't much different from the contractor handing over to the employer the amount of the bond to be returned on the date of its expiry.

JCT 05 Contract Particulars may be completed to require the contractor to procure, in favour of the employer, three types of bonds issued by sureties approved by the employer. They are designed to improve the contractor's cashflow. It is for the parties to work out the details of these instruments, including sureties approved by the employer, before the contract is executed. In particular, where the Contractor offers a discount on his tender price in return for these facilities, the priced Bills of Quantities need to be carefully amended before contract execution.

15.10.1 The Advance Payment Bond

A contractor will usually include in a tender for a building project the cost of working capital required to complete it. An employer who is able to pay part of the contract sum before commencement on site may therefore attract lower tenders. Such payment is referred to as 'advance payment' or 'mobilization payment'. The amount is reimbursed to the employer through deduction of instalments of the advance payment from amounts that should otherwise be certified as stage payment to the contractor. To protect the Employer against subsequent failure to perform, the contractor is usually required to furnish an advance payment bond designed to repay to the employer the money advanced in that event.

JCT 05 supports this practice by providing for the particulars of any advance payment to be completed in Contract Particulars. The items of information required to be completed are:

- whether or not Clause 4.8, the clause that entitles the Contractor to advance payment, applies;
- the amount of the Advance Payment;
- the date on which the Employer must make the Advance Payment;
- A schedule of instalments of the Advance Payment to be reimbursed at stated times.

15.10.1.1 The Advance Payment Bond Form

The bond is of the on-demand variety although its terms require the Employer's demand to be in the prescribed form attached as a schedule to the form of bond in Part 1. Upon receipt of a demand on the bond, the Surety must make the payment demanded within 5 Business Days. 'Business Day' is defined as 'day (other than a Saturday or a Sunday) on which commercial banks are open for business in London', a different definition from that in Clause 1.1 of the Conditions. Upon making such payment, the Surety will usually charge it against the Contractor's account with the Surety or recover it as debt.

The aggregate of amounts that may be demanded is to be stated in the bond document. In most cases it will be the same as the amount of the Advance Payment but it may be greater to reflect the time value of money. The aggregate is to be considered reduced by any instalment reimbursed by the Contractor but only if the Employer has advised the Surety in writing of the reimbursement.

The Surety's obligation to meet any demand ceases upon whichever of the following occurs first: (i) a longstop date stated in the bond document;[78] (ii) written certification by the Employer that the Advance Payment has been reimbursed in full through the certification process; (iii) written certification that the Contractor has paid off the Advance Payment or any balance of it outstanding.

15.10.2 The Off-site Materials Bond

JCT 05 allows the value of any item of off-site materials destined for the Works to be included in Interim Certificates before it is delivered to the site but only if it belongs to the Listed Items annexed to the Contract Bills. There are risks to the Employer that, for whatever reason, off-site materials paid for never actually get delivered to the site for incorporation into the Works. It is therefore a pre-condition for the inclusion of Listed Items of off-site materials in Interim Certificates that they are insured in the joint names of the Employer and the Contractor for their full value against the Specified Perils. By appropriately completing the Contract Particulars, the Employer may impose the provision of an on-demand bond as additional protection. The function of such a bond is to compensate the Employer up to the specified aggregate amount should such materials become permanently lost to the Employer. However, it is to be noted that, as it is an on-demand bond, the Employer may call the bond without any justification.

To impose an obligation to provide the bond, the aggregate amount that the Employer may demand from the Surety must be stated against Clause 4.17.4 or Clause 4.17.5 in the Contract Particulars. Clause 4.17.4 applies if the materials are uniquely identified; Clause 4.17.5 applies to materials not uniquely identified. Clause 4.17 reads as if they are intended as alternatives. Presumably the Clause 4.17.5 part is to be completed where some of the materials are uniquely identified whilst others are not.

15.10.2.1 The Off-site Materials Bond Form

The bond should be in the form provided in Part 2 of Schedule 6. It is stated expressly that any waiver by the Employer of his rights under a construction contract, variation of the Contract or extension of time granted to the Contractor is not to invalidate the bond. For example, although the Contract requires the Contractor to insure the materials against the Specified Perils, the Employer does not have to enforce the taking out of such insurance before he can call on the bond.

Any demand for payment on the bond must be in a prescribed form of demand provided as an attachment to the form. It must be signed for the Employer by two persons from the Employer's organization whose signatures are to be authenticated by the Employer's bankers. The Surety must pay the amount properly demanded up to the aggregate amount

78 In fixing this date the possibility of extensions of time should be borne in mind.

of the bond without entering into inquiries as to whether the Employer is entitled to call the bond.

The bond expires on whichever of the following occurs first: (i) a longstop date stated in it; (ii) the date certified by the Employer as the date on which all the Listed Items were delivered to the site.

15.10.3 The Retention Bond

The main purpose of retentions is to provide a fund from which the project owner may draw to make good any loss from the Contractor's breach of contract. From the owner's standpoint, this is infinitely to be preferred to having to recover damages by proceedings. However, the main problem with retention from the Contractor's standpoint is that his cashflow is adversely affected. An alternative is to provide a retention bond to the Employer in lieu of having retentions deducted. The owner then has the same degree of protection because he can call upon the bond in circumstances where he would otherwise have had the protection of the retention fund.

To adopt this alternative to deducting retention, the Contract Particulars in JCT 05 must be completed to indicate that Clause 4.19 applies. The amount of the bond and its expiry date are also to be stated. Clause 4.19.1 requires the Contractor to provide to the Employer by the Date of Possession the Retention Bond specified by the relevant entries in the Contract Particulars. The Architect is not to deduct the Retention in his calculation of the amount to be stated in an Interim Certificate provided the Contractor is in compliance with his obligation to provide the Retention Bond. However, the amount of Retention that should have been deducted in respect of each Interim Certificate is still to be calculated by the Architect or the Quantity Surveyor. If the Contractor defaults on his obligation to provide and maintain the required bond the Architect is to revert to deduction of Retention from the Interim Certificate following the default and continue with such deduction until the Contractor ceases the default. Upon the Contractor ceasing his default, the Employer is to release to the Contractor the Retention deducted during the period of the default. Presumably the Employer is to instruct the Architect not to deduct the Retention in his preparation of the next Interim Certificate.

If at any time the Retention calculated exceeds the aggregate amount in the Retention Bond either the Contractor arranges for the bond amount to be topped up or the Architect must deduct the difference before arriving at the amount of the relevant Interim Certificate.

There is no requirement for the Contractor to provide a performance bond, which provides protection in relation to the Contractor's performance of the Contract as a whole. However, such a requirement may be imposed through appropriate provisions in the Contract Bills or amendment of the Articles of Agreement or the Conditions. In respect of a Contractor's default for which the Employer may call on the performance bond or the Retention Bond, Clause 4.19.5 requires the Employer to call on the Retention Bond first.

15.10.3.1 The Retention Bond Form

The bond required should be in the form set out in Part 3 of Schedule 6. Its main provisions are as follows:

1. The Employer may call the bond up to the aggregate amount stated on it.

2. The Employer must notify to the Surety the date of the Interim Certificate immediately following the practical completion of the Works. The bond expires on this date.
3. The aggregate amount of the bond is reduced by a half upon issue of the Practical Completion Certificate.
4. Any demand on the bond should state the applicable Retention and the amount demanded, which must not exceed the former.
5. A demand on the bond must be signed on behalf of the Employer by a person(s) whose signature(s) should be authenticated by the Employer's bankers.
6. The Employer must produce certification that he gave the Contractor 14 days' notice of the demand and that the Contractor failed to pay the amount demanded within the 14 days.
7. The demand must state which of the following the demand is for: (i) cost incurred by the Employer by reason of the Contractor's failure to comply with an Architect's instruction; (ii) insurance premiums paid by the Employer as remedial action taken in response to the Contractor's default on his insurance obligations under the Contract; (iii) liquidated damages payable to the Employer; (iv) the Employer's expenses, direct loss or damage caused by termination of the Contractor's employment by the Employer; (v) any other costs actually incurred by the Employer which he is entitled under the Contract to recover from the Contractor.

15.11 The Local Democracy, Economic Development and Construction Bill, Payment and Notices

The most criticized provisions in the original Construction Act are easily those on notices to be served and their effectiveness in promoting clarity in communication about payment entitlements and quick identification of disputed issues for speedy resolution by adjudication.[79] Improvements have been sought through the Local Democracy, Economic Development and Construction Bill as follows.

15.11.1 Payment Notice

A new s. 110A requires a 'payer' or a 'payee' under a construction contract to take certain steps in relation to interim payment. These roles equate to the Employer and the Contractor, respectively, for the purposes of JCT 05. The Employer must serve a Payment Notice setting out the sum he believes to be due and the basis upon which it is calculated. Any payment already made is to be disregarded in the calculation. A certificate under JCT 05 attached to a summary of the underlying valuation and supporting calculations of the amount due would meet the content requirements of the notice. The notice must be served not later than 5 days after the 'payment due date', which is defined as 'the date provided for by the contract as the date on which the payment is due'. Under JCT 05, this date

79 For examples see comments in *Melville Dundas Ltd (in receivership) and Others* v. *George Wimpey UK Ltd* [2007] UKHL 18.

would be the date of the relevant payment certificate. A specified person may be designated under the Contract to serve the Notice on the Employer's behalf. For example, the Architect may be given the responsibility for serving these notices if JCT 05 is amended accordingly.

15.11.2 Default Payment Notice

Where no Payment Notice is served, the Contractor may, at any time before the final date for payment, serve a Default Payment Notice setting out the amount the Contractor believes to be due and the basis of its calculation. The sum stated in the Payment Notice or, where none has been served, the Default Payment Notice becomes the 'notified sum', a new concept used in place of the 'amount due' in the original Construction Act, which has been a matter of considerable controversy.

Subject to a valid Withholding Notice as described later, the Employer must pay the notified sum on or before the final date for payment, which is likely to be retained as 14 days after the date of the relevant Interim Certificate. However, where no Payment Notice is served and the Contractor serves a Default Payment Notice, the final date for payment of the sum in the latter is to be considered postponed by the number days between the latest date when the Employer's Payment Notice should have been served and the date of the Default Payment Notice.

15.11.3 Withholding Notice

The proposed payment procedures also allow the Employer to serve a Withholding Notice either after his Payment Notice or in reply to the Contractor's Default Payment Notice if he intends to pay less than the notified sum. It must be served after the appropriate Notice but not later than an agreed date before the final date of the notified sum. The interval between the latest date for service of the Withholding Notice and the final date is likely to be retained by the JCT as 5 days in the interest of continuity. It is important to note that whilst the Withholding Notice under the original Construction Act had to specify the amount to be withheld broken down to relate to individual stated reasons for the withholding, the notice under the new procedures is required to set out the sum he intends to pay and the basis upon which it is calculated. This change is unlikely to make much difference in practice since an acceptable statement of the basis of calculation should identify the reasons for the difference between the notified sum and the amount the Employer proposes to pay. The Employer needs to pay only the amount in the Withholding Notice. Where the Contractor disputes the entitlement to withhold, he may refer the matter to adjudication upon receipt of the Withholding Notice.

15.11.4 Disputes over notified sum and Withholding Notices

The Contractor may consider that the amount payable for actual progress on the job under the terms of the Contract is more than the notified sum in the Payment Notice, or that the Withholding Notice is defective, and commence adjudication as soon as he becomes aware of the problem. The new legislation requires the notified sum to be paid by the final date pending the outcome of any such adjudication. This express requirement avoids the debate over whether an Employer may refuse to pay by simply contending that the amount being claimed is not the amount due under the Contract. If the outcome of the adjudication is that the Contractor was entitled to payment of a greater sum either because of under-certification or wrongful withholding, the appropriate additional payment should be made to the Contractor by the later of: the final date for payment of the applicable notified sum or the 7th day after the adjudicator's decision. There is no express provision for the situation where the adjudicator's decision is that the notified sum overstates the amount payable under the Contract. The philosophy underlying the valuation and certification procedures under JCT 05 suggests that the appropriate remedial action is for the Architect or the Quantity Surveyor to take account of the decision in the next interim certificate.

16

Termination

The ideal outcome of a building contract is for both parties to perform their obligations in accordance with their contract. Whilst this ideal is achieved in the majority of building contracts, a contract can sometimes be brought to a premature end by one of the parties. When this happens, the contract is said to be 'determined' (i.e. terminated). A note of caution must be sounded here regarding terminology. The terms 'termination of a contract' and 'determination of a contract' are to be understood as shorthand for the ending of the primary obligations under the contract. These obligations consist of the contractor's obligation to carry out and complete the works and the employer's obligation to pay the contract price in accordance with the conditions of the contract. Strictly speaking, the contract itself does not come to an end because its secondary obligations (i.e. the contract-breaker's liability for damages) remain unaffected. Also, the right to refer to adjudication is not lost.[1]

A contract may be terminated either under the common law or by exercising rights of termination expressly provided for in it. This chapter examines termination under the terms of JCT 05. It will be seen that many of the rights to terminate under the Contract are expressed to be without prejudice to any other rights or remedies that the terminating party may possess. This means that the party concerned may choose to bring the Contract to an end on common law grounds. Before the provisions in the Contract are examined, the general nature of this choice is therefore explained.

16.1 Termination at common law for repudiation

Repudiation, also sometimes referred to as a 'repudiatory breach', of a contract arises when an act or omission of a party to the Contract is such a serious breach that the innocent party is entitled to treat it as evidence that the contract-breaker no longer intends to be bound by it. This situation can come about although the contract-breaker did his best to avoid the breach. When this happens, the innocent party has two choices. First, he can accept the repudiation and thereafter the party who repudiated can no longer, without the agreement of the innocent party, revert to the *status quo* before the repudiation. Not only is the

1 See, e.g. *A & D Maintenance and Construction Ltd* v. *Pagehurst Construction Services Ltd* (2000) 16 Const LJ 199; *Northern Developments (Cumbria) Ltd* v. *J & J Nichol* [2000] BLR 158; *Connex South Eastern Ltd* v. *MJ Building Services Group plc* [2004] EWHC 1518; [2004] BLR 333; 95 ConLR 43; *Melville Dundas Ltd and Others* v. *George Wimpey UK Ltd and Others* [2007] UKHL 18.

innocent party discharged from further performance of his obligations under the contract but he can also sue for damages immediately.[2] Second, the innocent party may choose to treat the contract as continuing despite the breach and claim damages instead. He is then said to 'affirm' the contract. Upon affirmation, the innocent party's right to accept that particular repudiatory breach is lost unless the breach is repeated or is of a continuing nature. Affirmation is very readily assumed if there is delay in accepting the repudiation. As the court is unlikely to order a party to perform a contractual obligation,[3] affirmation works only where the party in breach is still willing to continue performance.

The most easily understood form of repudiation is renunciation (i.e. an express statement by a party to the effect that he no longer intends to perform any of his obligations under the contract). For example, in *Multiplex Construction (UK) Ltd* v. *Cleveland Bridge UK Ltd and Another*,[4] which arose from a subcontract for the design and construction of the steel arch spanning the internal length of the Wembley Stadium, the defendant contended that the claimant main contractor's failure, as a consequence of under-valuation of work executed by the subcontractor, to make payment amounted to a repudiatory breach and served notice to terminate the subcontract by a specified date. Jackson J rejected the contention and held that it was rather the defendant who had, by serving the notice of termination and then stopping work, committed a repudiatory breach. It can also arise from a breach of a condition of the contract as opposed to a breach of a warranty. In addition, persistent and nonchalant breaches of a warranty may also constitute repudiation. For example, in *Sutcliffe* v. *Chippendale & Edmondson*[5] it was stated that persistent poor quality work could be treated as repudiation.

16.1.1 Conditions and warranties

A term is a condition if it is so important that its breach by a party entitles the other to treat the contract as repudiated. For this reason, a breach of a condition is often referred to variously as a 'repudiatory breach', 'fundamental breach', or a 'breach that goes to the root of the contract'. For example, it would be a condition of any construction contract that the employer will be able to grant possession of site without undue delay[6] and, further, that he will not expel the contractor from the site of the works without reasonable cause. Refusal to honour payment obligations or threats not to pay further sums due in accordance with the contract is capable of amounting to repudiation.[7] On the contractor's side, it would be a condition that he will not wholly abandon the works without lawful reason[8] or sub-let their entirety without the employer's consent.

South African authority suggests that where a construction contract requires the contractor to procure a performance bond, failure to comply would constitute a breach for which the Employer may terminate the contract at common law.[9] However, in *South Oxfordshire DC* v. *SITA UK Ltd*[10] Steel J rejected a contention by an employer that a contractor's failure to

2 *Heyman* v. *Darwins* [1942] AC 356; *Photo Production* v. *Securicor* [1980] AC 827.
3 The court is more likely to award damages for the breach.
4 [2006] EWHC 1341(TCC).
5 (1971) 18 BLR 149.
6 *Carr* v. *J. A. Berriman Property Ltd* (1953) 27 ALJR 273.
7 *CJ Elvin Building Services Ltd* v. *Noble* [2003] EWHC 837 (TCC).
8 *Marshall* v. *Mackintosh* (1898) 78 LT 750.
9 *Swartz & Son (Pty) Ltd* v. *Wolmaransstad Town Council* (1960) 2 SARL 1.
10 [2006] EWHC 2459 (Comm).

provide a performance bond in accordance with their contract amounted to repudiation. The judge explained that the fact that the parties had overlooked the need for the bond was indicative of the commercial unimportance of the bond in that particular transaction.

A warranty is a term of less importance than a condition. Its breach does not entitle the innocent party to terminate the contract; there is only entitlement to damages. For example, an isolated delay of a few days by an employer to pay on a certificate would not entitle the contractor to treat the contract as terminated. Similarly, an isolated defect that can easily be put right by the contractor would not result in a right of the employer to treat the contract as terminated.

For a long time, the position of the law was that a term was either a condition or a warranty. However, in modern times, it has been realized that some terms cannot be categorized in this way because whilst one type of breach of such a term could have only very minor consequences another breach of the same term could be serious enough to deprive the innocent party of substantially the whole benefit of the contract. For such a term, the legal consequence of its breach depends on the nature of the events arising from it. Where the legal effect of the breach of term is best judged by examining the nature of the events arising from the breach, the term is referred to as an 'innominate term' or 'intermediate term'.[11] It is submitted that failure to pay on a certificate falls into this category because, whilst delay by a few days would be tolerable, delays for months without explanation would be serious enough to entitle the Contractor to terminate the Contract at common law.

16.1.2 Wrongful termination at common law

A court may decide that a term is a condition or a warranty because it has been categorized as such by statute or binding judicial precedent. In the absence of relevant statutory provision and precedent, the court must determine from the Contract itself and the matrix of its surrounding circumstances which category was intended by the parties. Provided that the intention of the parties is clear enough that a term is to be a condition, the court will treat it as such even if the consequences of its breach are very minor.[12] As a corollary, a clear statement that breach of a very important term is not to give rise to a right to terminate would be accepted by the courts as evidence that the term is, by the common intention of the parties, only a warranty. However, in view of the draconian consequences of a breach of a condition, the courts lean against deciding that a term is a condition unless an intention to that effect is very clearly stated. Thus, merely stating that a term is a condition may not be conclusive that the parties intended that the innocent party shall have a right to terminate the Contract in the event of its breach.[13] This approach reflects recognition that the term 'condition' is often used to refer to terms in general without any intention that the terms shall be subject to the consequences of a condition in the legal sense. Indeed, many of the standard forms in the construction industry, JCT 05 included, are labelled 'conditions' although only a small proportion of their terms are intended as conditions in the legal sense.

There is a need to exercise due caution when faced with what appears to be repudiation. This need arises from the fact that where party A terminates for alleged repudiation by

---

11 *Hong Kong Fir Shipping Co. Ltd* v. *Kawasaki Kisen Kaisha Ltd* [1962] 1 All ER 474.
12 *Lombard North Central plc* v. *Butterworth* [1987] QB 527.
13 *Schuler AG* v. *Wickham Machine Tool Sales Ltd* [1974] AC 235.

party B and the conduct relied upon is not accepted in law as repudiation, A is guilty of wrongful termination. B may be entitled to treat the wrongful termination as repudiation by A which, if accepted by B, would result in a turning of the tables whereby A is liable to B for damages.[14] However, it was suggested in *Woodar Investment Development Ltd* v. *Wimpey Construction UK Ltd*[15] that a purported termination under the contract based on an honest but mistaken interpretation does not always amount to a repudiatory breach. Lord Wilberforce explained:

> So far from repudiating the contract, the appellants were relying on it and invoking one of its provisions, to which both parties had given consent. And unless the invocation of the provision was totally abusive, or lacking in good faith (neither of which is contended for), the fact that it has proved to be wrong in law cannot turn it into repudiation... Repudiation is a drastic conclusion which should be only held to arise in clear cases of refusal, in a matter going to the root of the contract, to perform.

16.2 Common law versus the Contract

It is a common law principle that contractual termination clauses will not preclude a party from terminating at common law for repudiation by the other party unless the Contract itself expressly or impliedly provides that it can only be terminated by exercise of the contractual right.[16] Clause 8.3.1 provides that the parties' rights under the termination clauses, including the right to terminate the Contractor's employment under the Contract, are without prejudice to any other rights and remedies that they may possess. The parties therefore have the choice to proceed under common law even where the contract may be terminated by invoking its terms. Regarding this choice, the factors that a party contemplating termination will usually consider include the following.

1. With termination at common law, assuming repudiation has really occurred, there are no special procedures to follow. A simple notice to the effect that the contract has been terminated for stated reasons would be sufficient. By contrast, contracts often lay down elaborate procedures to be followed. For this reason, a party who has failed, or is unable, to comply with procedures laid down in the contract may elect to bring its operation to an end by exercising his common law rights.
2. After termination at common law, neither party has any obligations under the contract except the contract-breaker's liability for damages and obligations under an arbitration agreement in the contract[17] or under the Construction Act.[18] However, where the employment of the contractor is terminated under the contract, both parties are still bound by the contract although most of its terms would not be applicable after the termination. A party entitled to terminate may therefore opt for a route because it offers greater advantages regarding his post-termination rights.

---

14 For example, see *Multiplex Construction (UK) Ltd* v. *Cleveland Bridge UK Ltd and Another*, [2006] EWHC 1341 (TCC). See also *Hayes (t/a Orchard Construction)* v. *Gallant* [2008] EWHC 2726 (TCC).

15 [1980] 1 All ER 571.

16 *Lockland Builders Ltd* v. *John Kim Rickwood* (1995) 77 BLR 38.

17 For explanation of the doctrine of separability of an arbitration agreement see Section 17.6.2.

18 As already stated, the right to refer to adjudication survives termination of the Contract.

3. Where the termination is at common law, the innocent party is entitled to damages assessed under common law principles, which are explained in Chapter 12, Sections 12.2 to 12.5. Most termination clauses state expressly what the innocent party may recover after the termination and how it is to be quantified (e.g. under Clause 8.7.5, 8.8.2 or 8.12.3). However, where the contractual remedy is not void (e.g. for being penal), the terminating party is entitled to only that remedy.[19]
4. If the contractor provided a performance bond, the employer may make a claim on it upon termination at common law. With termination under the contract, as explained in Section 16.20 in this chapter, the employer cannot call the bond until after completion of the outstanding work or, where the employer decides to abandon the Works, after submission of the employer's statement of accounts under Clause 8.8.1.

16.3 JCT 05 termination clauses: an overview

JCT 05 provides for termination of the employment of the Contractor by the Employer or the Contractor himself in a number of defined situations. This is done in the following ways:[20]

- termination by the Employer in defined situations: five specified defaults of the Contractor (Clause 8.4.1),[21] the Contractor's insolvency (Clause 8.5.1),[22] and corruption (Clause 8.6);[23]
- termination by the Contractor in defined situations: four specified defaults of the Employer (Clause 8.9.1),[24] suspension beyond the period of suspension in the Contract Particulars caused by the specified suspension events (Clause 8.9.2),[25] and the Employer's insolvency (Clause 8.10.1);[26]
- termination for withdrawal of Terrorism Cover (6.10.2.2);[27]
- termination by either party after major loss or damage to the Works from certain insured risks (para. C.4.4 of Schedule 3);[28]
- termination by either Party on account of *force majeure* and the like (Clause 8.11.1).[29]

16.4 Notices required by the termination procedures

The termination procedures laid down in the Contract require the giving of certain notices. It is stated in Clause 8.2.3 that notices of termination on any of the grounds in Section 8

19 *Thomas Feather & Co. (Bradford) Ltd* v. *Keighley Corporation* (1953) 52 LGR 30.
20 A clause of this type and intended to be operated by an employer is commonly referred to as a 'forfeiture clause'.
21 See Section 16.6 in this chapter.
22 See Section 16.7 in this chapter.
23 See Section 16.8 in this chapter.
24 See Section 16.9 in this chapter.
25 See Section 16.10 in this chapter.
26 See Section 16.11 in this chapter.
27 See Section 16.12 in this chapter.
28 See Section 16.13 in this chapter.
29 See Section 16.14 in this chapter.

of this standard contract should be in writing and served by actual delivery, registered post or recorded delivery. It is further provided in that clause that if a notice is sent by registered post or recorded delivery, subject to proof to the contrary, it is deemed to have been received on the second Business Day after the date of its posting. Business Day is defined under Clause 1.1 as 'any day which is not a Saturday, Sunday or a Public Holiday'. 'Public Holiday' is defined under the same clause as 'Christmas Day, Good Friday or a day which under the Banking and Financial Dealings Act 1971 is a bank holiday'.

Termination is such a drastic step that, if it is contested, the courts tend to construe relevant contractual provisions against the party seeking to terminate. It is therefore absolutely essential that the procedure for termination spelt out in the Contract is followed to the letter. Any attempt to terminate without compliance with the stipulated procedure may amount to termination at common law but only if the default relied upon is one for which there is entitlement to terminate at common law.[30] Two requirements are of paramount importance. First, the party terminating must comply meticulously with the timetable. Second, the notice must be clear as to what is being notified. Case law[31] suggests that a notice in general terms but which clearly directs attention to what is amiss is sufficient. However, it is recommended practice not only to state clearly the default in question but also to specify the applicable clauses of the Contract. Ideally, the notice should adopt the words used in the Contract to describe the default.

Whether a defective notice of termination under the Contract is capable of being rectified or replaced depends on the circumstances. Such remedial action would not be effective where the receiver is entitled to treat the notice as a repudiatory breach and he chooses to accept the repudiation before the remedial action is taken. Immediate remedial action must therefore be taken as soon as it is realized that a notice was defective. It follows from the above discussion that where the default relied upon in a notice also entitles the innocent party to terminate at common law, it may be prudent practice to serve the notice in the alternative. This way, if subsequently it is found to be invalid under the Contract, it can take effect at common law. For similar reasons, where the default complained of entitles the innocent party to terminate under the Contract but he prefers the termination to take effect at common law, service in the alternative may be advisable unless it is clearly a default for which termination at common law is available.

16.5 Definition of insolvency

In the context of a company, the term 'insolvency' is an omnibus word referring to various states in which the company is unable to pay its debts as they fall due. It is defined under s. 123 of the Insolvency Act 1986 by reference to a number of presumptions (e.g. failure to meet a statutory demand by a creditor and its liabilities exceeding its assets). These definitions are not particularly helpful in the operational environment of a construction contract. JCT 05 provides a more practical definition of insolvency for the purposes of termination of the employment of the Contractor. Under Clause 8.1, a party (either the

30 *Architectural Installation Services Ltd* v. *James Gibbons Windows Ltd* (1989) 46 BLR 91.

31 *Hounslow London Borough* v. *Twickenham Garden Developments Ltd* [1970] 3 WLR 538; *Supamarl Ltd* v. *Federated Homes Ltd* (1981) 9 ConLR 25.

Contractor or Employer in this particular contract) is to be considered Insolvent if any of the five following specified events occurs in relation to that party.

1. The Party enters into an arrangement, a compromise or a composition in satisfaction of his debts. It is recognized that companies often reorganize and restructure for operational reasons rather than for the reason of being unable to meet business debts as they fall due. This recognition is in the exclusion from this strand of insolvency situations where the Party, not being insolvent within the general meaning of that term, carries out a scheme of corporate restructuring and/or amalgamation with other business units.
2. He passes a resolution or makes other determination to go into liquidation or bankruptcy without a prior declaration of solvency. The qualification is designed to deal with situations where a solvent company is being liquidated. In such cases the directors of the company have to make a declaration of solvency before they commence liquidation of the company. The declaration is a statement by the directors that they have made full enquiry into the affairs of the company and have formed the opinion that the company will be able to pay all its debts in full within a stated period, which must not exceed 12 months, from the date of commencement of the liquidation.
3. A court has ordered that the life of the company as a business unit should be terminated or, where the Party is a natural person, that the person should be put into bankruptcy. The process of terminating the company's life, referred to as 'winding up' or 'liquidation', entails the collection and realization (often sale) of the company's assets and the distribution of the proceeds to its creditors. Thereafter the name of the company is struck off the list of companies.
4. An administrator or administrative receiver is appointed. This applies where the Party is a company. An administrator is an individual appointed with or without the assistance of the court to rescue the party from ultimate financial failure or to run the company until it can be wound up more cost-effectively. The appointment of an administrator is therefore a very strong signal of severe financial difficulties. An administrative receiver is an individual appointed by a secured creditor (very often a bank that has loaned money to the company) of a company to take possession of certain assets of the company, realize them and pay the debt owed to the creditor.[32] The assets affected are those that the company used as security for the loan. The appointment of an administrative receiver is therefore also a strong signal of financial problems.
5. Any of the events described above has occurred in any other jurisdiction. This strand of the definition is designed to capture the situation where the Party operates internationally and financial problems have arisen outside the UK. In such a case it would usually be just a matter of time before the financial difficulties affect the operation in the UK.

The events described above contemplate the Party being a company or a natural person. Where the Party is a partnership, Clause 8.1.6 provides that if any of the events arises in relation to any partner the business is considered Insolvent.

---

32 The Enterprise Act 2002 severely restricts the rights of a creditor to appoint an administrative receiver. Instead the creditor is given the power to appoint an administrator.

16.6 Termination by the Employer for specified defaults

Clause 8.4.1 identifies five grounds upon which the Employer may terminate the employment of the Contractor. These are referred to as the Contractor's 'specified defaults'. They are if, before practical completion, the Contractor:

1. wholly or substantially suspends the carrying out of the Works without a reasonable cause (Clause 8.4.1.1);
2. fails to proceed regularly and diligently with the Works (Clause 8.4.1.2);
3. refuses or neglects to comply with an AI requiring him to remove work, materials or goods and by such refusal or neglect the Works are materially affected (Clause 8.4.1.3);
4. assigns rights under the Contract or sub-lets any part of the Works without the required consent (Clause 8.4.1.4);
5. fails to comply with his contractual obligations in respect of the CDM Regulations (Clause 8.4.1.5).

It should be noted that some of these defaults might not be serious enough to constitute repudiation at common law. It is also to be noted that, although the Contract does not limit the time within which the Architect may give notice of default, it would be implied that any such notice must be given within a reasonable time of the default.[33]

It is proposed in the Local Democracy, Economic Development and Construction Bill passing through Parliament at the time of writing that the Contractor's right to suspend performance for non-payment should include a right to partial suspension (i.e. suspension of only some of his obligations, entirely in his discretion). Enactment of such a change would put in serious doubt the Employer's right, during a period of non-payment, to terminate the Contractor's employment under the Contract.

16.6.1 Suspension of the Works by the Contractor

This default is stated in Clause 8.4.1.1 as 'if the Contractor without reasonable cause wholly or substantially suspends the carrying out of the Works or the design of the Contractor's Designed Portion' (CDP). To qualify as a valid ground for termination, any suspension of the carrying out of the Works relied upon must therefore satisfy two conditions. First, the suspension should be a total or substantial cessation of work on the whole of the site. Second, there must be no reasonable cause for the suspension.[34] In practice, it is difficult to prove that these conditions apply in any given situation unless the Contractor is clearly minded to abandon the Works. As long as the Contractor does not move all his resources off the site, he cannot be said to have wholly suspended the carrying out of the Works. However, it may amount to substantial cessation of the work. Whether that is the case is a matter of fact and degree.

It would be even more difficult to prove that the Contractor has wholly or substantially suspended design of the CDP unless there is an agreed programme for this activity with specific deliverables by definite dates.

---

33 See *Architectural Installation Services Ltd* v. *James Gibbons Windows Ltd* (1989) 46 BLR 91.

34 Under Clause 4.14 the Contractor is entitled to suspend performance of his obligations on account of the Employer's failure to pay on an Interim Certificate by its final date for payment. See Chapter 15, Section 15.6.2 for discussion of the Contractor's right to suspend.

16.6.2 Failure to proceed regularly and diligently

Under Clause 2.4 the Contractor undertakes to proceed regularly and diligently with the Works until completion. Failure to do so is therefore a breach. Under the general law, it is doubtful if this type of breach goes to the root of the Contract. However, by Clause 8.4.1.2, the Employer is expressly entitled to terminate the employment of the Contractor on this ground. In a number of cases, very scathing judicial comments were made on the vagueness of the term 'regularly and diligently'. For example, in *Hounslow* v. *Twickenham Garden*[35] Megarry J described the same phrase in the JCT 63 as 'elusive words on which the dictionaries help little'. This remark was supported wholeheartedly by O'Connor J in *Lintest Builders Ltd* v. *Roberts*.[36] In *West Faulkner Associates* v. *London Borough of Newham*[37] Judge Newey QC described what amounts to proceeding regularly and diligently in the following terms:

> Contractors must go about their work in such way as to achieve their contractual obligations. This requires them to plan their work, to lead and to manage their workforce, to provide sufficient and proper materials and employ competent tradesmen, so that the works are fully carried out to an acceptable standard and that all time, sequence and other provisions of the contract are fulfilled.

When the case got to the Court of Appeal,[38] although the general approach was supported, Simon Brown LJ said that Judge Newey's definition of proceeding regularly and diligently could not be accepted in its entirety. He added that attendance and effort were not enough unless there was some measure of accomplishment. Pointing out that it would be unhelpful to seek to define the words 'regularly' and 'diligently' separately, he then offered an alternative definition in these terms:

> Taken together the obligation upon the contractor is essentially to proceed continuously, industriously and efficiently with appropriate physical resources so as to progress the works steadily towards completion substantially in accordance with the contractual requirements as to time, sequence and quality of work. Beyond that I think it impossible to give useful guidance. These are after all plain English words and in reality the failure of which clause 25(1) (b)[39] speaks is, like the elephant, easier to recognize than describe.

He concluded that, whichever of the definitions was applied, the Architect was not only entitled to give the notice but could not reasonably have done otherwise than give it. It would therefore appear that failure of the Contractor to comply with the master programme is some, although not conclusive, evidence of failure to proceed regularly and diligently unless compliance with the programme is a term of the Contract.

In *Sindall Ltd* v. *Abner Solland and Others*[40] HHJ Humphrey LLoyd QC pointed out that an implication of Lord Justice Simon Brown's reference to contractual requirements in the extract from his judgement is that the Architect must have regard to the date by when the Contractor is contractually bound to complete the Works. This observation suggests that the Architect

---

35 See Note 31.
36 (1978) 10 BLR 120; affirmed by Court of Appeal: (1980) 13 BLR 38.
37 (1993) 9 Const LJ 233, at 249; (1994) 71 BLR 1, at p. 13.
38 (1995) 11 Const LJ 157; (1994) 71 BLR 1.
39 The equivalent of JCT 05 Clause 8.4.1.2.
40 (2001) 3 TCLR 30; (2001) 30 ConLR 152.

must assess any outstanding entitlements to extension of time before considering the possibility of this default, as the proper yardstick is the operative date (i.e. not necessarily the completion date in the Contract Particulars) by when the Works must be completed.

Taking the judicial comments as a whole, it is concluded that the question whether the Contractor is proceeding regularly and diligently is a matter of fact and degree to be decided taking into account the master programme, the adequacy of the resources deployed for the purpose of performing the Contract, actual progress, outstanding work, productivity trends and the extension of time to which the Contractor is currently entitled.

16.6.3 Refusal to remove defective work and materials

Refusal or neglect by the Contractor to comply with an AI requiring the removal of work, materials and goods not in accordance with the Contract is a breach of the Contractor's duty to comply with all valid AIs imposed by Clause 3.10. To constitute a valid ground for termination under this Contract, the refusal or neglect must materially affect the Works.[41] Apart from the inherent vagueness of this qualification, it seems that termination on the ground that the Works are likely to be affected at a future date may not be valid.

It has to be pointed out that a more appropriate course of action might be for the Employer to employ third parties to carry out the removal and to recover the cost of that course of action from the Contractor under Clause 3.11.

16.6.4 Assignment and sub-letting

Under Clause 7.1 the Contractor undertakes not to assign the Contract without the written consent of the Employer.[42] Clause 3.7 prohibits sub-letting any portion of the Works without the written consent of the Architect.[43] Breach of any of these terms is a ground for termination by the Employer.

16.6.5 CDM Regulations[44]

Under Clause 3.25.2, for as long as the Contractor is also the Principal Contractor, he must properly discharge the duties of a Principal Contractor under the CDM Regulations. In particular, he must ensure that the Construction Phase Plan is developed and submitted to the Employer before any construction starts. Every subsequent amendment to the Plan must be notified to the Employer so that he can keep the CDM Co-ordinator and the Architect informed.

If a different Principal Contractor is appointed, the Contractor must comply with all reasonable requirements of the CDM Co-ordinator to the extent that such requirements are necessary for compliance with the CDM Regulations. The Contractor is to supply information reasonably required by the CDM Co-ordinator in writing for the purpose of

---

41 The Architect must have specifically required removal of the work or materials; condemning them as non-complying is not enough: *Holland Hannen & Cubitts (Northern) Ltd* v. *Welsh Health Technical Services Organisation* (1981) 18 BLR 80.

42 See Section 8.2 for explanation of the meaning of 'assign a contract' and a commentary on JCT 05 provisions on assignment.

43 For detailed commentary on sub-contracting under JCT 05 see Chapter 8, Section 8.3.

44 The CDM Regulations are covered in detail in Chapter 10.

preparing the Health and Safety File and to ensure that all sub-contractors do the same. Failure to comply with any of these obligations is a specified default for which the Employer may terminate the Contractor's employment.

It is submitted that any attempt to terminate for a default giving rise to only very minor consequences would be caught by the requirement under Clause 8.2.1 that notice of termination is not to be given 'unreasonably or vexatiously'.[45] Furthermore, there is Court of Appeal authority that such wide termination rights must be tempered with common sense. In *Rice (t/a Garden Guardian)* v. *Great Yarmouth BC*[46] the contract stated: 'if the Contractor commits a breach of any of its obligations under the contract...the Council may, without prejudice to any accrued rights or remedies under the contract, terminate the Contractor's employment under the contract by notice in writing'. The Court held that the Employer was entitled to terminate only where the breaches amounted to repudiation in the normal sense.

16.6.6 Procedure for termination for the Contractor's default

The procedure to be followed for termination on account of the specified defaults is shown in Figure 16.1. The Architect is to set the ball rolling by giving notice to the Contractor that the Contractor has committed a default for which the Employer is entitled to terminate his employment (Clause 8.4.1). This notice is referred to hereafter as the 'Default Notice'. Termination of the Contractor's employment is then conditional upon the Employer serving a valid notice of termination (referred to hereafter as 'Termination Notice') within the relevant period from receipt by the Contractor of the Architect's Default Notice.

The two-tier nature of the notices required is to be noted. The Employer must be particularly careful not to issue the second notice too early. He must also be certain that the Contractor has either continued the default for the relevant period or repeated it. It was explained in *J.M. Hill & Sons Ltd* v. *London Borough of Camden*[47] that the Architect is better placed than the Employer to issue the notice of default not only because of his independent status but also because of his greater expertise and knowledge in recognizing any occurrence of the defaults. However, the Architect would be liable for the Employer's loss if the Employer's right to terminate is lost because of the Architect's failure to give the notice.

West Faulkner Associates v. London Borough of Newham:[48] the claimant architectural design practice performed the role of Architect on a contract let on the JCT 63 form. Under the Contract, the defendants, the Employer under the Contract, were entitled to terminate the employment of the contractors if they failed to proceed regularly and diligently with the work. Before this right could be exercised, the Architect was required to issue to the contractors notice that they were failing to proceed regularly and diligently. During the course of work, the defendants, being concerned at the fact that the contractors were making very slow progress, asked the Architect whether they could issue the appropriate notice. On a part of the project programmed to be completed in 9 weeks, the contractors took 28 weeks. The Architect responded that, although progress was slow, it did not amount to failure to proceed regularly and diligently. The defendants eventually replaced the contractors and terminated the Architect's contract of engagement. When the Architect sued for their

45 For a commentary on the meaning of 'unreasonably and vexatiously' see Section 16.19.
46 [2003] TCLR 1; (2001) 3 LGLR 4; see also *Peregrine Systems Ltd* v. *Steria Ltd* [2005] EWCA Civ 239.
47 (1980) 18 BLR 31.
48 (1994) 71 BLR 1.

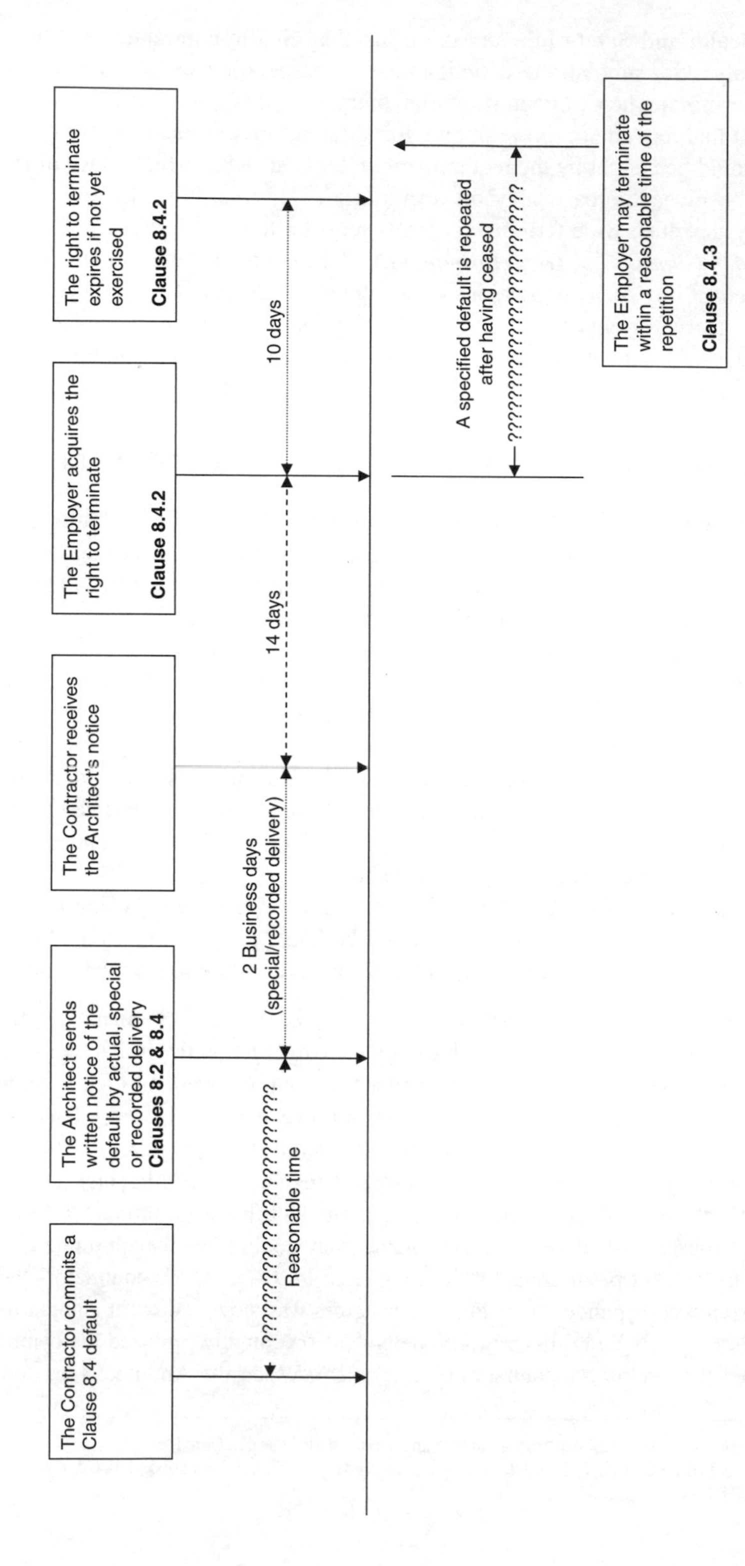

Fig. 16.1 Procedure for Terminating on Account of the Contractor's Specified Defaults

fees, the defendants counterclaimed damages for negligence in failing to issue the notice. Judge Newey held that there was no doubt that the contractors were failing to proceed regularly and diligently and that the Architect should have issued the notice. This decision was upheld by the Court of Appeal.

Another point that a prudent Employer must bear in mind is that the Architect's Default Notice is open to challenge as to its validity.[49] As pointed out by HHJ LLoyd QC in *Sindall* v. *Solland*,[50] it would be most unsatisfactory for the Employer to go ahead with the termination only for the appeal process afterwards to conclude that the Architect's notice had been invalid. Except in the clearest cases of default, the Employer would therefore be well advised not to act on the notice without independent legal advice.

If the Employer has not served the Termination Notice either because the Contractor ceased the specified default within the 14 days or for another reason and the Contractor repeats the default, the Employer may terminate the Contractor's employment within a reasonable time of the repetition (Clause 8.4.3). It was explained in *Robin Ellis Ltd* v. *Vinexsa International* Ltd,[51] which concerned the same termination procedure under the 1998 edition of the JCT Standard Form of Intermediate Contract, that no second Architect's Default Notice is required to trigger off the Employer's right to serve a Termination Notice for repetition of a specified default already covered by a Default Notice. In that case the Architect issued a Default Notice when the Contractor suspended work. The Contractor resumed work but suspended again. The Architect issued another Default Notice in relation to the second suspension. The Contractor challenged the validity of the Employer's termination notice on the grounds that it had been issued before expiry of 14 days from the second Default Notice. Deciding for the validity of the termination, HHJ Thornton QC explained that the Architect's second Default Notice lacked contractual validity, as no such notice was required.

A decision of the Court of Appeal suggests that, where there is no waiver of the right to terminate in reliance upon a valid Default Notice, any repetition of the default long after the notice would entitle the Employer to serve a Termination Notice. In *Reinwood Ltd* v. *L. Brown & Sons Ltd*[52] the Contractor served a Notice of Default (non-payment of VAT due) on 12 May 2005. On 26 January 2006 the Contractor served another Default Notice in respect of withholding of liquidated damages. On 28 June 2006 the Employer failed to pay an interim certificate, resulting in the Contractor serving on 4 July 2006 a Termination Notice, expressly relying on the Default Notice dated 26 January 2006. The House of Lords held that the Default Notice dated 26 January had been invalid because the Employer had been entitled to withhold that payment.[53] The Court of Appeal accepted the Contractor's assertion that, in reliance upon the notice of 12 May 2005, the termination was valid although there was no mention of the earlier Default Notice in the Termination Notice.

The use of the phrase 'within a reasonable time' to qualify when the Employer may terminate for repetition of a previous specified default creates unnecessary uncertainty where certainty is very necessary. A second area of uncertainty in Clause 8.4.3 concerns the use of the wording 'the Contractor repeats *a* specified default' instead of 'the Contractor

49 *Beaufort Developments (NI) Ltd* v. *Gilbert-Ash (NI) Ltd* [1999] 1 AC 266.
50 See Note 39.
51 [2003] EWHC (TCC); [2003] BLR 373; 93 ConLR 92.
52 Case No. 2007/2913 [2008] EWCA Civ 1090.
53 See Chapter 11, Section 11.10.2 for discussion of that litigation.

repeats *the* specified default'. It is arguable that the Employer may also be entitled to terminate for repetition if the second specified default is different from the first.

It is not unusual, after the first notice, for the parties to enter into negotiations towards avoidance of the termination. Where an agreement not to terminate is reached the Employer may no longer be entitled to rely on the first notice to terminate the Contractor's employment.[54]

16.7 Termination for the Contractor's insolvency

Clause 8.5.2 requires the Contractor to inform the Employer in writing of certain developments indicating that the Contractor is about to become Insolvent as defined in Clause 8.1. These include the commencement of relevant procedures and appointment of insolvency personnel to the Contractor's organization as well as proposals and notices of meetings the purpose of which is to initiate such procedures or appointments. The requirement for notice of such developments is calculated to give the Employer advance warning so that he can take appropriate steps to protect his interests (e.g. obtain professional advice as to how to deal with the situation). There are other sources of such warning independent of the Contractor. According to Powell-Smith and Sims,[55] the specified defaults listed under Clauses 8.4.1 are often the tell-tale signs of impending insolvency. They suggest the following additional warning signs:

- sudden disappearance of plant or materials from the site;
- high turnover in site management;
- excuses about late deliveries of materials;
- complaints by sub-contractors about non-payment;
- general lack of diligence in the carrying out of the works.

The effects of the Contractor becoming Insolvent are as follows.

1. The Employer acquires a right to terminate, at 'any time' while the Contractor is still Insolvent, the Contractor's employment under the Contract. The termination becomes effective upon the Contractor's receipt of written notice (Clause 8.2.2). The Clause 8.2.3 deeming provisions on receipt of notices applies to this notice.[56]
2. Pending final settlement under Clauses 8.7.4 and 8.7.5, any provision of the Contract that requires any sum to be paid to the Contractor (including release of retention) ceases to be applicable even if the Employer has not yet terminated the Contractor's employment (Clause 8.5.3.1).
3. The Contractor's right to carry out and complete the Works and carry out the design of any CDP is suspended. However, as explained in subsequent sub-sections, the Employer may enter into agreements requiring the Contractor to recommence performance of these obligations.

One of the main changes to the provisions on termination of the Contractor's employment on insolvency grounds has been the omission of the Contractor's employment being

54 See *Ellis Tylin* v. *Co-operative Retail Services Ltd* [1999] BLR 205; 68 ConLR 137.
55 Powell-Smith and Sims, *Determination and Suspension in Construction Contracts,* Collins, 1985, pp. 54–55.
56 For explanation of the deeming provisions see Section 16.4.

automatically terminated on the occurrence of certain insolvency events. The philosophy underlying this change appears to be that it is better to retain the contractual relationship but to give the Employer the right to terminate, which employers should exercise only after taking professional advice as to whether that course of action is the most appropriate in all the circumstances of the particular project and Parties. Alternatives open to the Employer include entering into continuation contracts or novation of the existing contract.

In most insolvency procedures, an administrative receiver, administrator or liquidator, hereafter referred to collectively as an 'Insolvency Practitioner', would be acting for the Contractor. Any agreement with the Contractor would therefore be with that person. To become a qualified Insolvency Practitioner, one must be a member of a professional body recognized by the Secretary of State for the Department of Business, Enterprise and Regulatory Reform and licensed to act in that capacity. To acquire such recognition, a professional body must satisfy the Secretary of State that its rules and regulations ensure that anybody licensed by it to practise as an Insolvency Practitioner has appropriate qualifications and training. Most Insolvency Practitioners are partners in the large legal and accountancy professions.

16.7.1 Instruction to the Contractor to continue

The Contractor's employment may be reinstated by agreement between the Employer and the Contractor (Clause 8.3.2). In practice, reinstatement is usually agreed only where either a novation can be arranged or it is clear that the Contractor, under the management of the Insolvency Practitioner, will be able to complete the project. If the project is near Practical Completion, the Insolvency Practitioner may well choose this course of action in order to recover retentions and avoid or minimize liquidated damages. The Employer may be attracted to such an arrangement to avoid the disruption of getting other contractors in.

16.7.2 Continuation contracts

A continuation contract entails the Employer entering into another contract whereby the Contractor, under the management of the Insolvency Practitioner, undertakes to complete the Works. Such a continuation contract is often in effect a variation to the original Contract (i.e. new terms are introduced). For example, the Employer may agree to advance payment or a new completion date. However, it is possible for the continuation contract to require the Contractor to continue and complete as he was originally obliged to. The value of a continuation contract contemplates a situation where the Employer, after examining the circumstances of the insolvency situation and alternative courses of action available to him, comes to a conclusion that his interests are best served by such an arrangement. A major factor that the Employer would usually consider is that the Insolvency Practitioner is personally liable for any non-performance of the continuation contract. This offers considerable protection since Insolvency Practitioners are normally from the major legal and accountancy firms and are therefore usually covered by appropriate professional indemnity insurance cover. On the part of the Insolvency Practitioner, he must also come to a view that such an arrangement is more advantageous to his appointor or the general body of creditors than termination.

16.7.3 Novation

The Employer and the Insolvency Practitioner may come to the view that novation is more in their mutual interests than termination of the Contractor's employment. In this context, a novation is an agreement involving the Employer, the Insolvent Contractor and a substitute contractor that the latter has replaced the Insolvent Contractor as the Contractor under the Contract. Thereafter the Substitute Contractor is governed by the Contract as if that has been the case from the start whilst the Insolvent Contractor ceases to have any further obligations or rights under the Contract. A novation is feasible only where the Contract is profitable and the Substitute Contractor is willing to share the surplus in the completed work with the Insolvency Practitioner. The surplus is the total value of the work executed less total payment made under the Contract. A conditional novation is a novation in which the terms of the Contract are varied by the agreement. This is a course of action that the Employer may consider where the Contract is not attractive enough to other contractors to bring about a pure novation. Examples of changes in terms include a new Completion Date, exclusion of the Substitute Contractor's liability for defects in the pre-novation work, and a new Contract Sum.

16.8 Corruption

The Employer may, by written notice, terminate the Contractor's employment if, in relation to this contract or any other contract between the Parties, the Contractor or any person employed by him or acting on his behalf commits an offence under the Prevention of Corruption Acts 1889 to 1916. A Local Authority Employer may also do so if the Contractor has given any fee or reward the receipt of which is prohibited by s. 117(2) of the Local Government Act 1972. The essence of these offences is the offering of inducement or reward for improper conduct in relation to the execution or performance of relevant contracts.

It is to be noted that acts of the Contractor's employees without his sanction or knowledge may amount to corruption. Also of significance is the provision that the clause covers similar acts committed on other contracts between the Employer and the Contractor.

16.9 Termination by the Contractor for Employer's defaults

The specified defaults of the Employer for which the Contractor may terminate his own employment are listed under Clause 8.9.1. These are that the Employer:

- fails to discharge to the Contractor any amount stated as payable under a certificate and/or VAT payable on the amount by the final date for payment;
- interferes with or obstructs the issue by the Architect of any certificate;
- assigns the Contract without the Contractor's consent;
- fails to discharge his contractual obligations in respect of the CDM Regulations.

16.9.1 Failure of the Employer to pay

The final date for payment pursuant to an Interim Certificate is 14 days from the date of its issue (Clause 4.13.1). The equivalent date for the Final Certificate is 28 days from the date of its issue (Clause 4.15.4). The Employer must pay the amount due under any certificate plus any VAT properly chargeable in respect of the certificate (Clause 4.6.1) by the relevant final date. Failure of the Employer to pay the amount properly due pursuant to a certificate entitles the Contractor to terminate his own employment. A less drastic course of action open to the Contractor is suspension under Clause 4.14.

In *Rupert Morgan Building Services (LLC) Ltd* v. *Jervis*[57] the English Court of Appeal held that, where a contract provides for a contract administrator as certifier of payment as in JCT 05, any amount stated on a certificate becomes the amount due. However, subject to serving appropriate notices as explained in Chapter 15, Section 15.2.7, the Employer may pay a lesser amount if he believes that it is the 'amount properly due' to the Contractor under the Contract. As stated by the House of Lords in *Melville Dundas Ltd (in Receivership)* v. *George Wimpey UK Ltd*[58] and *Reinwood Ltd* v. *L Brown & Sons Ltd*[59] the requirement for these notices is designed to ensure that a contractor knows immediately and with sufficient clarity where a set-off against payment otherwise due is being made and the grounds for it so that he may seek speedy redress by reference to adjudication.

16.9.2 Interference/obstruction by the Employer

It is explained in Chapter 2, Sections 2.1.3 and 2.4 that, even in the absence of an express term of the kind in Clause 8.9.1.2, the Employer would be under an implied duty not to interfere with the professional judgment of the Architect whenever he makes decisions in his capacity as an independent third party holding the balance fairly between the Employer and the Contractor.

For this condition in Clause 8.9.1.2 to apply, two facts must be established. First, the Employer, or his agents, must have had the intention either to prevent the Architect from performing his certification duties or to influence unduly the Architect's judgement in the performance of such duties. Second, there must have been actual interference or obstruction. In *R. B. Burden Ltd* v. *Swansea Corporation*[60] it was said that inadvertent errors, negligence or omissions of agents of the Employer who, at the request of the Architect, assisted in the certification process would not usually amount to interference or obstruction.

16.9.3 Assignment

Clause 7.1 prohibits any of the parties from assigning the Contract without the consent of the other.[61] Assignment by the Employer of any right under the Contract without the Contractor's consent is a default for which the Contractor has a right to terminate his own employment.

57 [2003] EWCA Civ 1563; [2003] BLR 18; [2004] TCLR 3; 91 ConLR. 81. In this case the English Court of Appeal approved a similar decision of the Scottish Sheriff's Court in *Clark Contracts Ltd* v. *Burrell Co (Construction Management) Ltd* (No. 1) (2002) SLT (Sh. Ct) 103.

58 [2007] BLR 257; 112 ConLR. 1; [2007] CILL 2469; [2007] 1 WLR 1136; [2007] 3 All ER 889.

59 [2008] UKHL 12.

60 [1957] 1 WLR 1167.

61 See Chapter 8, Section 8.2 for discussion on 'assigning a contract'; see Section 16.6.4 for termination for assignment by the Contractor.

16.9.4 CDM Regulations[62]

Under Clause 3.25, each Party undertakes to the other a contractual duty to comply with the CDM Regulations in relation to the Works and the site. The Employer is to ensure compliance with respect to two particular aspects of the Regulations. First, the Employer must ensure that the CDM Co-ordinator carries out his duties under the CDM Regulations. The same obligation applies in respect of the Principal Contractor but only where the Contractor is not also the Principal Contractor. The Employer commits a termination default if they fail to discharge their duties properly or he fails to appoint their replacements when they cease to act in those capacities.

As explained in Section 16.6.5 in this chapter, there is no entitlement to terminate for minor breaches of the Regulations.

16.9.5 Procedure for termination for Employer's defaults

The procedure for termination for the Employer's default is as follows. The Contractor gives the Employer notice specifying the default. It is important that the Contractor states expressly that he is giving a preliminary notice of termination. If the Employer continues the default for 14 days from the receipt of the notice then the Contractor may within a further 10 days give a final notice terminating his own employment. The right to terminate expires after the 10 days. However, if the Employer's default is repeated, the right becomes available again but it can only be exercised within a reasonable time after the repetition.[63] The procedure is shown in Figure 16.2.

16.10 Termination by Contractor for the specified suspension events

The Contractor may serve upon the Employer a preliminary notice of termination if any of specified events, referred to collectively as the 'specified suspension events', has the effect of suspending the carrying out of the whole, or substantially the whole, of the uncompleted Works for a continuous period beyond the maximum period of suspension specified in the Contract Particulars against Clause 8.9.2. The Employer must, at the pre-contract stage, consider very carefully the possibility of termination on this ground in deciding the period most appropriate to the particular contract and complete the Contract Particulars accordingly. The default maximum period of suspension is 2 months.

The specified suspension events are stated in Clause 8.9.2 as:

1. an AI under Clause 2.15 (discrepancies or divergences between documents), 3.14 (Variations and expenditure of Provisional Sums) or 3.15 (postponement of any part of the Works);
2. any impediment, prevention or default by the Employer, the Architect, the Quantity Surveyor of any of the Employer's Persons.

---

62 See Chapter 10 for a detailed commentary on the CDM Regulations.

63 See Section 16.6.6 for commentary on the right to terminate for a default repeated a year after an earlier notified default.

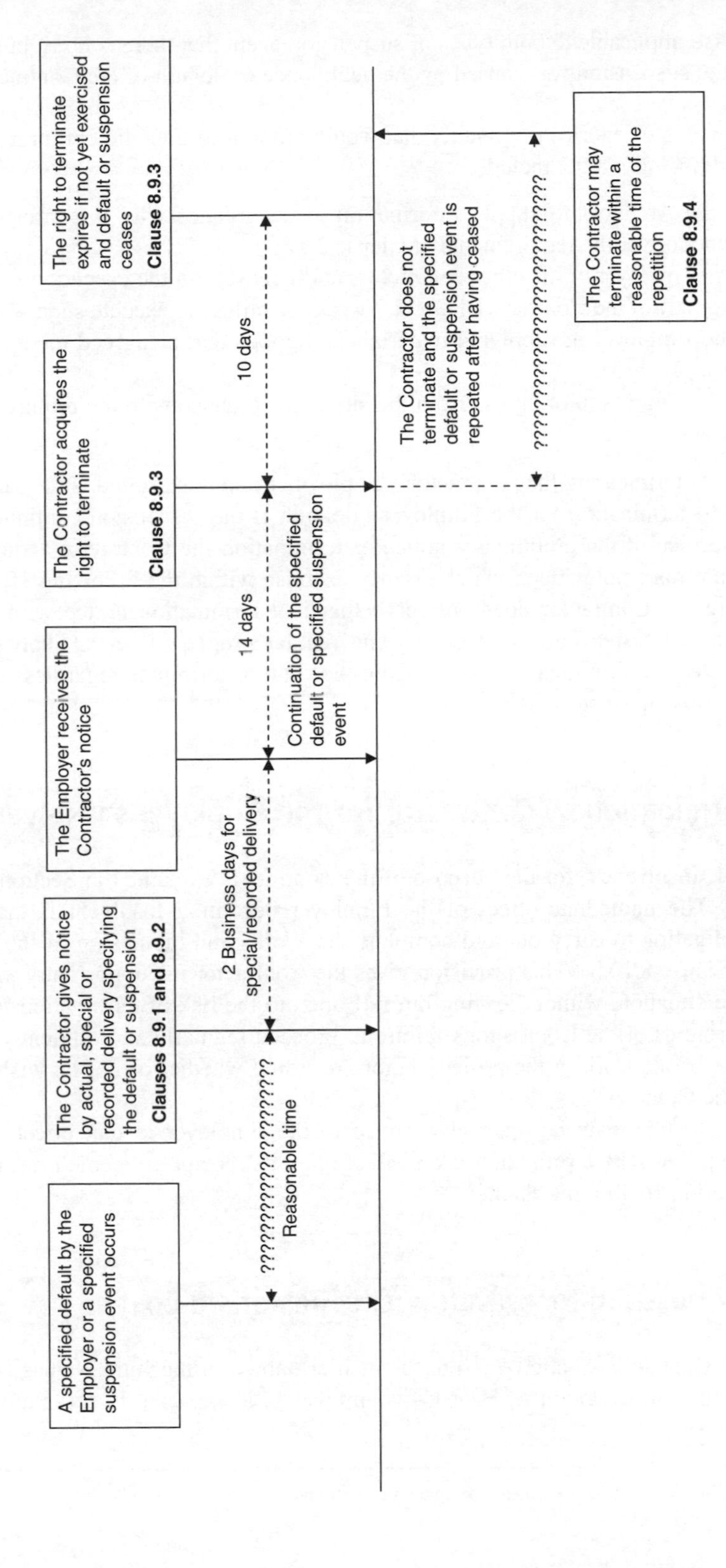

Fig. 16.2 Procedure for Terminating for the Employer's Specified Default or Specified Suspension

There is a proviso applicable to both types of suspension event that there is no right to terminate where the suspension was caused by the negligence or default of the Contractor or of the Contractor's Persons.

The most obvious categories of conduct that would amount to impediment, prevention or default[64] under Clause 8.9.2 include:

1. failure of the Architect to supply information in accordance with the Information Release Schedule or the requirements of Clause 2.12;
2. delay by the Employer or his other contractors and licensees in the execution of work not forming part of the Contract (Clause 2.7 work) or failure to execute such work;
3. delay by the Employer in supplying materials and goods that he agreed to supply or failure to supply them;
4. failure to give ingress to or egress from the site in accordance with the requirements of the contract.

The procedure for terminating the Contractor's employment on this ground is the same as that applicable to termination for the Employer's default. If the suspension continues for 14 days after service of the preliminary notice of termination the Contractor acquires a right to terminate his employment, which he may exercise within the following 10 days. Similarly, where the Contractor does not serve the final termination notice within the 10 days, a specified suspension event occurs and regular progress is or is likely to be affected materially, the Contractor acquires again the right to terminate regardless of the duration of the subsequent suspension.

16.11 Termination by Contractor for the Employer's insolvency

The meaning of 'insolvency' for the purpose of this contract is explained in Section 16.5 in this chapter. The immediate effect of the Employer becoming Insolvent is that the Contractor's obligation to carry out and complete the Works and the design of the CDP is suspended (Clause 8.10.3). This provision gives the Contractor the opportunity to stop and monitor the situation, without leaving himself open to the risk of termination by the Employer on grounds of the Contractor's failure to proceed regularly and diligently with the carrying out of the work in the project, before deciding whether or not he wishes to continue with the Contract.

If he decides to terminate he must give notice to the Employer to that effect. Only one notice is required. The termination takes effect upon its receipt or deemed receipt as explained in Section 16.4 in this chapter.

16.12 Termination for withdrawal of Terrorism Cover

As explained in Chapter 7, Section 7.3.4.6, the insurer named in the Joint Names Policy required under Insurance Option A, B or C, as applicable, is expected to serve advance

64 See Chapter 11, Section 11.4.6 for extension of time on this ground.

notice if the Terrorism Cover is to be withdrawn. The insurer's notice is referred to as the 'Insurer's Notification'. The date when the withdrawal is to come into effect is referred to as 'cessation date'.

The party responsible for taking out and maintaining the Joint Names Policy for the Works and Site Materials is to inform the other party immediately after receipt of the Insurer's Notification. The Employer has an option to terminate the Contractor's employment, which must be exercised within the time window between the date of the Insurer's Notification and the cessation date. A written notice to the Contractor stating the date from when the termination takes effect is required.

16.13 Termination by either party for loss/damage to the Works

Insurance Option C is intended for use where the Works entail alteration of, or extensions to, existing structures and their contents. Paragraph C.2 requires the Employer to take out and maintain insurance against loss and/or damage to the Works from the risks covered by 'All Risks Insurance' as defined in Clause 6.8.[64a]

Paragraph C.4.4 of Schedule 3 entitles either Party to terminate the employment of the Contractor in the event of the occurrence of loss or damage to the Works from the insured risks. However, there is an important proviso that it must be 'just and equitable' to terminate. It is submitted that useful considerations as to whether it is just and equitable to do so include the extent:

- of loss or damage to the existing structures;
- to which the nature of the Contract has been changed;
- to which the financial commitments of the Employer have increased;
- to which the Contractor can be adequately remunerated for carrying out the changed Works by the variation provisions in the Conditions.

It is likely that the type of loss and damage contemplated would be such as to frustrate the Contract at common law.

There are four key requirements to the procedure for termination under this clause.

1. Upon discovery of the loss/damage, the Contractor must forthwith give notice in writing to both the Employer and Architect.
2. The notice of termination must be served by either Party on the other within 28 days of the occurrence of the loss/damage.
3. The notice of termination must be sent by actual, recorded or special delivery. This requirement on method of delivery is only directory rather than mandatory (i.e. notice actually delivered by other methods within the time limit would be valid).[65] It is to be noted that the 'deeming' provisions in Clause 8.2.3 regarding the time of receipt of notices do not appear to apply to notices under this clause.

64a See Chapter 7, Section 7.3.4.

65 *Goodwin* v. *Fawcett* (1965) 175 EG 27; *J.M. Hill & Sons Ltd* v. *London Borough of Camden* (1980) 18 BLR

4. Where the notice of termination is to be contested, there is a limit of 7 days from receipt of the notice within which either Party may invoke the relevant procedures applicable to the resolution of disputes in order that it may be decided whether such termination is just and equitable.[66]

16.14 Termination by either party for *force majeure* and the like

Either Party may terminate the employment of the Contractor if, before Practical Completion, the carrying out of the Works, or substantially the whole of the Works, has been suspended for a continuous period exceeding the appropriate maximum stated in the Contract Particulars against Clause 8.11 and such suspension is caused by any of the following:

- *force majeure*;[67]
- an AI issued under Clause 2.15 (dealing with discrepancies and the like) as the result of the negligence or default of any Statutory Undertaker;
- an AI issued under Clause 3.14 (an instruction requiring a Variation) as the result of the negligence or default of any Statutory Undertaker;
- an AI issued under Clause 3.15 (an instruction postponing any work required to be carried out under the Contract) as the result of the negligence or default of any Statutory Undertaker;
- loss or damage to the Works from the Specified Perils;[68]
- civil commotion[69] and/or the activities of the relevant authorities in dealing with it;
- use or threat of terrorism[70] and/or the activities of the relevant authorities in dealing with it;
- exercise by Government of statutory power which directly affects the execution of the Works.

Clause 8.11.2 provides that, where the suspension arises from any of the Specified Perils, the Contractor has no right to terminate if the loss or damage from the Specified Peril was caused by the negligence or default of the Contractor or of any of the Contractor's Persons.

The procedure for termination under this clause, which is illustrated in Figure 16.3, is as follows. The Party wishing to terminate gives notice to the other that if the suspension is not terminated within 7 days of receipt of the notice, the Contractor's employment is to be terminated. If, after receipt of this notice, the suspension continues beyond that period, the termination takes effect upon its expiry.

---

66 For discussion on whether this limitation of the right to challenge termination contravenes the Construction Act see Chapter 17, Section 17.3.
67 See Chapter 11, Section 11.4.13 for explanation of this term.
68 See Clause 6.8 for the definition of this term.
69 See Chapter 11, Section 11.4.10.
70 See Chapter 7, Section 7.3.4.6 for explanation of this term.

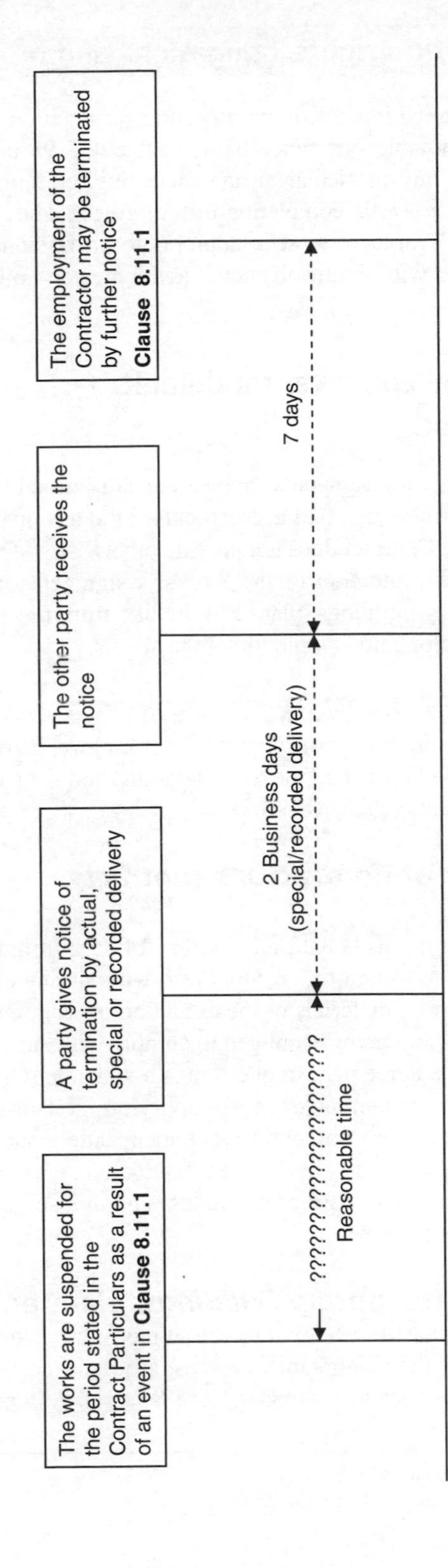

Fig. 16.3 Procedure for Terminating for Suspension Caused by the Clause 8.11.1 Events

16.15 Post-termination rights, obligations and procedures

The procedures, rights and obligations of the parties after termination vary according to the grounds relied upon. A notable difference between the JCT 98 and JCT 05 is that the latter does not anticipate any particular arrangement (e.g. continuation or novation agreements) by the Employer towards completion of the project. The JCT appears to be of the view that it is for the Employer to seek appropriate professional advice towards negotiation of the way forward with the Insolvency Practitioners and others acting for the Insolvent Party.

16.15.1 Termination by Employer for default, insolvency or corruption

The effect of termination is that the rights and obligations are limited to those stipulated in Clauses 8.7 and 8.8. Also, rights that had accrued before the termination may be exercised after the event where the Contract does not provide otherwise.[71] The rights and obligations of the Parties concern protection of the Works, assignment of the Contractor's contracts, removal of temporary buildings, plant and the like from the site, suspension of further payment to the Contractor, and completion of the Works.

16.15.1.1 Protection of the Works

Where the reason for the termination is Insolvency the Employer may take reasonable measures to protect the site, the Works and the Site Materials, and the Contractor is not to stand in the way of such measures (Clause 8.5.3.3).

16.15.1.2 Assignment of Contractor's contracts with third parties

The Employer, or the Architect on his behalf, may, within 14 days from the date of termination, request the Contractor to assign to the Employer, without any charge, the benefit of any agreement for the supply of materials or the execution of work entered into for the purposes of the Contract. The Contractor is obliged to comply with such a request only to the extent that the benefit of the agreement in question is assignable (Clause 8.7.2.3). For example, if the contract with a subcontractor or a supplier prohibits assignment without the consent of the other there can be no lawful assignment unless such consent is first obtained.[71a] Also, where the Contractor is in insolvent liquidation, such assignment would be in breach of insolvency law, which requires all creditors of an insolvent company to be treated with the same degree of fairness.

16.15.1.3 Removal of temporary buildings, plant and the like

The Contractor must ensure that any temporary buildings, plant, tools, etc., belonging to the Contractor or any of the Contractor's Persons, are removed from the Works as and when required by the Architect in writing (Clause 8.7.2.1). As explained later, the

71 *Lintest Builders Ltd* v. *Roberts* (1980) BLR 38 CA.
71a However, this right is limited by Clause 3.9.1 of JCT 05.

Employer has the right to use the Contractor's temporary buildings, equipment and materials on the site in completing the Works. Presumably, the Architect needs to consult the Employer concerning the resources he intends to use for the Works. He may even be under a duty either to advise the Employer on the resources he may lawfully use or draw to his attention the need to seek appropriate professional advice on the matter.

The JCT 98 authorized the Employer to remove and sell such property if the Contractor failed to remove them within a reasonable time of being properly requested to do so. The enforceability of this type of provision is fraught with difficulty, particularly where the Contractor is in insolvent liquidation. This power has not been brought forward into JCT 05.

16.15.1.4 Suspension of further payment to the Contractor

Clause 8.7.3 states that one of the effects of termination for reasons including Insolvency is that '(if not already applicable) Clauses 8.7.4, 8.7.5 and 8.8 shall thereupon apply and the other provisions of this Contract which require further payment or any release of Retention to the Contractor shall cease to apply'. Clause 8.7.4 requires the final account to be drawn up as an Architect's certificate within a reasonable time of completion of the project. Clause 8.7.5 provides for payment to be made by the appropriate party as indicated by the final account. Clause 8.8 contains alternative provisions on final accounting between the parties that are to apply where the Employer decides not to complete the project.

The purported effect of Clause 8.7.3 is therefore that the Contractor is not entitled to any further payment under the Contract other than pursuant to the final account under Clauses 8.7.5 and 8.8. Comment is called for on the enforceability of this provision against payment due that accrued prior to the termination and an adjudicator's decision directing the Employer to make payment to the Contractor.

The JCT 98 provided that the suspensory effect of the equivalent of Clause 8.7.3 did not apply to payment due under the Contract that had accrued 28 days or more before the suspension took effect.[72] This qualification has been omitted from JCT 05. There is pre-Construction Act case law indicating that the Employer would have a right to withhold further payment, including payment that accrued due before the commencement of the insolvency, where the Contractor is undergoing insolvent liquidation. For example, in *Willment Brothers Ltd* v. *North West Thames Regional Health Authority*[73] the Employer had issued a cheque to cover payment overdue on a certificate. On learning that a liquidator of the Contractor had been appointed, the Employer stopped the cheque. The Court of Appeal decided that, on what is now s. 323 of the Insolvency Act 1986, the Employer was entitled to set-off against the certificate the contingent liability of the Contractor to the Employer as a result of the termination. This case appears to be giving a green light to Employers to delay paying on certificates on the slightest suspicion of insolvency and thereby to bring about the situation where otherwise it could have been avoided. The impact of the Construction Act on this principle is examined later.

Clauses suspending payment to contractors and subcontractors after termination of their employment have been a common feature of construction contracts. As an effect of the

72 In *Melville Dundas Ltd and Others* v. *George Wimpey UK Ltd and Others* [2007] UKHL 18, the House of Lords treated this proviso in a similarly worded clause as supporting its construction whereby the suspension applied to not only payment that is yet to become due under the contract but also payment that had already become due but which the paying party had, in breach of contract, not yet made.

73 (1984) 26 BLR 51.

Construction Act, every qualifying construction contract, either expressly or by implied terms, provides for adjudication from which the adjudicator's decision is binding on the parties pending final resolution of the dispute by litigation, arbitration or agreement. This Contract provides that any dispute under it may be referred to and resolved in adjudication in accordance with Part I of the Scheme for Construction Contracts Paragraph 23(2) of which implements the binding effect of an adjudicator's decision. It is therefore arguable that, where an adjudicator makes a decision requiring an employer to make payment to a contractor, that requirement arises under their contract. However, the question whether termination clauses of the type referred to above include a right not to comply with the decision of an adjudicator directing payment has been a matter of some controversy. On the authority of the Court of Appeal in *Bouygues UK Ltd* v. *Dahl-Jensen UK Ltd*,[74] the court should not normally enforce an adjudicator's payment decision in favour of a contractor in insolvent liquidation where the employer has a contingent liability claim against the contractor.

The question that remains to be considered is whether the Employer is entitled to withhold payment that accrued prior to termination for insolvency other than liquidation. Some first instance decisions appeared to endorse the proposition that the obligation to comply with an adjudicator's decision arises under the relevant contract and that, therefore, it can be trumped by the suspensory effect of termination clauses.[75] However, the Court of Appeal took a different line in *Ferson Contractors Ltd* v. *Levolux A.T. Ltd.*[76] The subcontract at the centre of that case provided for termination of the employment of the subcontractor in terms similar to Clause 8.7.3 of JCT 05. By the time an adjudicator had made a decision on a payment dispute Ferson (the main contractor) had terminated the subcontractor's employment. Citing the subcontract's provision on the suspensory effect of the termination, Ferson refused to comply with the adjudicator's decision. HHJ Wilcox rejected this argument and decided that the subcontractor was entitled to enforce the decision without delay. In his opinion, the suspensory effect of the termination clause did not apply to payment required by reason of an adjudicator's decision. The Court of Appeal unanimously agreed with him.

The question was finally considered by the House of Lords in *Melville Dundas Ltd and Others* v. *George Wimpey UK Ltd and Others*[77] which arose from a contract incorporating the terms of the 1998 edition of the JCT Standard Form of Building Contract with Contractor's Design (WCD98). On 2 May 2003 the Contractor under the Contract applied for interim payment of about £400,000. Under the Contract the final date for payment of the application was 16 May 2003. Wimpey did not pay and on 22 May 2003 the Contractor went into administrative receivership. On 30 May 2003 Wimpey exercised its contractual right to terminate the Contractor's employment under the Contract. Clause 27.6.5.1 of WCD98 provided that upon such termination:

> Subject to clauses 27.5.3 and 27.6.5.2 the provisions of this contract [WCD98] which require any further payment or any release or further release of retention to the Contractor shall not apply; provided that clause 27.6.5.1 shall not be construed so as to prevent the enforcement by the Contractor of any rights under this contract in respect of amounts properly due to be paid by the Employer to the Contractor which the Employer has

---

74 [2000] BLR 522; [2001] 3 TCLR 2; 73 ConLR 135.

75 See, for example, *KNS Industrial Services (Birmingham) Ltd* v. *Sindall Ltd* (2000) CILL 1652; (2001) 17 Const LJ 170; *Bovis Lend Lease Ltd* v. *Triangle Development Ltd* (2002) CILL 1939; [2003] BLR 31.

76 [2002] EWCA Civ 11; [2003] BLR 118; [2003] TCLR 5; 86 ConLR 98.

77 [2007] UKHL 18.

unreasonably not paid and which, where clause 27.3.4 applies, have accrued 28 days or more before the date when under clause 27.3.4 the Employer could first give notice to determine the employment of the Contractor...

Wimpey contended that it was entitled under this clause not to pay the application even though no Withholding Notice had been served. One of the issues in the litigation was whether Part II of the Construction Act invalidated the clause. Their Lordships decided by a majority of three to two that the clause was enforceable against the Contractor (i.e. Wimpey did not have to pay the application).

One of the proposals in the Local Democracy, Economic Development and Construction Bill passing through Parliament at the time of writing is to adopt the principle in *Dundas* v. *Wimpey* as part of the Construction Act but to limit its application to only termination for the payee's insolvency.[78]

16.15.1.5 Supply of as-built drawings for CDP

Where there is a CDP, the Contractor must supply the Employer with two copies of the Contractor's Design Documents,[78a] including any such documents already supplied (8.7.2.2).

16.15.1.6 Completion of the Works

The Employer may complete the Works by employing other persons. In doing this, he is entitled to use, or authorize the other persons he has employed to use, the Contractor's temporary buildings, materials and equipment. However, where these resources do not belong to the Contractor, the Employer must obtain the consent of their actual owners before such use (Clause 8.7.1). The enforceability of the right to use the Contractor's resources is doubtful for at least two reasons. First, if the Contractor is in administration, leave of the court or the administrator would be required. Second, if the Contractor is in liquidation or bankrupt, the liquidator or trustee would be entitled to take into his possession any asset owned by the Contractor for the benefit of the general creditors.

16.15.2 Other termination under the Contract

The post-termination procedures described in this section are applicable to any termination by the Contractor under the Contract and termination by the Employer for:

- prolonged suspension on account of *force majeure* and the like (Clause 8.11.1);
- withdrawal of Terrorism Cover (Clause 6.10.2);
- loss or damage to the Works from any risk within the Clause 6.8 definition of 'All Risk Insurance' (paragraph C.4.4 of Schedule 3).

1. The Contractor is to remove all temporary buildings, plant, materials[79] and the like belonging to the Contractor or the Contractor's Persons from the site with reasonable

78 In any case, as suggested in *Westwood Structural Services Ltd* v. *Blyth Wood Park Management Co. Ltd* [2008] EWHC 3138 (TCC), *Dundas Wimpy* is likely to be limited at common law to insolvency situations.

78a This is defined under Clause 1.1 as 'the drawings, details and specification of materials, goods, and workmanship and other related documents prepared by or for the Contractor in relation to the Contractor's Designed Portion'.

79 Subject to retention of title clauses in favour of suppliers, the Employer owns materials he has paid for.

despatch (Clause 8.12.2.1). It is submitted that the Contractor's indemnities to the Employer in respect of personal injury, death and damage to property under Clauses 6.1 and 6.2 apply until the removal is complete. In these termination situations there is no provision for the Employer to use the Contractor's site resources in completing the Works.

2. The Contractor must provide the Employer with two copies of each item of as-built drawings produced for the purposes of the CDP (Clause 8.12.2.2).
3. The Contractor is to prepare a statement of the final financial settlement. In some of the termination situations the Employer may take over this task. The details on the preparation of this statement are provided in Section 16.16.3.

16.16 Financial settlement

The following final financial settlement regimes are discernible from the provisions: (i) where the termination is by the Employer for the Contractor's default, insolvency or corruption and the Works are completed; (ii) where the termination is by the Employer for the Contractor's default, insolvency or corruption and the Works are abandoned; (iii) termination in any other context.

16.16.1 Where the Works are completed after termination by the Employer for Contractor's default, insolvency or corruption

The financial settlement, which may be in the form of a statement by the Employer or a certificate issued by the Architect, must be prepared within a reasonable time of completion of the Works and the making good of defects (Clause 8.7.4). The settlement must be itemized to include: (i) the amount of 'expenses properly incurred by the Employer' as a result of the termination; (ii) 'direct loss and/or damage caused to the Employer' as a result of the termination; (iii) the amount of any payment made to the Contractor; (iv) the total amount which would have been payable for the Works in accordance with this Contract (referred to hereafter as the 'Notional Final Account').

16.16.1.1 Expense, loss and/or damage

There is the question whether 'direct loss and/or damage', as used under Clause 8.7.4.1 is the same as 'direct loss and/or expense' under Clauses 4.23 and 3.24. The decision in *F.G. Minter* v. *Welsh Health Technical Services Organisation*[80] suggests that this entitlement is damages under the first limb of *Hadley* v. *Baxendale.*[81] Allowable items include:

- amount payable to the completion contractor;
- additional professional fees payable on account of the termination;
- legal costs of the termination procedures;
- cost of managerial time expended in dealing with the termination;
- cost of work done to protect the uncompleted Works;
- cost of general site security;

80 43 (1980) 13 BLR 1.
81 4 (1854) 9 Ex. 341.

- cost of disposing of the Contractor's plant, temporary buildings, etc. (with credit for proceeds);
- cost of insuring the works for the period before the start of the completion contract;
- cost of additional finance;
- damages for delay in the completion of the Works.

16.16.1.2 Notional Final Account

As explained in Section 16.16.1 above, the term 'Notional Final Account' refers to 'the total amount which would have been payable for the Works in accordance with this Contract' stated in Clause 8.7.4.3. Many items would therefore have to be valued twice: first, in accordance with the completion contract and, second, in accordance with the original Contract. In the highly unlikely event of there being no variations and disruptions for which the Employer is responsible, this amount is very easy to determine. It is to be quantified in accordance with the relevant provisions in Section 4 of this Contract and on the assumption that the original Contractor carried out and completed the Works.

Unfortunately, there is some uncertainty as to how the amount is to be determined in situations where there are variations and loss and/or expense items in favour of the completion Contractor in circumstances where the same would have happened had the original Contractor completed the work. As an illustration of the uncertainty, consider a variation that would have been issued and for which the completion Contractor furnished an accepted Schedule 2 Quotation. There are several ways of determining the amount that would have been payable to the original Contractor in respect of the variation. The obvious and easiest way is to accept the amount in the quotation. An alternative is to use the estimates in the quotation but apply the original Contractor's unit prices for the estimated resources. As the Architect is usually not privy to such prices, this refinement is hardly workable. A third possibility is to price the variation in accordance with the Valuation Rules in Clause 5.6.1 and the original Contract Bills.

There are similar questions surrounding what to do with loss and/or expense under Clause 4.23. Entitlement to recovery under that clause is expressed to be conditional upon notice of disruption. Perhaps, the fiction that the original Contractor carried out the remainder of the Works is to be applied to Clause 4.23 notices served by the completion Contractor. Some practitioners favour a *pro rata* approach to variations, loss and/or expense and the like. This involves determining the two final accounts, ignoring variations and loss and/or expense items. The original Contractor is then credited for these items with sums that bear the same proportion to the completion Contractor's figures as the two final accounts bear to each other.

16.16.1.3 Calculations involved

The amount payable to or payable by the Contractor, D, is to be calculated as follows:

$$D = a_0 + a_1 + a_2 - A$$

where: a_0 = total amount paid or otherwise discharged to the Contractor before the termination; a_1 = the amount of expenses properly incurred by the Employer in completing the Works; a_2 = direct loss and/or damage caused to the Employer and for which the Contractor is liable; A = total amount that would have been payable to the Contractor if he had completed the Works in accordance with the Contract (Notional Final Account).

If D is positive the amount is to be paid by the Contractor. However, if it is negative it has to be paid to the Contractor.

16.16.2 Where the Works are abandoned after termination by the Employer for Contractor's default, insolvency or corruption

If the Employer, within 6 months of the termination, decides to abandon the carrying out of the Works, he must so notify the Contractor in writing (Clause 8.8.1). The Employer must then provide the Contractor with a statement of their final financial settlement within a reasonable time of the date of the notification. If the Employer abandons the Works but fails to serve the required notice within the 6 months, he must prepare the statement of settlement upon expiry of that period.

The amount payable to or payable by the Contractor, D, is to be calculated as follows:

$$D = a_0 + a_2 + a_4 - a_3$$

where: a_0 = total amount paid or otherwise discharged to the Contractor before the termination; a_2 = direct loss and/or damage caused to the Employer and for which the Contractor is liable; a_3 = the total value of work executed by the Contractor before termination plus any other amount due to the Contractor under the Contract; a_4 = the amount of expenses properly incurred by the Employer before the abandonment as a result of the termination.

If D is positive the amount is to be paid by the Contractor. However, if it is negative it has to be paid to the Contractor.

16.16.3 Other termination under the Contract

The financial settlement regime next outlined applies to the following types of termination by:

- the Contractor for the Employer's default or suspension of the Works (Clause 8.9.3);
- the Contractor for the Employer's Insolvency (Clause 8.10.1);
- either party for prolonged suspension on account of *force majeure* and the like (clause 8.11.1);
- the Employer for withdrawal of Terrorism Cover (Clause 6.10.2);
- either party for loss or damage to the Works from any risk within the Clause 6.8 definition of 'All Risk Insurance' (paragraph C.4.4 of Schedule 3).

Where the Contractor terminates his employment for prolonged suspension of the Works or the Employer's default or insolvency (the first two of the listed types of termination) the responsibility for preparing the statement of final financial settlement (referred to in the contract as the 'account') is on the Contractor, who must produce it as soon as reasonably practicable after the termination. A more definite timetable could have been imposed. Perhaps the drafters took the view that, as the Contractor would usually have a powerful cashflow-related incentive to act on this task with expedition, undue delay by the Contractor in producing the account is unlikely to arise. In addition, some may take the view that the Contractor should be given some latitude in when he chooses to deal with such situations for which the Employer is responsible.

With the remaining three types of termination the Employer has the option of preparing the account himself (Clause 8.12.3). If he elects to do this, the Contractor must supply him all the necessary documents within 2 months of the date of the termination. The Employer must then prepare the statement with 'reasonable despatch'. At the latest, he must do so within 3 months after receipt of the documents.

The account must state:

- the total value of work properly completed at the date of the termination (the method of determining this amount is the same as that used for the valuation of executed work for interim payment under Section 4);
- the Contractor's direct loss and/or expense under Clauses 3.24 and 4.23;
- cost of materials and goods ordered for the Works and for which the Contractor has incurred liability to pay (on payment they become the property of the Employer and the Contractor must not remove them);
- reasonable cost of removal of temporary buildings, plant, etc., as required by Clause 8.12.2.1.

The account is to include any direct loss and/or damage caused to the Contractor where the termination is for any of the following:

- the Employer's default;
- prolonged suspension under Clause 8.9.2;
- the Employer's Insolvency;
- loss or damage to the Works occasioned by any of the Specified Perils caused by the negligence or default of the Employer or of any the Employer's Persons.

The direct loss and/or damage item would include loss of profit on the remaining Works and, arguably, the Contractor's liability to sub-contractors arising directly from the termination.

The difference between their total and total payment already made on account represents the final settlement. If it is positive, it is payable to the Contractor and *vice versa*. Payment must be made by the Employer without deduction of Retention within 28 days after submission of the account. No definite timetable is provided for payment in the highly unlikely case where the final settlement is to be paid to the Employer. Paragraphs 7 and 8 of the Scheme would therefore apply by implication. The consequence of such an application is that the payment becomes due 7 days after the Employer makes a written demand for it whilst the final date is 17 days thereafter.

16.17 Contingent liability claims against an Insolvent Contractor

The final financial settlement upon completion may be years after the inception of the insolvency procedure. There is therefore the danger that the Insolvency Practitioner might have concluded his role and left. To counter this risk, a projected liability of the Contractor, referred to in the industry as a 'Contingent Liability Claim', is usually prepared and submitted to the Insolvency Practitioner pending the final accounts of the completion contract. As explained in Section 16.15.1.4 in this chapter, the Employer is also entitled to set-off for such a contingent liability claim against payment due.

16.18 Recoverability of liquidated damages

The general question whether a liquidated damages clause survives termination is difficult to answer from relevant case law. However, the correct starting point is that the answer depends on the terms of the particular contract. Where the contract provides expressly, clearly and unambiguously that the Contractor is to be answerable for liquidated damages for the overall delay in completion regardless of who achieved the completion, it will be given effect.[82] Where the Contract is silent on the issue, the general principle is that rights that accrued before termination are enforceable.[83] A contractor would therefore be liable for liquidated damages applicable to delay suffered before termination. On the issue of recoverability of liquidated damages for delay beyond the date of termination, the authorities conflict to some extent. It has been decided that liquidated damages for delay beyond the date of termination are recoverable only if there are express provisions in the contract to that effect.[84] However, as detailed in *Hudson's*,[85] different approaches have been followed in some common law jurisdictions.

The issue in relation to JCT 05 is clouded because it falls between the two extremes of clear express provisions and contractual silence. Two clauses may be relied upon in support of the proposition that the Employer is entitled to liquidated damages for the whole period of actual delay regardless of termination by the Employer for the Contractor's default, insolvency or corruption. Clause 8.3.1 provides that all the provisions on such termination are 'without prejudice to other rights and remedies of the Employer'. Some commentators[86] treated similar provisions in previous editions of this standard form as effectively maintaining the Employer's entitlement to liquidated damages regardless of the time of termination. There is support for this view in *Re Yeadon Waterworks Co. and Wright*[87] in which the employer was given the right to terminate the contractor's employment 'but without thereby affecting in any respects the liabilities of the said contractor'. That form of words was treated as retaining the employer's right to recover liquidated damages for the overall delay. In addition, Clauses 8.7.4 and 8.7.5 provide that in the settlement of accounts between the Employer and the Contractor, the Contractor is to be credited with 'the total amount which would have been payable for the Works in accordance with this Contract'. It is arguable that the amount which would have been payable must reflect the Employer's right to set-off for liquidated damages for delay under Clauses 2.32.1 and 4.13.5. It is submitted that both routes to recovery of liquidated damages are equally tenable. However, care must be exercised to avoid double recovery of damages for delay: as 'direct loss and/or damage' under Clause 8.7.4.1 and as liquidated damages in the preparation of the Notional Final Account.

---

82 See the New Zealand case of *Bayliss* v. *Wellington City* (1886) 4 NZLR 84.

83 *Bank of Boston Connecticut* v. *European Grain and Shipping Ltd* [1989] AC 1056.

84 *Re Yeadon Waterworks Co. and Wright* (1895) 72 LT 538; *British Glanzstoff Manufacturing Co. Ltd* v. *General Accident Fire & Life Assurance Corporation Ltd* [1913] AC 143.

85 *Hudson's Building and Engineering Contracts*, paragraphs 10.047 to 10.453.

86 Powell-Smith and Sims, *Determination and Suspension of Construction Contracts*, Collins, 1985, at p. 62; see also n. 19 in para. 10.047 of *Hudson's*.

87 Re *Yeadon Waterworks Co. and Wright* (1895) 72 LT 538.

16.19 Meaning of 'unreasonably or vexatiously'

The consequences of many of the grounds for termination vary in gravity. For example, the effects of sub-letting without consent may vary from very trifling to disastrous, depending upon the nature of the work sub-let and the calibre of the sub-contractor. The policy in the Contract is to provide a filter in the form of Clause 8.2.1 for minor defaults for which termination, although available technically, is not to take place. However, whilst the need for such filters is clear, the use of the phrase 'unreasonably or vexatiously' deprives the provision of the required certainty. In *J.M. Hill & Sons Ltd* v. *London Borough of Camden*,[88] which concerned notice by a contractor of termination for failure by the Employer to pay on a certificate, the Court of Appeal considered the meaning of that phrase. On the subject of the meaning of 'unreasonably' used in similar context in relation to the JCT 63, Ormrod LJ said:

> I imagine that it is meant to protect an employer who is a day out of time in payment, or whose cheque is in the post, or perhaps because the bank has closed or there has been a delay in clearing the cheque or something – something purely accidental or purely incidental so that the court could see that the contractor was taking advantage of the other side in circumstances in which, from a business point of view, it would be totally unfair and almost smacking of sharp practice.

In *J. Jarvis Ltd* v. *Rockdale Housing Association*[89] the Court of Appeal considered the same phrase used in Clause 28.1 of the JCT 80. Bingham LJ agreed with the views of Ormrod LJ as to the meaning of 'unreasonably'. He also suggested that it might be helpful to compare the benefit to the Contractor of terminating against the burdens to the Employer of that action. The Contractor's exercise of his right to terminate would not be unreasonable unless there is a gross disparity between the benefits and burdens. The use of 'vexatiously' is equally troublesome. According to Bingham LJ in the *Jarvis* v. *Rockdale* case, it suggests 'an ulterior motive to oppress, harass or annoy'.

16.20 Performance bonds and guarantees

A performance bond, also sometimes referred to as a 'performance guarantee', is an agreement by deed between an employer and a third party (bondsman or surety), usually a bank, insurance company or specialist bonding company, that if the contractor defaults in the performance of his obligations under the construction contract with the employer, the surety will pay the resultant loss of the employer up to a stated maximum sum. This is typically 10 per cent of the contract price. The contractor is usually required under the construction contract to obtain the bond for the employer. To procure the bond, the contractor pays a premium ranging typically from 1 to 3 per cent of the bond amount, depending upon the surety's assessment of the risk of default by the contractor. As a contractor would normally include this cost in his tender, the premium is ultimately paid for by the employer.

Although JCT 05 does not require the Contractor to procure a performance bond in favour of the Employer, experience suggests that it will often be amended to include such

88 (1980) 18 BLR 31.
89 (1986) 36 BLR 48 (hereafter *Jarvis* v. *Rockdale*).

a requirement. A particular problem highlighted in litigation arising from past editions of the contract, or contracts with similar provisions on termination and bonds, concerns when the employer may call on the bond.[90] The decisions suggest that, as the obligation of the surety to pay under the bond mirrors that of the contractor under the construction contract, the surety would not have to meet any payment until the contractor is obliged to do so. This means that in the event of termination, the surety will not have to pay until the work has been completed by alternative means, thus imposing upon the Employer the need to find additional funds to bring the project to completion.[91]

The decisions suggest that, where a bond is required, care must be exercised to ensure terms giving the Employer the right to immediate recovery of the estimated additional costs to complete, but allowing final settlement of accounts between the Employer and the surety on actual completion.[92]

16.21 Contesting Termination

Any step taken by either the Employer or the Contractor towards termination of the Contractor's employment is open to challenge by the other on procedural or substantive grounds. There are several contexts in which such a challenge may be raised.

1. The recipient of an Architect's notice of default may question whether what has happened amounts to the default provided for in the Contract. For example, in response to the Architect's notice that the Contractor has committed the default of failing to proceed regularly and diligently with the Works, the Contractor may admit to being behind programme but maintain that he is still proceeding regularly and diligently with the works.
2. The recipient of a Default Notice may be of the view that the wording of the notice is so unclear that he cannot tell exactly what the Architect is complaining of for remedial action.
3. As explained in Sections 16.6.6 and 16.9.5 in this Chapter, in most cases, a valid Termination Notice must be preceded by a valid Default Notice. A recipient of a Termination Notice may therefore dispute its validity on the grounds that no valid Default Notice had been served by the time of service of the Termination Notice.
4. As also explained in Sections 16.6.6 and 16.9.5 above, and illustrated in Figures 16.1 and 16.2, there are specified time windows within which valid Termination Notices by the Employer or the Contractor may be served. Similarly therefore, a purported Termination Notice may be challenged on the ground that it had been served outside the appropriate time frame.

---

90 For a review of the problems see Issaka Ndekugri, 'Performance Bonds and Guarantees in Construction: A Review of Some Recurring Problems' (1999) ICLR 294.

91 *Trafalgar House Construction* v. *General Surety* (1995) 73 BLR 32; see also *Paddington Churches Housing Association* v. *Technical and General Guarantee Company Ltd* [1999] BLR 244 in which His Honour Judge Bowsher held that, following termination under the JCT Standard Form of Building Contract With Contractor's Design 1981 edition for the Contractor's insolvency, the surety's obligation to pay on a performance bond did not arise until the Employer had not only completed the Works but also prepared the statement of accounts required under the equivalent of Clause 8.7 of this Contract.

92 For a sample of this type of bond see *The Use of Performance Bonds in Government Contracts,* a report published in 1996 by the Construction Sponsorship Directorate of the Department of the Environment, Transport and the Regions.

Depending on the parties' agreement, only the court or an arbitrator can finally determine the issue of the validity of the purported termination. Pending the decision, which could take months or even longer, there would be an impasse unless steps are taken to avoid it. To appreciate the seriousness of this risk, consider a situation where the Employer believes that he has validly terminated the Contractor's employment but the Contractor, believing otherwise, refuses to vacate the site. Only the court or the arbitrator can decide the question of the validity of the termination. What then is to happen in the interim? In the context of a failing Contractor, the situation could worsen for the Employer unless and until a way forward is found. An Employer in such circumstances would want to get the court to remove the Contractor from the project without any delay in order to bring in another contractor to complete the works. If the Contractor refuses to leave, the Employer may apply to the court for an interlocutory injunction against the Contractor continuing to remain on the site. Such an application will usually be accompanied by an undertaking to compensate the Contractor in damages in the event of the disputed termination being decided in the Contractor's favour.

The Contractor may respond to the pressure to eject him from the site or the application for the injunction with his own cross-application for an order to stop the Employer from ejecting him pending the determination of the disputed termination. Depending on the context, the desired intervention of the court can come in various other forms. For example, the court may be asked to order a party threatening termination not to go ahead and issue the final termination notice. An Employer insisting on the Contractor completing the works may ask the court to restrain the Contractor from removing his resources from the site or vacating it.

In each of these contexts, the court is required to deal with an application for an interlocutory injunction without knowing which of the parties will ultimately be found to be right. On the one hand, if it refuses the Employer's application and he is eventually found to have been entitled to terminate, the Employer could have suffered irretrievable loss from not being able to expel the Contractor at the earliest opportunity. On the other hand, if the court issues the injunction and it is finally decided that the termination had been without foundation, it would have sanctioned a serious breach of contract which could have disastrous consequences on the Contractor.

The problem of a disputed termination was considered in *London Borough of Hounslow* v. *Twickenham Garden Developments Ltd.*[93] The contract in that case, in the JCT 63 form, provided that if the Architect served notice that the Contractor had committed any of a number of specified defaults and the Contractor failed to remedy his default within 14 days after the notice, the Employer could, by notice, terminate the Contractor's employment. The Architect served notice that the Contractor was failing to proceed regularly and diligently with the works and, after expiry of 14 days, the Employer sought to terminate the Contractor's employment by written notice. The Contractor challenged the validity of the termination and refused to leave the site. The Employer commenced proceedings for an injunction to eject the Contractor and for damages for trespass. The Employer also applied for an interim injunction pending final determination of the dispute. Adopting an interesting analysis of the Contractor's entitlement to possession of site, Megarry J refused to issue the interim injunction. The judge stated that, in the context of a building contract, there was an implied term that the Contractor's licence granted by the contract to remain

93 (1970) 7 BLR 81 (hereafter *Hounslow* v. *Twickenham*).

on the site would not be revoked otherwise than in accordance with the contract while the contract was still in force. It was not therefore appropriate to issue an interim injunction to remove the contractor whilst the employer's right to terminate was still in doubt.

An implication of *Hounslow* v. *Twickenham* is that, where the Employer decides to abandon the project because of changes in business circumstances, the Contractor may continue with the works against the Employer's will. This proposition is described in Hudson's as absurd, thus raising serious doubts as to the correctness of the analysis of Megarry J. This doubt has been echoed in court decisions in Australia and New Zealand. Help in deciding applications for interlocutory injunctions of the kind in *Hounslow* v. *Twickenham* came after that case in the form of guidelines from the House of Lords in *American Cyanamid* v. *Ethicon Ltd.*[94] The court must first determine that there is serious question to be tried in the pending dispute. Once that trigger condition is satisfied, the next guiding principle is that an interlocutory injunction should be issued only where the court determines that, if the applicant eventually succeeds in the pending case, he would not be adequately compensated in damages for the loss he is likely to suffer as a consequence of the defendant not having been enjoined. In our example, the court should issue an injunction only where, if the Contractor were left on the site as a consequence of the court's refusal of the application, the Employer would not be adequately compensated in damages were it to be found that the Employer had been entitled to terminate at the first opportunity. If the Employer is able to prove that the Contractor will not be able to pay the promised damages because of insolvency or other good reason, the court should be inclined towards issuing the injunction.

The Contractor's position should also be considered. Will he be adequately compensated by damages if he is removed from the site but it is determined finally that the Employer had not been entitled to terminate the contract? In our example, if the Contractor was ordered off the site and it is finally determined that the Employer had not been entitled to terminate his employment, would damages for the wrongful termination be an adequate remedy for the contractor? The court must weigh the needs of the applicant against those of the defendant and determine where the balance of convenience lies. In the words of Lord Diplock:

> It is where there is doubt as to the adequacy of the respective remedies in damages available to either party or to both, that the question of balance of convenience arises. It would be unwise to attempt even to list all the various matters which may need to be taken into consideration in deciding where the balance lies, let alone to suggest the relative weight to be attached to them. These will vary from case to case. Where other factors appear to be evenly balanced it is a counsel of prudence to take such measures as are calculated to preserve the status quo.

The application of the *American Cyanamid* principles in the context of disputed termination of a contractor's employment is illustrated in *Tara Civil Engineering* v. *Moorfield Developments Ltd.*[95] The claimant, a contractor, was engaged by the defendant on a road works contract incorporating the 5th edition of the *ICE Conditions of Contract*. Under the contract, the Employer acquired a right to terminate the Contractor's employment if the Engineer certified that the Contractor had committed certain defaults. The Engineer issued such a certificate and the Employer notified the Contractor of his intention to expel the Contractor from the site. Contesting the validity of the Engineer's certificate,

94 [1975] AC 396.
95 (1989) 46 BLR 72 (hereafter *Tara* v. *Moorfield*).

the Contractor obtained an *ex parte* injunction restraining the Employer from expelling him. The Employer applied to discharge the injunction and for an order to the Contractor to vacate the site after completing work under another contract with a third party on the same site. The Contractor opposed this application. It was contended on the Contractor's behalf that *Hounslow* v. *Twickenham* was authority for the proposition that the Contractor was entitled to remain on the site until the validity of the termination was finally decided. Applying *American Cyanamid*, HHJ Bowsher QC held that the balance of convenience pointed towards supporting the Engineer's certificate, discharged the injunction and issued an order requiring the Contractor to vacate the site. The judge was of the opinion that, if it was finally determined that the termination was invalid, the damages payable by the Employer for the breach of contract would be an adequate remedy for the Contractor's loss from the invalid termination. He stated that it was not appropriate in these types of contested termination to look behind the contract administrator's notice of default unless, on the face of it, it was not the document required by the Contract.[96]

Wiltshier Construction (South) Ltd v. *Parkers Developments Ltd*[97] shows that, in exceptional circumstances, the balance of convenience could point towards restraining an employer from terminating the contractor's employment. The claimant was a management contractor for the construction of a supermarket for Tesco with the defendant development company as the employer under the 1987 edition of the *JCT Form of Management Contract*. Under this contract, the employer had the power to terminate the management contractor's employment if the contract administrator served notice of a specified default by the management contractor. 'Failure to carry out the works with due diligence and in an economical and expeditious manner' was the default stated in the contract administrator's letter that the employer sought to rely upon as a valid notice of default. The claimant commenced proceedings for an interim and permanent injunction to restrain the employer from terminating the claimant's employment. An *ex parte* injunction against service of the termination notice was granted and the developer applied to set it aside. HHJ Havery QC determined that, on the face it, the notice of default was invalid because the obligations of the management contractor did not include carrying out the works. He concluded that the balance of convenience pointed towards continuing the injunction. Factors considered included the invalidity of the notice of default, doubts as the developer's financial ability to compensate the contractor in damages, and the fact that Tesco was likely to step into the shoes of the developer to complete the project.

The conclusions from the analysis of the case law on disputed terminations are summarized as follows. As the Contractor's main reason for the contractual relationship is to make a profit, evidence of the Employer's ability to pay damages for wrongful termination should normally defeat the Contractor's application for an injunction to restrain the Employer from terminating the Contractor's employment or ejecting him from the site. It would be difficult to find circumstances where the balance of convenience would point towards granting an injunction to an Employer to restrain the Contractor from stopping the carrying out of the Works. Furthermore, such court intervention would amount to an order of specific performance, which is rarely appropriate in the context of a construction contract.

---

96 For support of the proposition that, in considering an application for an interim injunction, the court should not look behind a contract administrator's notice of default see also *The Attorney General of Hong Kong* v. *Ho Hon Mau (t/a KO Construction Company)* (1988) 44 BLR 144.

97 (1997) 13 Const LJ 129.

17

Dispute resolution

Certain aspects of the construction process make the performance of construction contracts intrinsically prone to disputes (e.g. new materials and construction techniques, multi-organizational participation, uncertainty in the physical and commercial environments within which they have to be performed, variability in human performance, and low margins). This reality of disputes is recognized in JCT 05 by way of provisions on how they are to be resolved if they arise. The aim in this chapter is to explain these provisions.

Generally, the court has the power to settle disputes between individuals. Furthermore, the parties often need the court to enforce settlements reached by the other techniques. Article 7 provides that any dispute arising under the Contract may be referred to adjudication in accordance with Clause 9.2.[1] By Article 8, completion of the Contract Particulars to indicate that 'Article 8 and Clauses 9.3 to 9.8 apply' has the effect that the parties have entered into an agreement in the terms of that Article to resolve their disputes by arbitration. The combined effect of Article 8 and the Contract Particulars[2] is that, if the phrase is deleted, there is no right or obligation to resolve any dispute by arbitration unless the parties enter into a separate agreement to that effect. The parties may also decide to resolve their disputes by negotiation or other techniques referred to collectively as 'Alternative Dispute Resolution' although the Contract mentions only mediation.

As Articles 7–9 stipulate that certain disputes or differences are to be resolved by the applicable resolution technique, a party may resist any of these procedures by denying that there is a 'dispute' or 'difference', thus raising the question of what these terms mean. A related source of contention concerns the types of disputes covered by the relevant dispute resolution clause, as the other party can resist the proceedings by arguing that, although there is a dispute, it is outside the ambit of the appropriate clause. The meaning of a 'dispute' and the ambit of the dispute resolution clauses in the Contract are therefore examined first.

---

1 It is also to be noted that if the form is to be used on a project not subject to the adjudication legislation, appropriate amendments must be made unless the parties wish to be bound by the adjudication provisions. See Chapter 1, Section 1.5.5 on amending the Contract.

2 It is stated in the Contract Particulars that if neither entry against Article 8 is deleted, Article 8 and Clauses 9.3 to 9.8 are not applicable to the Contract (i.e. there is no arbitration agreement by default).

17.1 The meaning of 'dispute' or 'difference'

Adjudication or arbitration proceedings cannot be properly commenced without a dispute between the parties. A party may therefore challenge a purported reference to adjudication or arbitration on the grounds that there is no dispute. Similarly, enforcement of a purported decision may be defended on the grounds that, without a dispute, the adjudicator or arbitrator had no jurisdiction to make a decision that is binding on the parties. Where such challenges are raised it is for the court to determine whether there is or was a dispute.

The existence of a dispute has also been litigated in the context of applications for stay of court proceedings to arbitration. In such an application a defendant in the court proceedings contends that an arbitration agreement covers the dispute the court is to try and that, therefore, the court proceedings should be suspended pending resolution by the arbitration. The response of the party wanting to avoid arbitration has often been that the matter before the court is not a dispute and that, therefore, it is outside any arbitration agreement that can apply to only disputes.

A major problem with the meaning of a 'dispute' is that a large body of case law has been built up from the litigation in all these different contexts. Some of the judicial comments were made in relation to specific legislation on court procedure or arbitration in force at the time of the relevant decision. As the legislation has changed over the last 60 years, some of the decisions are no longer relevant to the general question of what amounts to a dispute although they are still cited by litigants.[3] The context of each decision therefore needs to be determined to assess its true value as authority on the question before the tribunal of interest.

17.1.1 The arbitration cases

In *Tradax Internacional SA* v. *Cerrahogullari TAS, The M Eregli*[4] the defendant, owners of a ship, and the claimant entered into a charterparty which provided that all disputes from their contract were to be referred to arbitration. Under the contract a claim was barred unless it was made in writing and referred to an arbitrator within 9 months of final discharge of the cargo concerned. The defendant did not dispute the correctness of the claimant's invoices claiming dispatch money under the charterparty. It just ignored all communications with the claimant.

The claimant purported to appoint an arbitrator after expiry of the 9 months after final dispatch. For various reasons there was little progress with the arbitration. The claimant eventually brought proceedings and applied for summary judgment on the grounds that, as there was no arguable defence to the claim for dispatch money, there was no dispute to which the arbitration clause and the time-bar could apply. The defendant admitted that, apart from the time-bar, there was no defence to the claim as to liability or quantum. Kerr J held that a dispute had arisen even though the defendant had not denied liability. At 350c-d he said this of the claimant's argument that there was no dispute:

> ...The fallacy in the [claimant's] argument can be seen at once if one considers what would have been the position if the plaintiffs had in fact purported to appoint Mr Barclay [the

3 Some of the cases often cited but which we believe no longer have general application include: *Ellis Mechanical Services Ltd* v. *Wates Construction Ltd* (1976) 2 BLR 57; *R M Douglas Construction* v. *Bass Leisure Ltd* (1990) 53 BLR 119; *Monmouthshire County Council* v. *Costelloe & Kemple Ltd* (1965) 5 BLR 83.

4 [1981] 3 All ER 344 (hereafter *Tradax*).

> arbitrator] as their arbitrator within the time limit of nine months. They could clearly have done so, and indeed any commercial lawyer or businessman would say that this is what they should have done under the clause to enforce their claim. Arbitrators are appointed every day by claimants who believe, rightly or wrongly, that their claim is indisputable. However, on the [claimant's] own argument, Mr Barclay would have had no jurisdiction, since there was then, as they now say, no 'dispute' to which the arbitration clause could be applied. In my view this argument is obviously unsustainable.

In *Ellerine Bros (Pty) Ltd and Another* v. *Klinger*[5] the Court of Appeal, by unanimous decision, approved and applied *Tradax*. It arose from an agreement between the claimants (two South African companies) and the defendant which provided that the defendant should be the principal distributor of a film financed by the claimants and that the claimants should each receive 20% of the net receipts of the film. The defendant also undertook to keep proper records of the receipts and to make them available for inspection by the claimants. The agreement provided that all disputes were to be referred to arbitration.

After a year of not receiving anything from the defendant, the claimants wrote to him demanding to see the accounts of the film. The defendant ignored this and repeated requests. The claimants wrote a final letter demanding the accounts and payment of the monies due within 7 days. The letter stated that legal proceedings would be instituted if the defendant did not comply with this final request. The claimants eventually issued proceedings alleging breach of the agreement and seeking payment of monies due under it. The defendant applied to stay the court proceedings to arbitration under s. 1(1) of the Arbitration Act 1975. The claimants opposed the application on the grounds that, as all they were asking for was payment of monies to which they were clearly entitled, there was no dispute to be referred and that, therefore, the court had no power to stay the proceedings.

The Court of Appeal unanimously held that the defendant was entitled to stay of proceedings to arbitration. Templeman LJ stated at p. 741:

> Again by the light of nature, it seems to me that...if letters are written by the plaintiff making some request or some demand and the defendant does not reply, then there is a dispute. It is not necessary, for a dispute to arise, that the defendant should write back and say 'I don't agree'. If, on analysis, what the plaintiff is asking or demanding involves a matter on which agreement has not been reached and which falls fairly and squarely within the terms of the arbitration agreement, then the applicant is entitled to insist on arbitration instead of litigation.

At p. 743 he continued:

> But the fact that the plaintiffs make certain claims which, if disputed, would be referable to arbitration and the fact that the defendant then does nothing (he does not admit the claim, he merely continues a policy of masterly inactivity) does not mean that there is no dispute. There is a dispute until the defendant admits that a sum is due and payable, as Kerr J said in the *Tradax* case.

In *Hayter* v. *Nelson and Home Insurance Co.*[6] Saville J considered the proposition that there cannot be a dispute where the underlying claim is undisputable. It concerned a reinsurance contract between the claimant and the defendant. The defendant entered into a retrocession

5 [1982] 2 All ER 737 (hereafter *Ellerine* v. *Klinger*).
6 [1990] 2 Lloyd's Rep 265.

agreement with the second defendant in relation to the reinsurance contract whereby the latter agreed to accept by way of retrocession 100% of the reinsurance contract. The retrocession agreement incorporated the reinsurance contract and provided that the second defendant would follow the fortunes of the defendant in respect of the reinsurance contract. It also incorporated an arbitration agreement referring to arbitration any dispute that could not be settled amicably.

The defendant was ordered by an arbitrator and the court to pay the claimant certain sums under the reinsurance contract. The defendant then claimed an indemnity from the second defendant in respect of these sums. The second defendant applied to stay the claim for indemnity to arbitration. The second defendant did not agree that the defendant was entitled to the indemnity. It was argued on behalf of the first defendant that, as the second defendant had promised in the retrocession agreement to follow the fortunes of the first defendant, the claim for the indemnity was indisputable both as a matter of fact and as a matter of law and that, therefore, there was no dispute between the parties. At p. 268 Saville J rejected this argument in these terms:

> In my judgment in this context neither the word 'disputes' nor the word 'differences' is confined to cases where it cannot then and there be determined whether one party or the other is in the right. Two men have an argument over who won the University Boat Race in a particular year. In ordinary language they have a dispute over whether it was Oxford or Cambridge. The fact that it can be easily and immediately demonstrated beyond any doubt that the one is right and the other is wrong does not and cannot mean that that dispute did not in fact exist. Because one man can be said to be indisputably right and the other indisputably wrong does not, in my view, entail that there was therefore never any dispute between them.

These comments were referred to with approval by the majority of the Court of Appeal in *Halki Shipping Corporation* v. *Sopex Oils Ltd*,[7] the facts of which are outlined later.

In *Cruden Construction Ltd* v. *Commission For The New Towns*[8] the claimant was the contractor on a JCT 63 contract dated 22 May 1980 for the construction of dwellings for Central Lancashire New Town Development Corporation, who contracted to sell some of the dwellings to a housing association. On discovery of structural defects, the housing association made a claim against the Corporation. The defendants, who were the Corporation's successors in title, wrote to Cruden's solicitors on 7 October 1993 informing them of the alleged defects and inquired whether they would accept notices of arbitration on Cruden's behalf. On 11 October 1993, the solicitors replied that they would accept service on the assumption that details of the claim would be supplied. The same day the defendant gave notice of arbitration. HHJ Gilliland QC granted the claimant's application for a declaration that at the material time there was no dispute capable of being referred to arbitration. At p. 148C he explained:

> The reference [in *Ellerine* v. *Klinger*] to 'a matter on which agreement has not been reached' implies that an opportunity had to be given at some stage for an agreement to have been reached on the matter but where a person has not in fact been told and is unaware in what respects he is alleged to have broken his obligations it is in my judgment

7 [1998] 1 WLR 726 (hereafter *Halki*).
8 (1994) 75 BLR 134.

> quite impossible to say that the matter is one on which agreement has not been reached, at least where further information about the matter is being sought.

He distinguished *Tradax* as applicable where there is evidence that the claim was ignored whilst *Ellerine* v. *Klinger* is authority for the situation where there is evidence of prevarication in responding to a claim.

Halki concerned a charterparty between the claimant ship owners and the defendant for the carriage of the defendant's oils. The charterparty incorporated an arbitration agreement in respect of 'any dispute arising from or in connection with' the charterparty. The claimant claimed demurrage for delay by the defendants to load the ship and discharge the cargo at the agreed times. The defendant did not admit liability for demurrage.[9] The claimant brought proceedings on the claim and applied for summary judgment on the grounds that most of the demurrage claimed was indisputably due. The defendant applied to stay the proceedings to the arbitration.

After a review of the authorities on what amounts to a dispute, the Court of Appeal decided by a majority that they were bound by the earlier decision of the Court in *Ellerine* v. *Klinger*. At p. 761G Lord Justice Swinton Thomas stated:

> In my view, following those cases, Mr Waller's submission is correct, and in the words of Templeman LJ in *Ellerine Bros. (Pty)* v. *Klinger* [1982] 1 WLR 1375, 1383H there is a dispute once money is claimed unless and until the defendants admit that the sum is due and payable. The cases relied on by [counsel for the claimant] to the opposite effect resulted from particular interpretation that the courts placed on the words in section 1 of the Act of 1975 and its predecessors to which I have referred. In my judgment if a party refused to pay a sum which is claimed or has denied that it is owing then in the ordinary use of the English language there is a dispute between the parties.

In *Collins (Contractors) Ltd* v. *Baltic Quay Management (1994) Ltd*[10] the contract between the parties was practically the same as the JCT Agreement for Minor Works, 1998 edition. As under JCT 05, it provides that where the Employer (the defendant in this case) serves neither a Payment nor a Withholding Notice after receipt of a payment certificate, the Employer is bound to pay the amount stated on the certificate within 14 days after its receipt. The Contractor brought an action to enforce payment on a certificate to which the Employer had served none of the required notices. In reply to the defendant's application under s. 9(4) to stay the action to arbitration, the claimant contended that there was no dispute capable of being referred to arbitration. The Court of Appeal, by unanimous decision, held that the action should be stayed to arbitration because they were bound by *Halki*.

17.1.2 The adjudication cases

Comments by TCC (Technology & Construction Court) judges as to what constitutes a dispute for the purposes of statutory adjudication can be assigned roughly into one or the other of two contrasting approaches. Interestingly, the judges in both approaches derive support from *Halki* although they are in direct conflict.

---

9 This is equivalent to liquidated damages for the period during which the ship's owners were wrongfully deprived of the use of their ship.

10 [2004] EWCA Civ 1757.

The first approach, referred to in this work as the 'narrow definition' of a dispute, is based on the proposition that for a dispute to arise not only must a claim be made but also the recipient of the claim should have been given reasonable opportunity to consider and respond to it. *Sindall Ltd* v. *Abner Solland and Other*,[11] which illustrates this approach, arose from a contract in the form of the JCT Intermediate Form of Contract for the refurbishment of Lombard House in Mayfair, London. On 1 December 2000 the Contract Administrator issued a notice of default that the Contractor was failing to proceed regularly and diligently with the Works. The parties were by then in adjudication in relation to the Contractor's entitlement to extension of time due to various delays for which the Employer was responsible. The Contractor therefore challenged the default notice. On 21 December the Employer served notice of termination of Sindall's employment. The Contractor considered this termination as a repudiatory breach.

The adjudicator delivered his decision on 8 January 2001 ordering additional payment and extension of time to 29 August 2000. This adjudication had considered only relevant events up to 11 August 2000. On 11 January 2001 Sindall applied for further extension of time for delays suffered after 11 August 2000. On 1 Feb 2001 the Contract Administrator wrote requesting Sindall to provide such further information that it considered supportive of its claim. Sindall had by then commenced procedures to terminate its own employment for the alleged repudiatory breach by Solland.

On 9 February 2001 Sindall, in support of its second extension of time claim, delivered a substantial package including three lever files to the Contract Administrator. The covering letter requested a response within 7 days of receipt of the supporting documentation. On 15 February the Contract Administrator wrote that he needed time to consult the design team in relation to the claim and that, for that purpose, he needed three copies of the claim. On 16 February Sindall served Notice of Adjudication. The redress sought was stated to include a declaration on the validity of the termination by the Employer, the Contractor's entitlement to extension of time and payment orders in respect of financing charges for non-payment and any further monies due under the contract.

The adjudicator decided that: (i) the termination by Solland was invalid and that, therefore, Sindall was entitled to terminate its own employment; (ii) Sindall was entitled to further extension of time of 17 weeks. The claimant brought action to enforce the decision whilst the defendant applied for a declaration that the adjudicator had decided the issues relating to extension of time without jurisdiction to do so. HHJ LLoyd QC decided that no dispute in relation to the extension of time claim had crystallized by the time of service of the Notice of Adjudication and that, therefore, the decision on extension of time had been reached without jurisdiction. At paragraph 15 he explained:

> I do not accept, first, that Sindall was entitled to say 'either let us have the result within seven days or otherwise there will be a deemed dispute' or, secondly, and in any event that [the Contract Administrator's] failure to respond to the letter of 11 February by the time the adjudication notice was served constituted a deemed dispute…For there to be a

---

11 (2001) 3 TCLR 30; (2001) 80 ConLR 152 (hereafter *Sindall* v. *Solland*). See also *Fastrack Construction Ltd* v. *Morrison Construction Ltd & Anor* [2000] BLR 168; *Griffin (t/a K & D Contractors)* v. *Midas Homes Ltd* (2000) 18 Const LJ 67, 78 ConLR 152; *Edmund Nuttall Ltd* v. *R.G. Carter Ltd* [2002] BLR 312; *Hitec Power Protection BV* v. *MCI Worldcom Ltd* (2002) EWHC 1953 (TCC); *Beck Peppiatt Ltd* v. *Norwest Holst Construction Ltd* [2003] EWHC 822 (TCC); *Carillion Construction Ltd* v. *Devonport Royal Dockyard Ltd* [2003] BLR 79.

dispute for the purposes of exercising the statutory right to adjudication it must be clear that a point has emerged from the process of discussion or negotiation that has ended and that there is something which needs to be decided.

He did not think such a point had emerged in the instant case. He stated:

> Here...a person in the position of the Contract Administrator must be given sufficient time to make up its mind before one can fairly draw the inference that the absence of a useful reply means that there is a dispute.

The decision of HHJ Seymour QC in *R. Durtnell & Sons Ltd* v. *Kaduna Ltd*[12] suggests that a timetable provided in the contract for dealing with claims may be relevant to the question whether a dispute can arise from a delay in responding to a claim. The contract used in that case was in the JCT 80 form but it was so amended to comply with the Housing Grants, Construction and Regeneration Act 1996 that it was practically the same as JCT 98. Durtnell served a Notice of Adjudication contending that Kaduna was in breach of contract for a number of reasons and seeking a further extension of time under the contract. The adjudicator found that the Contractor had been delayed and ordered the contract period to be extended but in the subsequent enforcement proceedings HHJ Seymour QC decided that the adjudicator had not had jurisdiction to make such decision as the time allowed in the Contract for the Architect to make a determination in respect of Durtnell's application for an extension of time had not yet expired, and the Architect had not yet made a determination.

Durtnell argued that a decision of the Architect in relation to an application for an extension of time was not a condition precedent to the exercise of a right to adjudicate and submitted that that right was one exercisable at any time. Kaduna accepted that it was not a condition precedent to the jurisdiction of an adjudicator to determine whether, under the contract, Durtnell was entitled to an extension of time for completion of the Works that there should have been a decision of the Architect on the claim. It however submitted that where Durtnell had elected to seek a decision from the Architect, until there was such a decision or the Architect had failed to make it within the time allowed him, there was nothing for Durtnell to dispute and hence no 'dispute' capable of being referred to adjudication.

This is how at paragraph 42 the Judge dealt with the rival arguments:

> ...in my judgment it cannot be said that there is a 'dispute' as to entitlement to extensions of time, or as to valuation of loss and expense consequent upon a grant of extensions of time, at a time at which the question of whether there should be any extension of time, or any further extension of time, has been referred to the architect for the purposes of the standard form, the time allowed by the standard form for him to make a determination has not expired, and no determination has been made. I readily accept that it is not, expressly, a condition precedent to any reference to adjudication of a dispute as to entitlement to an extension of time under a contract in the standard form that the dispute should first have been referred to the architect. However, it is not easy to see how a dispute as to entitlement to an extension of time could arise until that had happened and the architect had made his determination or the time permitted for doing so had expired. The reason is that under the standard form it is not for the employer to grant an extension of time or not. That function is entrusted to the architect who is under an obligation to act impartially

12 [2003] All ER (B) 281; [2003] EWHC 517 (TCC).

> in making his assessment. Until the architect has made his assessment, or failed to do so within the time permitted by the standard form, there is just nothing to argue about, no 'dispute'. …It is nonsensical to suggest that a 'dispute' can exist between two parties as to a matter entrusted to a third party for independent decision in advance of the decision being known. For practical purposes, therefore, it seems to me that it is a condition precedent to the reference to adjudication of a 'dispute' as to entitlement to an extension of time and as to anything which is dependent upon such decision, such as a claim for payment of loss and expense in relation to an extension of time claimed but not granted, that the person to whom the making of a decision on the relevant issue is entrusted under the contract between the parties should have made his decision, or the time within which it should have been made has elapsed without a decision being made.

The Judge went on to hold that, in granting an extension of time for completion of the Works, the adjudicator had acted without jurisdiction.

The second approach, hereafter dubbed the 'wide definition' of a dispute, consists of the cases in which the court applied the proposition that there is a dispute once a claim is made unless and until the defendant admits that the claimant is entitled to the rights claimed. *Cowlin Construction Ltd* v. *CFW Architects (a firm)*[13] illustrates the general approach in the second category of cases. CFW had been appointed as the Architect by Cowlin, a design and build contractor on a Ministry of Defence project. On 27 February 2002 Cowlin submitted a claim to CFW for costs that were said to have been incurred as a result of delays by CFW. The claim totalled £672,395.29, with a breakdown into seven elements. At CFW's request, Cowlin wrote again to CFW on 11 March 2002 enclosing what was described as 'full supporting documentation of the monetary claim'. CFW rejected this, saying that there was insufficient detail. A meeting between Cowlin and CFW's loss adjusters took place but failed to produce a settlement. On 3 May 2002 Cowlin wrote to CFW providing a deadline of 17 May 2002 for a satisfactory offer of settlement, failing which 'immediate and substantive action' would be taken. On 14 May an agent of CFW responded:

> I refer to our recent meeting in connection with this matter and confirm, that in view of the various complexities of your allegations, it has been necessary for us to obtain and review all the files held by CFW, following our meeting with them, which unfortunately is taking longer than anticipated. Please be assured it is our intention to return to you as soon as possible and I would ask you to bear with us for the time being.

Cowlin served Notice of Adjudication on 18 May 2002. CFW argued that there was no dispute since the issue was still under discussion and their insurers had not had sufficient time to consider it.

In the action brought to enforce a payment decision in Cowlin's favour HHJ Kirkham, after considering a number of the authorities, expressed a preference for the test of a dispute adopted by Swinton Thomas LJ in *Halki* that 'there is a dispute once money is

---

13 [2003] EWHC50 (TCC); [2003] BLR 241. See also, for example, *Watkin Jones & Son Ltd* v. *Lidl UK GMbH* [2002] EWHC 183 (TCC); (2002) 86 ConLR 155; [2002] CILL 1847; [2005] TCLR 1; *Costain Ltd* v. *Wescol Steel Ltd* [2003] EWHC 312 (TCC); *Orange EBS Ltd* v. *ABB Ltd* [2003] EWHC 1187; *London & Amsterdam Properties Ltd* v. *Waterman Partnership Ltd* [2003] EWHC 3059 (TCC); *AWG Construction Services Ltd* v. *Rockingham Motor Speedway Ltd* [2004] EWHC 888 (TCC); *McAlpine PPS Pipeline Systems Joint Venture* v. *Transco plc* [2004] EWHC 2030 (TCC); *CIB Properties Ltd* v. *Birse Construction* [2004] EWHC 2365 (TCC).

claimed unless and until the defendants admit that the sum is due and payable'. At paragraph 86 she stated:

> Most disputes stem from claims. But the existence of a detailed claim is not necessary to give rise to a dispute. Many court and arbitral proceedings are begun before the nature of the dispute or difference between the parties has been explicitly set out. In any event [Counsel for CFW] does not rely on the lack of particularity. Absence of a reply gives rise to the inference that there is a dispute. That is what happened here. That conclusion is reinforced by the fact that Cowlin had delivered an ultimatum. I conclude that by 18 May 2002 a dispute had arisen. Cowlin had made an ultimatum. The nature of the claim had been outlined so that, although CFW did not know the detail, they were aware of the bare bones of it. Although CFW had not expressly rejected the claim, Cowlin made it clear that, unless CFW made their position clear by 17 May, Cowlin would assume that CFW did not accept the claim. In the absence of an acceptance of Cowlin's claim, CFW must be taken to have rejected it, so that a dispute had arisen. By the time the deadline passed, there was undoubtedly a dispute.

17.1.3 The current approach

AMEC Civil Engineering Ltd v. *The Secretary of State for Transport*[14] arose from a contract for the renovation of a viaduct over the M6 carried out by AMEC. The Works were certified complete on 23rd December 1996. In June 2002 the Employer informed AMEC of defects that had come to light. The defects were then believed to be attributable to deterioration and failure of roller bearings on some piers. At a meeting with the Employer and the Engineer acting as contract administrator on the project, AMEC denied responsibility for the defects. On 6 December 2002 the Employer sent AMEC a claim letter that alleged that the defects were the consequence of AMEC's breaches of their duties in contract and tort. AMEC replied that until their requests for details on the failure of the roller bearings were met, they were not in a position to comment.

The Contract provided that, as a condition precedent to a right to refer any dispute to arbitration, the dispute must be referred to and decided by the Engineer. On 11 December 2002 the Employer referred the matter to the Engineer, who decided that the defects were the consequence of AMEC's materials and workmanship not being in accordance with the Contract. As the Contract had been executed as a simple contract, the limitation period was to expire on 22 December 2002. Mindful of this date, the Employer sent AMEC a fax at 1.53 pm on 19 December imposing a deadline of 5 pm the same day by which AMEC was to accept the Engineer's decision. Not having received any response, the Employer served a notice of arbitration and had an arbitrator appointed.

AMEC brought proceedings challenging the jurisdiction of the arbitrator. Its principal ground was that on 11th December no crystallized dispute existed for a reference to the Engineer because of various reasons but particularly that: there were ongoing investigations to determine the precise cause of the failure of the roller bearings; the Employer's estimate of cost of repairs as being between £5 million and £20 million was too imprecise; it had not been given reasonable opportunity to consider the claim. Mr Justice Jackson,

14 [2004] EWHC 2339 (TCC) (hereafter *AMEC* v. *SS*).

after a review of the authorities on what amounts to a dispute for the purposes of arbitration or adjudication, derived these propositions (at para. 68):

1. The word 'dispute' which occurs in many arbitration clauses and also in s.108 of the Housing Grants Act should be given its normal meaning. It does not have some special or unusual meaning conferred upon it by lawyers.
2. Despite the simple meaning of the word 'dispute', there has been much litigation over the years as to whether or not disputes existed in particular situations. This litigation has not generated any hard-edged legal rules as to what is or is not a dispute. However, the accumulating judicial decisions have produced helpful guidance.
3. The mere fact that one party (whom I shall call 'the claimant') notifies the other party (whom I shall call 'the respondent') of a claim does not automatically and immediately give rise to a dispute. It is clear, both as a matter of language and from judicial decisions, that a dispute does not arise unless and until it emerges that the claim is not admitted.
4. The circumstances from which it may emerge that a claim is not admitted are Protean. For example, there may be an express rejection of the claim. There may be discussions between the parties from which objectively it is to be inferred that the claim is not admitted. The respondent may prevaricate, thus giving rise to the inference that he does not admit the claim. The respondent may simply remain silent for a period of time, thus giving rise to the same inference.
5. The period of time for which a respondent may remain silent before a dispute is to be inferred depends heavily upon the facts of the case and the contractual structure. Where the gist of the claim is well known and it is obviously controversial, a very short period of silence may suffice to give rise to this inference. Where the claim is notified to some agent of the respondent who has a legal duty to consider the claim independently and then give a considered response, a longer period of time may be required before it can be inferred that mere silence gives rise to a dispute.
6. If the claimant imposes upon the respondent a deadline for responding to the claim, that deadline does not have the automatic effect of curtailing what would otherwise be a reasonable time for responding. On the other hand, a stated deadline and the reasons for its imposition may be relevant factors when the court comes to consider what is a reasonable time for responding.
7. If the claim as presented by the claimant is so nebulous and ill-defined that the respondent cannot sensibly respond to it, neither silence by the respondent nor even an express non-admission is likely to give rise to a dispute for the purposes of arbitration or adjudication.

Applying these propositions to the facts, Jackson J concluded that a dispute existed between the parties on the date of reference to the Engineer for his decision. He referred to *Sindall* v. *Solland* and *Beck Peppiatt Ltd* v. *Norwest Holst Construction Ltd*[15] with approval. Referring to the former he said at paragraph 66:

> Two comments should be made about this case. First, MEA, the contract administrator, could not simply admit or deny the claim. By reason of its special position MEA was under

15 [2003] EWHC 822 (TCC).

> a duty to give a properly considered response to the claim for extension of time. Secondly, this case illustrates that failure to respond to a claim only gives rise to the inference of a dispute after a reasonable time. What constitutes a reasonable time depends critically upon the facts of the case and the contractual structure within which the parties are operating.

He however cautioned against a literal application of the proposition in *Ellerine* v. *Klinger* and *Halki* that there is a dispute once money is claimed unless and until the defendant admits that the sum is due and payable. In response to counsel for the Secretary of State's contention that the dispute arose at the very moment when AMEC received from the Highways Agency (HA) a letter of claim, he said at paragraph 71:

> It seemed to me that the third of these contentions was somewhat extreme and I pointed this out during argument. [Counsel for HA] relented only to this extent: AMEC needed to open the envelope and actually read the letter of claim before it could be said that a dispute within clause 66 had arisen.

The Court of Appeal, by unanimous decision, dismissed an appeal against the decision in *Amec* v. *SS*, with broad endorsement for the Jackson propositions.[16] Mr Justice Jackson's propositions also received the broad approval of the Court of Appeal in *Collins (Contractors) Ltd* v. *Baltic Quay Management (1994) Ltd.*[17] In particular, Clarke LJ, with whom the rest of the Court concurred, endorsed his statement that the mere making of a claim does not amount to a dispute. However, probably in response to Mr Justice Jackson's approval of *Sindall* v. *Solland*, he rejected the suggestion in that case that a dispute does not arise until negotiation and discussion have concluded or that a dispute could not be inferred.

17.2 Ambit of the dispute resolution clauses

The right to refer to adjudication is stated in Article 7 of JCT 05 as applying to any dispute or difference that 'arises under this Contract'. Article 8 makes arbitration available only in relation to a dispute of whatsoever nature 'arising out of or in connection with this Contract'. There is a long line of authorities concerning the meanings of similar phrases in dispute resolution clauses. For example, in *Mackender* v. *Feldia AG*[18] it was held that disputes 'arising under the Contract' included disputes concerning whether the contract had become illegal to perform or a voidable contract has been validly rescinded.[19] However, in *Fillite (Runcorn) Ltd* v. *Aqua-lift*[20] the Court of Appeal held that that form of words would not cover disputes concerning negligent misstatement under the principle in *Hedley Byrne Co. Ltd* v. *Heller & Partners Ltd*[21] or innocent misrepresentation under the Misrepresentation Act 1967.

In *Ashville Investments* v. *Elmer Contractors*[22] the Court of Appeal considered the meaning of 'any dispute or difference as to the construction of this contract or any matter or thing of whatsoever nature arising thereunder'. It was stated that 'any matter . . . arising thereunder' included disputes about the interpretation of the terms of the contract. It was

---

16 *Amec Civil Engineering Ltd* v. *Secretary of State for Transport* [2005] EWCA Civ 291.
17 [2004] EWCA Civ 1757.
18 [1967] 2 QB 590.
19 On frustration see *Kruse* v. *Questier & Co. Ltd* [1953] 1 QB 669 and *Government of Gibraltar* v. *Kenney* [1956] 2 QB 410.
20 (1989) 45 BLR 27.
21 [1964] AC 465.
22 (1989) 37 BLR 55.

also stated that 'any matter ... in connection therewith' covered disputes concerning rectification of the contract, negligent misstatement and misrepresentation.

The decision of the House of Lords in *Fiona Trust and Holding Corp and Others* v. *Privalov and Others*[23] suggests that these authorities may no longer be applicable, at least in arbitration clauses. That litigation arose from eight charterparties in the same form between entities within the Russian Sovcomflot group of companies and eight charterers. The dispute resolution clause stated:

> *Any dispute arising under this charter* [authors' emphasis] shall be decided by the English courts to whose jurisdiction the parties hereby agree. Notwithstanding the foregoing,..., either party may, by giving written notice of election to the other party, elect to have such dispute referred...to arbitration in London...[24]

The owners of the ships alleged that the charterparties had been procured by bribery of certain of their senior officers and purported to rescind them on that ground. When proceedings were commenced in the English court for a declaration that the charterparties had been validly rescinded, the charterers applied to stay them to arbitration. In their response to this application, the owners argued that the question whether the charterparties had been validly rescinded for bribery was outside the scope of the arbitration agreement. Morrison J declined to stay the claim for rescission. The Court of Appeal allowed the charterers' appeal. Referring to the long line of authorities on the ambit of arbitration and related clauses, Longmore LJ, who handed down the judgment of the Court, stated:[25]

> Not all the authorities are readily reconcilable but they are well known in this field and some or all are invariably cited by counsel in cases such as this. Hearings and judgments get longer as new authorities have to be considered. For our part we consider that the time has now come for a line of some sort to be drawn and a fresh start made at any rate for cases arising in an international commercial context. Ordinary businessmen would be surprised at the nice distinctions drawn in the cases and the time taken up by argument in debating whether a particular case falls within one set of words or another very similar set of words. If businessmen go to the trouble of agreeing that their disputes be heard in the courts of a particular country or by tribunal of their choice they do not expect (at any rate when they were making the contract in the first place) that time and expense will be taken in lengthy argument about the nature of particular causes of action and whether any particular cause of action comes within the meaning of the particular phrase they have chosen in their arbitration clause. If any businessman did want to exclude disputes about the validity of a contract, it would be comparatively simple to say so.

The appeal by the owners failed in the House of Lords. After applauding the opinion of Longmore LJ that the time had come to draw a line under the authorities to date and to make a fresh start, Lord Hoffman stated that arbitration agreements are to be construed with the presumption that the parties, as rational businessmen, are likely to have intended that 'any dispute arising out of the relationship' is to be decided by the same tribunal. According to this approach, each of the formulations analysed by the authorities, such as 'dispute under the contract', 'dispute arising out of', and 'dispute arising in connection

---

23 [2007] UKHL 40 (hereafter *Fiona Trust*).

24 It is to be noted that the use of the phrase 'such dispute' had the effect that the arbitration agreement covered 'any dispute arising under the charter'.

25 *Fiona Trust and Holding Corp and Others* v. *Privalov and Others* [2007] EWCA Civ 20, at para. 17.

with the contract', would now be wide enough to include disputes concerning the validity, rectification, rescission or frustration of the contract, or misrepresentation, negligent misstatement, or other tort claims.

17.3 Matters expressly not subject to review

Several clauses provide either that certain matters are not subject to review at all or that the relevant decisions are final. The provisions are as follows.

1. Under Clause 3.13, the Contractor may challenge the validity of an Architect's instruction by requesting the Architect to name the clause from which his power to issue it arises. As explained in Chapter 5, Section 5.3, if the Architect names a clause, the instruction is deemed authorized under the Contract if the Contractor complies with it before its validity is formally challenged by invoking the appropriate dispute resolution method. This means that whether such an instruction was in accordance with the Contract cannot amount to a dispute under the Contract.
2. Paragraph C.4.4 of Insurance Option C in Schedule 3 entitles either party to terminate the employment of the Contractor in case of loss or damage to the Works from the insured risks provided it is 'just and equitable' to do so. The other party may contest any purported termination on the grounds that it is not just and equitable to do so. There is a limit of 7 days from receipt of the notice of the challenge within which either party may invoke an appropriate dispute resolution mechanism to determine the issue. The effect of this limitation clause is that, after the 7 days, the matter cannot be a dispute under the Contract.
3. Clause 1.10.1 provides that the Final Certificate is final and conclusive evidence of a number of issues in 'any proceedings under or arising out of or in connection with this Contract (whether by adjudication, arbitration or legal proceedings)' unless the issue is properly challenged before expiry of 28 days after the issue of the Certificate.[26]
4. Fluctuations Options A, B and C in Schedule 7 provide in detail for how the amount of fluctuations under those clauses is to be determined. Paragraphs A.5, B.6 and C.4 in that Schedule provide that, as an alternative to calculating the amount of adjustment strictly in accordance with those provisions, the Quantity Surveyor and the Contractor can agree the amount of adjustment. Provided the Quantity Surveyor acts within his authority, any such agreement reached with the Contractor is not subject to review.[27]

It is interesting to note that some of these matters may qualify as disputes within the normal meaning of a 'dispute' as explained in Section 17.1. It is therefore arguable that their effect is that, in contravention of the Construction Act, the parties are not given the right to refer 'any dispute' 'at any time' for adjudication. The validity of these provisions is still to be challenged in the court.[28]

---

26 See Chapter 3, Section 3.9.

27 See Chapter 14, Section 14.1.5 for commentary on this subject.

28 In *John Mowlem & Co. plc* v. *Hydra Tight Ltd (t/a Hevilifts)* (2001) 17 Const LJ 358 the Contract provided that any disagreement between the parties was only a 'matter of dissatisfaction' and that, until the matter had been referred to and decided by the Contract Administrator, there was to be no dispute capable of being referred to adjudication. Both parties agreed that the provision was not compliant with the Construction Act. HJJ Toulmin stated that they were right in their agreement on the issue.

17.4 Adjudication

At the time of writing, a Local Democracy, Economic Development and Construction Bill to amend the then existing legislation on adjudication had just started its passage through Parliament. It is anticipated that the final changes will be implemented towards the autumn of 2009. This section outlines the law and procedure of dispute adjudication under the current JCT 05, which is based on the existing legislation. The main impact of the proposed changes on these matters concerns the cost of adjudication and the adjudicator's powers to correct errors in his decision. For this reason, we have explained the proposed provisions on these two matters in Sections 17.4.5.3 and 17.4.9. Other likely impacts on contracts in JCT 05 form are examined in the relevant parts of the book.[29]

Adjudication was first introduced into a UK standard form of construction contract in 1976 when it was incorporated into a form of domestic sub-contract intended for use with the JCT 63. The ambit of adjudication under the sub-contract covered only disputes between the main contractor and a sub-contractor about set-offs against monies otherwise payable to the sub-contractor. The process involved submission of the dispute to an independent third party, the adjudicator, appointed either in the sub-contract or after the dispute had arisen. The adjudicator was required to make a very speedy decision which would be binding unless and until reviewed by a court or an arbitrator after practical completion. Similar clauses were thereafter incorporated into other sub-contracts and the works contracts of the JCT family of contracts. From the 1980s wider adjudication clauses were introduced into standard forms of main contract. For example, the 1981 edition of the JCT's Standard Form of Building Contract with Contractor's Design contained an optional adjudication clause in respect of disputes arising in connection with a wide spectrum of matters referred to as 'The Adjudication Matters'. The first edition of the New Engineering Contract required any dispute to be referred to adjudication before practical completion.

Acting on the Latham Report's[30] recommendation that dispute resolution by adjudication should be incorporated into all relevant standard forms in the construction industry, the Government passed the Housing Grants, Construction and Regeneration Act 1996 (hereafter the 'Construction Act'). Subsections 108(1) to (4) of the Act require such qualifying construction contracts to contain, as a minimum, the following provisions on adjudication.

1. A party may give notice of his intention to refer any dispute arising under the Contract to adjudication at any time.
2. The timetable in the Contract has the object of securing the appointment of the adjudicator and referral of the dispute to him within 7 days of the notice to refer.
3. The adjudicator must make his decision within 28 days of the referral or such longer period as agreed by the parties after the dispute has arisen.
4. The adjudicator is allowed to extend the period for his decision by 14 days with the consent of the party who referred the dispute.
5. The adjudicator is under a duty to act impartially.
6. The adjudicator has authority to take the initiative to determine facts and law applicable to the issues in dispute.

---

29 See Chapter 15, Section 15.11 for proposed changes to the provisions on payment. New provisions on the right to suspend are examined in Chapter 11, Section 11.4.5 and Chapter 15, Section 15.6.2.

30 Sir Michael Latham, *Constructing the Team*, HMSO, 1994.

7. The decision of the adjudicator is binding until the dispute is finally determined by arbitration or litigation proceedings or by agreement of the parties.
8. The adjudicator, his employees and agents are not liable to the Parties for actions and omissions in furtherance of the adjudication unless they were in bad faith.

Article 7 provides that either Party may refer any dispute under the Contract for adjudication in accordance with Clause 9.2, which incorporates Part 1 of the Scheme for Construction Contracts (England Wales) Regulations 1998 (The Scheme).[31] Part 1 is an adjudication procedure, which includes the provisions made mandatory by the Construction Act. As the JCT 98 provided for a bespoke adjudication procedure, the adoption of the Scheme must be one of the major changes made in JCT 05.

Whether the liberal approach in *Fiona Trust*[32] will be adopted in the construction of clauses referring disputes 'under' construction contracts to adjudication remains to be seen. A common context in which the issue of the scope of an adjudication clause arises is where there has been a supplementary or settlement agreement and there is a subsequent dispute concerning the agreement. In *Shepherd Construction Ltd* v. *Mecright Ltd*[33] there were variations to a contract between the parties. After completion of the contract works the parties entered into a compromise agreement on a specified sum in final settlement of the amount due. HHJ LLoyd QC stated that a dispute as to the terms of the settlement agreement was not a dispute under the original construction contract.[34] Similar cases have been decided on the different basis that the subsequent agreement constitutes a variation to the original contract and that, therefore, a dispute as to the terms of the agreement constitutes a dispute under the original contract.[35] In some of these cases, *Shepherd v. Mecright* has been distinguished as decided on the special facts of an agreement independent of the original contract made after completion of the works and in full and final settlement.

This analysis of the appellate courts in *Fiona Trust* was carried out in the specific context of arbitration, particularly international commercial arbitration often driven by an imperative to avoid the risk of delay or partiality of national jurisdiction. Different considerations may well apply to clauses referring disputes to adjudication under the Construction Act. There are many in the construction industry who would be horrified at the adjudicators, most of whom are not legally qualified, determining disputes concerning matters other than contractual claims and other technical issues.

The key stages of an adjudication under the Contract are:

- service of a Notice of Adjudication by the Referring Party;
- appointment of the adjudicator;
- referral of the dispute to the adjudicator;
- submission of the response to the Referral Notice;
- decision by the adjudicator;
- enforcement of the decision if it is not complied with voluntarily.

---

31 The Scheme for Construction Contracts (England and Wales) Regulations 1998, SI 1998 No. 649; The Scheme for Construction Contracts (Scotland) Regulations 1998, SI 1998 No. 687, applies in Scotland.

32 *Fiona Trust and Holding Corp and Others* v. *Privalov and Others* [2007] UKHL 40.

33 [2000] BLR 489.

34 For similar analysis of a settlement agreement, see *Lathom Construction Ltd* v. *Cross* (1999) CILL 1568.

35 For example, see *Quarmby Construction Co. Ltd* v. *Larraby Land Ltd* (Leeds TCC 14 April 2003); *Westminster Building Co Ltd* v. *Beckingham* [2004] EWHC 138 (TCC); [2004] BLR 163; *L Brown & Sons Ltd* v. *Crosby Homes (North West) Ltd* [2005] EWHC 3503 (TCC); *McConnell Dowell Construction (Aust) Pty* v. *National Grid Gas plc* [2006] EWHC 2551 (TCC).

17.4.1 Notice of Adjudication

The function of this document is to confirm to the other party that a dispute has arisen as a consequence of the latter's response, or failure to respond to a claim[36] made by the author of the notice and that the latter demands reference of the dispute to adjudication. The recipient of the claim may also serve this notice first since, as explained in Section 17.1, a dispute crystallizes after rejection of the claim. For example, the Employer may serve the Notice of Adjudication after the Architect's rejection of a financial or extension of time claim by the Contractor. The author of this notice is hereafter referred to as the 'Referring Party' whilst its recipient is the 'Responding Party'.

There are two particular reasons why the Notice of Adjudication must provide a reasonable amount of detail on the dispute. Firstly, it may constitute the main document upon which the applicable Adjudicator Nominating Body (ANB)[37] would rely to match the expertise and geographical area of practice of the nominee with the equivalent particulars of the dispute. A person approached by the ANB to take on the role of adjudicator would also normally check as to his suitability for the dispute in terms of professional expertise and skills required, independence from the parties, the likely duration of the adjudication, and his or her other commitments. Secondly, the Notice of Adjudication establishes the jurisdiction of the adjudicator. There is the clearest possible recognition of this principle by the Court of Appeal in *Carillion Construction Ltd* v. *Devonport Royal Dockyard Ltd.*[38] The adjudicator would not therefore have jurisdiction to decide any issue not covered by the notice. Although the Referring Party may provide more detail at subsequent stages, such details may amount to a new dispute and, therefore, be outside the jurisdiction of the adjudicator.[39]

Paragraph 1(2) of Part I of the Scheme requires the Notice of Adjudication to be served on every party to the adjudication whilst paragraph 1(3) requires the following information to be set out in it:

- the names and addresses of the parties to the Contract;
- addresses specified by the parties for the giving of notices;
- brief description of the nature of the dispute and of the parties involved;
- details of where and when the dispute has arisen;
- the nature of the redress sought.

Steps may be taken in the Notice of Adjudication towards appointment of the adjudicator. To this end, if an adjudicator is named in the Contract, the notice may enclose a copy of a written request to the named person to act. If the adjudicator is not so named the notice may either state an intention to apply to an appropriate ANB for a nomination or enclose such an application.

---

36 The term 'claim' is used here in a wider context than contractors' claims. It includes other assertions as to the terms of the Contract.

37 An ANB is a body that provides a function as a nominator of people to act as adjudicators upon request from parties to disputes under a construction contract.

38 [2005] EWCA Civ 1358 at para. 21 of the judgment of Lord Justice Chadwick.

39 For examples of adjudicators' decisions that could not be enforced for this reason see *Edmund Nuttall Ltd* v. *RG Carter Ltd* [2002] BLR 312; *London & Amsterdam Properties Ltd* v. *Waterman Partnership* [2003] EWHC 3059 (TCC); [2003] BLR 179; (2004) 20 Const LJ 215; *AWG Construction Services Ltd* v. *Rockingham Motor Speedway Ltd* [2004] EWHC 888 (TCC); *McAlpine PPS Pipeline Systems Joint Venture* v. *Transco plc* [2004] EWHC 2030 (TCC); [2004] BLR 352; (2004) 96 ConLR 69.

17.4.2 Appointment of the adjudicator

The procedure for appointing an adjudicator for a dispute that has arisen depends on the provision in the Contract Particulars on this issue. However, it is open to the parties to agree an adjudicator outside the contractual adjudicator appointment procedure. Indeed, if the relationship between the parties is good enough, in relation to a particular dispute, it would be in their mutual interests to agree on an adjudicator whom they both believe to possess the professional expertise required to understand the issues in the dispute and the procedural skills to make a fair determination without undue delay. In contemplation of such an agreement the Referring Party may, in the Notice of Adjudication, put forward name(s) of suitable candidates. The Responding Party may also suggest alternatives. Such communication could continue until an adjudicator is agreed. In most cases, this consensual method is not even attempted because of the risk of failure to achieve the appointment of an adjudicator and referral of the dispute to him within 7 days after the Notice of Adjudication as required by the mandatory adjudication procedural rules.

The Contract anticipates two alternative methods of specifying the person to act as adjudicator when a dispute under the Contract arises and the Parties are unwilling or unable to agree on the person to act as adjudicator. First, either Party applies to an ANB to nominate a suitable person for appointment by the Parties. To use this method the ANB must be identified in the Contract Particulars by making a selection from the following existing ANBs:

- the Royal Institute of British Architects (RIBA);
- the Royal Institution of Chartered Surveyors (RICS);
- the Construction Confederation;
- the National Specialist Contractors Council;
- the Chartered Institute of Arbitrators (CIArb).

The selection is to be made by deleting the ANBs that are not applicable. If no selection is made, the Referring Party may apply to any of the listed ANBs for a nomination.

The second contractual method for appointing the adjudicator anticipates a specific person being retained for the entire duration of the project to act as adjudicator whenever a dispute arises. The name of the person is to be completed in the Contract Particulars against Clause 9.2.1. Having an adjudicator named in the Contract has the advantage of allowing proactive action towards greater certainty on the quality of the adjudicator, speedier references, and greater familiarity with the project and mutual understanding between the parties and the named person. The main shortcoming is that the named adjudicator would be unlikely to possess all of the types of expertise and skills appropriate to every dispute. Furthermore, because of being on holiday, illness and the like, the named adjudicator might be unable to act when a dispute arises. In *AMEC Capital Projects Ltd* v. *Whitefriars City Estates Ltd*[40] the named individual had unfortunately died when the need for his services arose. To get over these problems an ANB must also be selected from the list to make nominations where, for whatever reason, the named adjudicator is unable to act.

Where there is a named adjudicator the appointment procedure specified in paragraph 2 of Part I of the Scheme involves the following steps:

1. service of a Notice of Adjudication;
2. written request to the named person to act;

40 [2004] EWCA Civ 1418.

3. response by the named person to both parties within 2 days;
4. approach to the ANB to make the appointment if the response is that the named adjudicator is unable or unwilling to act.[41]

The importance of compliance with this procedure is illustrated by *Ide Contracting Ltd* v. *RG Carter Cambridge Ltd.*[42] The Contract in that case provided that any dispute under the Contract was to be referred to a Mr X and that, in the event of he being unable or unwilling to act, the Chartered Institute of Arbitrators (CIArb) was to nominate an adjudicator. An agent of Ide, the Referring Party, phoned Mr X and was informed that he would not be available to act as adjudicator because of other commitments during the relevant period. Ide then served a Notice of Adjudication in which it stated that it would be approaching the CIArb to nominate an adjudicator because Mr X would not be available. R G Carter objected to this and put forward two names from which Ide were to choose the adjudicator. Ide ignored this and had an adjudicator nominated by the CIArb. HHJ Havery QC held that the adjudicator had acted without jurisdiction because of departure from the applicable appointment procedure. He pointed out two particular departures. First, Paragraph 2 of Part I of the Scheme requires the approach to be made to the person named as adjudicator only after service of the Notice of Adjudication; otherwise a party would be able to avoid the named adjudicator by finding out when he/she would not be available and then serving the notice. Second, an approach is to be made to the ANB only after either communication from the named adjudicator to both parties that he/she cannot act or failure to make any communication within 2 days.

The importance of compliance with the applicable procedures for appointing the adjudicator cannot be overemphasized. In *Pegram Shopfitters Ltd* v. *Tally Wiejl (UK) Ltd*[43] the Court of Appeal stated that an adjudicator nominated by an ANB other than that specified in the contract lacks jurisdiction to determine a dispute under the contract. Depending on the terms of the contract, as shown by *Ide Contracting*, an application for a nomination must not be made before service of the Notice of Adjudication.

It is good practice for the adjudicator and the parties to enter into an agreement regulating their respective rights and obligations in relation to the adjudication proceedings. The JCT had produced two versions of such agreements for use for adjudications from the JCT family of contracts.

17.4.3 Referral of the dispute

The Referral Notice states the Referring Party's case and should therefore be prepared with the aim of gaining the tactical advantage of a cogent and fully explained case. It should be structured in such way that the adjudicator can very easily understand the dispute in terms of the events from which it arose, the underlying issues, the position of each party on the issues, and the facts and law relied upon. A well-structured submission not only reduces the risk of the adjudicator misunderstanding the dispute, and thereby making

41 Failure by the named adjudicator to respond within the 2 days after the request to act would amount to inability or unwillingness to act.

42 [2004] EWHC 36 (TCC); [2004] BLR 172 (hereafter *Ide Contracting*).

43 [2003] EWCA Civ 1750; [2004] BLR 65; 91 ConLR 173; see also *John Mowlem & Co. plc* v. *Hydra-Tight Ltd (t/a Hevilifts)* (2001) 17 Const LJ 358; CILL 1649.

a decision that cannot be enforced, but also keeps the time taken by the adjudicator, and therefore costs, to a minimum.

As already explained, the Notice of Adjudication establishes the adjudicator's jurisdiction. The adjudicator would not therefore have jurisdiction to consider issues raised in only the Referral Notice. Where possible, the Referring Party would be well advised to have the Referral Notice almost ready for submission before the Notice of Adjudication is served. Such a strategy has the advantage that it encourages consistency between the two documents as to the description of the dispute.

Case law suggests that many a Referring Party starts the adjudication process and then realizes that it lacks the necessary expertise, skills or even time to formulate the legal basis of the claim underling the dispute or to produce the Referral Notice. Experts are then brought in, usually at the time of preparing the Referral Notice. A Referral Notice based on expert advice given very late may well depart from the assertions made in the Notice of Adjudication.[44] For these reasons, where external support is necessary, the expert should be brought in as early as possible to review the circumstances in which the dispute crystallized and to prepare the Referral Notice and the Notice of Adjudication concurrently.

Paragraph 7(2) requires the Referral Notice to be accompanied by copies of the construction contract (or relevant extracts from it) and of any other documents to be relied upon by the Referring Party. Paragraph 7(1) requires it to be served on the adjudicator within 7 days after the Notice of Adjudication and copied simultaneously to every other party to the dispute. There has been judicial inconsistency on the effect of failure to serve a valid Referral Notice within the specified timetable. The decisions in two TCC cases suggest such failure does not prevent the adjudicator acquiring jurisdiction to make a binding decision.[45] However, in *Hart Investments Ltd* v. *Fidler*,[46] HHJ Coulson QC stated that an adjudicator cannot acquire jurisdiction unless the referral is completed within the 7-day timetable. The trend appears to be towards treating the timetable requirements under s. 108 of the Construction Act as mandatory.[47] Prudence therefore suggests fresh service of the Notice of Adjudication, ensuring that the second time around the timetable is complied with, unless both parties agree to waive the timetable requirement.

17.4.4 The Response to the Referral Notice

This document is the reply of the Responding Party to the issues, arguments and facts relied upon in the Referral Notice. If the Responding Party is questioning the jurisdiction of the adjudicator that fact should be clearly stated. After raising doubts as to the adjudicator's jurisdiction, there should be the clearest reservation of the right to raise the jurisdictional challenge against enforcement of any decision eventually reached by

44 For example, see *Edmund Nuttal Ltd* v. *RG Carter Ltd* [2002] BLR 312 in which a claims consultant brought in after service of the Notice of Adjudication to assist with an adjudication treated as causes of delay some events stated in the Notice of Adjudication not to have caused delay and *vice versa*.

45 *William Verry* v. *North West London Communal Mikvah* [2004] EWHC 1300 (TCC); [2004] BLR 308; *Tracy Bennett* v. *FMK Construction Ltd* [2005] EWHC 1268 (TCC).

46 [2006] EWHC 2857 (TCC); [2007] BLR 30.

47 See *Hart Investments Ltd* v. *Fidler and Another* [2006] EWHC 2857 (TCC); *Cubitt Building and Interiors Ltd* v. *Fleetglade Ltd* [2006] EWHC 3413 (TCC); *Epping Electrical Company Ltd* v. *Briggs & Forrester (Plumbing Services) Ltd* [2007] BLR 1126; *Aveat Heating Ltd* v. *Jerram Falkus Construction Ltd* [2007] EWHC 121 (TCC).

the adjudicator unless the Responding Party is agreeing to be bound by the adjudicator's determination of the question.[48]

The format of this document depends, to a large extent, on that of the Referral Notice. It would be a good strategy to follow the format of the Referral Notice and to deal with the issues in the same general order. Such a strategy would be of assistance to the adjudicator in his identification of the disputed matters and the positions of the parties on them.

17.4.5 The powers of the adjudicator

The powers of the adjudicator are stated in paragraphs 12–18. Subject to a requirement that he must act impartially and avoid unnecessary expense, the adjudicator has almost unbridled powers. He may set his own procedure and take any steps to ascertain the facts and law as they pertain to the dispute.[49] The only possible limitation is that he is entitled to only reasonable fees and expenses. This means that, for example, if he is too extravagant regarding the level of, and remuneration for hired skills and expertise, he may not be able to recover some of the expenses.

Specific powers include to:

- draw up the timetable for the adjudication including deadlines by when specified steps are to be taken by the parties;
- fix limits to the length of any written submission;
- fix time limits for oral representations;
- decide the language or languages to be used in the adjudication;
- decide whether any document is to be translated into another language and the party responsible for the translation;
- request any of the parties to supply him with any document he reasonably requires to make his decision;
- request from any party a written statement clarifying or supplementing documents already submitted;
- meet with any of the parties and their representatives;
- make site visits and other inspections, with or without the parties, as he considers appropriate provided he obtains any necessary third party consents;
- carry out any tests or experiments subject to obtaining necessary third party consents;
- obtain and consider any representations and submissions he requires;[50]
- appoint experts, assessors and legal advisers provided he notifies the parties of his intention to do so;[51]
- issue such other directions relating to the conduct of the adjudication.

48 For cases in which it was held that, from the content of the challenging party's communication, he had accepted that the adjudicator should determine the question of his own jurisdiction and that he would be bound by the determination, see *Northern Developments (Cumbria) Ltd* v. *J. & J Nichol* [2000] BLR 158; *Whiteways Construction (Sussex) Ltd* v. *Castelli Construction UK Ltd* (2000) 16 Const LJ 453; *Nordot Engineering Services Ltd* v. *Siemens plc* (unreported, 14 April 2000, HHJ Gilliland QC); *Hortimax Ltd* v. *Hedon Salads Ltd* (2008) 24 Const LJ 47. *Nolan Davis* v. *Steven P Catton* (unreported, 22 February 2000, TCC, No 590).

49 See s. 108(2)(f) of the Construction Act and para. 13 of the Scheme.

50 This would include a power to invite representations from the Architect, sub-contractors, and employees of the parties.

51 The JCT98 required the adjudicator to provide estimates of the cost of the advice sought. Although there is no such express provision in the Scheme, it would be good practice to do so as it could influence the attitude of the parties to the remainder of the proceedings.

The adjudicator must be careful to comply with the rules of natural justice in the exercise of these powers. To the extent possible within the adjudication timetable, he must give the parties the opportunity to comment on any information from whatever source that he will rely on in reaching his decision. This obligation applies to information from meetings with the parties, site visits, legal and other expert advisors, and the adjudicator's own knowledge. The court has also recognised that, within the tight adjudication timetable, an adjudicator may not be able to invite comment on every piece of information from these sources. The decision may therefore be enforceable where the failure to invite comment either was minor or did not affect the outcome.[52]

The extent of the adjudicator's power to award the costs of an adjudication calls for more comment. These costs fall into two categories: (i) the adjudicator's fees and expenses; (ii) costs incurred by the parties in the course of the adjudication.

17.4.5.1 Fees and expenses of the adjudicator

Paragraph 25 of the Scheme states:

> The adjudicator shall be entitled to the payment of such reasonable amount as he may determine by way of fees and expenses reasonably incurred by him. The parties shall be jointly and severally liable for any sum which remains outstanding following the making of any determination on how the payment shall be apportioned.

This paragraph is generally considered sufficient to give the adjudicator the power to apportion liability for payment of his fees and expenses.[53] There is no doubt that, by this provision, after stating how the parties are to pay towards his fees and expenses, he may pursue any or both parties to recover any payment that has not been made by any party in accordance with the determination. Thus, if the adjudicator directs that the losing party is to pay all his fees and expenses, he is entitled to be paid the amount by the winning party in full if the losing party fails to make the payment. The second sentence of the paragraph envisages the adjudicator receiving payment of part of his fees and expenses before reaching a decision on the dispute referred.

17.4.5.2 Costs incurred by the parties

These are the costs incurred by the parties themselves. They usually cover the costs of the services of the ANB that nominated the adjudicator, legal and expert advisers engaged by the parties, legal or other representatives in the conduct of the adjudication, and expert witnesses. The Scheme is silent on the adjudicator's power to award this element of cost. Parties have sought to recover their cost of adjudication on the following grounds:

- an implication from the Scheme;
- both parties claiming costs in their submissions to the adjudicator;

---

52 *Carillion Construction Ltd* v. *Devonport Royal Dockyard* [2005] EWHC 778 (TCC). In *Carillion Construction Ltd* v. *Devonport Royal Dockyard* [2005] EWCA Civ. 1358 the Court of Appeal stated that the first instance court was plainly right on this issue and, therefore, refused leave to appeal against its decision. On this issue see also: *Karl Construction (Scotland) Ltd* v. *Sweeney Civil Engineering (Scotland) Ltd* (2002) 18 Const LJ 55; *Martin Girt* v. *Page Bentley* [2002] EWHC 2434 (TCC).

53 For example, see *Northern Developments (Cumbria) Ltd* v. *J & J Nichol* [2000] BLR 158, at para. 38; [2000] 2 TCLR 261.

- express power given by the contract to the adjudicator to award costs;
- damages where the matter submitted to the adjudicator constitutes a breach of contract.

In *John Cothliff* v. *Allen Build (North West) Ltd*[54] the Referring Party, in the Referral Notice, asked the adjudicator to order the other party to pay its costs. The adjudicator considered that he had such power and ordered the Responding Party to pay 70% of the Referring Party's costs. Judge Marshall Evans QC in the Liverpool County Court held that the adjudicator in that case did have the power to award costs. He explained such power may be implied from paragraphs 13(h) and 16 of the Scheme.

Judge Evans also stated that the adjudicator's power to award costs arose as an implication from the fact of the Referring Party seeking recovery of its cost and the Responding Party's legal representatives making representations in the adjudication on the matter of costs without questioning his jurisdiction to award them. The issue of the adjudicator's power to award costs on the basis of the parties' submissions was considered in a slightly different factual context in *Northern Developments.*[55] The Contract from which it arose was not compliant with s. 108 of the Act. The Scheme therefore applied. Nichol, a subcontractor of Northern Developments, referred a dispute and requested the adjudicator to order Northern Developments to pay Nichol's costs of the adjudication. In its response, Northern Developments also requested the adjudicator to order Nichol to pay its costs. HHJ Bowsher QC held that the parties had, by their exchanges on costs during their adjudication, agreed to confer upon the adjudicator the power to award costs. The adjudicator's decision on costs was therefore binding. He stated, however, that in the absence of such agreement there is no power under the Scheme to award costs. He disagreed expressly with HHJ Evans in *Cothliff* v. *Allen Build* on the proposition that an adjudicator has implied power under the Scheme to award costs.[56]

The doctrine of freedom of contract suggests that, where the parties, as part of their contract, specify their financial responsibilities in respect of any adjudication under the contract, the adjudicator would have the power to determine liability for the costs of the adjudication as specified in the contract. In *Bridgeway Construction Ltd* v. *Tolent Construction Ltd*,[57] which arose from a contract between a main contractor and a sub-contractor, the contract provided that the party who served a Notice of Adjudication was to bear not only its own costs but also the other party's costs in any event. In his decision in an adjudication initiated by the claimant, the adjudicator directed that the claimant was to pay his fees and expenses and the defendant's costs of the adjudication. HHJ Mackay QC, in the Liverpool District Registry, refused the claimant's application for a declaration that the contract's provision on costs was void for being designed to circumvent the Construction Act. This decision was based on an application of the doctrine of freedom of contract. One of the suggested amendments to the Construction Act that has received considerable support is a provision in the Act rendering null and void terms of the type highlighted by *Bridgeway* v. *Tolent*.

---

54 (1999) CILL 1530 (hereafter *Cothliff* v. *Allen Build*).

55 *Northern Developments (Cumbria) Ltd* v. *J & J Nichol* [2000] BLR 158; [2000] 2 TCLR 261.

56 As explained in Section 17.4.5.4, the Court of Appeal confirmed the principle of submission to the jurisdiction of the adjudicator where the Responding Party answers to the Referring Party's claim for costs without reservation of the issue of jurisdiction.

57 (unreported, Liverpool Registry, 11 April 2000) (hereafter *Bridgeway* v. *Tolent*).

In *John Roberts Architects* v. *Parkcare Homes (No. 2) Ltd*[58] the Contract gave the adjudicator power in his discretion to direct payment of the legal costs and expenses of one party by the other as part of his decision. The Court of Appeal held that the adjudicator retained the contractual power to award costs even after the claimant had discontinued its claim.

JCT 05 does not give the adjudicator any power to award party costs.

The entitlement to costs of an adjudication was argued on the basis of breach of contract in *Total M and E Services Ltd* v. *ABB Building Technologies Ltd*,[59] which arose from a contract to which the Scheme applied. The claimant, the Referring Party in the adjudication, contended that if a party fails to make payment due under a construction contract and the other invokes adjudication to obtain payment, the cost of the adjudication proceedings should be recoverable as damages for breach of the payment obligation. HHJ Wilcox stated that, absent agreement by the parties to confer jurisdiction on an adjudicator to award the cost of adjudication proceedings, there was no such jurisdiction where the Scheme applied. The Judge went on to decide that, as the Construction Act envisages that contractual parties may refer disputes to adjudication without providing for recovery of the cost of such proceedings, such costs were not recoverable as damages for breach of contract. In his opinion, allowing such a claim would have the effect of undermining the purpose of the Construction Act.

17.4.5.3 The Local Democracy, Economic Development and Construction Bill and adjudication costs

The Bill will insert into the original Construction Act a new s. 108A which provides that any agreement by the parties to a dispute under a construction contract about the allocation between them of costs relating to an adjudication (including the fees and expenses of the adjudicator) is ineffective unless such agreement is made in writing after the giving of notice of intention to refer the dispute to adjudication. Agreements on adjudication costs of the type highlighted by *Bridgeway* v. *Tolent* made after the Bill becomes law would therefore be void.

17.4.5.4 Interest

Paragraph 20(c) provides that an adjudicator may 'having regard to any term of the contract relating to the payment of interest decide the circumstances in which, and the rates at which, and the periods for which simple or compound rates of interest shall be paid'. In *Carillion Construction Ltd* v. *Devonport Royal Dockyard Ltd*[60] Jackson J construed it as granting a freestanding power to adjudicators under the Scheme to award interest (i.e. there is such power even where the contract itself does not provide for interest on monies outstanding and there is no agreement by the parties that the adjudicator should consider entitlement to interest as part of his decision). The Court of Appeal took a

58 [2006] EWCA Civ. 64; [2006] BLR 106.
59 [2000] EWHC 348 (TCC).
60 [2005] EWHC 778 (TCC).

different view.[61] In their Lordships' view, an adjudicator under the Scheme may consider questions of interest:

> if, but only, if (i) those questions are 'matters in dispute' which have been properly referred to him or (ii) those are questions which the parties have agreed should be within the scope of the adjudication or (iii) those are questions which the adjudicator considers to be 'necessarily connected with the dispute'.

Where the dispute properly referred concerns the amount due under the contract and the contract itself provides for inclusion of interest for delayed payment, the adjudicator would have jurisdiction to consider award of interest. Where a party claims interest and the other party challenges such entitlement on quantum or principle but without denying jurisdiction the adjudicator would, by such submission of the parties to his jurisdiction, acquire power to consider award of interest. This latter scenario of acquisition of jurisdiction was what the Court of Appeal determined had arisen in *Carillion* v. *Devonport*. In its Notice of Adjudication Carillion advanced a claim for certain sums as due plus interest. Devonport replied that no sum was due and owing and that therefore the question of interest did not arise. Lord Justice Chadwick, who delivered the unanimous judgment of the court, described as 'irresistible' the conclusion that the parties to the dispute had agreed that the question whether interest should be paid on monies outstanding was to be within the scope of the adjudication.

Where questions of interest are within the scope of the adjudication, the adjudicator is to determine first whether there is entitlement in principle, and then subsidiary issues such as the rate of interest, whether compound or simple interest and the dates from which interest is payable. If the details (the legal basis, rates of interest, periods, whether simple or compound interest) of the claim are not specified in the Referral Notice the adjudicator would be acting properly to invite the Referring Party to make representations on these matters to him and allow the other party to comment to the extent possible within the adjudication timetable.

It is explained in Chapter 15, Section 15.6.1 that interest may be claimed for delayed payment of debt: (i) on a compound basis at common law; (ii) on a simple or compound basis under the Late Payment of Commercial Debts (Interest) Act 1998 (LPCDIA) but only where both parties entered into the contract in the course of a business; (iii) on a simple basis under Clause 4.13.6 of JCT 05. As explained in that part of the book, Clause 4.13.6 probably prevails over the LPCDIA and the common law claim. However, whichever basis is adopted by an adjudicator is, at worst, only a mistake within his jurisdiction and, therefore, binding until final resolution by agreement, arbitration or litigation.

17.4.5.5 VAT

The general principle is that any amount decided by the adjudicator to be payable under the terms of the contract may attract VAT. An amount decided to be payable as damages for breach of contract or interest on such damages does not attract VAT. This distinction

61 *Carillion Construction Ltd* v. *Devonport Royal Dockyard Ltd* [2005] EWCA Civ 1358 (hereafter *Carillion* v. *Devonport*).

can be confusing because of loose use of terms. For example, as pointed out by HHJ Humphrey LLoyd QC in *Pring & St Hill Ltd* v. *C J Hafner T/A Southern Erectors*,[62] 'loss and/or expense' is payable under the terms of the JCT family of contracts and will therefore attract VAT. A claim for damages at common law for delay and disruption, although often mistakenly labelled 'loss and/or expense', should not attract VAT.

17.4.6 The adjudicator's decision

The adjudicator must define the dispute over which he possesses jurisdiction to act. He also has to determine the facts and the law relevant to the dispute in compliance with the rules of evidence. He then applies the law to the facts to arrive at his decision, which must address liability in relation to interest, costs and VAT only if they form part of the dispute.

The adjudicator must make the decision within 28 days after his appointment and referral of the dispute. With the permission of the Referring Party, the adjudicator may extend this timetable by up to 14 days. He must obtain the agreement of both parties if he needs further extension. There has been judicial inconsistency on the question of the effect of a decision made outside the statutory or agreed timetable. In one group of cases late decisions were enforced.[63] However, after the Scottish Court of Appeal decided in *Ritchie Brothers (PWC) Ltd* v. *David Philp (Commercials) Ltd*[64] that a late adjudicator's decision was a nullity, the English TCC has tended to adopt the same principle to refuse enforcement.[65]

17.4.6.1 Reasons

Paragraph 22 imposes upon the adjudicator an obligation to provide reasons for his decision if they are requested by a party to the proceedings. Unfortunately, there is no limit to when a party may make such a request. To avoid embarrassment by a late request, an adjudicator would be well advised to give directions on the latest date by when such a request may be made. Some adjudicators take the view that they ought to provide reasons even if not requested as they are likely to be of value in assisting the parties to accept the decision as finally resolving their dispute or work out other negotiated settlement. The provision of reasons will often involve some additional costs. A prudent adjudicator would therefore consult the parties before committing them to such additional expense.

Challenges to enforcement of adjudicators' decisions on grounds of poor quality of the reasons given have so far failed, the reason for such failure being that any errors brought out by the reasons are within the adjudicator's jurisdiction.[66] However, there has been

---

62 [2002] EWHC 1775 (TCC).

63 *Barnes & Elliott Ltd* v. *Taylor Woodrow Holdings Ltd* [2003] EWHC 3100 (TCC), [2004] BLR 111; *St Andrews Bay Development Ltd* v. *HBG Management Ltd and Another* [2003] ScotCS 103; *Simons Construction Ltd* v. *Aardvark Developments Ltd* [2003] EWHC 2474 (TCC).

64 [2005] ScotCS CSIH 32 (Inner House).

65 *Hart Investments Ltd* v. *Fidler and Another* [2006] EWHC 2857 (TCC); *Cubitt Building and Interiors Ltd* v. *Fleetglade Ltd* [2006] EWHC 3413 (TCC); *Epping Electrical Company Ltd* v. *Briggs & Forrester (Plumbing Services) Ltd* [2007] BLR 1126; *Aveat Heating Ltd* v. *Jerram Falkus Construction Ltd* [2007] EWHC 121 (TCC).

66 *Gillies Ramsay Diamond and Others* v. *PJW Enterprises Ltd* (2004) BLR 131; 2004 SC 430; 2004 SLT 545 (hereafter Gillies Ramsay); *Carillion* v. *Devonport* [2005] EWCA Civ. 1358.

some judicial recognition that, in principle, failure to provide any reasons or providing unintelligible reasons could render an adjudicator's decision unenforceable.[67]

17.4.7 Immunity of the adjudicator

Paragraph 26 provides that the adjudicator, his employees and other agents shall not be liable to the contracting parties for any act or omission in furtherance of the adjudication provided the act or omission was not in bad faith. This implies that neither the Contractor nor the Employer is entitled to claim against these people for mistakes made in the adjudication process. However, this immunity does not bind third parties to the Contract who may be able to establish liability for negligent misstatement under *Hedley Byrne.*[68]

17.4.8 Enforcing the decision of an adjudicator

Leading up to the passing of the Construction Act, the appropriate procedure for enforcing an adjudicator's decision was a matter of great debate. The concern centred around the effect of an arbitration agreement on access to the court for such enforcement. Section 9 of the Arbitration Act 1996 provides that, where a party to an arbitration agreement commences court proceedings in relation to a matter within the scope of the agreement, any other party to the agreement and the legal proceedings may apply to the court to stay them in favour of arbitration. As explained in Section 17.6.2, on such an application, the court must stay the proceedings unless extremely narrow exceptions apply. As the parties are under a contractual obligation to comply with the decision of an adjudicator, the matter of compliance with such a decision, and therefore the issue of enforcement, would be a matter within the standard arbitration agreements in construction contracts. It was, therefore, feared that the response to any attempt to enforce an adjudicator's decision through the court would be met with an application to stay the enforcement proceedings to arbitration.

The JCT responded to this concern by excluding the matter of enforcing adjudicator decisions from the scope of the arbitration agreements in its contracts, including JCT 05. This precaution proved needless, as applications to stay enforcement proceedings have so far failed.[69] From the experience of adjudication so far, the most appropriate method of enforcing an adjudicator's decision through the court is to apply for summary judgment/ interim payment. Other methods that have been tried are through an application for a mandatory injunction, making a statutory demand, and instituting Part 8 proceedings for a declaration by the court on an issue on which, by the agreement of the parties, the question of compliance with the decision hangs.

---

67 See *Gillies Ramsay Diamond and Others* v. *PJW Enterprises Ltd* (2004) SC 430; (2004) SLT 545; [2004] BLR 131; *Carillion Construction Ltd* v. *Devonport Royal Dockyard* [2005] EWHC 778 (TCC); *Balfour Beatty Northern Ltd* v. *Modus Corovest (Blackpool) Ltd* [2008] EWHC 3029 (TCC).

68 *Hedley Byrne & Co. Ltd* v. *Heller & Partners Ltd* [1964] AC 465; [1964] 3 WLR 101.

69 See for example: *Macob Civil Engineering Ltd* v. *Morrison Construction Ltd* [1999] BLR 93; *Absolute Rentals Ltd* v. *Gencor Enterprises Ltd* (2001) 17 Const LJ 322; *The Construction Group Centre Ltd* v. *The Highland Council* [2002] BLR 476; *David McLean Housing Contractors Ltd* v. *Swansea Housing Association* [2002] BLR 125.

17.4.8.1 Summary judgment/interim payment

The standard practice in enforcing the decisions of adjudicators has been to commence proceedings in the appropriate court and to apply for summary judgment under Part 24 of the Civil Procedure Rules (CPR).[70] Rule 24.2 of the CPR states:

> The court may give summary judgment against a claimant or defendant on the whole of a claim or on a particular issue if –
> (a) it considers that –
> (i) that claimant has no real prospect of succeeding on the claim or issue; or
> (ii) that defendant has no real prospect of successfully defending the claim or issue; and
> (b) there is no other compelling reason why the case or issue should be disposed of at a trial.

In the context of enforcement of the decision of an adjudicator, the basis of such an application is that: (i) the defendant has no real prospect of successfully defending his failure to comply with it; and (ii) there is no other compelling reason why the case should go for trial. The court is therefore asked to order entry of judgment in the applicant's favour without a full trial. Such judgment may then be enforced as any other court judgment. At the centre of the application is a need to establish that the Responding Party has 'no real prospect of succeeding' in his defence if the case goes to a full trial. According to Lord Woolf MR in *Swain* v. *Hillman*,[71] the phrase does not need any amplification as it speaks for itself. What is required for dismissal of the application is 'realistic' rather than 'fanciful' prospects of success at trial.

In *Glencot Development and Design Co. Ltd* v. *Ben Barrett & Son (Contractors) Ltd*[72] HHJ Humphrey LLoyd QC advised that a cautious applicant for summary judgment to enforce the decision of an adjudicator should always consider making an interim payment application at the same time. He stated that an interim payment order is open to the court where a summary judgment application is unsuccessful but the court is nevertheless satisfied that, if the claim on the decision went to trial, the applicant would obtain judgment for a substantial amount of money against the defendant.[73] The court may, in such circumstances, order interim payment of a reasonable proportion of the amount likely to be obtained at full trial. The legal basis of an interim payment order has been doubted.[74]

17.4.8.2 Mandatory injunction

In the pre-Construction Act case of *Drake and Scull Engineering Ltd* v. *McLaughlin and Harvey plc*[75] the court granted a mandatory injunction requiring compliance with the award of an adjudicator. In *Macob Civil Engineering Ltd* v. *Morrison Construction Ltd*[76] Dyson J,

---

70 For commentary on the CPR see Section 17.5.
71 [2001] 1 All ER 91, at p. 92.
72 [2001] BLR 207.
73 See Rule 25 of the CPR. For an example of a case in which a summary judgment application failed but the applicant was held entitled to interim payment under CPR 25.7 see *Ken Griffin and John Tomlinson* v. *Midas Homes Ltd* (2002) 18 Const LJ 67.
74 See *RSL (South West) Ltd* v. *Stansell Ltd* [2003] EWHC 1390 (TCC).
75 (1992) 60 BLR 102 (hereafter *Drake and Scull*).
76 [1999] BLR 93 (hereafter *Macob* v. *Morrison*).

as he then was, stated that a mandatory injunction was not appropriate for payment of contractual debts between contractual parties. He obviously had in contemplation the fact that failure to comply with such an injunction constitutes contempt of court, a criminal offence carrying possible sanctions of a fine or imprisonment. *Drake and Scull* was distinguished in *Macob* v. *Morrison* on the grounds that the earlier case involved a decision requiring payment of the amount in dispute to a third party for safe-keeping as a trustee stakeholder pending final determination of the dispute by arbitration.[77] He cited,[78] as examples of situations where an injunction might be appropriate, decisions of an adjudicator requiring a party to return to the site to recommence work, to provide access or inspection facilities, to open up work, or to carry out specified work.

17.4.8.3 Statutory demand

Under s. 122(1)(f) of the Insolvency Act 1986 a company may be wound up on the grounds that it is unable to pay its debts. The court does not need to see the company's accounts to determine that it is unable to pay its debts as s. 123 lists situations that the court may accept as sufficient evidence of such inability. One such event is 'if a creditor (by assignment or otherwise) to whom the company is indebted in a sum exceeding £750 then due has served on the company, by leaving it at the company's registered office, a written demand (in the prescribed form) requiring the company to pay the sum due and the company has for 3 weeks thereafter neglected to pay the sum or to secure or compound for it to the reasonable satisfaction of the creditor'. The written demand is commonly referred to as a 'statutory demand'. The prescribed form is available from law stationers.

In a number of cases[79] the court decided that an adjudicator's payment decision creates a debt which can be the subject matter of a statutory demand. If payment is not received the beneficiary of the decision may present a petition for the winding up of the debtor company or bankruptcy of the individual debtor. For this reason payment will usually be made where the paying party is solvent. However, a statutory demand carries the risk that if the paying party successfully defends the demand, the beneficiary of the decision would be liable for the costs of the legal proceedings.

17.4.8.4 Part 8 proceedings (Civil Procedure Rules)

These are appropriate where the only difference between the parties regarding the adjudicator's decision relates to a question of law (e.g. whether the adjudicator acted with jurisdiction), and the court's declaration on that question will resolve the matter of compliance. The court is authorized, under Part 8(3), at any stage to order the claim to continue as if the claimant had not used the Part 8 procedure. The court may also give any other direction it considers appropriate (e.g. orders for summary judgment or full trial of any remaining issues). The evidence on which the court is required to make the declaration is usually limited to that served with the claim form or the defendant's acknowledgement of service.

---

77 The particular contract expressly gave the adjudicator the power to make such a decision. It is doubtful whether an adjudicator acting under the Construction Act would have such power if the Contract does not make a similar provision.

78 At p. 100.

79 *George Parke* v. *The Fenton Gretton Partnership* (2001) CILL 1712; *William Oakley and David Oakley* v. *Airclear Environmental Ltd and Airclear TS Ltd* (2002) CILL 1824; *Jamil Mohammed* v. *Dr Michael Bowles*, 11 March 2003, unreported; *Guardi Shoes Ltd* v. *Datum Contracts* (2003) CILL 1934.

17.4.9 The adjudicator's powers to correct errors in his decision

Generally, after making and publishing his decision, the adjudicator has no power to review it to reflect any changes in his views on the merits of the dispute. This proposition flows from the provision in s. 108(3) of the Construction Act that his decision is to be binding on the parties until the dispute is finally determined by agreement, arbitration or litigation. In England and Wales it has been decided by the court that the adjudicator may however revise his decision: (i) to correct clerical or other accidental errors; (ii) to provide clarification; (iii) to remove ambiguity; (iv) to make good an omission provided the correction is made soon after publication of the original decision. Case law suggests that such corrections may be made up to 7 days after publication of the decision.[80]

One of the proposed amendments to the Construction Act is the insertion of a new Section 108(5A) which requires a construction contract subject to Scottish law to include in writing a provision giving the adjudicator the power to correct a clerical or typographical error arising by accident or omission. Such a requirement of a construction contract subject to English law has been considered unnecessary because the English common law has the same effect.

17.4.10 Advantages of statutory adjudication

1. Its most valued feature is the speediness of the decision. It is often argued that such a decision, albeit temporarily binding, enables the parties to devote their energy to execution of the project rather than to the demanding procedures of more permanent resolution methods.
2. It avoids the conflict of interest inherent in the traditional role of the project architect/engineer as the first-tier tribunal for resolution of disputes between employers and contractors.
3. It is intended as a relatively cheap process, with minimum lawyer or expert witness involvement. In practice, some adjudications have involved considerable teams of lawyers and experts on both sides.
4. In the vast majority of cases the parties accept the decision as a final determination of their dispute (i.e. there is rarely any reference of the same dispute to arbitration or litigation). Anecdotal evidence and some reported cases[81] suggest that the step of simply invoking adjudication often compels the parties to agree a compromise before any further steps are taken.
5. If either party cannot accept the decision of the adjudicator as final, it can always be reviewed in arbitration or litigation.

---

80 In *Bloor Construction (UK) Ltd* v. *Bowmer & Kirkland (London) Ltd* [2000] BLR 314, HHJ Toulmin CMG QC held that correction within a reasonable time of giving the decision was an implied term. In that case he upheld correction within 2 days of the decision. In *Edmund Nuttall Ltd* v. *Sevenoaks District Council,* 14 April, 2000, TCC Case No. HT/00/119, the adjudicator made his decision on 9th March 2000. On 19th March he acknowledged by letter that his decision contained an error although he stated that he had no jurisdiction to make a correction. Mr Justice Dyson treated the letter as a correction within the adjudicator's jurisdiction.

81 See, for example, *Outwing Construction Ltd* v. *H. Randell & Son Ltd* [1999] BLR 156, 64 ConLR 59, (1999) 15 Const LJ 308; *Rentokil Ailsa Environmental Ltd* v. *Eastend Civil Engineering Ltd* (1999) CILL 1506.

17.4.11 Disadvantages of statutory adjudication

1. Its most serious criticism is that, with the complexity of construction disputes, the restrictive timetable for the process can result in serious injustice, which can only worsen adversarialism.
2. It tips the balance in favour of large and well-resourced parties who can more easily accommodate the demanding timetable requirements.
3. It affords the opportunities for one party to ambush the other (i.e. one party prepares a full and sophisticated submission of his case over a long period which then has to be responded to by the other party in 1–2 weeks).
4. The legislation fails to recognize that not all disputes can be suitably resolved, even on a temporary basis, by adjudication.
5. Where the party in whose favour a payment decision is given subsequently becomes insolvent, the availability of the final review in arbitration or litigation is only academic.
6. It is not particularly suitable for multi-party disputes.
7. It undermines the well-understood role of the project architect/engineer.
8. Concerns have been expressed that the costs of adjudication have been increasing at such a rate that it is becoming less and less appropriate for disputes in relation to sums less than £50,000.
9. Adjudication is being used in some wholly inappropriate circumstances (e.g. disputed final accounts of large and complex projects and allegations of professional negligence).

17.5 Litigation

Article 9 provides that the English courts have jurisdiction to decide any dispute arising out of or in connection with the Contract. This Article therefore has to be amended where the parties, particularly multinational companies, do not wish to litigate in England. However, such a course of action must not be taken lightly as the ability of the English courts to deal with disputes under a JCT contract can hardly be matched by the courts of other countries.

Article 9 is a non-exclusive jurisdiction clause (i.e. the parties are not limited to resolving their disputes in only the English courts). A party may choose to bring proceedings in the court of another country provided the jurisdiction of that other court can be established by the law of that country. Here again, amendment is called for if there is reason to contemplate foreign proceedings as a serious possibility.

Article 9 is expressed to be subject to Articles 7 and 8. This means that, as explained in the next Section, there may be a concurrent right to refer disputes to arbitration depending upon whether the Contract Particulars are completed to incorporate the arbitration agreement. The policy underlying English arbitration law is that involvement of the court in a dispute within a valid arbitration agreement is to be kept to only as much as is necessary to support the parties' choice of arbitration as their preferred method of resolving their disputes.

Procedure in civil litigation in the County Court, High Court and the Court of Appeal is governed by the so-called Civil Procedure Rules (CPR). They came into force in 1999 as the final outcome of Lord Woolf's investigation into necessary reform in civil litigation. They have been supplemented by Practice Directions and pre-action protocols for specific types of disputes. The protocol applicable to disputes in connection with a building contract is the Pre-Action Protocol for Construction and Engineering Disputes.[82] There is also the Technology and Construction Court Guide, which describes procedures specific to the TCC, the division of the High Court in which most litigation from construction and engineering takes place.

A party who decides on litigation as the method of resolving a dispute must therefore familiarize himself with the relevant parts of the Rules and the supplementary documents. The defendant must also develop such awareness to be able to respond correctly to steps taken by the claimant. Although the parties' legal advisors will usually be aware of these documents and the importance of complying with them, and would therefore provide appropriate guidance, there are advantages to the parties having reasonable awareness. Not the least of these advantages is the fact that communication with the legal advisors at an informed level reduces delay and the cost of resolving the dispute.

17.5.1 The Civil Procedure Rules

The overriding objective of the CPR, as stated in Rule 1.1, is to ensure that cases are dealt with justly. Treating a case justly includes: ensuring that the parties are on an equal footing; saving of expense; ensuring proper proportionality in terms of the amount involved, the importance of the case, the complexity of the issues and the financial positions of the parties; ensuring that the case is dealt with expeditiously and fairly.

The court is required to promote the overriding objective by managing cases actively. In particular, the court must not only encourage parties to use alternative dispute resolution (ADR) to resolve their dispute but also facilitate such procedure. Steps available to the court in this respect include proposing ADR to the litigants and staying court proceeding in favour of ADR. The parties themselves are under a duty to assist the court in furthering the overriding objective by seriously considering resolution of their dispute without litigation. To underline the importance of this duty, the court may impose cost sanctions against a party who unreasonably refused to consider ADR or failed to comply with the Protocol.

17.5.2 Compliance with the Pre-Action Protocol

This protocol is designed to encourage greater contact between the parties at the earliest opportunity for the purpose of sharing information relevant to the dispute, thereby promoting settlement without litigation or speedier court proceedings in relation to the issues on which settlement could not be reached. It requires parties at the pre-action stage of their dispute to follow a procedure involving a Letter of Claim, Letter of Response and Pre-Action Meeting as vital signposts. The purpose of these steps is to ensure that, before court proceedings commence, the claimant and the defendant have a reasonable amount of information on their respective positions on the issues in the dispute. It also encourages them to meet and, if necessary, to carry out further pre-action investigation to plug

82 There is a separate protocol for professional negligence cases.

any gaps in the information necessary to dispose of the dispute without the need for the proceedings.

17.5.2.1 Letter of Claim

Section 3 of the Protocol requires a claimant, prior to commencing court proceedings, to send this document to the proposed defendant. The Letter must contain the following information:

- the claimant's full name and address;
- the proposed defendant's full name and address;
- a summary of the facts on which the claim is based;
- the legal basis of the claim;
- the nature of relief to be claimed;
- whether the claim has been rejected previously and the reasons for the rejection;
- name of any expert already instructed and the issues on which the expert evidence will be directed.

17.5.2.2 The defendant's response

The defendant must acknowledge the Claim Letter within 14 calendar days after its receipt. In the acknowledgement document, the defendant may raise any objection to the commencement of the proceedings on account of lack of jurisdiction of the court, the existence of an arbitration agreement covering the dispute or other valid ground. If the defendant fails to serve any acknowledgement, the claimant may curtail further compliance with the Protocol and commence the proceedings.

Within 28 days after receipt of the Letter of Claim the defendant must serve a Letter of Response and Counterclaim, if any. With the agreement of the claimant, this period may be extended to a maximum of 3 months. If the Letter of Response is not served on time, the claimant may go ahead with the proceedings without further compliance with the Protocol.

17.5.2.3 Pre-action meeting

It is an expectation that the parties will meet within 28 days after the claimant's receipt of the Letter of Response. This meeting should normally be attended by a senior representative of each party, their legal advisors and, where relevant, insurers. It affords them the final opportunity, with the knowledge of the dispute outlined in the correspondence, to craft a solution to the problem that does not require litigation. They should now know in some detail what their differences are. Alternatives to litigation available to resolve the dispute should then be explored. In the event that they conclude that litigation is inevitable, they should consider the most sensible and cost-effective way of managing the impending proceedings.

Notes of the meeting must be taken and kept, as any party who attended it is allowed to inform the court what happened in the meeting so that the court may consider appropriate sanctions where there was unreasonable conduct. The meeting is, however, 'without prejudice' in that any admissions of liability must not be disclosed and, if disclosed, must be ignored by the court in deciding the case. The rationale for this principle is that a party who compromised in the interest of amicable settlement should not be penalized by the court for doing so.

17.5.3 Non-compliance with the Protocol

It has been reported that, as a consequence of how the courts are driving the ADR agenda, whilst it was once considered a sign of weakness to suggest mediation or settlement, it is increasingly being thought stupid not to do so.[83] The courts are required to encourage resolution by ADR in several ways.

The court may stay proceedings to give the parties a chance to attempt ADR or even order the parties to use ADR. *Cundall Johnson & Partners LLP* v. *Whipps Cross University Hospital NHS Trust*[84] illustrates exercise of the court's jurisdiction to stay proceedings commenced without compliance with the Protocol. The defendant NHS Trust engaged the claimant for engineering services in respect of construction projects. The action was brought by the claimants for outstanding fees. The defendant applied to stay the proceedings on the grounds that, although the claimants had made piecemeal attempts at compliance with the Protocol, the information provided in the correspondence from the claimant was confusing and inconsistent. Jackson J determined that there was a real possibility of settlement if the parties went through the processes of the Protocol, and granted the application. He explained that a clear and concise summary of the parties' respective cases was a crucial requirement of compliance with the Protocol.

A related course of action available to the court is to order pre-action disclosure of documents if there is a real prospect of the disclosure resulting in speedy resolution without recourse to litigation.[85]

The court may also impose sanctions adverse to a party who failed to comply with the pre-Action Protocol or unreasonably declined invitations to consider ADR. For example, the court may decline to order the defendant to pay the costs of a claimant who, although a winner in the proceedings, is found to have been guilty of unreasonable pre-action conduct.[86] A claimant unsuccessful in the proceedings may be ordered to pay the defendant's costs on an indemnity basis.[87] Similarly, a recalcitrant party entitled to recovery of interest may be awarded the interest at a lower rate or forfeit it completely. Correspondingly, such a party liable for interest may be ordered to pay it at a higher rate.

There is judicial recognition that it is not every refusal of an offer of mediation or other ADR technique that amounts to unreasonable pre-action conduct that should attract the cost sanctions of the court. In *Halsey* v. *Milton Keynes General NHS Trust*[88] the Court of Appeal identified six factors for assessing the reasonableness of refusal to take part in ADR:

- the nature of the dispute;
- the merits of the case;

---

83 Penny Brooker and Anthony Lavers, 'Construction lawyers' experience with mediation post-CPR' (2005) 18 Const LJ 97-116.

84 [2007] EWHC 2178 (TCC); [2007] BLR 520; [2008] TCLR 1; 115 ConLR 125.

85 See for example *Birse Constuction Ltd* v. *HLC Engenharia e Gestao de Projectos SA* [2006] EWHC 1258 (TCC).

86 See *Dunnett* v. *Railtrack plc* [2002] EWCA Civ. 303.

87 See *Paul Thomas Construction Ltd* v. *Hyland and Another* (2002) 18 Const LJ 345; (2001) CILL 1848.

88 [2004] EWCA Civ 576; [2004] 1 WLR 3002. See also *Hurst* v. *Leeming* [2002] EWHC 1051 (Ch) (the refusal to mediate was reasonable because the mediation had no real prospect of success; other grounds for the refusal accepted by the court were that: (i) the party inviting the mediation was an undischarged bankrupt; (ii) his attitude and character had negative elements); *Hickman* v. *Blake Lapthorn* [2006] EWHC 12 (QB) (adverse cost order refused because the refusal to mediate was not shown to have been unreasonable); *Nigel Witham Ltd* v. *Smith* [2008] EWHC 12 (TCC) (uncompromising attitude of the other party meant that the mediation had no real prospect of success).

- the extent to which other settlement methods were attempted;
- whether it would have been disproportionately costly;
- whether delay in setting up or attending the ADR had prejudicial effects;
- whether the ADR had reasonable prospect of success.

In that case the claimant brought proceedings against the NHS Trust for allegedly negligent treatment of her husband. The first instance judge found that the claimant's conduct was 'somewhat tactical' and dismissed the claim at trial with costs to the defendant. The claimant appealed against the award of costs on the grounds that the defendant had refused her invitations to take part in mediation. It was the defendant's position that the cost of mediation would have been disproportionately high in comparison to the cost of the trial. The Court of Appeal dismissed the appeal because the claimant had failed to discharge the burden of proving that the mediation had reasonable prospect of success.

Earl of Malmesbury v. *Strutt and Parker*[89] suggests that adoption of an unreasonable stance in mediation may amount to refusal to mediate. In that case the court determined that the claimant's offer in final settlement of its damages claim was unreasonably high. Its entitlement to recovery of the costs of a hearing on the damages issue was reduced by 20% to reflect the unreasonableness of its position in the mediation. A limitation to the importance of this case concerns the principle that settlement negotiations are normally on a 'without prejudice' basis (i.e. the negotiations are not to be disclosed to the court). But for the fact that the parties had waived this rule, the evidence of unreasonable conduct might not have been admissible in the court proceedings.

17.5.4 Recovery of cost of compliance with Protocol

Considering the documents that have to be prepared, the cost of compliance with the Protocol can be substantial. Such costs are incidental costs of any subsequent legal proceedings and the court has discretion to award them in the normal way. Where the basis of the claim, as pursued in the legal proceedings, is completely different from those argued at the pre-action stage, the question arises whether the court should award the pre-action costs in favour of the winning party in the proceedings. In *McGlinn* v. *Waltham Contractors and Others*[90] the court answered this question in the affirmative. The justification given for this position was that it would be contrary to the whole purpose of the Protocol if claimants are penalized for not pursuing in litigation claims they had included in the pre-action correspondence. For the same reason, a defendant who lost in the proceedings is not entitled to recover its cost of defending allegations not pursued in the proceedings. In recognition that this position is open to abusive use to oppress, the court stated that gross misconduct in pursuing highly speculative claims would be liable to be punished.

17.6 Arbitration

Arbitration may be defined as a private procedure for settling disputes whereby a dispute between parties is decided judicially by an impartial individual or a panel of individuals

89 [2008] EWHC 424 (QB).
90 [2005] EWHC 1419 (TCC); BLR 432.

appointed for that purpose. An individual so appointed is referred to as an 'arbitrator' whilst his decision is referred to as his 'award'. An arbitrator's award is legally binding on all the parties to the arbitration proceedings to whom it is addressed.

An arbitrator does not have to possess any particular skills or qualifications unless they are specified in the agreement to resolve disputes by arbitration. In most cases, a person is appointed to act as an arbitrator on the strength of his expertise and experience in the subject matter of the dispute. In addition, as he is expected to act in a judicial capacity, he must possess some knowledge of the law, particularly the law of contract, tort and evidence. Many practising arbitrators are fellows of the Chartered Institute of Arbitrators, which undertakes the training of arbitrators in all fields. Most construction arbitrators are professionally qualified in a construction field and/or the law. Research[91] suggests that, with construction disputes, arbitrators qualified in the law and a construction discipline are preferred to those qualified in only one field.

Most of the law on arbitration is enshrined in the Arbitration Act (AA) 1996, Part I of which applies where the seat of the arbitration is England and Wales or Northern Ireland.[92] The seat of an arbitration, as defined in s. 3 of AA 1996, is the place designated as such by:

- the parties to the arbitration agreement; or
- any person or institution vested with the power to do so; or
- an arbitral tribunal authorized by the parties.

If the seat is not specified in any of these ways, it is to be determined objectively by examining the arbitration agreement in the light all the relevant circumstances. For practical purposes, it is the place where the arbitration is conducted but it is possible to designate a seat on which the parties and the Arbitrator never set foot.

References to section(s) in the rest of this chapter are to sections of that Act unless the contrary is expressly indicated.

17.6.1 JCT 05 arbitration agreement

Article 8 constitutes the arbitration agreement under JCT 05. As explained in the introduction to this chapter, it does not apply unless the Contract Particulars are completed indicating that 'Article 8 and Clauses 9.3 to 9.8 (Arbitration) apply'. If they do not apply, it means that the parties have decided that there will be neither a right to refer any dispute to arbitration nor any obligation to participate in such proceedings unless they subsequently enter into a free-standing arbitration agreement. Such an arbitration agreement not embedded in the substantive contract between the parties is often referred to as an 'ad hoc' arbitration agreement.

The scope of JCT 05 arbitration agreement is stated as 'any dispute or difference between the Parties of any kind whatsoever arising out of or in connection with this Contract' except:

- disputes in connection with the enforcement of any decision of an adjudicator appointed to determine a dispute arising under the Contract;[93]

91 Ndekugri, I. and Jenkins, H., 'Construction Arbitration: A Survey' (1994) ICLR 388, pp. 366–83.
92 See s. 2(1) of AA 1996. Scotland has adopted the Model Arbitration Law of the United Nations Commission on International Trade Law most of which is also adopted in the AA 1996.
93 For explanation of the rationale behind this exclusion see Sections 17.4.8 and 17.6.2 in this chapter.

- disputes in respect of the Construction Industry Scheme to the extent that legislation provides another method for their resolution;
- disputes in respect of VAT to the extent that legislation provides another method for their resolution.

In addition, as explained in Section 17.3, the Contract provides that, in stated circumstances, some matters are not subject to review.

17.6.2 Enforcement of the arbitration agreement

As an arbitration agreement is a contract, a party to the agreement who commences litigation proceedings would be acting in breach of it. However, the common law remedies are either inappropriate (damages) or rarely granted (specific performance). The party preferring arbitration may apply to the court before which the litigation proceedings have been commenced to stay them in favour of arbitration (i.e. suspend the court proceedings so that the arbitration can go ahead). Section 9(4) of the AA 1996 provides that, upon such an application, the court must grant a stay unless it is satisfied that 'the arbitration agreement is null and void, inoperative, or is incapable of being performed'.

It is difficult to see any circumstances in which the arbitration agreement in JCT 05 could be 'null and void, inoperative, or is incapable of being performed'. It is already established at common law that an arbitration agreement is a contract separate from the main contract in which it is embedded and that, therefore, an arbitration agreement would normally survive termination of the underlying contract.[94] This doctrine of the separability of the arbitration agreement is now codified in s. 7 of the AA 1996. On account of this doctrine, an arbitration agreement is not therefore rendered 'null and void' simply by the fact of the underlying contract being null and void.[95] Also, there is authority in relation to previous arbitration legislation suggesting that neither the incapacity of a party to settle the award[96] nor lack of funds to finance the proceedings[97] has the effect of rendering an arbitration agreement 'incapable of being performed'.

17.6.3 Arbitration procedure

Under s. 33, the Arbitrator has a general duty to act fairly and impartially as between the parties, giving each party reasonable opportunity to put forward his case and to respond to that of his opponent. In particular, he must adopt appropriate procedures for achieving this objective. On the question of procedure, Article 8 of JCT 05 provides that it is to be in accordance with Clauses 9.3 to 9.8 and JCT 05 edition of the Construction Industry Model Arbitration Rules (CIMAR).[98] Rule 6 of CIMAR requires the Arbitrator to consult the parties before deciding on the procedure. Subject to the above considerations, the procedure to follow is entirely at the Arbitrator's discretion. CIMAR comprises 14 Rules

94 *Heyman* v. *Darwins Ltd* [1942] AC 356.
95 *Fiona Trust & Holding Corp.* v. *Privalov* [2007] UKHL 40; 114 ConLR 69.
96 *The Rena K* [1979] QB 377.
97 *Paczy* v. *Haendler* [1981] 1 Lloyd's Rep. 302.
98 These were produced at the instigation of the Society of Construction Arbitrators who viewed as unsatisfactory the use of different arbitration rules within the construction industry.

governing various aspects of arbitration procedure. The matters needing special attention include:

- choice from three recognized types of procedure;
- commencement of the proceedings;
- appointment of the Arbitrator;
- multi-party disputes.

17.6.3.1 Choice of three types of procedure

CIMAR describes three types of procedures that the Arbitrator may decide to adopt: (1) Short Hearing Procedure; (2) Documents Only Procedure; and (3) Full Procedure.

(1) The Short Hearing Procedure (Rule 7) is appropriate where the Arbitrator can determine the dispute by inspecting work, materials or the operation of machinery. It requires the parties to submit to the Arbitrator written statements of their case. The hearing should not take more than a day unless the parties agree to extend it. An example of a dispute for which this procedure would be appropriate is whether work is in accordance with the Contract.

(2) The Documents Only Procedure (Rule 8) involves the Arbitrator making his award only on the basis of written statements of claim. This is appropriate where the issues in dispute are such that there is no need for oral evidence or the amount involved does not warrant it. As an example, consider a dispute over whether an instruction issued by the Architect is authorized under the Conditions. There is no reason why the Arbitrator cannot determine such a dispute by examining the instruction, the Conditions and the parties' written submissions.

(3) The Full Procedure is appropriate where neither of the two procedures described above is appropriate. With this procedure, the parties must exchange statements of claim and defence before the hearing. Rule 9 provides guidelines with which the statements must comply. An example of dispute for which this type of procedure would be suitable is a complex claim for delays and disruption from a variety of different causes.

17.6.3.2 Commencement of arbitration proceedings

The Limitation Acts also apply to arbitration (i.e. if the proceedings are not commenced within the appropriate limitation period, the matter can no longer be pursued unless the defendant decides not to invoke the limitation defence).[99] Although the courts have power under s. 12(1) to extend time, it is exercised only very sparingly. This makes it of paramount importance that parties know what type of act amounts to commencement of the proceedings. Under s. 14(1), the parties may agree when proceedings are to be considered commenced. Sub-rule 2.1 of CIMAR states that proceedings are commenced in respect of a dispute when one party serves on the other a written notice of arbitration. The notice should specify the dispute or disputes and require the other party to agree to the appointment of an arbitrator. The notice may further specify names of persons proposed for appointment as the Arbitrator but this may be left to subsequent correspondence.

---

99 See Section 17.6.5.2 of this chapter.

17.6.3.3 Appointment of the Arbitrator

The parties have 14 days from the date of the notice of arbitration, or of a previously appointed arbitrator ceasing to act, to agree the person to act as the Arbitrator. Obviously, this period may be extended by agreement between the parties. On receipt of the notice, the other party will often respond by proffering his own list of names of possible arbitrators. Fourteen days is a very short time to check CVs and availability of arbitrators. CIMAR recognizes the possibility of the parties failing to agree the Arbitrator by providing that, in such event, either party may apply to a person empowered under the Contract to make the appointment on the parties' behalf. With JCT 05, this is the person named as the Appointor of Arbitrator in the Contract Particulars.[100] The options in the Contract Particulars arc the President or Vice-President of: (i) the Royal Institute of British Architects (RIBA) or (ii) the Royal Institution of Chartered Surveyors (RICS) or (iii) the Chartered Institute of Arbitrators (CIArb). If no selection is made the Appointor is the RIBA by default. Research[101] indicates that most arbitrators are appointed by the designated institutions rather than the parties themselves. This reflects the considerable distrust created by disputes. If the designated appointor fails to make the appointment, either party may apply to the court to exercise its jurisdiction under s. 18(3) of AA 1996 to make an appointment.

17.6.3.4 Multi-party disputes

The reality of construction disputes is that there are often several parties with some interest in the subject matter of a dispute. As an illustration, consider the cause of delay. It could be the Contractor, the Architect, or a sub-contractor. Furthermore, because of sub-contracting, a party who is found liable under one contract may wish to pass that liability down the chain of contracts to the party ultimately responsible. However, there is the problem that, with an arbitration from one contract, third parties to that contract have neither the right nor the obligation to be a party to the proceedings. The Architect therefore has no right or obligation to join in an arbitration between the Employer and the Contractor. This thus raises the problem of separate proceedings on the same or related disputes and, consequently, the risk of inconsistent outcomes (e.g. the finding in the Employer–Contractor proceedings is that a defect is attributable to a design error whilst the Employer–Architect proceedings conclude that it was caused by poor workmanship of the Contractor). The risk of inconsistent arbitration awards used to be a ground for the court to refuse to stay proceedings to arbitration, the rationale being that, as the court has powers to join all the relevant parties in the same action, it was better for the court to decide the matter.[102] Section 9 of AA 1996, which governs the court's consideration of applications for stay of litigation proceedings, does not allow it any discretion to refuse a stay for this reason.

Section 35(1) provides that the parties are free to authorize the Arbitrator to consolidate related proceedings or conduct concurrent hearings. Under s. 35(2), in the absence of such agreement, the Arbitrator has no power to do so on his own initiative. There is some attempt in a JCT 05 Contract to reduce the risk of inconsistent findings on issues common to two or more related arbitrations. Clause 9.4.2 provides that Rules 2.6, 2.7 and 2.8 of CIMAR apply where two or more arbitral proceedings in respect of the Works arise from

100 See Chapter 1, Section 1.5.2.4 dealing with entry for Clause 9.4,1.
101 Ndekugri, I. and Jenkins, H. 'Construction Arbitration: A Survey' (1994) ICLR 388.
102 See *Tauton-Collins* v. *Cromie* [1964] 1 WLR 637.

different contracts. Rule 2.6 requires any person with responsibility for appointing the Arbitrator (the parties and the Appointor of Arbitrator) to consider referring the disputes to the same arbitrator unless there are sufficient reasons for not doing so. Parties may also add disputes not included in the original notice of arbitration (see Clause 9.4.3 and Rule 3.3 of CIMAR).

Risk of prejudice against one or more of the parties could provide sufficient reason for not appointing the same arbitrator to related disputes. There has been judicial notice of this risk.[102a] *Abu Dhabi Gas Liquefaction Co. Ltd* v. *Eastern Bechtel Corporation*[103] involved a contract for the construction of liquefied natural gas tanks with the claimant as the employer and the defendant as the main contractor. Leaks in the tanks were detected after completion for which the employer made a substantial claim against the defendant, who in turn claimed against sub-contractors it considered responsible for the defects. The defendant suggested the appointment of the same arbitrator for the two disputes but both the employer and the sub-contractors opposed this suggestion. The sub-contractors feared that if a common arbitrator formed a view as to the cause of the leak in the first arbitration in which they could not participate, he would be inclined to take the same view on causation in the subsequent arbitration with them. To avoid the risk of inconsistent findings on the same facts, the Court of Appeal appointed the same arbitrator. However, in recognition of the possibility of the prejudice referred to, the Court indicated that it would consider any application to appoint a second arbitrator for particular issues in a separate arbitration.

To complete the circle, the Arbitrator may consolidate proceedings and conduct concurrent hearings in specified circumstances. All disputes between the Contractor and the Employer notified before the appointment of the Arbitrator must be consolidated (Rule 3.2). The Arbitrator may consolidate additional disputes from the same arbitration agreement notified after his appointment with the proceedings if appropriate (Rule 3.3). However, where the separate proceedings involve different parties, the Arbitrator *may* order concurrent hearings only where the disputes arise from the same project and raise a common issue (Rule 3.7). In this type of situation, he must not consolidate the proceedings without the agreement of all concerned (Rule 3.8). With consolidated proceedings, the Arbitrator, unless the parties otherwise agree, must deliver a single award. Separate awards are required where there have only been concurrent hearings.

It is obvious from the description of the scheme for consolidation that it will be workable only if CIMAR, or other appropriate arbitration rules, is incorporated into the related contracts and the appointing bodies cooperate with each other in the interest of consistent awards. Other obstacles to effective implementation of the joinder provisions are discussed in Section 17.7.14 of this chapter.

17.6.4 Enforcement of an arbitrator's award

The successful party has two ways of enforcing the award: (i) he can sue on it; (ii) he can register it as a judgment or order of a court under s. 66 and enforce it as such. It is often an express term of the arbitration agreement that the award will be binding on the parties. For example, under Clause 9.6, the award of the Arbitrator is final and binding on the parties

---

102a See *Pring & St Hill Ltd* v. *C J Hafner T/A Southern Erectors* [2002] EWHC 1775 (TCC) which concerned the same adjudicator deciding the same question in disputes in a chain of contracts.

103 (1982) 21 BLR 117.

subject to any court determination of preliminary points of law under s. 45 or determination of points of law arising out of an award under s. 69. Even in the absence of such express provisions, a term to the same effect would normally be implied. Failure to abide by the award would therefore constitute an actionable breach of contract. Under s. 66, an arbitration award may, with the leave of the High Court, be entered as a judgment of the court and enforced as such. This alternative, which has the advantage of convenience, is available where the award is clear and without doubt as to its validity. The unsuccessful party may seek to prevent this method of enforcement by asking the court to set it aside for lack of substantive jurisdiction, serious irregularity or a point of law arising from the award.[104]

17.6.5 Court intervention into arbitration

The aim of an arbitration agreement is to provide for the ability to choose a private tribunal, in preference to the courts, to settle any disputes between the parties to the agreement. However, for two main reasons, arbitration agreements do not completely remove the possibility of the courts getting involved in their dispute. First, any agreement that parties to a contract should never go to court in connection with disputes under the contract would be null and void on public policy grounds. The only exception is a form of agreement commonly referred to as a *Scott* v. *Avery*[105] clause, after the name of the case in which it first appeared. This form, which does not actually oust the jurisdiction of the courts, makes the award of an arbitrator a condition precedent to litigation. A defendant who is party to this form of agreement is entitled, as a right, to stay of proceedings, unless he has prejudiced this right either by a waiver or improper conduct. However, s. 9(5) provides that the court may ignore a *Scott* v. *Avery* clause and hear the case where the arbitration agreement is null and void, inoperative or incapable of being performed. Second, in addition to the court's power to enforce arbitration awards as explained in Section 17.6.4 above, various parts of the AA 1996 invest the courts with other supervisory powers over arbitrations. However, these powers have to be looked at in the light of the general principle stated in s. 1(c) of Part I of the Act that 'in matters governed by this Part the court should not intervene except as provided by the Part'. Generally, the policy in the legislation is to keep interference by the court to a minimum and to support the arbitration where it is allowed to intervene. The more important of these powers are next explained.

17.6.5.1 Appointment of arbitrators

Most arbitration agreements expressly state the manner in which the Arbitrator is to be appointed. If the contractual method proves unsuccessful, the court can appoint the Arbitrator upon an application from either party to do so.[106] In the event of an appointed Arbitrator being unable or refusing to act, dying, or otherwise failing to act after he has been validly appointed, a replacement has to be appointed. Subject to the express provisions of the arbitration agreement on this eventuality, this vacancy can be filled along the same lines as used in the initial appointment (i.e. agreement, third-party appointment or appointment by the High Court).

104 See Section 17.6.5.8 of this chapter.
105 (1856) 5 HL Cas 811.
106 Section 18.

17.6.5.2 Extension of time for commencement of arbitration

The statutes of limitation apply to both litigation and arbitration.[107] This means that arbitration would be statute-barred if a notice to refer to arbitration were not given within 6 years (for simple agreements) or 12 years (for agreements executed as deeds) from when the right to arbitrate accrued. However, an arbitration agreement may stipulate different limitation periods. The notice to refer to arbitration is the equivalent of the claim form in litigation.

Under s. 12, the High Court has discretion to extend time if a claimant is out of time. The section expressly states that the discretion is to be exercised only if the court is satisfied that: (i) the circumstances were outside the reasonable contemplation of the parties when they agreed the provision in question and it would be just to extend time; or (ii) the conduct of a party makes it unjust to hold the other party to the strict term of the provision limiting the period for commencement of arbitration. In practice, the courts are generally reluctant to interfere with arbitration agreements in this manner.

17.6.5.3 Removal of an arbitrator

Section 24 empowers the courts to remove an arbitrator in four situations.

1. There are justifiable doubts as to his impartiality.
2. He does not possess the qualifications required by the arbitration agreement.
3. He is physically or mentally incapable.
4. He has refused or failed properly to conduct the proceedings or to use all reasonable dispatch in conducting the proceedings or making an award, thus causing substantial injustice.

17.6.5.4 Resignation of an arbitrator

Under s. 25, the parties may agree with the Arbitrator his entitlement to his fees and expenses and his liability if he resigns. If they do not so agree, the Arbitrator may apply to the court for relief from any liability and an order in respect of his fees.

17.6.5.5 Interlocutory orders

These are orders issued by the Arbitrator on procedural matters. For example, under s. 39, if both parties agree, the Arbitrator can order, on a provisional basis, payment of a sum of money or the disposition of property between the parties or an interim payment in respect of costs. Any such order can be subject to the Arbitrator's final assessment in his award. If the party to whom an order is addressed fails to comply, it can be enforced by the court unless the parties excluded such jurisdiction of the court. JCT 05 does not contain such an exclusion.

17.6.5.6 Determination of recoverable costs

This is the process whereby the make-up of the amount demanded by a successful party as costs is assessed as to their validity. Where they are excessive, they will be determined on

---

107 Section 13(1).

the basis of criteria set out in s. 63(5). The main guideline is that recoverable costs shall be determined on the basis of a reasonable amount in respect of all costs reasonably incurred and that any doubt as to reasonableness is to be resolved in favour of the paying party.

17.6.5.7 Determination of preliminary points of law

Under s. 45, unless the parties otherwise agreed, the court has jurisdiction to determine any question of law arising in the course of the proceedings if a party applies to it and the court is satisfied that the question of law substantially affects the rights of one or more of the parties. Even where the court is so satisfied, it must not consider the application unless either of two conditions is satisfied:

1. the other parties to the proceedings consent that the court may consider it; or
2. the Arbitrator consents and the court is convinced that the application was brought without delay and that the determination is likely to produce a substantial saving of costs.

Section 45(2) requires the applicant to state in the application the question to be determined and, unless all the parties to the proceedings consent to the application, why the court must determine it. It follows from the above discussion that the application must contain the information required by the court to decide whether the conditions have been met.

Clause 9.7 of JCT 05 provides that the parties have given their consent to this type of application to the court. The validity of this type of term (i.e. terms giving consent to appeals and applications in the arbitration agreement itself) has been upheld in a long line of authorities.[108] There is therefore an automatic right under JCT 05 to apply to the court to determine preliminary points of law.

Section 45 is non-mandatory (i.e. the parties are free to agree that there will be no appeals on points of law to the High Court). Section 45(1) also provides that where the parties agree that the Arbitrator shall not give reasons for his award, that is equivalent to excluding the court's jurisdiction. This is the case irrespective of whether or not a domestic arbitration agreement is involved.

Leave of the court is required to appeal against its decision as to whether or not any of the conditions for the exercise of its jurisdiction to determine the question has been met. There is a similar condition on appeal against the court's determination of the question of law, but with the added requirement that such leave is not to be given unless the court considers that the question is one of general importance or that there are other special reasons why the Court of Appeal should consider it. It is to be noted that the Court of Appeal itself has no discretion to grant leave if the lower court declines to do so.

17.6.5.8 Appeals against an arbitration award

Under Ss 67 (jurisdiction), 68 (serious irregularity affecting the Arbitrator, the proceedings or the award) and 69 (point of law arising out of the award), a party may appeal against an

---

108 *How Engineering & Services Ltd* v. *Lindner Ceiling and Floors plc* (unreported, 17 May 1995, HHJ Thornton QC); *Vascroft Constructors Ltd* v. *Seeboard plc* (1996) 78 BLR 132; *Taylor Woodrow Civil Engineering Ltd* v. *Hutchinson Development* (1998) 75 ConLR 1; *Poseidon Schiffahrt GmbH* v. *Nomadic Navigation Co. Ltd (The Trade Nomad)* [1998] 1 Lloyd's Rep. 57; *Fence Gate Ltd* v. *NEL Construction Ltd* (2001) 82 ConLR 41; *Robin Ellis Ltd* v. *Vinexsa International Ltd* [2003] EWHC 1352 (TCC); *B.R. Cantrell and Another* v. *Wright & Fuller* [2003] EWHC 1545 (TCC); *Hallamshire Construction Plc* v. *South Holland DC* [2004] EWHC 8 (TCC); 93 ConLR 103.

arbitrator's award. Ss 67 and 68 are mandatory but s. 69 applies subject to the agreement of the parties otherwise. By Clause 9.7 of JCT 05, the parties have given their consent in advance to appeals on points of law arising from an award.[109]

Upon appeal, the court may confirm, vary, or set aside an award, or remit it to the Arbitrator for reconsideration together with the court's opinions on the question of law that was the subject of the appeal. However, with challenges for serious irregularity, under s. 68, the court will normally avoid setting aside an award if it can do so. This is because setting aside the award would entail re-commencement of the arbitration with a new Arbitrator, with all the implications for costs and delays. Where an award has been remitted, the Arbitrator, unless the order otherwise directs, must make his award within 3 months of the date of the remission order.

All appeals are subject to s. 70, which sets out that the appeal is first subject to the requirement that:

- any available process of appeal to, or review by the Arbitrator has been exhausted;
- any necessary application for correction of the award under s. 57 has been made;
- the appeal is brought within 28 days of the date of the award.

There are additional conditions depending on the grounds for the appeal.

Substantive jurisdiction or serious irregularity

The substantive jurisdiction of the Arbitrator is defined under s. 30(1) as:

- whether there is a valid arbitration agreement;
- whether the Arbitrator has been properly appointed;
- whether the dispute is within the arbitration agreement;
- whether the agreed procedure was used in referring it to arbitration.

Serious irregularity is defined in s. 68(2) as any of the following that the court considers to have caused or will cause substantial injustice to the applicant:

1. failure of the Arbitrator to comply with his s. 33 obligations (e.g. failing to be impartial and following an unfair procedure);
2. the Arbitrator exceeding his powers;
3. failure of the Arbitrator to follow agreed procedure;
4. failure of the Arbitrator to deal with the issues submitted to him;
5. the Appointor exceeding his powers;
6. uncertainty or ambiguity as to the effect of the award;
7. the award was obtained by fraud or other means unacceptable on public policy grounds;
8. failure of the award to comply with agreed form;
9. any irregularity in the proceedings that is admitted by the Arbitrator or the Appointor.

Where the appeal is on any of these grounds, s. 73 imposes the additional condition that the objection as to jurisdiction or serious irregularity must have been made as soon as it became known. If that was not done, a party may not raise the objection later unless he can show that, at the time he took part or continued to take part in the proceedings, he did not know and could not, with reasonable diligence, have discovered the grounds for this objection.

---

109 As explained in Section 17.6.5.7 of this chapter, the parties therefore have automatic rights of appeal on points of law arising out of an award.

Questions of law

For appeals on a point of law, leave of the court is required first, unless all parties to the proceedings consent that the appeal must go ahead.[110] Where leave of the court has to be obtained, s. 69(3) lists the following factors that must apply before the court can grant it.

1. The determination of the question will substantially affect the rights of one or more of the parties.
2. The Arbitrator was asked to determine the question of law.
3. From the facts in the award, either the Arbitrator's determination was obviously wrong or the question is one of general importance and the determination of the Arbitrator is at least open to serious doubt.
4. It is just and fair for the court to interfere by determining the question.

These requirements have been derived in part from guidelines formulated by the House of Lords in *Pioneer Shipping Ltd* v. *BTP Tioxide Ltd: The Nema*[111] regarding the exercise of this discretion under superseded legislation.[112] There is a restricted right of appeal to the Court of Appeal from the High Court. The restriction arises from the requirement that the High Court must grant leave to appeal.

17.7 Arbitration or litigation

Pre-1998 versions of the JCT 80 provided for arbitration as a mandatory dispute resolution technique. Arbitration is now optional. The change was made partly to avoid the effect of the decision in *Northern Regional Health Authority* v. *Derek Crouch Construction Ltd*[113] that the court had no power to review certificates, opinions, decisions and the like of an architect where an arbitration agreement invests arbitrators with such powers. To make the decision whether to opt for arbitration, it is necessary to understand its advantages and disadvantages when compared with litigation. They are therefore next compared on a number of issues.

17.7.1 Expertise

Proponents of arbitration, as against litigation, advance the argument that an arbitrator is normally selected for his expert knowledge of the subject matter of the dispute, whereas a judge rarely has any expert knowledge of the technical complexities of the construction industry. There are three counter-arguments commonly put forward. The first is that the TCC, which is the division of the High Court that hears the vast majority of construction

110 Section 69(2).

111 [1982] AC 724; the guidelines, referred to as the 'Nema' Guidelines, were reaffirmed in *Antaios Compania Naviera SA* v. *Salen Rederiena AB: The Antaios* [1985] AC 191.

112 Under the Nema Guidelines, the test on a point of law of general public importance required the judge to form the view that the Arbitrator's determination of the point was probably wrong. Under s. 69(3)(c) the judge need only form the view that the Arbitrator's determination is open to serious doubt. In *CMA CGM SA* v. *Beteiligungs KG MS Northern Pioneer Schiffahrsgesellschaft mbH & Co* [2002] EWCA Civ 1878 the Court stated that this statutory provision opens the door to an appeal on a point of law of general public importance a bit wider.

113 (1984) 26 BLR 1; this decision was overruled in *Beaufort Developments (NI) Ltd* v. *Gilbert-Ash (NI) Ltd* (1998) 88 BLR 1; [1999] 1 AC 266.

contract disputes, has accumulated expertise in this area over the years. Second, it is claimed that the issues usually involved are more of a legal nature than of construction contract management and construction practices. Third, technical experts can always be called as witnesses to assist the judges in coping with the technical issues arising, although it has to be pointed out that such a course of action can involve extra costs. Finally, anecdotal evidence suggests that, because of the inevitable disparity in familiarity with the relevant procedures, some non-lawyer arbitrators may have trouble in controlling proceedings where very experienced professional advocates represent the parties.

17.7.2 Advocacy

Only practising barristers and solicitors obtaining a 'rights of audience' certificate have the right to appear before a High Court as advocates unless the litigant wishes to appear and represent himself in person. Anybody can appear before an arbitrator on behalf of the parties unless they agree otherwise.[114] There is the counter-argument that arbitration has spawned a new type of professional, the lay advocate, who offers specialist skills in the presentation of cases on behalf of parties to arbitration proceedings. The lay advocate is not necessarily cheaper than the traditional legal representatives and will normally not be as proficient in the law.

17.7.3 Costs

It is claimed that the cost of arbitration is usually much less than the cost of litigation. Indeed, Ss 1 and 33 make it an obligation of an Arbitrator to avoid unnecessary delay and expense. The other school of thought is that this belief is based purely on fiction rather than fact. The points of the counter-argument as to costs are:

- parties to arbitration usually have exactly the same type and level of legal representation as in litigation;
- the Arbitrator and the hearing venue have to be paid for by the parties whilst the court and judges come 'free';
- secretarial and other support to record the proceedings of the arbitration is frequently a source of cost.

17.7.4 Simplicity

Procedure in the Court is now governed by the CPR. Although they are intended to be simpler than the Rules of the Supreme Court that they replaced, they are still quite complex. By contrast, arbitration is intended to be simplicity itself. By s. 34, unless the parties agree their own procedure, the procedure to be followed in arbitration is at the discretion of the Arbitrator subject to his mandatory obligations under s. 33 to adopt fair, impartial and appropriate procedures. It is often argued that the Arbitrator can adopt a procedure that best fits the case he is called upon to settle and thus avoid the complexities of litigation procedures.

114 Section 36.

There are two arguments against this point. First, it is said that an arbitrator's discretion on procedure is not as unfettered as is often imagined and that where an arbitrator fails to follow proper procedure his award would be appealable to the courts. For example, in *Modern Engineering (Bristol) Ltd* v. *C. Miskin & Son Ltd*[115] the Arbitrator was faced with a point of law as to whether an architect's certificate could be reviewed. After listening to one side's argument on the point, the Arbitrator made up his mind without giving the other side the opportunity to put forward their argument. Though his opinion was the correct position, the Court of Appeal held that he should have listened to the counter-argument and that his failure to do so constituted misconduct (now referred to as 'serious irregularity' under the 1996 Act). He was therefore removed for misconduct and his award set aside under the 1950 Arbitration Act. Second, many arbitration agreements incorporate model procedures published by various bodies (e.g. CIMAR and the Institution of Civil Engineers (ICE) Arbitration Procedure), some of which are quite complex.

17.7.5 Expedition

The law's delays are notorious but, following the reforms instigated by Lord Woolf, many people believe that the position has improved under the new Civil Procedure Rules. Arbitration can be quicker. The reasons usually given for this relate to the simplicity issue already discussed. Much depends on the attitude of the parties and the robustness and experience of the arbitrator.

17.7.6 Convenience

It is common practice for an Arbitrator to arrange for the arbitration proceedings to be held at times and places to suit the convenience of both parties. An action in court is listed subject to the availability of a judge and a court primarily and then, possibly, the availability of the legal representatives of the parties.

17.7.7 Courtesy

Rarely will an arbitrator forget the fact that, in effect, he has been appointed by the parties and that, therefore, he owes them a moral duty to extend to them the normal courtesies required in commercial and professional communication. By contrast, some believe that judges can be abrupt. However, judges are professionally trained and well experienced in controlling proceedings whereas arbitrators are, by contrast, amateurs who, if inexperienced, can be dominated by the parties' professional advocates.

17.7.8 Privacy/confidentiality

A lawsuit must normally be heard in open court with the Press and public free to attend. Litigation may therefore involve unpleasant publicity or result in the disclosure of trade secrets. With arbitration, the entire hearing takes place in private, thus avoiding the publicity

---

115 (1981) 15 BLR 82; [1981] 1 Lloyd's Rep. 135.

and the disadvantages of the nervousness commonly induced by proceedings in court. However, if an arbitrator's award is appealed then the appeal is heard in open court and is liable to be reported in the Press.

Confidentiality applies to not only an arbitration award but also the accompanying pleadings, submissions and proofs of evidence as an implied term of the relevant arbitration agreement.[116] Court decisions are often reported in not only the Press but also formal law reports.

17.7.9 Future business relations

Litigation is inherently adversarial, thus giving rise to risks of reduction in the chances of future contracts between the parties. Even where they enter into other contracts, the distrust resulting from the experience of litigation could easily poison such subsequent contractual relationships. Contractual relationships in general may even suffer because a reputation for litigious behaviour is very easily established within the industry as a whole. The counter-argument is that arbitration can be equally adversarial.

17.7.10 Powers of the Arbitrator

In *Northern Regional Health Authority* v. *Derek Crouch Construction Ltd*,[117] which arose from a JCT 63 contract, it was held by the Court of Appeal that, where an arbitration agreement expressly conferred powers on an Arbitrator to open up, review or revise any decision, opinion, direction, certificate or valuation of the Architect/Engineer, the High Court did not possess similar powers. This meant that many cases could be resolved only by arbitration. However, the House of Lords overruled *Crouch* in *Beaufort Developments (NI) Ltd* v. *Gilbert-Ash (NI) Ltd*.[118] Their Lordships stated that, unless a contractual provision was expressed to be final and conclusive on an issue, the court always had jurisdiction to determine the contractual entitlements of the parties. They explained further that it was the Arbitrator who had no such power unless the parties positively gave it to him, thus necessitating arbitration clauses of the type considered in *Crouch*.

17.7.11 Summary relief

In litigation, a party may apply for Summary Judgment and/or Interim Payment under the CPR. The essence of a Summary Judgment application under Rule 24 is that: (i) the defendant has no real prospect of successfully defending the claim or a particular issue; and (ii) there is no other reason why the case or issue should go for full trial. The court is therefore asked to decide for the applicant without a full trial. Usually both the evidence to support the application and that rebutting it are provided in the form of sworn statements. An Interim Payment order (under Rule 25), which is applicable to only monetary

---

116 *Ali Shipping Corp.* v. *Shipyard Togir* [1999] 1 WLR 314; [1998] 2 All ER 136; [1998] 1 Lloyd's Rep 643; *Michael Wilson and Partners Ltd* v. *Emmott* [2008] EWCA Civ 184.

117 (1984) 26 BLR 1 (hereafter *Crouch*).

118 (1998) 88 BLR 1.

claims, is made where the court is satisfied from the sworn evidence that, although a full trial is necessary, the applicant would be awarded a substantial amount. In such cases, the court may order immediate payment of the part of the claim indubitably due pending a final trial. However, if the outcome at trial is that the amount paid was not owed in part or at all, the court can order an appropriate refund.

Contractors usually applied for these orders to enforce payment obligations under the contract. The customary response of employers was to apply for stay of proceedings to arbitration if the contract contained an arbitration agreement, suggesting that arbitrators would not be able to grant equivalent relief. With Interim Payment applications, the courts had no difficulty ordering Interim Payment and staying the proceedings to arbitration.[119] Similarly, on an application for Summary Judgment, the court could grant it for sums for which there is no reasonable prospect of being successfully defended and stay the proceedings to arbitration.[120]

As explained in Sections 17.1.1 and 17.6.2 in this chapter, the *Halki* litigation highlighted the fact that the court's discretion to refuse stay of proceedings is now severely limited. It is therefore argued that parties may refuse to use arbitration because the court would be unable to grant these types of summary relief.[121] The facts of *Collins (Contractors) Ltd* v. *Baltic Quay Management (1994) Ltd*[122] illustrate the application of *Halki* to prevent Summary Judgment for sums indubitably due. The contract from which that litigation arose provided for certification and service of notices in identical terms to those in JCT 05. The contract administrator under the contract issued an interim certificate which the Employer failed to pay by the final date but without service of any of the required notices. When the contractor brought court proceedings for Summary Judgment to enforce payment the employer applied to stay them to arbitration under s. 9 of the Arbitration Act 1996. The Court of Appeal, by a unanimous decision, held that, on the authority of *Halki*, the employer was entitled to such a stay.

However, there is the counter-argument that the AA 1996 gives arbitrators powers to grant equivalent relief (e.g. a provisional award under s. 39, where the parties have agreed that the Arbitrator is to have such power). Also, Rule 10 of CIMAR gives the Arbitrator the power to grant provisional relief.

17.7.12 Finality

It is often argued that the scope for challenging the awards of arbitrators is much less than that for the decisions of the courts as the statutory framework places a number of restrictions on appeals from arbitrations.[123] Whilst this finality may be attractive to some parties, Wallace[124] cautioned that any misuse of the extensive powers of arbitrators under the Act could lead to serious injustice against which the courts would be powerless to intervene.

---

119 See *Imodco Ltd* v. *Wimpey Major Projects and Taylor Woodrow International Ltd* (1987) 40 BLR 1.

120 *Ellis Mechanical Services Ltd* v. *Wates Construction Ltd* (1976) 2 BLR 57; *Associate Bulk Carriers* v. *Koch Shipping* (1978) 7 BLR 18; *R. M. Douglas Construction Ltd* v. *Bass Leisure Ltd* (1990) 53 BLR 119.

121 All applications to stay proceedings to enforce adjudicators' decisions by summary judgment to arbitration have so far failed.

122 [2004] EWCA Civ 1757; [2005] BLR 63; [2005] TCLR 3; 99 ConLR 1.

123 See Section 17.6.5.8 of this chapter.

124 Wallace, I.N. D, 'First Impressions of the 1996 Arbitration Act' (1997) ICLR 71, pp. 71–116.

17.7.13 National sovereignty

It is often argued that many foreign governments would rather use the private forum of arbitration to resolve their disputes than submit to the jurisdiction of the courts of another country.

17.7.14 Multiple parties

The duty to submit a difference to arbitration is imposed by the contract between the parties. From the doctrine of privity of contract, it follows that no third party is obliged to, nor has a right to take part in the proceedings even if the dispute concerns it.[125] The reality of construction disputes is that there are often several parties involved in the cause of the dispute. For example, who is responsible for delay in a construction contract? It could be the architect, the contractor, or a sub-contractor. In litigation, the court has jurisdiction under the Civil Liability (Contribution) Act 1978 and the CPR[126] to join any party concerned for either an indemnity or for a contribution. In arbitration, an arbitrator has no such powers. This inability to join third parties was a proper ground on which a court could refuse stay of proceedings when it had discretion whether or not to stay to arbitration.[127] The position now is therefore that there is an increased risk of parallel proceedings from the same facts where the dispute involves multiple parties.

Attempts have been made to allow consolidation of disputes in the same arbitration proceedings by having suitable joinder provisions in all relevant contracts. Such provisions are notoriously problematical not only in their drafting but also in their operation by parties and arbitrators.[128] For example, in *Dredging & Construction Co. Ltd* v. *Delta Civil Engineering Co. Ltd (No.1)*[129] the dispute concerned whether, under the joinder provisions in the *ICE Conditions of Contract, 6th Edition* and the *Federation of Civil Engineering Contractor's Form of Subcontract, 1991 Edition*, appointment of an arbitrator and reference of a dispute under the main contract to him constituted a condition precedent to the main contractor's right to require a dispute under the sub-contract to be referred to an arbitrator under the main contract.

The provisions in JCT 05 aimed at supporting multiparty arbitrations are explained in Section 17.6.3.4. However, because of the fragmentation of the industry, multiplicity of appointing institutions leading to gaps in knowledge of ongoing proceedings and frequent

125 Under s. 8 of the Contracts (Rights of Third Parties) Act 1999, a person on whom the Contract confers a third-party right is considered a party to the arbitration agreement. That party may therefore invoke arbitration to enforce the right. Correspondingly, any court proceedings commenced by the party to enforce the right may be stayed to the arbitration. For detailed commentary see Chapter 9, Sections 9.2.3, 9.3.3 and 9.5.

126 See CPR, Part 20.

127 See *Tauton-Collins* v. *Cromie* [1964] 1 WLR 637.

128 For judicial notice of this difficulty see: *Trafalgar House Construction (Regions) Ltd* v. *Railtrack plc* (1995) 75 BLR 55, at p. 80 (the judge in this case also criticized the way the Arbitrator had sought to implement the joinder provisions in the JCT 80 and the related nominated sub-contract); *Lafarge Redland Aggregates Ltd* v. *Shephard Hill Civil Engineering Ltd* [2000] BLR 385 (the sub-contract provided that the main contractor had the right to require any related sub-contract dispute to be 'dealt with jointly with the dispute under the main contract in accordance with the provisions of Clause 66 thereof'. Clause 66 of the main contract required any dispute under the Contract to be referred in the first instance for the decision of the project engineer and then to arbitration. The House of Lords was split 3/2 on the question whether the sub-contract provided for tripartite arbitration).

129 (2000) CLC 213; 68 ConLR 87.

amendments to contracts, it may not always be possible to achieve this end. Possible sources of difficulty include gaps in the contractual and organizational framework, conflicting claims, compromise settlements, and disagreements as to whether the disputes are related.

Fragmentation of the industry works against the necessary dovetailing of the relevant contracts to enforce consolidation. Contractors and sub-contractors are not always in a position to impose contracts with relevant dovetailing joinder provisions on all suppliers. Assuming industry-wide availability of such joinder provisions, there is still the problem of frequent amendments to the standard forms, which may result in loss of a vital piece of the contractual jigsaw for enabling multiparty arbitrations. Furthermore, the multiplicity of appointing institutions and rivalry among them not only promotes gaps in knowledge of ongoing proceedings but also stands in the way of the desired level of cooperation. For example, research[130] suggests that the professional institutions tend to appoint only their members regardless of the nature of the dispute to be referred.

It is not unusual for the common party in related disputes to make a claim in one dispute which conflicts with its claims in the other. For example, in *Monk (A) & Co. Ltd* v. *Devon County Council*[131] the main contractor (Monk) denied liability for extra costs to their sub-contractors. As the disputes with the sub-contractors had not yet been resolved, Monk had to recognize the possibility that eventually the disputed liability might attach to them. With this in mind, they claimed in respect of this liability in their dispute with the employer. A related difficulty is that sometimes there may be a compromise in one dispute which turns out to be completely unacceptable to the other party in the second dispute. For instance, the main contractor may reach a compromised settlement under the main contract for reasons of future business relations with the employer. Such a settlement may be totally unsatisfactory to the sub-contractor.

Lafarge Redland Aggregates Ltd v. *Shephard Hill Civil Engineering Ltd*[132] illustrates the problems that can arise from differences in the needs and priorities of the parties. The dispute resolution clause in the main contract required any dispute to be referred for the decision of the project engineer and then to arbitration if either party was dissatisfied with the engineer's decision. The sub-contract gave the main contractor the right to require a related sub-contract dispute to be 'dealt with jointly with the dispute under the main contract in accordance with the provisions of clause 66 thereof'. When related disputes arose the main contractor preferred pursuit of a negotiated settlement with the Employer prior to reference to arbitration whilst the sub-contractor demanded immediate reference of the sub-contract dispute. The House of Lords held that, as there was delay in commencement of the arbitration under the main contract, the sub-contractor was entitled to appointment of an independent arbitrator for the sub-contract dispute.

A not uncommon obstacle to achieving joinder of parties or consolidation of proceedings is disagreement as to whether the conditions for appointing the same arbitrator are satisfied. Where the disagreement prevents appointment of an arbitrator there is failure of the procedure for appointing the arbitrator. Section 18 provides that, unless the parties have agreed a method of getting over the impasse, either party may apply to the court to give directions as to the appointment of an arbitrator or even make the appointment itself.

---

130 Ndekugri, I. and Jenkins, H., 'Construction Arbitration: A Survey' (1994) ICLR 388.
131 (1978) 10 BLR 9.
132 [2000] BLR 385.

In *City & General (Holborn) Ltd* v. *AYH plc*[133] an arbitrator was appointed to decide a main contractor's extension of time and related loss and expense claims against the employer. The employer blamed the causes of the claims on the defendant, who had performed the role of project manager and quantity surveyor under separate contracts with the employer. The employer's claim against the defendant included allegations of breaches of duty in the provision of pre-contract advice on costs and contractual arrangements. In the employer's notice of arbitration against the defendant it was stated that this second dispute would be referred to the arbitrator already appointed. The defendant objected to such a reference.

One of the issues in the ensuing court proceedings brought by the employer for the court to exercise its power under s. 18 to appoint the arbitrator was whether the two disputes were substantially the same or connected, the condition for appointing a common arbitrator. Jackson J. concluded from his examination of the disputes that the issues in the second dispute were substantially the same or connected with those in the first dispute. The judge, however, described as 'difficult' the general question as to the proportion of the issues in the two disputes that must converge to trigger the appointment of a common arbitrator. He stated that, as the commercial purpose of joinder provisions is to avoid multiplicity of proceedings generating excessive costs and involving risk of inconsistent findings, the threshold did not have to be set too high. In his opinion it was not necessary for the majority of the issues in the two disputes to be the same or connected.

17.7.15 Compound interest

The effect of s. 49 and Sub-Rule 12.8 of CIMAR is that an arbitrator has unfettered discretion to award simple or compound interest on money payable but not paid. In contrast, the court will only do so in very particular circumstances.[134] In the light of *Sempra Metals Ltd (formerly Metallgesellschaft Ltd)* v. *Her Majesty's Commissioners of Inland Revenue & Anor*,[135] the difference between arbitration and litigation in terms of recovery of interest is likely to diminish. In that case the House of Lords stated that the court has jurisdiction to award compound interest not only on non-payment of a debt but also on damages in contract or tort.

17.7.16 Legal aid

If a party qualifies for legal aid, he is likely to prefer litigation to arbitration because such aid is not available for arbitrations. A party not entitled to legal aid but who is confident of winning may prefer arbitration because he would be entitled to recover costs against the losing party. In litigation this may not be possible where the losing party was assisted by legal aid.

17.8 Mediation

The essence of mediation is that a third party, the mediator, assists the parties to come to a mutually acceptable solution to their dispute. This type of dispute resolution is referred

133 [2005] EWHC 2494.
134 *Westdeutsche Landesbank Girozentrale* v. *Islington* [1996] 2 WLR 802.
135 [2007] UKHL 34; [2007] 3 WLR 354.

to as 'facilitative mediation'. It may be abandoned at any point upon request by any party. The facilitative mediator does not express any views on the relative merits of the parties' positions. There is also evaluative mediation, where the approach is generally the same but the evaluative mediator is authorized to submit a recommended settlement in the event that the parties fail to reach an agreed settlement of their dispute. The mediator's recommendation is non-binding but often forms the foundation for a negotiated settlement after the 'failure' of the mediation.

There is no universally accepted procedure for mediation. However, there are a number of common practices.

1. The first step is for the parties to agree to resolve their dispute by mediation. In the early days of mediation, it was often very difficult for any party to take the first step towards mediation because of fear that such an overture could be misconstrued by the other side as indicative of lack of confidence in the strength of the opposing case. However, this must have changed because, as explained in Section 17.5, the parties are under an obligation to make reasonable effort to resolve their dispute without legal proceedings.

   When the parties are in agreement to undergo mediation, it is good practice to enter into a mediation agreement governing its conduct. *JCT Practice Note 28*[136] contains a sample of a mediation agreement. Matters addressed include: the parties' common intention to undergo mediation; the disputed issues to be referred to the mediator; the name of the mediator if already appointed or otherwise the method for appointing one; confidentiality of the proceedings; the right of any party to withdraw from the proceedings at any time without giving reasons; that the proceedings are 'without prejudice'; how the costs of the mediation are to be met;[137] the parties' undertaking to draw up any settlement as a binding and enforceable contract.[138]
2. It is standard practice to enter into a contract with the mediator. This contract should address issues such as procedure to be followed, payment, the parties' liabilities for the costs of the mediation, confidentiality, privilege, and whether the mediator can be called as a witness in subsequent proceedings. *JCT Practice Note 28* also contains a sample of this type of agreement.
3. Each party must prepare his statement of claim at the earliest possible moment after agreeing to attempt settlement by mediation. The contents of such a statement would normally include: a summary of the issues in dispute; a factual narrative which may contain references to the contract and other documents (e.g. correspondence, instructions from the contract administrator, specifications, minutes of meetings); the legal principles relied upon; details of the remedy being sought (e.g. breakdown of any financial claim, and time-impact analysis to support any extension of time claim); documents relied upon (these may be omitted at this stage if they are too bulky).
4. The first meeting is normally a joint session. The purposes of such a meeting include: (i) to introduce the mediator and the representatives of the disputants, (ii) to acquaint the mediator with the general nature of the dispute through brief oral statements of each party's case; (iii) to agree a timetable for the rest of the mediation process.

136 Joint Contracts Tribunal, RIBA Publications Ltd, 1995.

137 Common practice is for each party to bear his own costs whilst the parties share equally the fees and expenses of the mediator.

138 *JCT Practice Note 28* also contains a sample of an agreement following settlement.

5. The mediator then engages in a process of shuttle diplomacy between the parties who, at this stage, are often in separate locations in the same premises. The mediator plays a variety of roles, the most important of which include: (i) as a facilitator of communications; (ii) assisting parties to find common grounds in their positions; (iii) assisting with problem-solving by joint creative searches for alternative solutions; (iv) assisting the parties to evaluate the legal or technical merits of their case; and (v) as a healer of emotional wounds.
6. If the outcome of the mediation process is an agreed solution, the parties are brought together in a joint meeting to adopt the common position. At this point the agreement is non-binding. However, the parties may agree to convey their decision as a formal enforceable agreement for purposes of certainty.

Table of Cases

Table of Clause References

JCT STANDARD BUILDING CONTRACT

Articles of Agreement

Conditions – Clause No.

Section 1

Section 2

Section 3

Section 4

Section 5

Section 6

Section 9

Schedules

Schedule 1

Schedule 2

Schedule 3

Schedule 4

ASSOCIATED DOCUMENTS

Construction Industry Model Arbitration Rules (CIMAR)

JCT Formula Rules

Standard Method of Measurement

7th edition

Table of Statutes and Statutory Instruments
(Including supporting publications and Parliamentary Bills)

Note – for abbreviated titles of Statutes see Abbreviations – General

Bills in passage through Parliament, and other publications:

Subject Index